Geographic Information Systems: Future Trends and Technologies

Geographic Information Systems: Future Trends and Technologies

Editor: Marina De Lima

NY RESEARCH PRESS
New York

Published by NY Research Press
118-35 Queens Blvd., Suite 400,
Forest Hills, NY 11375, USA
www.nyresearchpress.com

Geographic Information Systems: Future Trends and Technologies
Edited by Marina De Lima

International Standard Book Number: 978-1-63238-571-0 (Hardback)

Cataloging-in-Publication Data

Geographic information systems : future trends and technologies / edited by Marina De Lima.
p. cm.
Includes bibliographical references and index.
ISBN 978-1-63238-571-0
1. Geographic information systems. I. De Lima, Marina.
G70.212 .G46 2018
910.285--dc23

Printed in China.

Contents

Preface

The main aim of this book is to educate learners and enhance their research focus by presenting diverse topics covering this vast field. This is an advanced book which compiles significant studies by distinguished experts in the area of analysis. This book addresses successive solutions to the challenges arising in the area of application, along with it; the book provides scope for future developments.

Geographic information systems gather and store geographical data. Some examples are desktop GIS, web servers, spatial database management systems, etc. A number of disciplines benefit from this technology, specially the fields of geography, forestry and topography. Some of the main applications of geographic information systems include national security, landscape architecture, archaeology, etc. This book includes some of the vital pieces of work being conducted across the world on various topics related to geographic information systems. As this field is emerging at a rapid pace, the contents of this book will help the readers understand the modern concepts and applications of the subject.

It was a great honour to edit this book, though there were challenges, as it involved a lot of communication and networking between me and the editorial team. However, the end result was this all-inclusive book covering diverse themes in the field.

Finally, it is important to acknowledge the efforts of the contributors for their excellent chapters, through which a wide variety of issues have been addressed. I would also like to thank my colleagues for their valuable feedback during the making of this book.

Editor

Geospatial Analysis on the Distributions of Tobacco Smoking and Alcohol Drinking in India

Sze Hang Fu[1], Prabhat Jha[1,2]*, Prakash C. Gupta[3], Rajesh Kumar[4], Rajesh Dikshit[5], Dhirendra Sinha[6]

1 Centre for Global Health Research, Li Ka Shing Knowledge Institute, St. Michael's Hospital, University of Toronto, Toronto, Ontario, Canada, **2** Dalla Lana School of Public Health, University of Toronto, Toronto, Ontario, Canada, **3** Healis-Sekhsaria Institute of Public Health, Navi Mumbai, India, **4** School of Public Health, Post Graduate Institute of Medical Education and Research, Chandigarh, India, **5** Department of Epidemiology, Tata Memorial Hospital, Mumbai, India, **6** South East Asia Regional Office, World Health Organization, New Delhi, India

Abstract

Background: Tobacco smoking and binge alcohol drinking are two of the leading risk factors for premature mortality worldwide. In India, studies have examined the geographic distributions of tobacco smoking and alcohol drinking only at the state-level; sub-state variations and the spatial association between the two consumptions are poorly understood.

Methodology: We used data from the Special Fertility and Mortality Survey conducted in 1998 to examine the geographic distributions of tobacco smoking and alcohol drinking at the district and postal code levels. We used kriging interpolation to generate smoking and drinking distributions at the postal code level. We also examined spatial autocorrelations and identified spatial clusters of high and low prevalence of smoking and drinking. Finally, we used bivariate analyses to examine the spatial correlations between smoking and drinking, and between cigarette and bidi smoking.

Results: There was a high prevalence of any smoking in the central and northeastern states, and a high prevalence of any drinking in Himachal Pradesh, Arunachal Pradesh, and eastern Madhya Pradesh. Spatial clusters of early smoking (started smoking before age 20) were identified in the central states. Cigarette and bidi smoking showed distinctly different geographic patterns, with high levels of cigarette smoking in the northeastern states and high levels of bidi smoking in the central states. The geographic pattern of bidi smoking was similar to early smoking. Cigarette smoking was spatially associated with any drinking. Smoking prevalences in 1998 were correlated with prevalences in 2004 at the district level and 2010 at the state level.

Conclusion: These results along with earlier evidence on the complementarities between tobacco smoking and alcohol drinking suggest that local public health action on smoking might also help to reduce alcohol consumption, and vice versa. Surveys that properly represent tobacco and alcohol consumptions at the district level are recommended.

Editor: Natalie Walker, The National Institute for Health Innovation, New Zealand

Funding: The authors have no support or funding to report.

Competing Interests: The authors have declared that no competing interests exist.

* Email: prabhat.jha@utoronto.ca

Introduction

The World Health Organization has identified tobacco smoking and binge alcohol drinking as two of the leading risk factors for premature mortality worldwide [1,2]. In India, 14% per cent of the population above age 15 smoke tobacco (24% for men and 3% for women) [3]. Smoking is already responsible for about 1 in 5 deaths for men and 1 in 20 deaths for women at ages 30–69 [4]. India has among the lowest alcohol use per capita in the world, but reports growth in alcohol sales [5,6]. More importantly, male alcohol drinking in India is characterized by problematic drinking (e.g. binge-drinking), which has negative health and social impacts on the consumers and their families [5,7].

Most studies examined tobacco smoking and alcohol drinking behaviours separately among the Indian population. Few studies which examined the concurrent use of tobacco and alcohol found that smokers have a higher likelihood to drink alcohol than non-smokers [8], and vice versa [9,10].

Understanding the geographic distributions of tobacco smoking and alcohol drinking, along with the determinants contributing to their use, is essential for developing targeted tobacco and alcohol control policies. Previous studies have examined the distributions of tobacco smoking [11–13] and alcohol drinking [14] in India at the state level separately. State level analysis can potentially mask the sub-state spatial heterogeneity, especially in the bigger states (e.g. 17 Indian states have populations greater than 25 million [15]). Implementation of tobacco and alcohol control policies is the responsibility of the state government [6,16], thus studying the distributions of tobacco smoking and alcohol drinking as well as their spatial association at the sub-state level is desirable.

Advances in geographic information system (GIS) technology enable analysis of public health phenomena by taking "space" into account [17]. GIS is an integrated set of tools that allow both the analytical manipulation and the visual presentation of spatial data [17,18]. Epidemiologic studies have employed GIS methods for disease mapping and cluster detection [19–21], and although these types of research often focus on diseases, GIS methods can also study risk factors for diseases.

Here, we utilized GIS methods to analyze the distributions of tobacco smoking and alcohol drinking in India. Our objective was to identify sub-state regions with significantly high and low prevalence of smoking and drinking. We also explored whether smoking and drinking were spatially correlated.

Methods

Dataset

The Special Fertility and Mortality Survey (SFMS) was a one-time study conducted in February 1998 [22]. The SFMS covered over one million nationally-representative households, based on the Registrar General of India's Sample Registration System (SRS) [22,23]. The SRS aims to provide reliable estimates of fertility and mortality at the state and national levels for rural and urban areas separately [24]. The SFMS interviewed household heads to provide various health indicators for the household residents, including their tobacco smoking and alcohol drinking behaviours. They were asked "does 'household member' smoke/drink alcohol", "age at which 'household member' started smoking/drinking", "what does he/she smoke: cigarette, bidi", and "how frequently 'household member' drinks alcohol: daily, 3–4 times a week, once a fortnight, once a month, less frequently". These definitions are similar to the National Family Health Survey (NFHS) 1998–1999 ("Does 'household member' smoke tobacco/drink alcohol?") [25] and the Global Adult Tobacco Survey (GATS) 2009–2010 ("Do you currently smoke tobacco on a daily basis, less than daily, or not at all?") [3]. The GATS survey interviewed respondents directly while the earlier surveys relied on proxy responses, usually from the head of the household.

We focused on males between 30 and 69 years old, as this age group comprises the majority of smokers (80%) and drinkers (83%), similar to the results from other surveys [11,26,27]. Among Indian females, only about 3% smoke regularly [3,11] and 1% drink regularly [27,28], thus females were excluded from the analysis. We analyzed the following smoking and drinking variables: any smoking, early smoking, cigarette smoking, bidi smoking, and any alcohol drinking. Excluding the missing records or those reporting initiation of smoking before age 10 years, the mean age of initiating any smoking, cigarette smoking, and bidi smoking was 21 (SD 4.8), 23 (SD 4.7), and 20 (SD 4.6), respectively. Therefore, we defined early smoking as those who initiated smoking before age 20. The mean age of initiating any alcohol drinking was 24 (SD 5.4).

The SFMS covered 445 out of the 467 districts from the 1991 Census. Three island districts were excluded. The geographic locations of the SRS units were further geocoded to the associated Indian postal codes, allowing analyses at the district and postal code levels.

Geospatial analyses

Three GIS methods were used to analyse the geographic patterns of tobacco smoking and alcohol drinking and identify areas with high and low prevalence. 1) ordinary kriging interpolation was used to describe the prevalence at the postal code level, 2) Moran's I was used to measure the overall spatial autocorrelation at the district level, 3) and the local indicators of spatial association (LISA) statistic was used to identify spatial clusters at the district level. These methods are described next, and detailed explanations can be found in Appendix S1.

Kriging is a technique that generates an estimated interpolation surface from a set of data points by incorporating the inference from the spatial structure of data points [29,30]. First, the study population at the SRS units was aggregated to the Indian postal code locations. Crude prevalences (i.e. without age standardization) were calculated for the smoking and drinking variables at the postal code locations. The kriging procedure used the calculated crude prevalence to create an interpolation surface for the entire country. Kriging was implemented using ArcGIS software (ESRI, 2011. Environmental Systems Research Institute, http://www.esri.com/software/arcgis/). Results from kriging interpolation allowed us to validate the calculated prevalence at the district level using the SFMS data, which was designed for state level analysis [24].

Moran's I is an index that measures the "global" spatial autocorrelation of values over the entire dataset [29]. An index score derived from the calculation indicates the strength of spatial autocorrelation of the values, with a score of 0.3 or more and −0.3 or less indicating relatively strong positive and strong negative autocorrelations, respectively [29,31]. While Moran's I determines the overall spatial autocorrelation of a variable, it does not provide a measure for spatial clusters. Hence, we used the LISA statistics to identify spatial clusters of the smoking and drinking variables. The LISA statistic gives an indication of the extent of significant spatial clustering of similar or dissimilar values around a spatial feature [32]. Four types of spatial associations can be derived from this statistic, with high-high (HH) and low-low (LL) types for spatial

Table 1. Prevalence of smoking and drinking among men age 30–69 years.

	District level			Postal code level		
	# of observations	Population (n)	Prevalence (%)	# of observations	Population (n)	Prevalence (%)
Study population	–	1,031,625	100	–	959,411	100
Any smokers	422,353	1,031,625	40.9	394,719	959,411	41.1
Cigarette smokers†	66,244	837,433†	7.9	60,908	783,426†	7.8
Bidi smokers†	245,274	837,433†	29.3	231,164	783,426†	29.5
Early smokers (before age 20)‡	170,162	419,419‡	40.6	158,835	391,980‡	40.5
Any drinkers	192,515	1,031,625	18.7	179,392	959,411	18.7

†Six states (Haryana, Bihar, Sikkim, Meghalaya, Mizoram, and Andhra Pradesh) were excluded from the analysis for smoking type as responses were not provided for majority of the smokers (>20%) from these states.

‡2934 and 2739 records from the smoking population at district and postal code level, respectively, were excluded due to invalid information for the resident's age at which he started smoking.

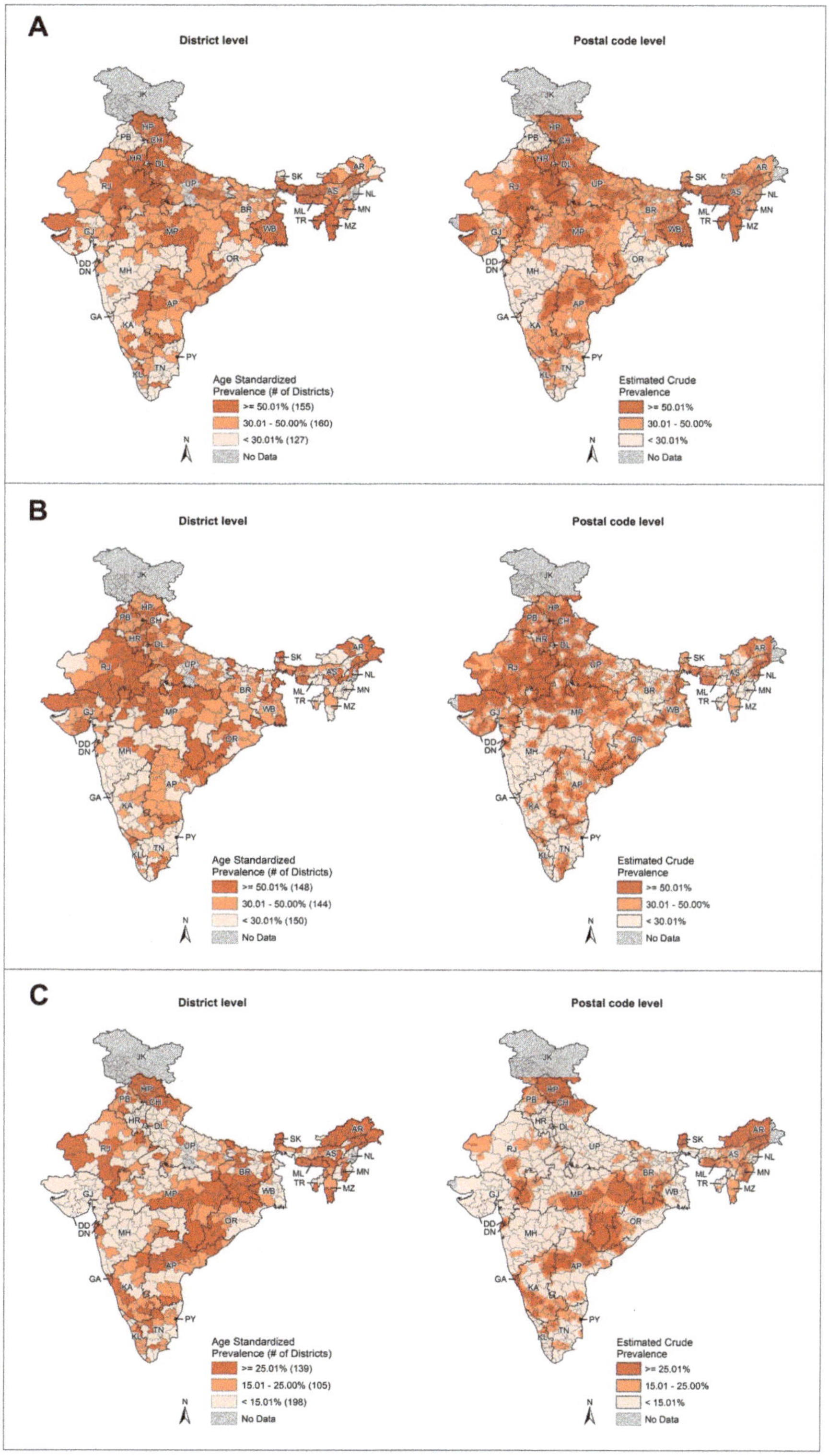

Notes on prevalence calculations:
Prevalence at the district level was age standardized using Census of India 2001 national population;
prevalence at the postal code level was estimated crude prevalence from kriging interpolation

Figure 1. Smoking and drinking prevalence at the district and postal code levels. Prevalence at district level was age standardized using Census of India 2001 national population; prevalence at postal code level was estimated crude prevalence from kriging interpolation. State abbreviations: AP - Andhra Pradesh, AR - Arunachal Pradesh, AS - Assam, BR - Bihar, CH - Chandigarh, DD - Daman and Diu, DL - Delhi, DN - Dadra & Nagar Haveli, GA - Goa, GJ - Gujarat, HP - Himachal Pradesh, HR - Haryana, JK - Jammu & Kashmir, KA - Karnataka, KL - Kerala, MG - Meghalaya, MH - Maharashtra, MN - Manipur, MP - Madhya Pradesh, MZ - Mizoram, NL - Nagaland, OR - Orissa, PB - Punjab, PD - Pondicherry, RJ - Rajasthan, SK - Sikkim, TN - Tamil Nadu, TR - Tripura, UP - Uttar Pradesh, WB - West Bengal. **A.** Any smoking prevalence at district and postal code levels. **B.** Early smoking (before age 20) prevalence at district and postal code levels. **C.** Any drinking prevalence at district and postal code levels.

Table 2. Univariate global Moran's I for smoking and drinking variables.

Variable	Moran's I
Any smoking	0.4204**
Cigarette smoking	0.5849**
Bidi smoking	0.4620**
Early smoking (before age 20)	0.2798**
Any drinking	0.3707**

**p<0.01.

clustering of similar values, and high-low (HL) and low-high (LH) types for spatial clustering of dissimilar values, that is, spatial outliers [31]. For example, the high-high type indicates that a district with high prevalence is surrounded by neighbouring districts also with high prevalence; while low-high type indicates that a district with low prevalence is surrounded by neighbouring districts with high prevalence. We used the first order queen spatial weight for identifying neighbours for the spatial features. The first order queen weight defines neighbours as districts that share either a common border or a vertex with a given district. Preliminary analysis showed that the first order queen weight produced superior results to higher order weights (data not shown).

Two types of Moran's I and LISA analyses were performed: univariate and bivariate. Using the district-level age standardized prevalence (Appendix S1), we performed univariate Moran's I and LISA to examine the spatial autocorrelation and spatial clusters, respectively, for each smoking and drinking variable. We used bivariate Moran's I and LISA to examine the spatial associations between smoking and drinking, as well as between cigarette and bid smoking.

A permutation approach was used to assess the statistical significance of the Moran's I and LISA results ($p<0.05$). The permutation approach produced a reference distribution for significance testing by randomly shuffling the actual data values over space a given number of times [33]. Moran's I and LISA statistics were calculated using GeoDa software [34]. Results from both kriging interpolation and LISA analysis were visualized using ArcGIS software.

Results

Among men aged 30–69 years in 1998, 40.9% were smokers and 18.7% were drinkers; among smokers, 40.6% were early smokers (before age 20) (Table 1). For smoking types, six states were excluded due to greater than 20% missing data on smoking types–Haryana, Bihar, Sikkim, Meghalaya, Mizoram, and Andhra Pradesh. In the remaining states, 39.7% of men aged 30–69 years were smokers, among which 7.9% were cigarette smokers and 29.3% were bidi smokers, while 2.5% of the records had missing data for smoking type. At the postal code level, about 7% of the study population was excluded as they could not be aggregated to postal code locations. Nevertheless, the prevalence for both smoking and drinking status at the two geographic levels remained similar.

The geographic distributions for both smoking and drinking were similar between the district and postal code levels in India (Figure 1A–C, Figure S1D–E). Kriging interpolation identified the sub-district level patterns that were concealed in the district level maps. It also helped to provide estimations for districts not covered by the SFMS, such as those in Uttar Pradesh. The prevalence of smoking was high in the central, northern (except Punjab), and the northeastern states, while the prevalence was lower in Punjab, Maharashtra and Orissa (Figure 1A). Some neighbouring states showed contrasting results (e.g. high prevalence in Madhya Pradesh vs. low prevalence in Maharashtra). Prevalence of early smoking (before age 20) was high in the central and northern states (Figure 1B). Although Punjab had low smoking prevalence, it had high prevalence of early smoking (before age 20). Cigarette and bidi smoking showed distinctly different geographic patterns: high cigarette smoking prevalence was found in the southern and northeastern states (Figure S1D); while high bidi smoking prevalence was found in the central and northern (except Punjab) states (Figure S1E). West Bengal had high prevalence for both cigarette and bidi smoking. The geographic pattern of bidi smoking was similar to early smoking (before age 20). High prevalence of any drinking was located in Himachal Pradesh, parts of the northeastern states, and the central eastern part of the country (Figure 1C).

In Moran's I analysis, all smoking and drinking variables were statistically significant ($p<0.01$) and had relatively strong autocorrelations (index scores close to or more than 0.3) (Table 2). Bivariate Moran's I results showed that cigarette smoking was positively correlated to any drinking (index score = 0.119, $p<0.01$) and negatively correlated to bidi smoking (index score = −0.112, $p<0.05$) (Table 3). However, any smoking did not show any significant spatial correlation with any drinking (index score = 0.031, $p = 0.14$) or drinking more than 3 days a week (data not shown).

The univariate LISA results highlighted districts with significantly high prevalence as high-high clusters and low prevalence as low-low clusters (Figure 2A–C, Figure S2D–E). Spatial outliers, specifically high-low and low-high types, also existed in the maps. Compared to the high-high and low-low clusters, fewer districts belonged to the high-low and low-high types, and these outliers did not often neighbour other outliers in clusters. For early smoking (before age 20) and cigarette smoking, outliers were found near the boundary shared by Rajasthan, Uttar Pradesh, and Madhya Pradesh, and they were surrounded by spatial clusters in the adjacent regions (Figure 2B, Figure S2D).

The results from bivariate LISA analysis showed that compared to univariate LISA, sizes of high-high and low-low clusters were smaller and the number of high-low and low-high outliers was higher (Figure 3A–B, Figure S3C–D). This indicates that the bivariate spatial correlations between smoking and drinking were not as strong as those found in the univariate analysis. The bivariate LISA results were supported by the bivariate Moran's I results as most of the bivariate pairs had weak and non-significant results (Table 3). Nonetheless, it can be observed that for smoking

Table 3. Bivariate global Moran's I for smoking versus drinking, and cigarette smoking versus bidi smoking.

Variable	Moran's I
Any smoking *vs.* any drinking	0.0308
Cigarette smoking *vs.* any drinking	0.1190**
Bidi smoking *vs.* any drinking	−0.0525
Cigarette smoking *vs.* Bidi smoking	−0.1115*

*p<0.05;
**p<0.01.

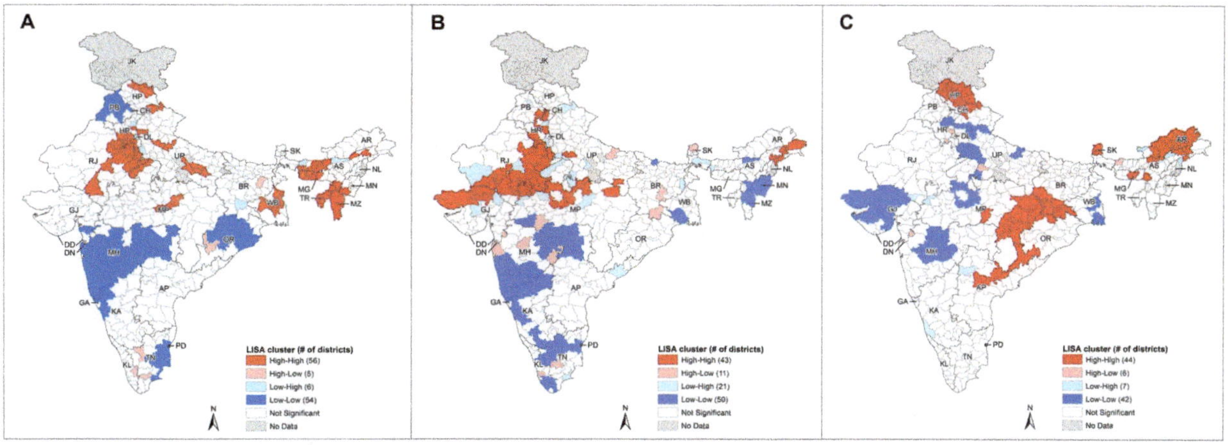

Figure 2. LISA cluster maps for smoking and drinking prevalence at the district level. State abbreviations: AP - Andhra Pradesh, AR - Arunachal Pradesh, AS - Assam, BR - Bihar, CH - Chandigarh, DD - Daman and Diu, DL - Delhi, DN - Dadra & Nagar Haveli, GA - Goa, GJ - Gujarat, HP - Himachal Pradesh, HR - Haryana, JK - Jammu & Kashmir, KA - Karnataka, KL - Kerala, MG - Meghalaya, MH - Maharashtra, MN - Manipur, MP - Madhya Pradesh, MZ - Mizoram, NL - Nagaland, OR - Orissa, PB - Punjab, PD - Pondicherry, RJ - Rajasthan, SK - Sikkim, TN - Tamil Nadu, TR - Tripura, UP - Uttar Pradesh, WB - West Bengal. **A.** LISA cluster map for any smoking prevalence. **B.** LISA cluster map for early smoking (before age 20) prevalence. **C.** LISA cluster map for any drinking prevalence.

versus drinking, high-high clusters exist in Himachal Pradesh, and low-low clusters exist in Gujarat and Maharashtra (Figure 3A). Cigarette smoking had the strongest spatial correlation with drinking (Figure S3C). For cigarette smoking versus bidi smoking, high-high clusters were found in West Bengal, and low-low clusters were found in Punjab, Maharashtra, and Orissa (Figure 3B). The low-high outliers found in the central states clearly showed that low prevalence of cigarette smoking coexisted with high prevalence of bidi smoking, and vice versa for the states of Kerala and Tamil Nadu. Overall, there was high prevalence for both cigarette and bidi smoking in West Bengal, and low prevalence for both smoking and drinking in Gujarat and Maharashtra (Table 4).

We compared our results on smoking prevalences with the living respondents in the Million Death Study (MDS) [23] and adults interviewed in the Global Adult Tobacco Survey (GATS) 2009–2010 [3]. We found correlations of smoking prevalences between SFMS in 1998 and MDS in 2004 at the district level (any smoking: $R^2 = 0.58$, $p < 0.01$; cigarette smoking: $R^2 = 0.52$, $p < 0.01$; bidi smoking: $R^2 = 0.56$, $p < 0.01$), and high correlations between SFMS in 1998 and GATS in 2010 at the state level (any

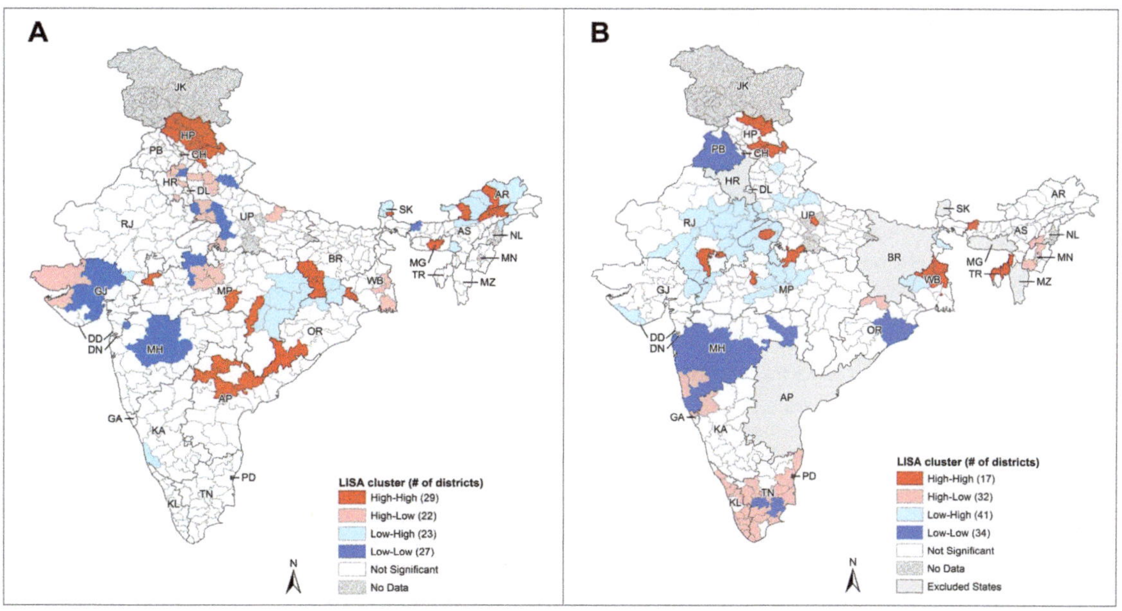

Figure 3. Bivariate LISA cluster maps. State abbreviations: AP - Andhra Pradesh, AR - Arunachal Pradesh, AS - Assam, BR - Bihar, CH - Chandigarh, DD - Daman and Diu, DL - Delhi, DN - Dadra & Nagar Haveli, GA - Goa, GJ - Gujarat, HP - Himachal Pradesh, HR - Haryana, JK - Jammu & Kashmir, KA - Karnataka, KL - Kerala, MG - Meghalaya, MH - Maharashtra, MN - Manipur, MP - Madhya Pradesh, MZ - Mizoram, NL - Nagaland, OR - Orissa, PB - Punjab, PD - Pondicherry, RJ - Rajasthan, SK - Sikkim, TN - Tamil Nadu, TR - Tripura, UP - Uttar Pradesh, WB - West Bengal. **A.** Bivariate LISA cluster map for any smoking prevalence versus any drinking prevalence. **B.** Bivariate LISA cluster map for cigarette smoking prevalence versus bidi smoking prevalence.

Table 4. Broad geographic patterns by regions across India for smoking and drinking variables.

Region	Variable (State name if applicable)	
	with high prevalence	with low prevalence
North	Any drinking (Himsachal Pradesh)	Any smoking (Punjab); Bidi smoking (Punjab)
Northeast	Any smoking; Cigarette smoking; Any drinking (Arunachal Pradesh, Sikkim)	n/a
East	Any smoking (West Bengal); Bidi smoking (West Bengal); Cigarette smoking (West Bengal)	Any smoking (Orissa); Bidi smoking (Orissa)
Central	Any smoking; Bidi smoking; Early smoking (before age 20)	Cigarette smoking
West	n/a	Any smoking (Maharashtra); Cigarette smoking (Gujarat, Maharashtra); Bidi smoking (Maharashtra); Early smoking (before age 20) (Maharashtra); Any drinking (Gujarat, Maharashtra)
South	Cigarette smoking (Kerala)	Any smoking (Tamil Nadu); Bidi smoking (Tamil Nadu)

smoking: $R^2 = 0.67$, $p<0.01$; cigarette smoking: $R^2 = 0.64$, $p< 0.01$; bidi smoking: $R^2 = 0.70$, $p<0.01$).

Discussion

Our results confirmed that smoking prevalence is significantly higher in the northeastern and central states, while it is significantly lower in Punjab, Maharashtra, and Orissa. These state level results of smoking are comparable to the findings from previous research studies [11,12] and GATS [3]. Punjab had significantly lower smoking prevalence as the majority of its population practices Sikhism, which prohibits tobacco use [11,12]. The high prevalence in the northeastern states is consistent with the GATS report [3]. Moreover, high prevalence has been reported among students in this region [35].

There was a high prevalence of any drinking in Himachal Pradesh, south Bihar, eastern Madhya Pradesh (presently Chhattisgarh state), and in the northeastern states (Sikkim and Arunachal Pradesh in particular). Many of these states have high proportion of tribal population [36] and liquor drinking is common among the tribal persons [37], as well as a low proportion of Muslim population [38], in whom the reported alcohol drinking prevalence is low [14,39]. The low prevalence in Gujarat and Maharashtra could be attributed to the alcohol prohibition in these states, with Gujarat under complete prohibition (theoretically) on the production and sale of alcohol [7]. Surprisingly, we found that Tamil Nadu had low drinking prevalence even though other research has found high drinking prevalence [40,41].

Our results showed the sub-state variations of smoking and drinking prevalence and the coexistence of high-high and low-low clusters with outliers within a state. For example, in Madhya Pradesh low-low clusters for cigarette smoking were found in the western and eastern part of the state, with high-low outliers found in the central part of the state. These spatial heterogeneities of tobacco and alcohol use at the sub-state levels were also found in previous studies [12,14].

Cigarette smoking was high in the northeastern and the southern states, while bidi smoking was high in the central states and in West Bengal. These two smoking types are characterized by different users in India. Cigarette smoking is common among the educated and people with skilled occupations, while bidi smoking is common among the rural population, the urban poor, and the less educated [9,42]. The relatively inexpensive bidis are more affordable and attractive to young people [43]. Our results show that the geographic patterns of bidi smoking closely resemble those of early smoking (before age 20). Control policies on bidi smoking thus may reduce smoking uptake among young people.

Since 1998, data has shown an increase in sales and the affordability of tobacco [44], despite the implementation of several tobacco control policies [45,46]. Limited data suggested an increase in smoking prevalence, in particular among younger people [44]. Evidence also suggests that cigarettes are displacing bidis among younger males [47]. Meanwhile, increases in alcohol sales and use have been observed [5,48], along with the more relaxed restrictions on consumption in many states [49]. For example, alcohol prohibition is slowly being liberalized and consumption is increasing in Maharashtra. The increase in popularity of both cigarette smoking and drinking required our attention as we found an association between the two. Previous studies also found complementarities between tobacco smoking and alcohol drinking among Indians [8–10]. Smoking cessation may enhance alcohol abstinence [50] and certain types of smoking cessation medication can help reduce the frequency of alcohol consumption [51]. Thus, local public health action on smoking might help reduce alcohol consumption, and vice versa. Tobacco and alcohol control is currently implemented at the state level [6,16], and thus reliable monitoring of changes in consumption patterns and complementarities between cigarette smoking and alcohol drinking at the district level is required to enable effective control.

This study has several limitations. First, the SFMS was aimed to provide information representative at the state level [24], similar to the NFHS 1998–1999 [25] and GATS 2009–2010 [3]. But its large sample size yielded over 1 million individuals for this study, which provide sufficient number for stable district level analysis. Results from the kriging interpolations also confirmed the geographic patterns at the district level. Second, the SFMS was conducted over a decade ago and the current level of smoking and drinking among Indians may now differ. However, smoking prevalences in 1998 were correlated with prevalence in 2004 at the district level (SFMS *vs.* MDS) and with prevalences in 2010 at the state level (SFMS *vs.* GATS). Notwithstanding the different survey designs between SFMS and GATS (i.e. the use of household informants in SFMS and selected individuals in GATS), conclusions derived from our results at sub-state level are still likely to apply to the current Indian population. Third, the use of household informants may lead to under-reporting of smoking and drinking behaviours as they may not be aware of these consumption behaviours of the household residents (such as number of days alcohol was used), especially among younger males. Under-reporting could also be caused by the social desirability bias. For example, in states where prohibitions on tobacco and alcohol use were in effect, respondents may be less likely to report the smoking and drinking behaviours of their household residents. Fourth, our results are nationally representative and applicable to Indian males aged 30–69, which comprises the majority of smokers in India [11,26]. The exclusion of women was reasonable as smoking and drinking prevalences among women remained low in the last decade [3,11,27,28]. Similarly, the exclusion of six states (with 19% of study population) from the analysis of cigarette and bidi geographic patterns should not greatly alter the national results for the remaining 23 states. Finally, although the prevalence for the various smoking and drinking variables were not markedly different between the district level and postal code level study populations, it is possible that the 7% of records excluded in the postal code level were spatially clustered. Therefore, the kriging interpolations only provide the approximate distributions of smoking and drinking at the sub-district level.

Conclusion

India has over 100 million current tobacco smokers and it accounts for approximately one-fifth of the world's tobacco-related deaths [52,53]. There is an urgent need of ongoing research to determine the current level of smoking in the population. Our study informs Indian tobacco control policy by filling the research gap of tobacco use and the use of cigarette and bidi smoking at the sub-state level in India. We found that cigarette smoking dominated the northeastern and southern states and bidi smoking dominated the central states, with early smoking (before age 20) closely resembling the geographic patterns of bidi smoking. We also identified high-burden areas of drinking and its spatial association with cigarette smoking. Finally, the sub-state variations of smoking and drinking require the attention of state policy-makers. In order to control tobacco and alcohol use more effectively, state governments should implement the existing national acts of tobacco and alcohol control at the district level. Surveys that properly represent their consumption at the district level are recommended. Reductions in smoking may simultaneously reduce alcohol consumption, and vice versa.

Supporting Information

Figure S1 Cigarette smoking and bidi smoking prevalence at the district and postal code levels. Prevalence at district level was age standardized using Census of India 2001 national population; prevalence at postal code level was estimated crude prevalence from kriging interpolation. State abbreviations: AP - Andhra Pradesh, AR - Arunachal Pradesh, AS - Assam, BR - Bihar, CH - Chandigarh, DD - Daman and Diu, DL - Delhi, DN - Dadra & Nagar Haveli, GA - Goa, GJ - Gujarat, HP - Himachal Pradesh, HR - Haryana, JK - Jammu & Kashmir, KA - Karnataka, KL - Kerala, MG - Meghalaya, MH - Maharashtra, MN - Manipur, MP - Madhya Pradesh, MZ - Mizoram, NL - Nagaland, OR - Orissa, PB - Punjab, PD - Pondicherry, RJ - Rajasthan, SK - Sikkim, TN - Tamil Nadu, TR - Tripura, UP - Uttar Pradesh, WB - West Bengal. **D.** Cigarette smoking prevalence at district and postal code levels. **E.** Bidi smoking prevalence at district and postal code levels.

Figure S2 LISA cluster maps for cigarette smoking and bidi smoking prevalence at the district level. State abbreviations: AP - Andhra Pradesh, AR - Arunachal Pradesh, AS - Assam, BR - Bihar, CH - Chandigarh, DD - Daman and Diu, DL - Delhi, DN - Dadra & Nagar Haveli, GA - Goa, GJ - Gujarat, HP - Himachal Pradesh, HR - Haryana, JK - Jammu & Kashmir, KA - Karnataka, KL - Kerala, MG - Meghalaya, MH - Maharashtra, MN - Manipur, MP - Madhya Pradesh, MZ - Mizoram, NL - Nagaland, OR - Orissa, PB - Punjab, PD - Pondicherry, RJ - Rajasthan, SK - Sikkim, TN - Tamil Nadu, TR - Tripura, UP - Uttar Pradesh, WB - West Bengal. **D.** LISA cluster map for cigarette smoking prevalence. **E.** LISA cluster map for bidi smoking prevalence.

Figure S3 Bivariate LISA cluster maps. State abbreviations: AP - Andhra Pradesh, AR - Arunachal Pradesh, AS - Assam, BR - Bihar, CH - Chandigarh, DD - Daman and Diu, DL - Delhi, DN - Dadra & Nagar Haveli, GA - Goa, GJ - Gujarat, HP - Himachal Pradesh, HR - Haryana, JK - Jammu & Kashmir, KA - Karnataka, KL - Kerala, MG - Meghalaya, MH - Maharashtra, MN - Manipur, MP - Madhya Pradesh, MZ - Mizoram, NL - Nagaland, OR - Orissa, PB - Punjab, PD - Pondicherry, RJ - Rajasthan, SK - Sikkim, TN - Tamil Nadu, TR - Tripura, UP - Uttar Pradesh, WB - West Bengal. **C.** Bivariate LISA cluster map for cigarette smoking prevalence versus any drinking prevalence. **D.** Bivariate LISA cluster map for bidi smoking prevalence versus any drinking prevalence.

Acknowledgments

The authors thank Chinthanie Ramasundarahettige for her support in data preparation, suggestions on the methodology, and comments on the manuscript, and Samir Gambhir and Peter Rodriguez for their support in GIS analysis. We also thank Marvin Hsiao for his valuable comments on the manuscript.

Author Contributions

Conceived and designed the experiments: SHF PJ. Performed the experiments: SHF. Analyzed the data: SHF PJ PG RK. Contributed reagents/materials/analysis tools: SHF PJ. Wrote the paper: SHF PJ PG RK. Reviewed the manuscript: SHF PJ PG RK RD DS.

References

1. Mathers C, Stevens G, Mascarenhas M (2009) Global Health Risks: Mortality and burden of disease attributable to selected major risks. Geneva: WHO.
2. Jha P, Nugent R, Verguet S, Bloom D, Hum R (2012) Chronic disease prevention and control. Challenge Paper for the Copenhagen Consensus 2012.
3. International Institute of Population Sciences (IIPS) (2010) Global Adult Tobacco Survey: India 2009–2010. New Delhi: Ministry of Health and Family Welfare, Government of India.
4. Jha P, Jacob B, Gajalakshmi V, Gupta PC, Dhingra N, et al. (2008) A nationally representative case-control study of smoking and death in India. N Engl J Med 358: 1137–1147.
5. Prasad R (2009) Alcohol use on the rise in India. Lancet 373: 17–18.
6. Benegal V (2005) India: alcohol and public health. Addiction 100: 1051–1056.
7. Mahal A (2000) What works in alcohol policy? Evidence from rural India. Economic Political Weekly 35: 3959–3968.
8. Mohan D, Chopra A, Sethi H (2002) The co-occurrence of tobacco & alcohol in general population of metropolis Delhi. Indian J Med Res 116: 150–154.

9. Narayan KM, Chadha SL, Hanson RL, Tandon R, Shekhawat S, et al. (1996) Prevalence and patterns of smoking in Delhi: cross sectional study. BMJ 312: 1576–1579.
10. Gupta PC, Maulik PK, Pednekar MS, Saxena S (2005) Concurrent alcohol and tobacco use among a middle-aged and elderly population in Mumbai. Natl Med J India 18: 88–91.
11. Rani M, Bonu S, Jha P, Nguyen SN, Jamjoum L (2003) Tobacco use in India: prevalence and predictors of smoking and chewing in a national cross sectional household survey. Tob Control 12: e4.
12. Subramanian SV, Nandy S, Kelly M, Gordon D, Davey Smith G (2004) Patterns and distribution of tobacco consumption in India: cross sectional multilevel evidence from the 1998–9 national family health survey. BMJ 328: 801–806.
13. John RM (2005) Tobacco consumption patterns and its health implications in India. Health Policy 71: 213–222.
14. Subramanian SV, Nandy S, Irving M, Gordon D, Davey Smith G (2005) Role of socioeconomic markers and state prohibition policy in predicting alcohol consumption among men and women in India: a multilevel statistical analysis. Bull World Health Organ 83: 829–836.
15. Census of India (2011) Primary Census Abstract Data Tables. New Delhi: Registrar General of India, Government of India.
16. Mehrotra R, Mehrotra V, Jandoo T (2010) Tobacco control legislation in India: past and present. Indian J Cancer 47 Suppl 1: 75–80.
17. Ricketts TC (2003) Geographic information systems and public health. Annu Rev Public Health 24: 1–6.
18. Beale L, Abellan JJ, Hodgson S, Jarup L (2008) Methodologic issues and approaches to spatial epidemiology. Environ Health Perspect 116: 1105–1110.
19. Singh A, Pathak PK, Chauhan RK, Pan W (2011) Infant and child mortality in India in the last two decades: a geospatial analysis. PLoS One 6: e26856.
20. Kandwal R, Augustijn EW, Stein A, Miscione G, Garg PK, et al. (2010) Geospatial analysis of HIV-Related social stigma: a study of tested females across mandals of Andhra Pradesh in India. Int J Health Geogr 9: 18.
21. Srivastava A, Nagpal BN, Joshi PL, Paliwal JC, Dash AP (2009) Identification of malaria hot spots for focused intervention in tribal state of India: a GIS based approach. Int J Health Geogr 8: 30.
22. Registrar General of India (2005) Special Fertility & Mortality Survey, A Report of 1.1 Million Indian Households. New Delhi: Registrar General.
23. Jha P, Gajalakshmi V, Gupta PC, Kumar R, Mony P, et al. (2006) Prospective study of one million deaths in India: rationale, design, and validation results. PLoS Med 3: e18.
24. Registrar General of India (1996) Sample Registration System Statistical Report 1996. New Delhi: Registrar General of India.
25. International Institute of Population Sciences (IIPS) (2000) National family health survey–II (1998–1999). Bombay: International Institute of Population Studies.
26. Registrar General of India (2004) Sample Registration System (SRS): Baseline Survey Report -2004. New Delhi: Office of the Registrar General of India.
27. Neufeld KJ, Peters DH, Rani M, Bonu S, Brooner RK (2005) Regular use of alcohol and tobacco in India and its association with age, gender, and poverty. Drug Alcohol Depend 77: 283–291.
28. International Institute for Population Sciences (IIPS) and Macro International (2007) National Family Health Survey (NFHS-3), 2005–06: India: Volume I. Mumbai: IIPS.
29. O'Sullivan D, Unwin D (2010) Geographic Information Analysis. Hoboken, New Jersey: John Wiley & Sons, Inc.
30. Oliver MA, Webster M (1990) Kriging: a method of interpolation for GIS. International Journal of Geographical Information Systems 4: 313–332.
31. Pfeiffer DU, Robinson TP, Stevenson M, Stevens KB, Rogers DJ, et al. (2008) Spatial Analysis in Epidemiology: Oxford University Press.
32. Anselin L (1995) Local Indicators of Spatial Association - Lisa. Geographical Analysis 27: 93–115.
33. Anselin L (2005) Exploring Spatial Data with GeoDa: A Workbook. Spatial Analysis Laboratory, Department of Geography University of Illinois and Center for Spatially Integrated Social Science, IL, USA.
34. Anselin L, Syabri I, Kho Y (2006) GeoDa: An Introduction to Spatial Data Analysis. Geographical Analysis 38: 5–22.
35. Sinha DN, Gupta PC, P G (2007) Tobacco use among students and school personnel in India. Asian Pac J Cancer Prev 8: 417–421.
36. Census of India (2001) Census Reference Tables, A – Series: General Population Tables. New Delhi: Registrar General of India, Government of India.
37. Joshi RA (2010) Role of NGOs In Removing Tribal Problems. In: Kanmony JC, editor. Dalits and Tribes of India. New Delhi: Krishan Mittal.
38. Census of India (2001) Census Reference Tables, C – Series: Social and Cultural Tables. New Delhi: Registrar General of India, Government of India.
39. Gupta PC, Saxena S, Pednekar MS, Maulik PK (2003) Alcohol consumption among middle-aged and elderly men: a community study from western India. Alcohol Alcohol 38: 327–331.
40. Chakravarthy C (1990) Community workers estimate of drinking and alcohol-related problems in rural areas. Indian Journal of Psychological Medicine 13: 49–56.
41. John A, Barman A, Bal D, Chandy G, Samuel J, et al. (2009) Hazardous alcohol use in rural southern India: Nature, prevalence and risk factors. The National Medical Journal of India 22: 123–125.
42. Rahman M, Fukui T (2000) Bidi smoking and health. Public Health 114: 123–127.
43. Gupta PC, Asma S (2008) Bidi Smoking and Public Health. New Delhi: Ministry of Health and Family Services, Government of India.
44. John RM, Rao RK, Rao MG, Moore J, Deshpande RS, et al. (2010) The Economics of Tobacco and Tobacco Taxation in India. Paris: International Union Against Tuberculosis and Lung Disease.
45. Kaur J, Jain DC (2011) Tobacco control policies in India: implementation and challenges. Indian J Public Health 55: 220–227.
46. Government of India (2000) Cable Television Networks (Amendment) Act.
47. Jha P., Guindon E., Joseph R A., Nandi A., John R M., et al. (2011) A Rational Taxation System of Bidis and Cigarettes to Reduce Smoking Deaths in India. Economic & Political Weekly 46: 44–51.
48. Jacob KS (2010) Alcohol and public health policies in India. The National Medical Journal of India 23: 224–225.
49. Das SK, Balakrishnan V, Vasudevan DM (2006) Alcohol: Its health and social impact in India. The National Medical Journal of India 19: 94–99.
50. Kalman D, Kim S, DiGirolamo G, Smelson D, Ziedonis D (2010) Addressing tobacco use disorder in smokers in early remission from alcohol dependence: the case for integrating smoking cessation services in substance use disorder treatment programs. Clin Psychol Rev 30: 12–24.
51. McKee SA, Young-Wolff KC, Harrison EL, Cummings KM, Borland R, et al. (2013) Longitudinal associations between smoking cessation medications and alcohol consumption among smokers in the International Tobacco Control Four Country survey. Alcohol Clin Exp Res 37: 804–810.
52. Tiwari R, Deb P, Debbarma A, Chaudhuri R, Chakraborty A, et al. (2006) Tobacco use and cardiovascular disease: a knowledge, attitude and practice study in rural Kerala. Indian J Med Sci 60: 271–276.
53. Jha P, Peto R (2014) Global effects of smoking, of quitting, and of taxing tobacco. N Engl J Med 370: 60–68.

2

A System for Household Enumeration and Re-identification in Densely Populated Slums to Facilitate Community Research, Education, and Advocacy

Dana R. Thomson[1]*, Shrutika Shitole[2], Tejal Shitole[2], Kiran Sawant[2], Ramnath Subbaraman[2,3], David E. Bloom[4], Anita Patil-Deshmukh[2]

1 Department of Global Health and Social Medicine, Harvard Medical School, Boston, Massachusetts, United States of America, **2** Partners for Urban Knowledge, Action, and Research, Mumbai, India, **3** Division of Infectious Diseases, Brigham and Women's Hospital, Boston, Massachusetts, United States of America, **4** Department of Global Health and Population, Harvard School of Public Health, Boston Massachusetts, United States of America

Abstract

Background: We devised and implemented an innovative Location-Based Household Coding System (LBHCS) appropriate to a densely populated informal settlement in Mumbai, India.

Methods and Findings: LBHCS codes were designed to double as unique household identifiers and as walking directions; when an entire community is enumerated, LBHCS codes can be used to identify the number of households located per road (or lane) segment. LBHCS was used in community-wide biometric, mental health, diarrheal disease, and water poverty studies. It also facilitated targeted health interventions by a research team of youth from Mumbai, including intensive door-to-door education of residents, targeted follow-up meetings, and a full census. In addition, LBHCS permitted rapid and low-cost preparation of GIS mapping of all households in the slum, and spatial summation and spatial analysis of survey data.

Conclusion: LBHCS was an effective, easy-to-use, affordable approach to household enumeration and re-identification in a densely populated informal settlement where alternative satellite imagery and GPS technologies could not be used.

Editor: Thomas Eisele, Tulane University School of Public Health and Tropical Medicine, United States of America

Funding: This project was funded by the Rockefeller Foundation. RS's contribution was funded by the Fogarty International Clinical Research Scholars and Fellows Program at Vanderbilt University (R24 TW007988) and more recently by a Harvard University T32 Post-Doctoral Fellowship (NIAID AI 007433). Financial support was also provided by The Weatherhead Center for International Affairs at Harvard University. No funding bodies had any role in the study design, data collection and analysis, decision to publish, or preparation of the manuscript.

Competing Interests: The authors have declared that no competing interests exist.

* E-mail: dana_thomson@hms.harvard.edu

Background

Conducting research and outreach in slum communities is challenging for several reasons, including that the absence of a street address system makes it difficult to uniquely identify and follow up with households. Alternative household enumeration methods, including use of GPS units and satellite imagery, do not work in densely populated slums. With one in six people living in slums worldwide, including one out of every three city dwellers [1], household enumeration and identification poses an enormous challenge for researchers, community organizers, and advocates working with the urban poor. Slum environments are also challenging because they change rapidly. Residents move frequently, and structures are rebuilt often due to poor construction, natural disaster, change in tenant configurations, local politics, and forced eviction [2]. In these dynamic settings, enumeration of the community's residents, physical structures, and boundaries aids in recognition of the community's existence which can serve as a tool to lobby officials for pro-poor housing policies and participatory slum redevelopment [2]. Adaptive, easy-to-implement housing enumeration methods that can be used by slum dwellers and researchers alike in high-density settings are sorely needed.

Various methods have been used for slum enumeration, including hand-drawn maps on paper or in a geographic information system (GIS) [3], mapping with GPS units [4], enumerating from satellite imagery [3], or some combination of the three [5]. Advanced technologies, such as spatially encoded video [6], have also been used to capture community health risks. Each enumeration method poses challenges. Drawing maps by hand is time-intensive and requires many enumerators. Use of GPS units and spatially encoded video require a technical team to collect, clean, and maintain spatial data, and technically trained workers or volunteers may not be readily available [7]. Members of our team who worked with organizations in other slum settlements found that locating a household with a GPS unit and a printed map (a common way to identify households coded by GPS) requires 20–30 minutes in the field per household to ground-truth the GPS reading and converse with residents in nearby homes. Furthermore, use of GPS is problematic in densely populated slums because current GPS units have a spatial error of up to 33 feet (10 meters), and building cover prevents satellite connection

[7,8]. High building density is especially problematic in a city such as Mumbai, where over half of the population lives in slums on six percent of its land [9], making Mumbai's slums some of the highest-density settlements on earth. For example, Dharavi, one of Mumbai's largest slums, is estimated to have a population density greater than 400,000 people per square kilometer, which is nearly ten times denser than daytime Manhattan [10]. Enumeration from recently collected satellite imagery can be very efficient and accurate [11], though in very densely populated slums like Dharavi, structures are stacked several stories high and share walls so that the roofs belonging to different dwellings are indistinguishable.

In 2008, researchers from Partners for Urban Knowledge, Action and Research (PUKAR); Harvard School of Public Health (HSPH); and New York University (NYU) formed a collaboration to perform a series of health studies, and health and education campaigns in Kaula Bandar, a slum of approximately 14,000 people in Mumbai, India. The slum is located on a former shipping pier one-tenth of a square kilometer in size with one road passable by motor vehicle down the center, and a labyrinth of pedestrian-only lanes on either side, most of which are covered by buildings. Many of the lanes are so narrow that individuals can only pass through in a single-file configuration, and larger individuals may need to turn sideways to navigate some of the narrow spaces (Figure 1). Approximately half of the land space in Kaula Bandar is residential - the rest is industrial or open space - which means that residential density is approximately 280,000 people per square kilometer (see Figure 2). Dwellings in Kaula Bandar range in size from 20 to 90 square feet (2 to 8 square meters), and are home to 5 people on average.

Kaula Bandar is a "non-notified" slum, which means it is not recognized by local and state governments. Non-notified status greatly limits residents' ability to pressure officials for basic water and sanitation infrastructure, schools, or health services [12]. As a result, all water, sanitation, and power infrastructure is improvised by residents except for a few public toilet facilities. PUKAR is unique in that the organization is staffed by a small national research team that hires and trains a cadre of what Arjun Appadurai calls "Barefoot Researchers" from the community to collect data and perform community education and outreach [13]. Barefoot Researchers are often young (ages 18–23), and many have limited formal education. Engagement and training of Barefoot Researchers was conducted in the spirit of what Kretzmann and McKnight call Community Asset Mapping; that is, taking inventory of the capacities, skills, and assets of low-income people and their communities to reinforce community pride and internally generated development, rather than document needs and deficiencies which reinforces dependency on outsiders [14]. In this way, research, education, advocacy, and community building in Kaula Bandar are integrated and informed by one another.

Nearly half of Kaula Bandar's adult residents have lived in the slum for more than 20 years, and its population continues to grow as new residents move in and young adults start their own families. As a result, new lanes are formed by the construction of new dwellings. Typically, a dwelling's ground level is occupied by the owner, while 2nd-level units are rented. A large portion of Kaula Bandar's residents move away seasonally for work or to care for relatives in rural villages. The majority of dwellings in Kaula Bandar have existed for years, though we estimate that the Bombay Port Trust demolished at least 5% of homes in the last five years targeting new construction in open spaces and on the main road, and large fires in 2010 and 2013 destroyed approximately 9% and 3% of dwellings, respectively. After these fires, government support to rebuild was promised, though only a few politically connected families received funds, and only after months of delay. Most burned dwellings were rebuilt and reoccupied by the same tenants using private resources, and those households that started rebuilding first slightly increased the size and quality of their dwellings.

In this densely populated community with regular changes to housing and population, none of the typical enumeration methods are feasible; GPS units cannot make satellite or cell phone network contact from narrow, covered lanes, and dwelling rooftops are indistinguishable in satellite imagery. While hand-mapping is an option, it requires extensive human-power and time. In this paper, we describe a method developed by the PUKAR research team to code and find households efficiently, present how we extended the coding system for mapping and spatial analysis, and discuss the utility of this method for research and practice in other densely populated or dynamic slum settings.

Methods

Location-based Household Coding

PUKAR's three lead field researchers (authors SS, TS, and KS), two of whom are from communities near Kaula Bandar, created a simple, innovative Location-Based Household Coding System (LBHCS) to uniquely identify households. This system allowed them and each of the Barefoot Researchers to find their way to a specific household in a matter of minutes. The PUKAR team started with a simple hand-sketched map of Kaula Bandar that included the main road and each of the community's 25 main lanes connecting the central road with the edge of the pier (see Figure 1). At the time that the system was developed, each of the three researchers led a team of Barefoot Researchers in a section of the community. Main lanes were assigned codes like T4, S3, and K7, indicating it was the 4th, 3rd, or 7th main lane in Tejal, Shrutika, or Kiran's section. Each sub-lane was coded with the direction (left or right) from which it originated from the main lane (from the perspective of where the pier connects to land), and the order in which it appeared (1st, 2nd, 3rd sub-lane). Additional branches of sub-lanes and smaller by-lanes were accordingly coded, resulting in such codes as K9-LSL2-LBL1, which represented the 9th main lane in Kiran's section, 2nd left sub-lane, 1st left by-lane. Special codes were created for homes near to the sea (NS) and the road outside of Kaula Bandar (RO). Specific home locations were indicating by adding a side of the lane and house number to the lane code, as in K9-LSL2-LBL1-R3D, the 3rd house on the right, downstairs, on lane K9-LSL2-LBL1. Second-story households were coded as U for upstairs level. In this way, household codes doubled as detailed, easy-to-follow walking directions and the basis for a relational database of all addresses at a particular point in time.

For each study, PUKAR maintained a database of each participant and their household code. Each time the participant was visited in their home for data collection, medical follow up, or educational outreach, the field researcher verified the accuracy of the household code, and noted on paper any corrections resulting from previous recording error or because the participant had moved. Updated household codes were entered into in an excel database by the database manager who tracked all old or incorrect codes in a separate column.

GIS Mapping

As noted, Kaula Bandar is so densely populated that individual households are indistinguishable in satellite imagery. And most lanes, sub-lanes, and by-lanes are covered by housing so they do

Figure 1. Kaula Bandar, Mumbai, India: A densely populated slum.

not receive GPS signals and they are not visible in satellite imagery. PUKAR researchers (authors SS, TS, KS) with extensive knowledge of the community mapped all lanes in a GIS with the help of a GIS student (author DT) using detailed satellite imagery and a few key GPS coordinates for reference. GPS coordinates were taken at the beginning and end of each main lane, and the satellite imagery provided additional cues about lane locations, such as "seams" over lanes formed by rooftops.

After the lane network was mapped, it was straightforward to map all households based on a household census with LBHCS. A household census using the coding system indicated the total number of households located on each side of each lane segment. The GIS student generated a latitude/longitude coordinate for each ground-level dwelling (upstairs dwellings were assigned the same coordinate as downstairs dwellings), which resulted in a map of all addresses. Although the coordinates do not identify dwelling centroids (geographic center of a shape), the resulting spatial data can be thought of as a household entrance, similar to an address geocoding service such as GoogleMaps. The household locations are sufficiently accurate for property identification (for example, locating a claim to rebuild after a fire), spatial analysis (for example, hot spot analysis), calculation of distance variables (for example, distance from home to nearest water tap), and establishing the relative location of neighbors.

Several additional spatial datasets were generated from the satellite imagery by a PUKAR researcher (author KS) and the GIS student (author DT) that were not dependent on LBHCS, but were essential for community organizing and advocacy (Table 1). These spatial datasets included information on basic infrastructure such as toilets, trash bins, trash piles, and water distribution sites, as well as community resources such as temples, religious pre-schools, and informal health clinics and pharmacies.

Resources

Our tool kit included a 25 square kilometer, 0.5 meter resolution GeoEye-1 satellite image purchased with an academic license for US $230 from a commercial provider (eMap), and a walking census of households using LBHCS. In 2010 and 2012, PUKAR researchers (authors TS, SS, KS) collected unique household codes for all households in the community in just four days, a database manager entered these data within two weeks, and a PUKAR researcher (author KS) and a GIS graduate student (author DT) mapped thousands of features and their attributes

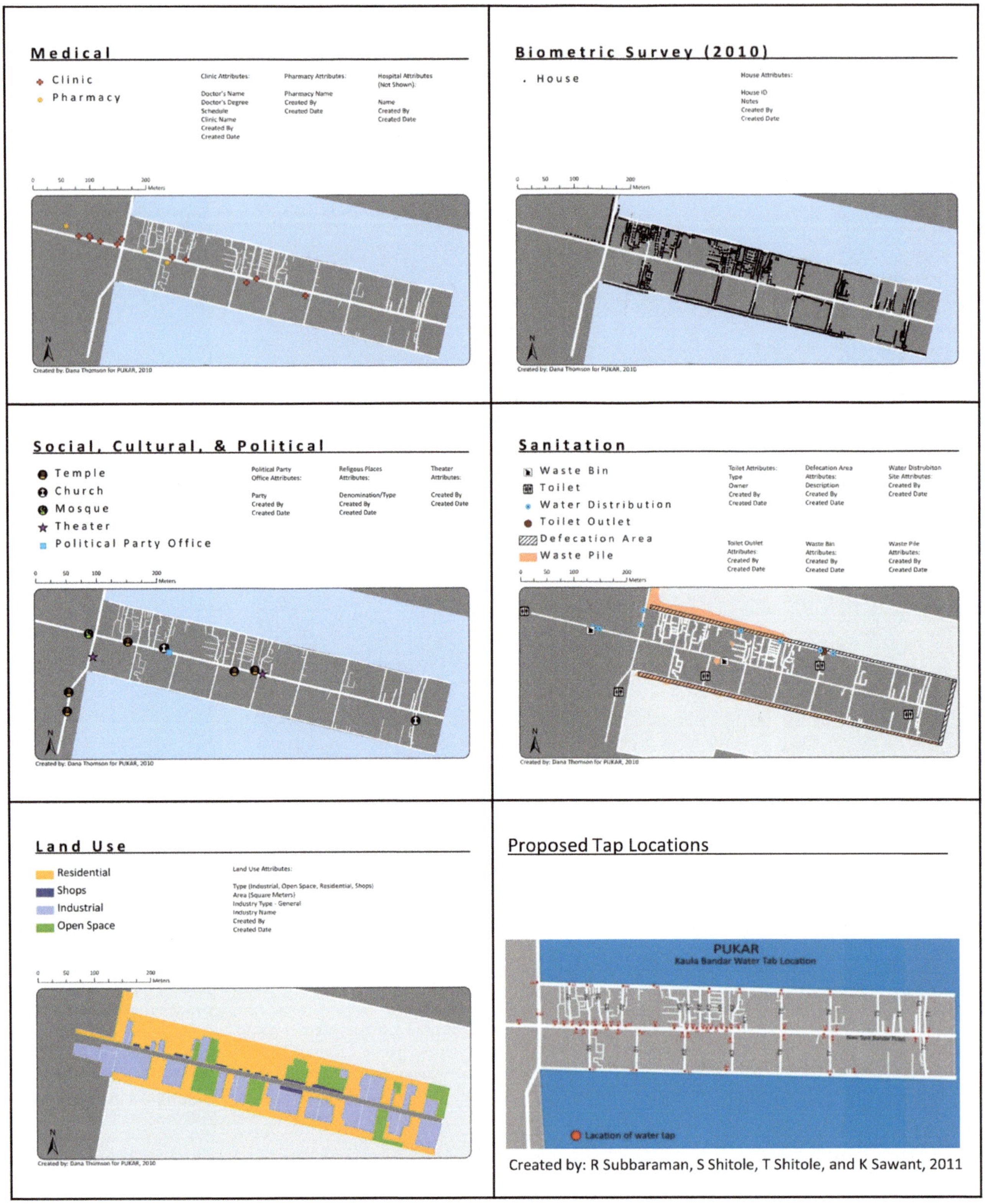

Figure 2. Example community maps.

directly in a GIS within three weeks. The success of this project hinged on a small team with strong local place knowledge, database management skills, and GIS skills. We used a computer with Excel (a spreadsheet program) and ArcGIS (a geographic information system), which at the time of this writing cost US $1500 for a single ArcGIS license plus a few hundred dollars for a computer with the MS Office Suite. Free spreadsheet and GIS programs such as Apache OpenOffice and QGIS were also

Table 1. Spatial data layers and attributes collected.

#	Spatial Data Layer	[Vector type] Attributes
Base Layers		
1	Health "Clinics"	[point], clinic type, clinic name, degree and name of service provider, hours
2	Pharmacies	[point], name
3	Industries	[polygon], name, type
4	Land use	[polygon], land use type, square meter
5	Lanes and Sub-lanes	[line], name
6	Police Stations	[point], name
7	Political Party Office	[point], party name
8	Ration Stations	[point]
9	Religions Places	[point], denomination
10	Religious Pre-Schools	[point], name, type, highest level, funding source
11	Theaters	[point], type
12	Toilet outlets	[point]
13	Trees	[point]
14	Waste bins	[point]
15	Waste piles	[polygons]
16	Water distribution sites	[point]
Advocacy & Research Layers		
17	Defecation Areas	[polygon]
18	Health Zones	[polygon], name
19	High Tide Flood Areas	[polygon]
20	Hospital	[point], name, type
21	House Addresses	[point], household id
22	Monsoon Season Flood Areas	[polygon]
23	Proposed Water Distribution Sites	[point]

available. We borrowed one GPSMAP 76CSx GPS device to collect the geographic coordinates of key landmarks, including the start and end of covered lanes.

LBHCS can be used alone for data collection and outreach by minimally trained staff at low cost. If using the household coding system alone, it is important to have at least one team member who is familiar with data entry and data cleaning, and to have a computer. If advocacy or quantitative research are key activities, we recommend integrating the household coding system with mapping activities as we present here. See Figure 3 for a summary of this toolkit.

Maintenance of the Data

LBHCS can be used in one of two ways. Users can either carry out a household census at one point in time and update it periodically, or users can regenerate household censuses periodically creating new unique household codes in each census that are not linked to previous censuses. The ideal approach will depend on the community and the reason for using a household coding system. In slums in which structures are knocked down or replaced regularly, it may be advantageous to regenerate household codes periodically so they remain useful as walking directions and capture an accurate snapshot of the community's residents and layout. In slums with a stable set of structures, it may make sense to maintain and update a longitudinal database of household codes and locations.

If new households are built between existing dwellings, the function of the codes as walking directions decreases; therefore, we recommend recoding households periodically (for example, annually), keeping track of previous household codes and their dates of use. However, for minor additions, users can add half numbers (as in K9-LSL2-LBL1-R*3.5*D) or a letter for new dwellings (as in K9-LSL2-LBL1-R*N*D). Notably, in Kaula Bandar, the pre-existing density of dwellings was very high (with nearly all dwellings being immediately adjacent to other ones), so we rarely ran into situations in which new dwellings could be built in between others. Most new homes could only be built at the very ends of existing lanes or in completely new lanes, which did not interfere with the utility of the coding system. As such, the coding system was very useful for longitudinal follow-up of homes in Kaula Bandar; however, in less dense settlements, this may not be the case.

Maintaining a longitudinal dataset of household IDs requires substantially more data infrastructure and skills than simple cross-sectional uses of the coding system, and may not be worthwhile for advocacy, education, or research teams that operate on shorter time scales. This observation applies to, for example, education programs staffed by students, or researchers working on shorter grant cycles. In those slums in which household IDs are not maintained, the IDs may still facilitate follow-up for many months, which is often the duration of an intervention, educational campaign, or study.

	Location-based coding system	Location-based coding system + Maps	Geo-locate households in Location-based coding system + Maps
Purpose	Community research; education/outreach	Community education/ outreach; advocacy	Advocacy; spatial analysis
Resources			
Place knowledge 	X	X	X
Data management skills 	X	X	X
GIS skills 		X	X
Computer 	X	X	X
Satellite Imagery 		X	X
GIS (ESRI ArcGIS)		X	X
GPS unit 		X (optional)	X (optional)

Figure 3. Location-based household coding system toolkit and applications.

Ethics Statement

A number of studies were conducted in Kaula Bandar using LBHCS including a biometric census, longitudinal water quality study, cost of diarrheal illness study, and mental health study. The biometric census included adults and children, while the other three studies included only adult participants. Residents were willing to participate in these studies largely because of PUKAR's community-based participatory research model, which involved youth from the Kaula Bandar community in data collection. In addition, findings were used to directly advocate for improved outreach to residents by the local government.

In the biometrics study, signed consent, or thumbprint consent for individuals who could not or did not know how to spell their names, was obtained from all adults and guardians of minors. During the study, the PUKAR team received feedback that many participants were hesitant to sign or provide thumbprints, even if they wanted to participate in the study, as these actions are associated with government or police interactions. As a result, verbal consent was obtained for all subsequent studies with the signature of the researcher. For interviews that required audio recording, the participant's consent was also digitally recorded. All study protocols, including the consent forms, received approval from either the Institutional Review Board of the Harvard School of Public Health (17740) or from the PUKAR Institutional Ethics Committee (FWA00016911).

Results

A Household Coding System for Community Research and Outreach

From March to December 2010, the PUKAR team engaged in an extensive biometrics survey in Kaula Bandar, in which basic demographic and health information, height, weight, blood pressure, and a photo (holding a chalkboard with the individual's household ID) were collected from community residents at a central research station that shifted to different parts of the community on the main road. All households in Kaula Bandar were coded by PUKAR researchers immediately prior to starting the survey. Approximately 1200 new residents moved to Kaula Bandar during the 10 months of the study. Every weekend, Barefoot Researchers were provided with a list of household codes. With minimal training about the coding system, the Barefoot Researchers were able to (1) independently identify each household in the community, (2) perform a household census, (3) recruit individuals to the main research station, and (4) perform home-based follow-up with nearly 100 individuals who had hypertension at the research station and whose blood pressure was re-measured so the individual could be referred to the hospital for appropriate evaluation, if needed. Moreover, each study subject's health information and photo were printed on a card, and these health cards were returned to residents to provide them with a basic health record. Using the coding system, Barefoot Researchers returned roughly 96% of health cards to residents within a week of initial data collection, as repeat identification of homes was easily and quickly performed by any of the research team members.

In 2011, PUKAR performed a study of water issues in the community, in which a consistent set of 21 homes located throughout the slum were followed longitudinally for a year to assess water cost, quantity, and quality, using serial microbiological testing of water samples [12]. On days when water testing was performed, it was important to locate all homes very rapidly, as water samples had to be transported in bulk to the laboratory within a couple of hours of collection. Without the household coding system, such rapid repeat identification of homes throughout the community on the same day would not have been feasible. The coding system also showed remarkable stability over a year of follow-up, largely because the community is so dense that new homes cannot be constructed between pre-existing homes.

In 2012, PUKAR engaged in a study of mental health in Kaula Bandar, for which a random sample of households was required. Since the previous coding of community households in 2010, a few new lanes had been added to the community, and new households had been added to the ends of lanes. As such, we decided to re-code the entire community to ensure an accurate roster for randomized sampling. With a team of ten people, all households in the settlement were re-coded in less than four days. This new registry allowed us to select a subset of households for the mental health survey using a random number generator.

In 2011 and 2012, the PUKAR team engaged in extensive door-to-door education of community residents on critical health topics such as child immunizations and diarrheal disease. The household coding system allowed Barefoot Researchers to record homes that were missed in initial rounds of the educational efforts, such that repeat efforts could be made to find the residents of these households to make sure they received individualized education.

Maps for Advocacy and Outreach

The PUKAR research team worked with a GIS student to map 23 spatial layers of existing infrastructure and services, and produce an atlas of maps that the team has since used extensively for community outreach and advocacy (Figure 2). In October 2010, for example, PUKAR was invited to present its work to Mumbai's Municipal Health Commissioner as part of a consultation of the city government with non-governmental organizations (NGOs). In preparation for this presentation, PUKAR researchers mapped nearby government health centers to emphasize the distance of Kaula Bandar from these resources. These maps graphically highlighted the community's isolation from public health centers, and they were critical in motivating the municipality of Mumbai to extend twice-monthly health camps to Kaula Bandar for provision of immunizations and other basic medical care. These efforts, along with door-to-door education efforts by PUKAR, helped to improve the fraction of children who are fully immunized with the basic set of vaccinations in Kaula Bandar from 29% to 80% in just three years [Unpublished results presented by PUKAR at the 2012 NIH Summit on the Science of Eliminating Health Disparities].

After PUKAR completed a study on water poverty in Kaula Bandar [15], PUKAR researchers presented their data to Mumbai's Municipal Commissioner for Water in late 2011, including a map of Kaula Bandar's limited water and sanitation infrastructure. As a result of this dialogue, the Water Commissioner took steps to facilitate extension of a new water supply to the Kaula Bandar community. The development of a new water supply is still in process, as fundamental water infrastructure needs to be extended to the entire nearby area before the local water supply in Kaula Bandar can be developed further. The Water Commissioner and the local corporator (equivalent to an alderman) suggested that once this main supply is in place, approximately 60 new community water taps can be placed in Kaula Bandar.

PUKAR researchers have concerns regarding equitable placement of these water taps, as studies have shown that the distance of a household from a water source may have a major impact on health outcomes [16–18]. Given these concerns, PUKAR researchers proposed a plan for community water tap placement and mapped it (Figure 2). Tap placement on the map was decided based on the population density in each lane, as well as PUKAR researchers' knowledge about open spaces in the slum to facilitate ease of access and mobility within the community. PUKAR presented this plan to the Water Commissioner and local corporator, and both officials agreed on this plan for tap placement. We hope that this map remains a shared roadmap for the community and government officials, and facilitates government accountability for equitable water tap placement.

Spatial Data for Research and Advocacy

The research team at Harvard used the household coordinates to reaggregate data from a biometric study to sub-community health zones to perform spatial summaries and analyses of exposures and outcomes. Health zone boundaries corresponded to landmarks. The team distinguished households located near the center or at the periphery of the pier because households on the periphery were believed to have higher exposure to contaminated water due to high tide and monsoon flooding. As noted, Kaula Bandar has a small geographic footprint, covering less than one-tenth of a square kilometer of land. We found little spatial variability of health outcomes and exposures within the slum; all of our exploratory spatial analyses produced null results. In larger slums with greater variability in terrain, resources, or within-slum socioeconomic disparities, use of the household coding system with geolocated addresses would facilitate extensive spatial analysis including spatial regression and cluster analysis (also called "hot spot analysis").

Maintaining a Household Coding System Over Time

The coding system worked well with the fluid, malleable nature of the community. Kaula Bandar is fluid in two ways. First, people in the community migrate seasonally to rural areas and residents who accrue resources migrate to "better" slums. They also move because of demolition of their homes by the government and eviction by owners of the dwelling. The dwellings also change periodically. Homes disappear and reappear due to fire, and some of the homes on the main road get periodically demolished. As such, the coding system allowed us to re-enumerate the community in four days and account for changes in living structures due to expansion, fires, and demolition. By integrating LBHCS with spatial data collection we were able to keep a permanent record of former household locations and lane configurations.

Discussion

Urban slums comprise one-sixth of the global population [1], but can be difficult to work in or study because lack of an address system limits follow-up with households over time. Household enumeration systems are needed in slum communities that comprise a substantial portion of the global population, particularly in high-density slums in which GPS units, satellite imagery, and other common enumeration tools do not work. We created an innovative Location-Based Household Coding System that facilitated several community-based research projects and door-to-door health interventions in a densely populated, non-notified slum in Mumbai, India. The coding system doubled as walking directions, which allowed rapid identification of individual in their households by a research team of youth from Mumbai, many of whom live in the slum community itself. For the cost of a few hours of training and one computer for data management, LBHCS allowed a household census and reliable household re-identification for over one year. Re-enumeration of all households was performed for a new study in just four days by ten people.

By combining a LBHCS with GIS, two people mapped several thousand households in just a few days, which allowed spatial analysis of detailed household survey data. Mapping at this scale by other methods would not have been possible, or it would have been prohibitively time-intensive and expensive. By generating a few additional spatial datasets, PUKAR researchers produced base maps of community infrastructure that added leverage in several important meetings between Kaula Bandar residents and city officials to extend city services.

Although Kaula Bandar is geographically small, these methods should easily scale to slums with larger populations covering larger geographic areas. The simplicity of the coding system means rapid training for any sized research or outreach team, including teams composed of members who are young or who have limited technology skills. LBHCS is extremely effective for following the same households over weeks and months because the codes double as clear walking directions. These methods are particularly amenable to high-density slums in which new development occurs on the periphery of the community, and new dwellings are not frequently added between existing dwellings. LBHCS is an efficient, easy-to-learn, cost-effective, scalable approach to household enumeration and re-identification in densely populated settings.

Acknowledgments

Preliminary versions of this manuscript were presented at the 10th International Conference on Urban Health in Belo Horizonte, Brazil, and at the 2012 Conference on the Science of Eliminating Health Disparities in National Harbor, MD, USA.

Author Contributions

Conceived and designed the experiments: SS TS KS DRT. Wrote the paper: DRT RS. Conducted research, education, and advocacy using LBHCS: APD DEB RS SS TS KS DRT.

References

1. United Nations Population Fund (UNFPA) (2007) State of the world population 2007: Unleashing the potential of urban growth. New York: UNFPA. 99 p.
2. Patel S, Baptist C, D'Cruz C (2012) Knowledge is power - informal communities assert their right to the city through SDI and community-led enumerations. Environ Urban 24: 13–26. Available: http://eau.sagepub.com/content/25/1/177. Accessed 2013 November 30.
3. Karanja I (2010) An enumeration and mapping of informal settlements in Kisumu, Kenya, implemented by their inhabitants. Environ Urban 22: 217–239. Available: http://eau.sagepub.com/content/22/1/217. Accessed 2013 November 30.
4. Livengood A, Kunte K (2012) Enabling participatory planning with GIS: A case study of settlement mapping in Cuttack, India. Environ Urban 24: 77–97. Available: http://eau.sagepub.com/content/24/1/77. Accessed 2013 November 30.
5. Vuksanovic-Macura Z (2012) The mapping and enumeration of informal Roma settlements in Serbia. Environ Urban 24: 685–705. Available: http://eau.sagepub.com/content/24/2/685. Accessed 2013 November 30.
6. Curtis A, Blackburn JK, Widmer JM, Morris JG Jr (2013) A ubiquitous method for street scale spatial data collection and analysis in challenging urban environments: Mapping health risks using spatial video in Haiti. Int J Health Geogr 12: 21. Available: http://www.ij-healthgeographics.com/content/12/1/21. Accessed 2013 November 30.
7. Militino AF, Ugarte MD, Iribas J, Lizarraga-Garcia E (2013) Mapping GPS positional errors using spatial linear mixed models. J Geodesy 87: 675–685.
8. Navstar Global Positioning System (2008) Global positioning system standard positioning service performance standard, 4th ed. Washington, DC: United States Department of Defense. 160 p.
9. Municipal Corporation of Greater Mumbai (2010) Mumbai Human Development Report 2009. New Delhi: Oxford University Press. 207 p.

10. Sharma K (2000) Rediscovering Dharavi: Stories from Asia's largest slum. New Delhi: Penguin. 209 p.
11. Makau J, Dobson S, Samia E (2012) The five-city enumeration: The role of participatory enumerations in developing community capacity and partnerships with government in Uganda. Environ Urban 24: 31–46. Available: http://eau.sagepub.com/content/24/1/31. Accessed 2013 November 30.
12. Subbaraman R, O'Brien J, Shitole T, Shitole S, Sawant K, et al. (2012) Off the map: The health and social implications of being a non-notified slum in India. Environ Urban 24: 643–663. Available: http://eau.sagepub.com/content/24/2/643. Accessed 2013 November 30.
13. Appadurai A, PUKAR (2008) Mumbai's barefoot researchers. Mumbai: PUKAR. 213 p.
14. Kretzmann J, McKnight JP (2007) Assets-based community development. Natl Civic Rev 85(4): 23–29.
15. Subbaraman R, Shitole S, Shitole T, Sawant K, O'Brien J, et al. (2013) The social ecology of water in a Mumbai slum: Failures in water quality, quantity, and reliability. BMC Public Health 13: 173. Available: http://www.biomedcentral.com/1471-2458/13/173. Accessed 2013 November 30.
16. Howard G, Bartram J (2003) Domestic water quantity, service level, and health. Geneva: World Health Organization. 33 p.
17. Gorter AC, Sandiford P, Smith GD, Pauw JP (1991) Water supply, sanitation and diarrhoeal disease in Nicaragua: Results from a case-control study. Int J Epidemiol 20: 527–533.
18. Water and Environmental Health at London and Loughborough (WELL) (1998) Guidance manual on water supply and sanitation. Loughborogh, UK: Water Engineering and Development Center. 302 p.

3

Modeling Spatial Patterns of Soil Respiration in Maize Fields from Vegetation and Soil Property Factors with the Use of Remote Sensing and Geographical Information System

Ni Huang[1], Li Wang[1]*, Yiqiang Guo[2], Pengyu Hao[1], Zheng Niu[1]

1 The State Key Laboratory of Remote Sensing Science, Institute of Remote Sensing and Digital Earth, Chinese Academy of Sciences, Beijing, China, **2** Land Consolidation and Rehabilitation Center, Ministry of Land and Resources, Beijing, China

Abstract

To examine the method for estimating the spatial patterns of soil respiration (R_s) in agricultural ecosystems using remote sensing and geographical information system (GIS), R_s rates were measured at 53 sites during the peak growing season of maize in three counties in North China. Through Pearson's correlation analysis, leaf area index (LAI), canopy chlorophyll content, aboveground biomass, soil organic carbon (SOC) content, and soil total nitrogen content were selected as the factors that affected spatial variability in R_s during the peak growing season of maize. The use of a structural equation modeling approach revealed that only LAI and SOC content directly affected R_s. Meanwhile, other factors indirectly affected R_s through LAI and SOC content. When three greenness vegetation indices were extracted from an optical image of an environmental and disaster mitigation satellite in China, enhanced vegetation index (EVI) showed the best correlation with LAI and was thus used as a proxy for LAI to estimate R_s at the regional scale. The spatial distribution of SOC content was obtained by extrapolating the SOC content at the plot scale based on the kriging interpolation method in GIS. When data were pooled for 38 plots, a first-order exponential analysis indicated that approximately 73% of the spatial variability in R_s during the peak growing season of maize can be explained by EVI and SOC content. Further test analysis based on independent data from 15 plots showed that the simple exponential model had acceptable accuracy in estimating the spatial patterns of R_s in maize fields on the basis of remotely sensed EVI and GIS-interpolated SOC content, with R^2 of 0.69 and root-mean-square error of 0.51 μmol CO_2 m^{-2} s^{-1}. The conclusions from this study provide valuable information for estimates of R_s during the peak growing season of maize in three counties in North China.

Editor: Ben Bond-Lamberty, DOE Pacific Northwest National Laboratory, United States of America

Funding: This work was supported by the National Natural Science Foundation of China (41301498), the Public Service Sectors (Ministry of Land and Resources) Special Fund Research (201311127), the Special Foundation for Young Scientists of the State Laboratory of Remote Sensing Science (13RC-07), and the Major State Basic Research Development Program of China (2013CB733405). The funders had no role in study design, data collection and analysis, decision to publish, or preparation of the manuscript.

Competing Interests: The authors have declared that no competing interests exist.

* Email: wangli@radi.ac.cn

Introduction

Soil CO_2 efflux from terrestrial ecosystems to the atmosphere has been considered the second largest global carbon flux and is a vital component of ecosystem respiration [1]. In recent decades, significant progress has been made in identifying the biophysical factors that influence soil respiration (R_s) to predict soil CO_2 emission accurately in time and space [2–4].

The majority of R_s arises from root and microbial tissue. Therefore, understanding the spatial and temporal changes of these sources will facilitate the modeling of R_s. However, the large spatial and temporal heterogeneity of root and microbial activity within the landscape and the covariation of potentially important factors (i.e., temperature and water content) pose great challenges to the development of mechanistically based models that account for spatial and temporal variability in R_s [2]. Thus, many different statistical models of R_s have been developed on the basis of data collected from different ecosystems [5]. Numerous studies have established R_s models based on soil temperature, soil moisture, or both [6,7]. Aside from soil temperature and moisture, plant productivity proxies [e.g., leaf area index (LAI), canopy chlorophyll content (Chl_{canopy}), and plant biomass] [8–10] and soil properties [e.g., soil organic carbon (SOC) content, soil total nitrogen (STN) content, and soil C and N ratio (soil C/N)] [11,12] also potentially influence R_s and are often included in models of R_s. However, most of the factors that affect variations in R_s tend to be derived through field measurements [13]. Furthermore, direct observation of these variables across long time spans or large spatial scales is expensive because of the required manpower and material resources. A simple method to derive data related to variations in R_s is necessary to facilitate the determination of the spatial and temporal distribution of R_s.

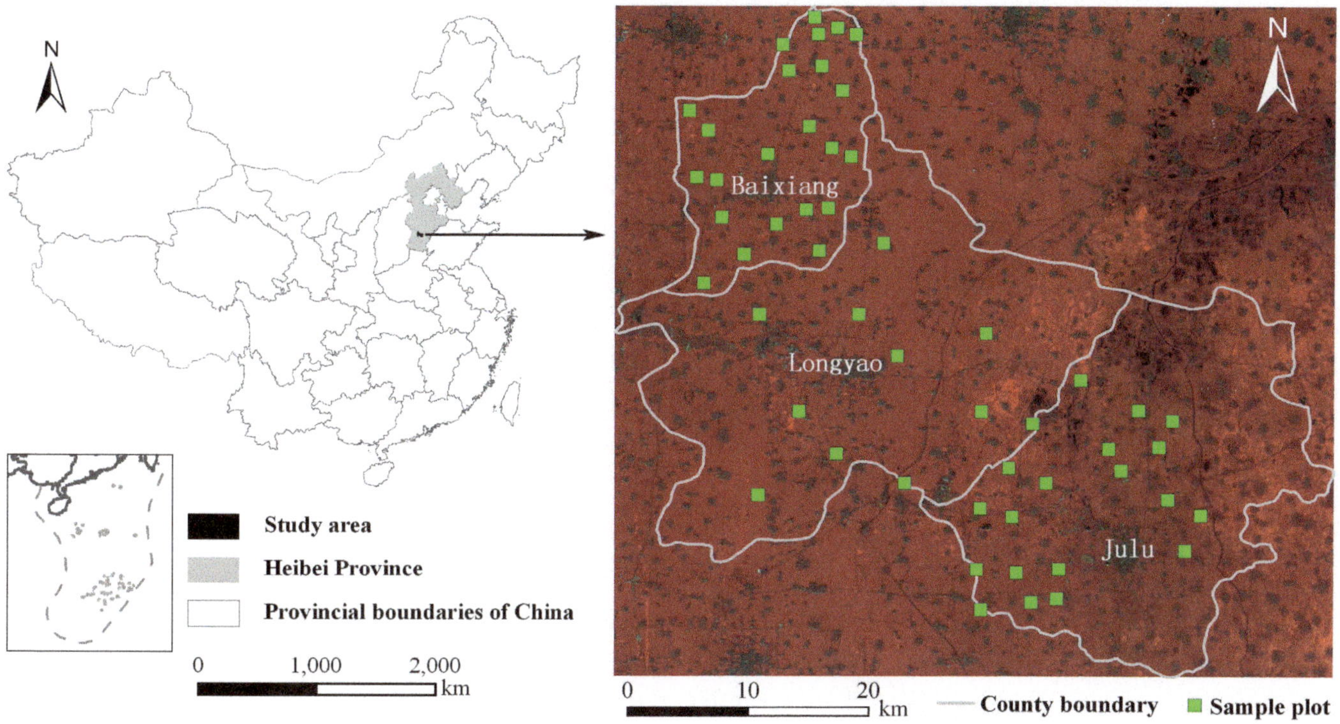

Figure 1. Spatial location of the sample plots for field experiments in three counties in North China. The box in the bottom left corner of Figure 1 shows the South China Sea islands.

Remote sensing and geographical information system (GIS) provide powerful tools for data acquisition, spatial analysis, and graphical display [14–16]. In the field of global change research, significant advances have been made in the development and application of remote sensing and GIS. These advances include land cover and land-use changes [17,18], environmental vulnerability and risk assessment [19,20], ecological restoration and management [21–23], and terrestrial ecosystem carbon cycle [24–26]. However, applying the data derived from remote sensing and GIS into R_s modeling remains controversial, especially for remote sensing data, because remotely sensed data in principle are independent measurements of site properties, not functionally important variables (e.g., soil temperature, soil moisture, and plant growth variables) that control R_s [3,27,28]. On the basis of statistical analysis of field experiments, previous studies found that remotely sensed vegetation indices (VIs) correlate with R_s in crop sites that lack drought stress [10] and can be used to model the spatial patterns of R_s during the peak growing season of alpine grasslands in the Tibetan Plateau [26]. However, few studies explore the potential of remote sensing and GIS data for estimating the spatial patterns of R_s in agricultural land, which may be affected by more complex factors than natural grasslands because of the influence of human activity. Although modern agriculture has successfully increased food production, the processes involved have profoundly affected the global carbon cycle through tillage, drainage and conversion of natural to agricultural ecosystems [29,30]. Therefore, a simple method should be identified to study the spatial characteristics of R_s in agricultural ecosystems.

This study aims to examine a potential new approach for estimating the spatial patterns of R_s during the peak growing season of maize by using remote sensing and GIS technology in Baixiang, Longyao and Julu Counties, which are typical agricultural areas in the north plain of China. Studying the spatial characteristics of soil CO_2 efflux in maize fields will contribute to eco-agricultural development.

Materials and Methods

Ethics Statement

No specific permissions were required for the 53 sample plots in this study. We confirmed that the field studies did not involve endangered or protected species, and the specific location of the sample plots was provided in the manuscript (Fig. 1).

Study Site

The study site is situated within three counties (Baixiang, Longyao and Julu) in Southern Hebei Province of North China (Fig. 1). The total area of the study site is 1.64×10^3 km^2. This area is located in the North China Plain with a flat open terrain, single landform type, and a mean elevation of 30 m above sea level. Calcareous alluvial soil with high capacity to retain water and fertilizer is the main soil type in the study area. The study site is suitable for farming, and maize is the main crop. The climate is continental monsoon with four distinct seasons and adequate light and heat resources. Long records of meteorological data near the study site (http://cdc.cma.gov.cn) indicate that the mean annual temperature is 13.5°C with the coldest temperatures in January and the hottest in July. The mean annual precipitation is 502.8 mm, but precipitation is distributed unevenly in the four seasons with the greatest precipitation occurring in summer (362.5 mm). Therefore, drought influences agricultural development, and agriculture mainly involves irrigation in this study site.

Fifty-three sample plots located in the maize fields were identified within the study site (Fig. 1). The distance between any two sample plots was larger than 2 km. Each sample plot (greater than 100 m×100 m) has a large maize area, flat terrain, and maize under uniform growing conditions. All measurements were performed from August 11, 2013 to August 20, 2013, which corresponded to the tassel stage and peak growing period of maize. During the 10 days of field measurements, continuous measurements were performed, except on August 12 because of a minor precipitation event. Therefore, all field measurements required 9 days.

Field measurements

Soil respiration measurements. In each sample plot, R_s was measured by using a soil respiration chamber (LI-6400-09; LiCor, Lincoln, Nebraska, USA) connected to a portable photosynthesis system (LI-6400; LiCor, Lincoln, Nebraska, USA). The soil respiration chamber was mounted on a PVC soil collar that was sharpened at the bottom. Each PVC collar (5 cm long, 11 cm inside diameter) was inserted 2 cm to 3 cm into the ground and was installed at least 24 h prior to performing any measurements. To reduce the difference in root biomass, soil collars were placed in three locations on the basis of their distance to the maize plant: near a maize plant, inter-plant, and inter-row. Two collars were placed in each of the three positions for each R_s measurement. At least three to four consecutive measurements on each collar were performed to prevent any systematic error in the R_s estimates. An average R_s value was used for each collar, and the average value from six collars was used to represent the R_s value at plot level. Each R_s measurement was conducted between 09:00 h and 15:00 h (local time) because fluxes measured during this time interval are usually representative of the daily mean flux.

Soil temperature and soil moisture measurements. After the soil respiration measurement on a PVC soil collar in each plot, soil temperature and soil moisture were measured in this collar to minimize sample difference. Soil temperature was measured at a 10 cm depth (T_{s10}) by using a ground thermometer. Volumetric soil moisture at a depth of 0 cm to 20 cm (SM_{20}) was determined by using a portable time domain reflectometry probe (HydroSense, Campbell, USA). Thus, six soil temperature and moisture measurements were performed in each plot. The average value was used to represent soil temperature or soil moisture at the plot level.

Maize biophysical parameter measurements. LAI was measured by using an LAI-2000 (LI-COR Inc., Lincoln, Nebraska). In each plot, six representative positions were selected for LAI measurement, and in every position, two repeated measurements were performed. Leaf chlorophyll content (Chl_{leaf}) was determined by using a portable chlorophyll meter (SPAD-502, New Jersey, USA). Fully expanded leaves, which depended on the height of the maize plant, were randomly selected from three locations that corresponded to the upper, middle, and lower parts of the maize plant. For each leaf location, 10 SPAD values were randomly collected. The vertical leaf area distribution in maize canopy was analyzed by measuring the area of each green leaf from the bottom to the top of eight randomly distributed maize plants with the use of an area meter (LI-3100, LI-COR, Lincoln, Nebraska). The area-weighted mean SPAD reading was used to derive Chl_{leaf}. However, the SPAD reading was in arbitrary units rather than in actual amounts of chlorophyll per unit area of the leaf tissue. A transform relationship exists between the SPAD readings and the actual chlorophyll content in maize [31]. To convert the SPAD readings to chlorophyll content per unit leaf area (μg cm^{-2}), this study used the transform relationship ($Chl_{leaf} = 0.95 \times \text{SPAD reading} - 3.25$) derived by Wu et al. [32] in maize plots, and the same SPAD meter was employed in this study. Chl_{canopy} was then determined by using the following equation:

$$Chl_{canopy} = Chl_{leaf} \times GLAI \quad (1)$$

where Chl_{canopy} is the canopy chlorophyll content (g m^{-2}), Chl_{leaf} is the leaf chlorophyll content of maize (g m^{-2}), and GLAI represents the green leaf area per unit ground area.

In each sample plot, three representative maize plants were harvested for aboveground biomass (AGB) measurement. These fresh maize plants were sealed in a plastic bag and immediately transported to a nearby laboratory for subsequent analysis. Thereafter, fresh samples were oven dried at 65°C until the mass of the sample reached a constant weight. The AGB in each plot can be derived by multiplying the average dry weight per plant (g plant^{-1}) and the average plant density of maize (plants m^{-2}).

Soil property measurements. Soil within the six PVC collars in each plot was destructively sampled after measuring R_s, soil temperature and soil moisture. Soil was sampled to a depth of approximately 20 cm by a cylindrical soil driller (4 cm diameter, 20 cm height), in which fine root biomass and microbial activity are the highest [33,34]. These collected soil samples were sealed in plastic bags and stored at room temperature while being transported to the laboratory. Six collected soil samples in each plot were uniformly mixed to form a composite sample for laboratory analysis. The composite sample was air-dried in the laboratory to a constant weight for soil chemical analyses. The air-dried soil samples were ground to pass through a 0.2 mm sieve after any visible plant tissues and debris were manually removed. The SOC content was estimated by using the standard Mebius method [35]. The STN content was analyzed by using the Kjeldahl digestion procedure [36]. In this study, soil C/N was calculated by the ratio of SOC and STN contents.

Table 1. Calculation for vegetation indices [a].

Vegetation index	Formula	Reference
Normalized difference vegetation index	$NDVI = \frac{R_{Nir}\text{-}R_{Red}}{R_{Nir}+R_{Red}}$	Rouse et al. [47], Gamon et al. [48]
Modified soil adjusted vegetation index	$MSAVI = \frac{2R_{Nir}+1-\sqrt{(2R_{Nir}+1)^2-8(R_{Nir}\text{-}R_{Red})}}{2}$	Qi et al. [49]
Enhanced vegetation index	$EVI = 2.5 \times \frac{R_{Nir}\text{-}R_{Red}}{1+R_{Nir}+6\times R_{Red}-7.5\times R_{Blue}}$	Huete et al. [50]

[a] R_{Blue}, R_{Red}, and R_{Nir} are reflectance of blue, red, and NIR band in the HJ-1A CCD optical image, respectively.

Table 2. Pearson's correlation among soil respiration and factors affecting soil respiration in maize fields during the peak growing season in three counties in North China.

	R_s	T_{s10}	SWC_{20}	Chl_{canopy}	LAI	AGB	SOC content	STN content	Soil C/N
R_s	1.00	−0.27	−0.18	0.54***	0.75***	0.59***	0.76***	0.59***	−0.23
T_{s10}		1.00	0.18	−0.15	−0.28	−0.27	−0.49**	−0.66***	0.51**
SWC_{20}			1.00	−0.17	−0.05	−0.07	0.16	−0.00	0.06
Chl_{canopy}				1.00	0.83***	0.81***	0.26	0.20	−0.18
LAI					1.00	0.76***	0.44**	0.38*	−0.28
AGB						1.00	0.45**	0.34*	−0.15
SOC content							1.00	0.78***	−0.29
STN content								1.00	−0.79***
Soil C/N									1.00

R_s is the daily mean soil respiration rate (μmol CO_2 m^{-2} s^{-1}), T_{s10} is the soil temperature at 10 cm depth (°C), SWC_{20} is the soil water content at 0 cm to 20 cm depth (m^3 m^{-3}), Chl_{canopy} is the canopy chlorophyll content (g m^{-2}), LAI is the leaf area index, AGB is the aboveground biomass (kg m^{-2}), SOC content is the soil organic carbon content (g kg^{-1}), STN content is the soil total nitrogen content (g kg^{-1}), and soil C/N is the soil C: N ratio. Significance levels:
*$p<0.05$,
**$p<0.01$,
***$p<0.001$.

Spatial data acquisition

Maize classification data. This study aimed to derive the spatial distribution of R_s in maize fields based on the field measurements at the plot scale. Maize classification data is necessary to spatially extrapolate R_s at the plot scale to the whole study area. Multi-temporal normalized difference vegetation index (NDVI) data collected over the growing season were used to classify maize at the study site [37–39]. Clouds are common occurrences in the study area during the growing season. Thus, obtaining a time sequence of cloud-free scenes is difficult. Two types of satellite data were used to establish the time-series NDVI data. One was the Operational Land Imager (OLI) image of Landsat 8, and the other was the small constellation for environmental and disaster mitigation (HJ-1A and B) charge coupled device (CCD) image [40–42]. Five scenes of OLI images acquired on May 3, 2013, May 19, 2013, July 6, 2013, October 10, 2013, and October 26, 2013 were downloaded from the U.S. Geological Survey (http://earthexplorer.usgs.gov/). Three HJ-1A and B CCD optical images acquired on June 6, 2013, August 17, 2013, and September 15, 2013 were downloaded from the China Center for Resource Satellite Data and Applications (http://www.cresda.com). The two types of remote sensing images exhibit same spatial resolution (30 m). The 30 m spatial resolution is appropriate for classifying maize patterns in the study area given the relatively large field in the region, which could spatially corresponded to five or more 30 m pixels. The strong relationship of the NDVI with biophysical vegetation characteristics, such as LAI and green biomass [43,44], enables the discrimination of land cover types on the basis of their unique phenological responses. Before land-use classification, pre-processing (i.e., radiometric calibration, atmospheric correction and geometric correction) of OLI images and HJ-1A and B CCD optical images was accomplished by using the Environment for Visualizing Images (ENVI) software (Version 4.7, Research Systems Inc., Boulder, Colorado, USA) [45,46]. This process ensured the consistency between the two types of remote sensing data and the seasonality of the NDVI time series. The maximum likelihood classification method, integrated in the ENVI software, was applied to the eight-date NDVI time series that spanned one maize growing season of the study site.

Spectral vegetation index for vegetation biophysical parameter estimation. Three greenness indices, namely, NDVI, enhanced vegetation index (EVI), and modified soil adjusted vegetation index (MSAVI), were derived from the HJ-1A CCD optical image acquired on August 17, 2013 (Table 1) for vegetation biophysical parameter estimation. Previous studies reported that greenness VIs offer important and convenient measures for vegetation biophysical parameters, such as LAI and Chl_{canopy} [51–54]. Meanwhile, LAI and Chl_{canopy} are also found to be good indicators of plant canopy photosynthesis [55–57] and are used in the modeling of R_s [58]. To obtain the spatial patterns of vegetation biophysical parameters in maize fields, the spatial distribution of vegetation biophysical parameters over the whole study area was overlapped with the maize classification data.

Quantifying the spatial pattern of SOC content. Statistics and geostatistics have been widely applied to quantify the spatial distribution patterns of SOC at a regional scale [59–61]. Based on the theory of regionalized variables, geostatistics provides advanced tools to quantify the spatial features of soil parameters and to conduct spatial interpolation [62,63]. In this study, geostatistical analyses were performed by using the geostatistical analyst module of ArcGIS software (Version 9.3, 2008) to quantify the spatial pattern of SOC content. To obtain the spatial pattern of the SOC content in the maize fields, the spatial distribution of

Table 3. Spatial characteristics of soil respiration (R_s, μmol CO_2 m^{-2} s^{-1}), soil temperature at 10 cm depth (T_{s10}, °C), soil water content at 0 cm to 20 cm depth (SWC_{20}, m^3 m^{-3}), canopy chlorophyll content (Chl_{canopy}, g m^{-2}), leaf area index (LAI), aboveground biomass (AGB, kg m^{-2}), soil organic carbon content (SOC content, g kg^{-1}), soil total nitrogen content (STN content, g kg^{-1}) and soil C: N ratio (soil C/N) in maize fields during the peak growing season in three counties in North China.

Variables	Mean	Maximum	Minimum	CV (%)
R_s	5.43	7.33	2.64	15.45
T_{s10}	28.32	30.93	25.78	4.73
SWC_{20}	27.54	33.27	19.54	12.48
Chl_{canopy}	0.18	0.21	0.16	6.54
LAI	3.75	4.53	2.81	8.64
AGB	0.94	1.89	0.44	31.93
SOC content	11.86	17.26	6.40	16.71
STN content	1.25	1.78	0.53	24.47
Soil C/N	9.82	14.38	7.07	18.53

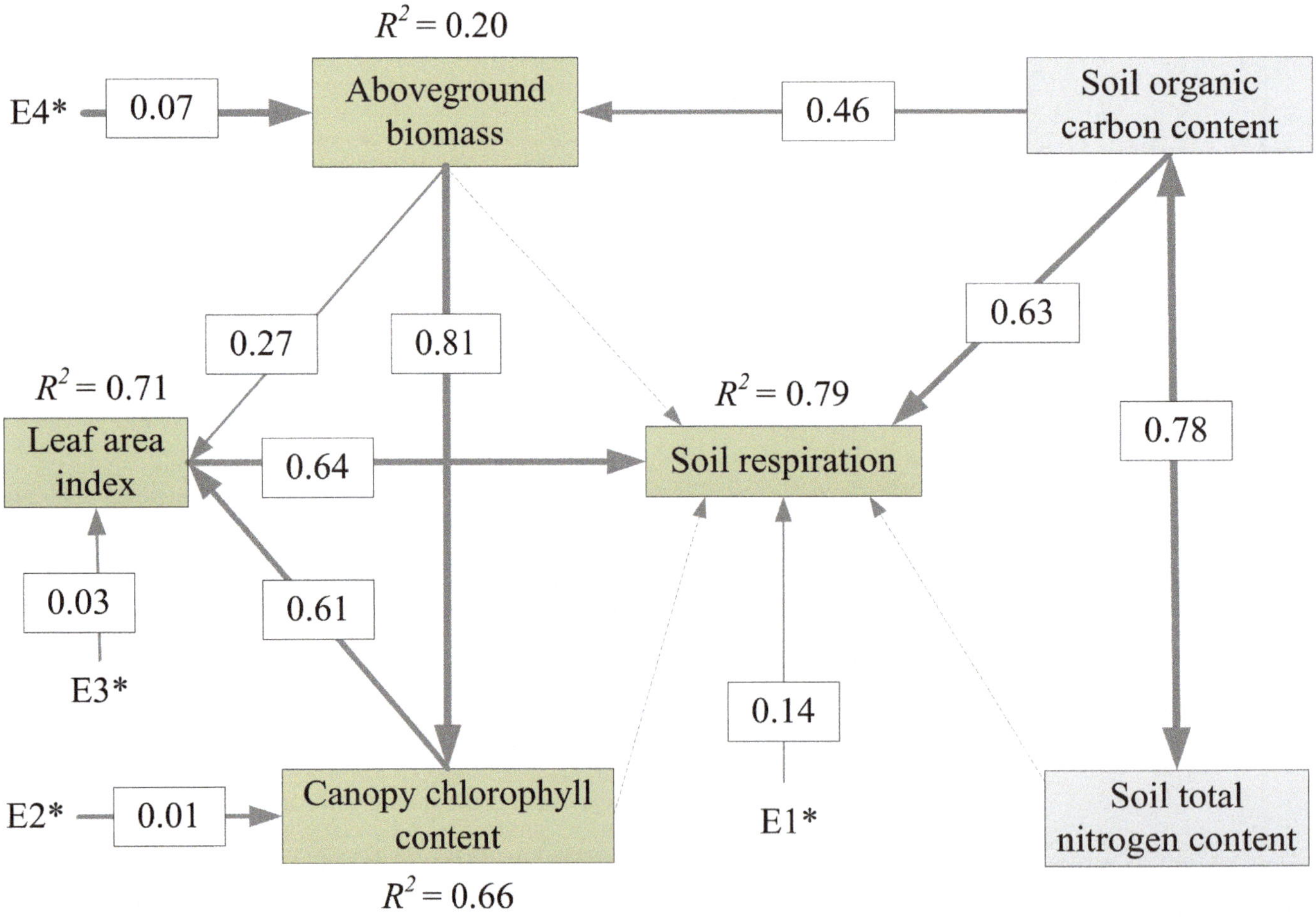

Figure 2. Final structural equation modeling (SEM) for soil respiration. Non-significant paths are shown in dashed line. The thickness of the solid arrows reflects the magnitude of the standardized SEM coefficients. Standardized coefficients are listed on each significant path. * represents error terms for observed variables, among them, E1, E2, E3, and E4 represent measurement errors for soil respiration, canopy chlorophyll content, leaf area index, and aboveground biomass, respectively.

Table 4. Total, direct, and indirect effects in the structural equation modeling.

Variable	Direct effect	Indirect effect	Total
Soil respiration			
Aboveground biomass	−0.10*ns*	0.46	0.36
Soil organic carbon content	0.63	0.16	0.79
Soil total nitrogen content	−0.09*ns*	0.30	0. 21
Leaf area index	0.64	-	0.64
Canopy chlorophyll content	−0.04*ns*	0.39	0.35
Aboveground biomass			
Soil organic carbon content	0.46	-	0.46
Soil total nitrogen content	−0.01*ns*	-	−0.01*ns*
Leaf area index			
Aboveground biomass	0.27	0.50	0.77
Soil organic carbon content	-	0.35	0.35
Soil total nitrogen content	-	−0.01*ns*	−0.01*ns*
Canopy chlorophyll content	0.61	-	0.61
Canopy chlorophyll content			
Aboveground biomass	0.81	-	0.81
Soil organic carbon content	-	0.37	0.37
Soil total nitrogen content	-	−0.01*ns*	−0.01*ns*

These effects were calculated using standardized path coefficients. Non-significant effects are indicated by "*ns*".

the SOC content over the whole study area was overlapped with the maize classification data.

Modeling spatial patterns of soil respiration

Identifying factors affecting spatial variability of soil respiration. The variables that explain the spatial variability of R_s are as follows: (1) soil properties, measured by SOC content, STN content and soil C/N; (2) environmental factors, encompassing T_{s10} and SM_{20}, and (3) plant photosynthesis proxy factors, including AGB, LAI and Chl_{canopy}. Pearson's correlation requires variables to be normally distributed and mutually independent. Each variable was tested for normal distribution by using the Shapiro–Wilk normality test and for randomness by the runs test of the Statistical Package for the Social Sciences (SPSS, Chicago, Illinois, USA). The results of the statistical analysis showed that each of these measured variables followed a normal distribution (Shapiro-Wilk, $p>0.05$) and showed randomness (runs test, $p>0.05$). Thus, Pearson's correlation analysis, as implemented in the SPSS software, was used to screen important variables that influence R_s. Five variables with statistically significant correlation ($p<0.05$) with R_s, namely, SOC content, STN content, LAI, AGB, and Chl_{canopy}, were screened out (Table 2). However, these variables were cross-correlated [64–66] and included both direct and indirect effects. To solve this problem, structural equation modeling (SEM) was used to evaluate explicitly the causal relationships among these interacting variables [67–69] and to divide the total effects of variables on R_s into direct and indirect effects. On the basis of the theoretical knowledge on the major factors that influence spatial patterns of R_s at regional scales [8,13,26], we developed an SEM model to relate R_s to SOC content, STN content, LAI, AGB, and Chl_{canopy}. This SEM model was used to identify the direct effect factors for R_s estimation. The SEM model was fitted by using AMOS 18.0 for Windows [70]. After using the SEM, the fit indices, namely, comparative fit index = 0.984 and goodness-of-fit index = 0.946. Thus, the theoretical model showed a good fit with the sample data.

Quantifying the spatial patterns of soil respiration in maize fields. In this study, the direct effect factors of R_s identified by SEM were used to estimate R_s. The spatial distribution data of these direct effect factors were first obtained on the basis of remote sensing or GIS to quantify the spatial patterns of R_s in maize fields. A simple exponential model that used the proxy data was then employed to estimate the spatial pattern of R_s during the peak growing season of maize. The accuracy of this method was examined by separating the observed data into two datasets through a random generator. One dataset consisted of 38 sample plots for analysis, whereas the other consisted of 15 for testing the accuracy of the R_s estimation.

Result

Spatial characteristics of soil respiration

Based on field-measured data at 38 plots, the daily mean R_s of maize during the peak growing season was 5.43 μmol CO_2 m^{-2} s^{-1} with a range of 2.64 μmol CO_2 m^{-2} s^{-1} to 7.33 μmol CO_2 m^{-2} s^{-1} and a coefficient of variation (CV) of 15.45% (Table 3). The spatial variability of soil temperature at 10 cm depth (T_{s10}) was relatively small at the study site with a CV of 4.73% and was far less than the spatial variation in soil water content at 0 cm to 20 cm depth (SWC_{20}). The AGB of maize showed greater spatial variability (CV = 31.93%) than LAI (CV = 8.64%) and Chl_{canopy} (CV = 6.54%).

Mean SOC content, STN content, and soil C/N at 0 cm to 20 cm depth in maize fields of the study site were 11.86 g kg^{-1} (ranged from 6.40 g kg^{-1} to 17.26 g kg^{-1}), 1.25 g kg^{-1} (ranged from 0.53 g kg^{-1} to 1.78 g kg^{-1}), and 9.82 (ranged from 7.07 to 14.38), respectively. Their CVs were not similar with the STN content which showed greater spatial variability than the SOC content and soil C/N.

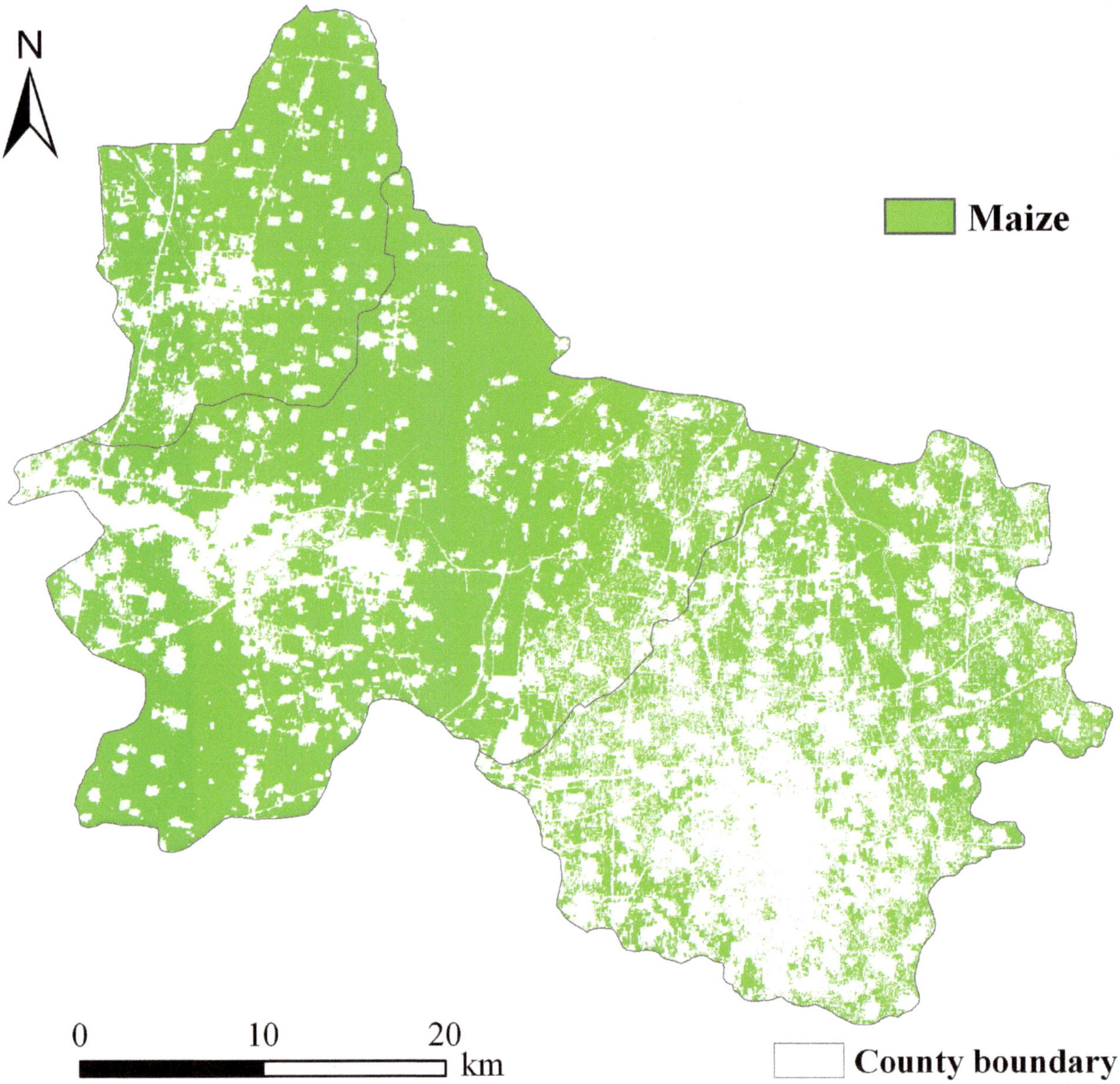

Figure 3. Maize classification map in three counties in North China.

Factors driving spatial variability of soil respiration

Based on Pearson's correlation analysis, five variables with significant correlation with R_s, namely, Chl_{canopy}, LAI, AGB, SOC content, and STN content, were selected (Table 2). However, the five selected variables were intercorrelated (Table 2), and their relationships with R_s combined both direct and indirect correlations. Thus, an SEM model was further used to evaluate the causal relationships among these interacting variables. The final SEM explained 79% of the variation in R_s (Fig. 2). The direct, indirect, and total effects of the variables are shown in Table 4. Among the five selected variables, LAI and SOC content directly affected R_s and can be used to predict R_s with relatively high accuracy ($R^2 = 0.79$). The other three variables (i.e., Chl_{canopy}, ABG, and STN content), despite having a significant correlation with R_s, only affected R_s indirectly through their direct relationship with SOC content and LAI. Thus, the two direct effect factors were used to estimate R_s, and the spatially distributed data proxies of these two factors were used to quantify the spatial patterns of R_s in maize fields during the peak growing season.

Spatial data used for soil respiration estimation

Maize classification. The maize classification map of the study area is shown in Figure 3. The classification accuracy for maize at the study site could not be quantitatively assessed because of the limitation of the sample data. However, 53 sample plots were all located in the maize classification map, and the county-level maize patterns classified in the map were consistent with the general maize patterns across the three counties. In addition, the classified maize area was close to the maize area reported by the China County Statistical Yearbook [71]. Thus, the classification accuracy of maize was believed to be reasonable, and the maize classification map was then used to predict the spatial pattern of R_s during the peak growing season of maize.

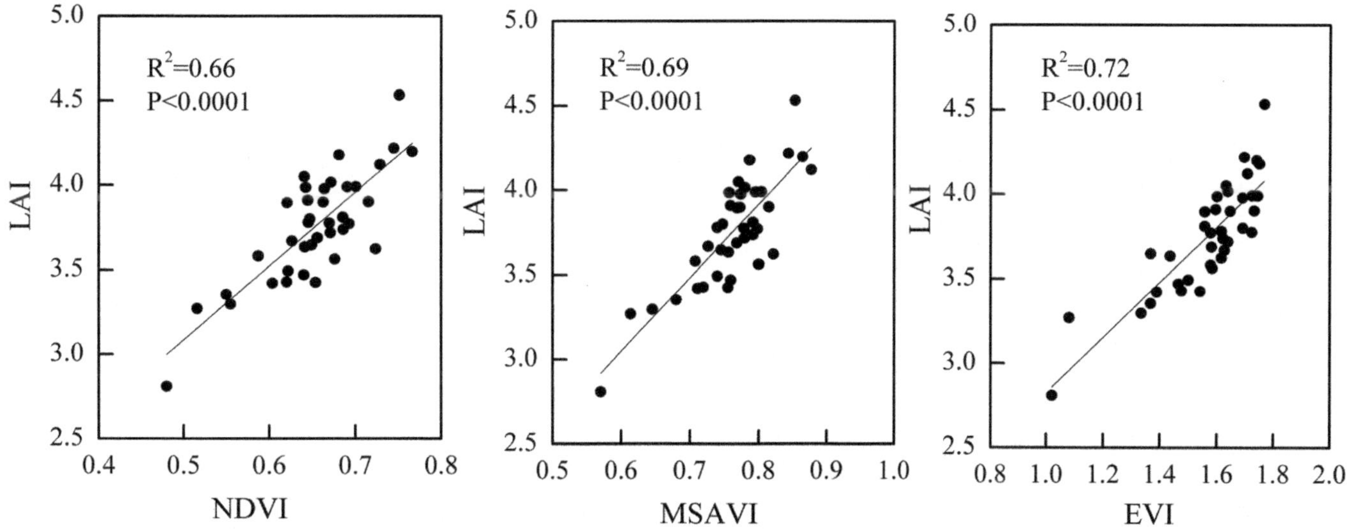

Figure 4. Linear relationships between three vegetation indices (VIs) and leaf area index (LAI) during the peak growing season of maize in three counties in North China (n = 38). The VIs are: normalized difference vegetation index (NDVI), enhanced vegetation index (EVI), and modified soil adjusted vegetation index (MSAVI).

LAI estimation from spectral vegetation index. Among the three greenness indices calculated from the optical image of HJ-1A satellite, EVI showed the best linear relationship with LAI, with a determination coefficient (R^2) of 0.72, followed by MSAVI and NDVI (Fig. 4). The explanation of LAI variance increased from 66% to 72% when EVI was used instead of NDVI for LAI estimation, and this increase was statistically significant ($p<0.05$). However, EVI and MSAVI did not exhibit a significant difference in explaining the variation in LAI, despite EVI having a slightly better relationship with LAI than MSAVI. Thus, EVI was used as a proxy for LAI to estimate R_s during the peak growing season of maize for simplicity. The spatially distributed EVI during the peak growing season of maize exhibited relatively small variability (Fig. 5). Overall, the EVI in the north and southwest parts of the study site (i.e., Baixiang and Longyao Counties) showed a high value. Relatively low EVI values mainly occurred in the southeast parts of the study site (i.e., Julu County), especially the northwest Julu County (Fig. 5).

Spatial distribution of SOC content. Kriging interpolation was performed by using ArcGIS 9.3 software to produce the spatial distribution map of the SOC content in maize fields of the study area. A cell size of 30 m×30 m was selected for the spatial interpolation to match the spatial resolution of images from OLI and HJ-1A/B. The final result of this spatial interpolation process is shown in Figure 6. Based on the spatial distribution map of the SOC content in maize fields, SOC content values were higher in the northwest and southwest parts of the study area than in the southeastern part.

Spatial distribution of soil respiration

The EVI and SOC content were used to estimate the spatial pattern of R_s during the peak growing season of maize on the basis of a simple exponential model. The geo-location information (latitude and longitude) of the 38 sample plots was used in the extraction of pixels. Pixels that contained these plots from the spatial distribution maps of EVI and SOC content data (Figs. 5 and 6) were extracted. These data were used to determine the model parameters by least-squares fitting. The resulting model was as follows:

$$R_s = 1.57 \times \exp(0.44 \times \text{EVI} + 0.05 \times \text{SOC content}) \quad (2)$$

$$(n = 38,\ R^2 = 0.73)$$

where R_s refers to the daily mean soil respiration rate in μmol CO_2 m^{-2} s^{-1}; EVI refers to enhanced vegetation index, as a proxy for LAI; and SOC content is the soil organic carbon content (g kg^{-1}) in maize fields of the study area. Eq. (2) was employed to predict the spatial pattern of R_s from spatially distributed EVI and SOC content data during the peak growing season of maize (Figs. 5 and 6). The spatial variation in R_s showed a pattern similar to that in SOC content (Figs. 6 and 7).

Figure 8 shows the accuracy assessment result of the R_s prediction model. The field measured R_s was comparable with the spatial data predicted R_s. Based on the independent test dataset, EVI and SOC content accounted for 69% of the spatial variation in ground-measured R_s, and the RMSE was 0.51 μmol CO_2 m^{-2} s^{-1}. The result of the accuracy assessment suggests that the prediction model, which used EVI and SOC content as the dependent variables, was effective in estimating R_s in maize fields during the peak growing season.

Discussion

Relationships between LAI and three VIs

In this study, in situ measured data were obtained during the peak growing period of maize (corresponding to the tassel stage of maize). The effect of soil background on the spectral reflectance of remote sensing images was negligible during this period because the maize cover was higher with LAI ranging from 2.81 to 4.53. The difference in the capability of spectral vegetation index (VI) responding to LAI variation mainly depended on the sensitivity of VI to the canopy structural variation of maize. Thus, the VI modified the effect of soil reflectance (i.e. MSAVI) did not exhibit a significantly greater advantage than NDVI, which is strongly affected by soil reflectance in sparsely vegetated areas [50]. EVI, which is more sensitive to variation in dense vegetation than

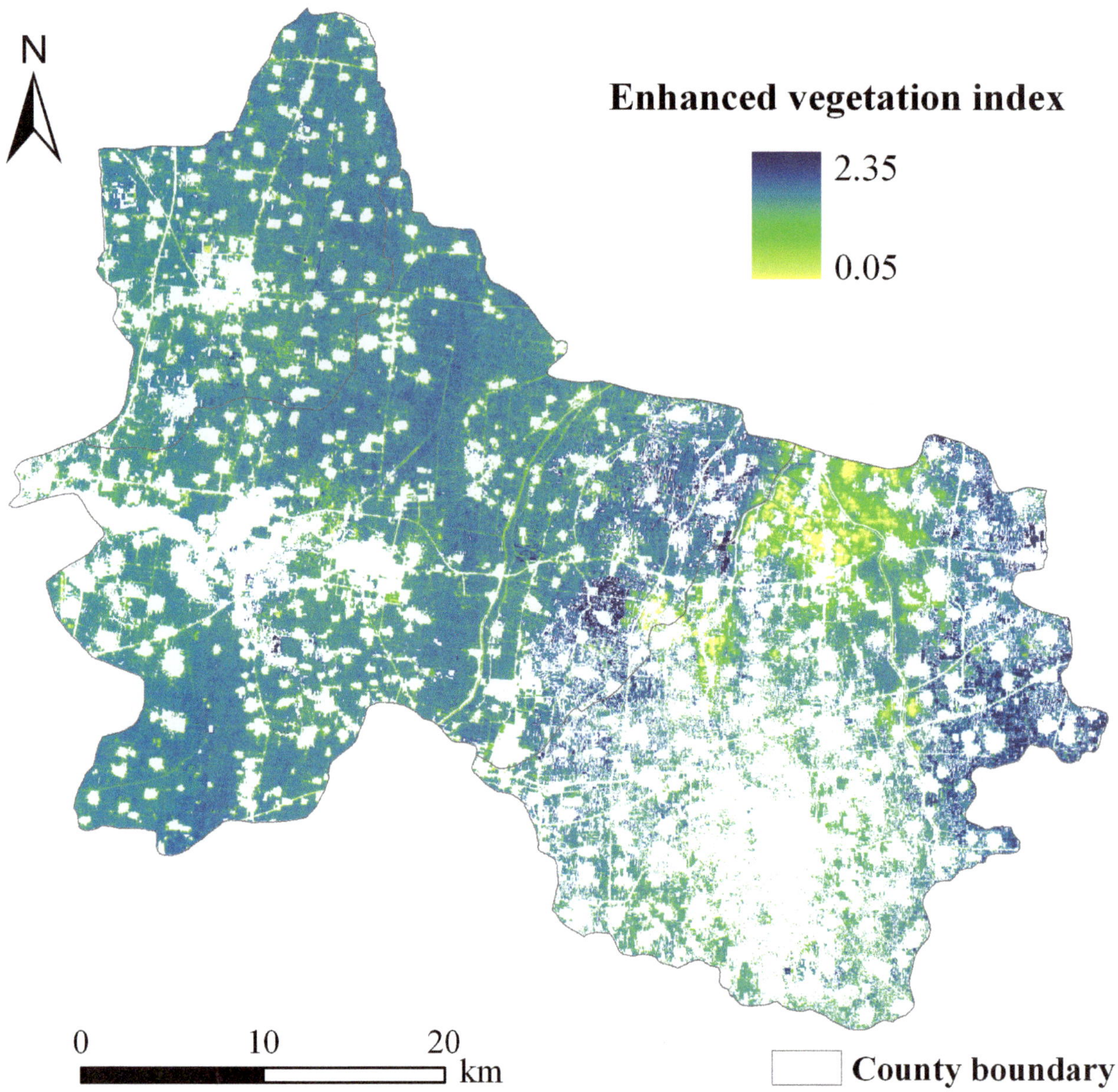

Figure 5. Spatial distribution map of enhanced vegetation index in maize fields in three counties in North China.

NDVI [50], showed the best relationship with the LAI of maize. This result was consistent with our previous study [58] that was conducted in irrigated and rainfed maize fields located at the University of Nebraska, Agricultural and Research Development Center, Mead, Eastern Nebraska, USA.

Measurement accuracy of SOC content

Field measurement data revealed that the SOC content at 0 cm to 20 cm depth in the maize fields ranged from 6.4 g kg^{-1} to 17.3 g kg^{-1}, and the mean value was 12.01 g kg^{-1}. For the mean dry land SOC content in North China, the value appeared to be higher than the previous estimate (0.83 from the average of 268 sample points) [72]. This difference was partly attributed to the fact that only the SOC content in maize fields, not in all dry land types, was considered. Most maize fields in the study site were on a winter wheat/maize rotation, and wheat straw was returned to the soil. The high productivity of maize crops contributed to the development of a thick A horizon and high SOC content [73,74]. Additionally, only the SOC content in maize fields at 0 cm to 20 cm depth was analyzed, whereas previous studies estimated the SOC content on the basis of organic carbon content to a depth of 1 m [72,75,76]. In agricultural land, soil depth at 0 cm to 20 cm is located in the cultivation layer and has a higher SOC content than the SOC content at the deeper soil layers [34]. This condition contributed to the higher SOC content from the measured soil property data than the previous estimate.

Factors affecting spatial pattern of soil respiration

The spatial differences in R_s at the study site can be mainly attributed to the differences in vegetation productivity and soil property factors among the sample plots, whereas soil temperature and soil moisture served a minor function in regulating the spatial pattern of R_s. A previous study also demonstrated that site variables that reflect site productivity (e.g., LAI or aboveground

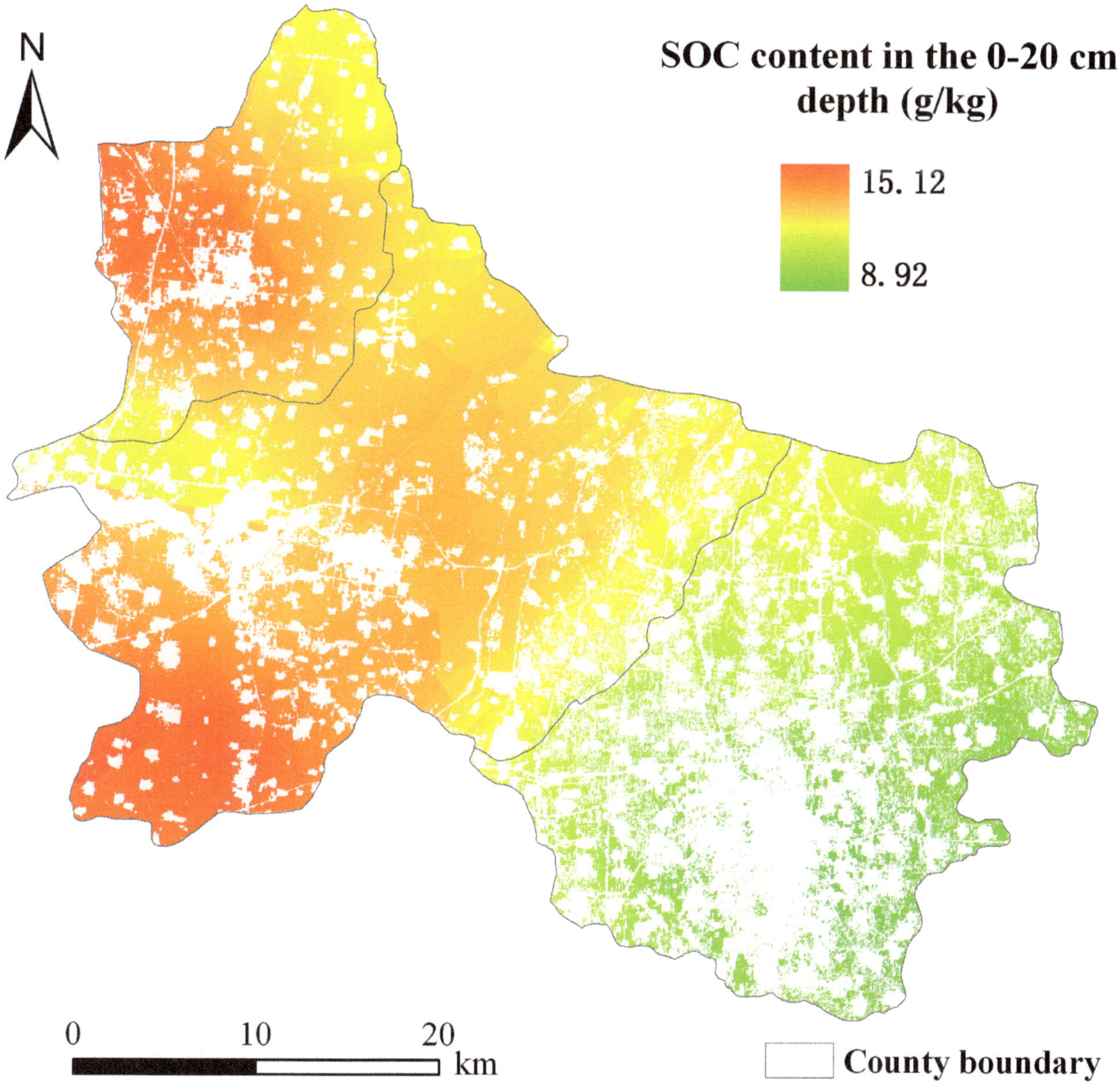

Figure 6. Spatial distribution map of soil organic carbon (SOC) content in the 0–20 cm depth in maize fields in three counties in North China.

net primary productivity) will provide a useful approach for large-scale estimates of regional R_s in terrestrial ecosystems [8]. Soil temperature evidently serves a predominant function in the spatial variations of R_s across sites of climatically contrasting environments [4]. However, at a local scale or under similar climatic conditions, other biological and biophysical factors, such as vegetation productivity and the size of organic carbon pools, may prevail as dominant drivers of R_s [4,77]. At a local scale, the spatial variation in T_{s10} in the study site was small (CV = 4.73%). Thus, soil temperature did not affect the spatial pattern of R_s. Although soil moisture in the maize fields showed a relatively large spatial variation (CV = 12.48%), this variation did not reach a degree that will affect the spatial dynamics of R_s. The soil C quantity and substrate quality factors (i.e., SOC and STN contents) were consistently and strongly correlated with one another and significantly affected the variation in R_s [5,12,13]. However, SEM results showed that the STN content only affected R_s indirectly through the direct effect on the SOC content at the study site.

During the peak growing season of maize, biophysical parameters, such as LAI, Chl_{canopy}, and AGB, were important variables that determined the size of the photosynthetic capacity [56,78]. However, these variables are not truly independent, and a correlation between one of them and R_s may lead to a correlation of the other with R_s. In this study, R_s was strongly correlated with LAI, Chl_{canopy} and AGB of maize fields, whereas LAI was the only variable directly related to R_s during the peak growing season of maize on the basis of SEM analysis.

The direct effect factors of R_s were used to estimate the spatial variability of R_s during the peak growing season of maize in three counties in North China. A simple exponential model, which included the corresponding spatial proxies from remote sensing

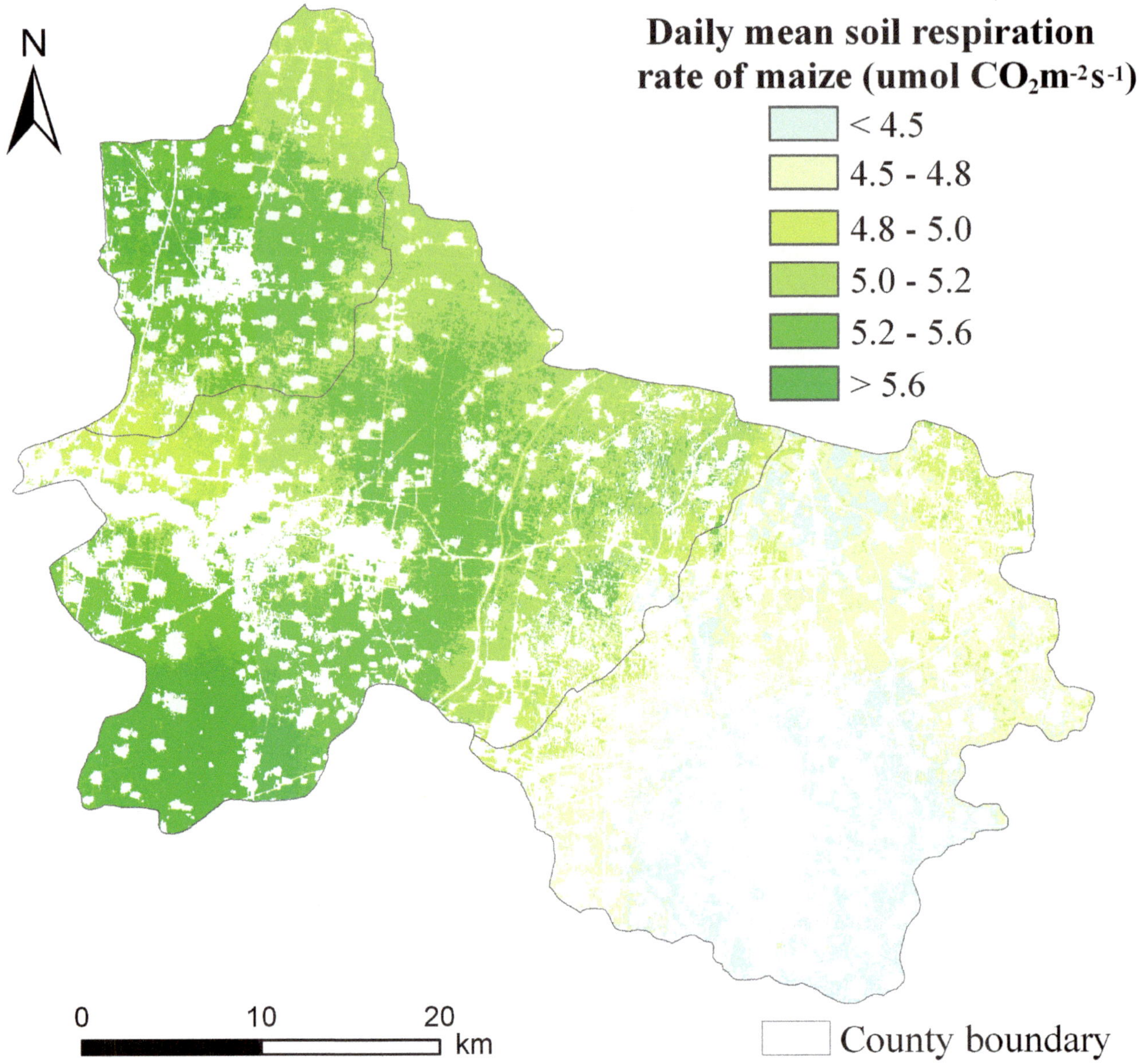

Figure 7. Spatial pattern of daily mean soil respiration rate during the peak growing season of maize in three counties in North China.

and GIS (i.e., EVI and spatially interpolated SOC content), was employed. A similar method was applied to a deciduous broadleaf forest site in the Midwest USA [79]. The independent test data also demonstrated the rationality of this method at the study site to a certain extent (Fig. 8). Regardless of the form of the R_s model, the relationship between LAI and EVI, as well as the kriging interpolation precision of the SOC content, affected the predictive accuracy of the R_s model. A moderate correlation between EVI and LAI (Fig. 4) affected the test accuracy of the exponential model with an R^2 value of 0.69 and an RMSE value of 0.51 μmol CO_2 m^{-2} s^{-1} (Fig. 8). The tendency of kriging to overestimate small values is supported by previous studies [80–82]. This tendency may help explain the bias toward overestimating R_s at low values (Fig. 8). Therefore, improving the accuracy of input parameters from remote sensing or GIS will increase the predictive capability of the R_s model.

Notably, the R_s model developed in this study was applicable to maize fields during the peak growth period in the three counties in North China. However, the model employed in this study does not consider temperature, a main driver of R_s that has high spatial variability. This model may be not used anywhere else or in other stages of the growing season. Furthermore, when spatially distributed data were used in the R_s model, a simple alternative method was employed to estimate the maize LAI by using the remotely sensed EVI, which may be problematic. Verstraeten et al. [83] highlighted that the assimilation of remotely sensed geophysical products into a carbon model is a complex process, and simply exchanging conventional input data for their remotely sensed counterparts is insufficient. Therefore, future research should focus on an integrating spatially distributed R_s datasets and geophysical products from remote sensing and GIS by using the data assimilation method, which has been extensively applied in

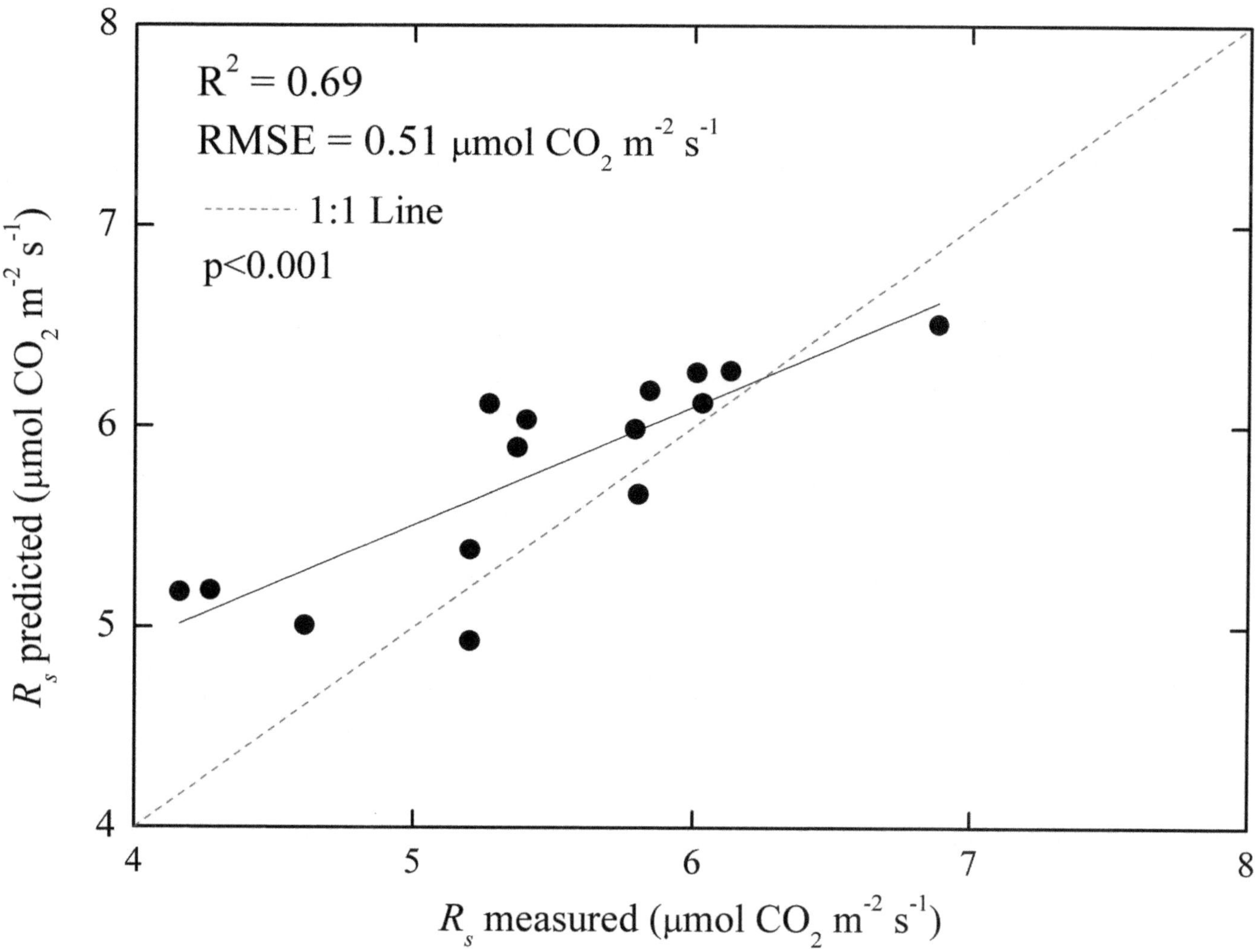

Figure 8. Spatial data predicted soil respiration (R_s) and corresponding ground-based measurements with R^2 and RMSE (μmol CO_2 m^{-2} s^{-1}) during the peak growing season of maize in three counties in North China ($n = 15$). The predicted soil respiration was attained with an exponential model that used EVI and SOC content as dependent variables.

terrestrial carbon cycle research [84–86]. However, this method lack the integration of R_s and spatially distributed data.

Conclusions

This study investigated the potential of spatial data from remote sensing and GIS for estimating the spatial patterns of R_s during the peak growing season of maize in three counties in North China. Based on in situ measurements, plant productivity (i.e., LAI) and soil property (i.e. SOC content) factors were identified as the most important determinants of spatial variability in R_s during the peak growing season of maize, and R_s was weakly related to soil temperature and soil moisture. Spectral VIs calculated from an HJ-1A CCD optical image were used to estimate LAI and EVI was found to be the best proxy for LAI. To derive the spatial pattern of R_s during the peak growing season of maize, a simple exponential model, which included remotely sensed EVI and GIS spatially interpolated SOC content, was employed. This method was tested by using an independent sample dataset and was shown to be reasonable at the study site.

Acknowledgments

We sincerely thank the anonymous reviewers for their important and constructive revision advices on the manuscript.

Author Contributions

Conceived and designed the experiments: NH LW YQG. Performed the experiments: NH LW PYH. Analyzed the data: NH LW PYH. Contributed reagents/materials/analysis tools: NH LW YQG PYH ZN. Wrote the paper: NH.

References

1. Raich JW, Potter CS (1995) Global patterns of carbon dioxide emissions from soils. Global Biogeochemical Cycles 9: 23–36.
2. Davidson EA, Belk E, Boone RD (1998) Soil water content and temperature as independent or confound factors controlling soil respiration in a temperature mixed hardwood forest. Global Change Biology 4: 217–227.
3. Buchmann N (2000) Biotic and abiotic factors controlling soil respiration rates in Picea abies stands. Soil Biology and Biochemistry 32: 1625–1635.
4. Campbell JL, Sun OJ, Law BE (2004) Supply-side controls on soil respiration among Oregon forests. Global Change Biology 10: 1857–1869.

5. Webster KL, Creed IF, Skowronski MD, Kaheil YH (2009) Comparison of the performance of statistical models that predict soil respiration from forests. Soil Science Society of America Journal 73: 1157–1167.
6. Gaumont-Guay D, Black TA, Griffis TJ, Barr AG, Jassal RS, et al. (2006) Interpreting the dependence of soil respiration on soil temperature and water content in a boreal aspen stand. Agricultural and Forest Meteorology 140: 220–235.
7. Phillips SC, Varner RK, Frolking S, Munger JW, Bubier JL, et al. (2010). Interannual, seasonal, and diel variation in soil respiration relative to ecosystem respiration at a wetland to upland slope at Harvard Forest. Journal of Geophysical Research: Biogeosciences 115: G02019. doi: 10.1029/2008JG000858.
8. Reichstein M, Rey A, Freibauer A, Tenhunen J, Valentini R, et al. (2003) Modeling temporal and large-scale spatial variability of soil respiration from soil water availability, temperature and vegetation productivity indices. Global Biogeochemical Cycles 17: 1104. doi: 10.1029/2003GB002035.
9. Geng Y, Wang Y, Yang K, Wang S, Zeng H, et al. (2012) Soil respiration in Tibetan alpine grasslands: belowground biomass and soil moisture, but not soil temperature, best explain the large-scale patterns. PloS one 7: e34968. doi: 10.1371/journal.pone.0034968.
10. Huang N, Niu Z, Zhan YL, Tappertc MC, Wu CY, et al. (2012) Relationships between soil respiration and photosynthesis-related spectral vegetation indices in two cropland ecosystems. Agricultural and Forest Meteorology 160: 80–89.
11. Chen ST, Huang Y, Zou JW, Shen QR, Hu ZH, et al. (2010) Modeling interannual variability of global soil respiration from climate and soil properties. Agricultural and Forest Meteorology 150: 590–605.
12. Almagro M, Querejeta JI, Boix-Fayos C, Martínez-Mena M (2013) Links between vegetation patterns, soil C and N pools and respiration rate under three different land uses in a dry Mediterranean ecosystem. Journal of Soils and Sediments 13: 641–653.
13. Martin JG, Bolstad PV, Ryu SR, Chen J (2009) Modeling soil respiration based on carbon, nitrogen, and root mass across diverse Great Lake forests. Agricultural and Forest Meteorology 149: 1722–1729.
14. Longley P (Ed.) (2005) Geographic information systems and science. John Wiley & Sons.
15. Weng Q (2001) Modeling urban growth effects on surface runoff with the integration of remote sensing and GIS. Environmental Management 28: 737–748.
16. Lin ML, Chen CW (2011) Using GIS-based spatial geocomputation from remotely sensed data for drought risk-sensitive assessment. International Journal of Innovative Computing, Information and Control 7: 657–668.
17. Shalaby A, Tateishi R (2007) Remote sensing and GIS for mapping and monitoring land cover and land-use changes in the Northwestern coastal zone of Egypt. Applied Geography 27: 28–41.
18. Dewan AM, Yamaguchi Y (2009). Land use and land cover change in Greater Dhaka, Bangladesh: using remote sensing to promote sustainable urbanization. Applied Geography 29: 390–401.
19. Wang XD, Zhong XH, Liu SZ, Liu JG, Wang ZY, et al. (2008). Regional assessment of environmental vulnerability in the Tibetan Plateau: development and application of a new method. Journal of Arid environments 72: 1929–1939.
20. Ceccato P, Connor SJ, Jeanne I, Thomson MC (2005) Application of geographical information systems and remote sensing technologies for assessing and monitoring malaria risk. Parassitologia 47: 81–96.
21. Franklin J (1995) Predictive vegetation mapping: geographic modelling of biospatial patterns in relation to environmental gradients. Progress in Physical Geography 19: 474–499.
22. Raup B, Kääb A, Kargel JS, Bishop MP, Hamilton G, et al. (2007) Remote sensing and GIS technology in the Global Land Ice Measurements from Space (GLIMS) project. Computers & Geosciences 33: 104–125.
23. Keane RE, Burgan R, van Wagtendonk J (2001) Mapping wildland fuels for fire management across multiple scales: integrating remote sensing, GIS, and biophysical modeling. International Journal of Wildland Fire 10: 301–319.
24. He C, Wang S, Xu J, Zhou C (2002) Using remote sensing to estimate the change of carbon storage: a case study in the estuary of Yellow River delta. International Journal of Remote Sensing 23: 1565–1580.
25. Yuan W, Liu S, Yu G, Bonnefond JM, Chen J, et al. (2010) Global estimates of evapotranspiration and gross primary production based on MODIS and global meteorology data. Remote Sensing of Environment 114: 1416–1431.
26. Huang N, He JS, Niu Z (2013) Estimating the spatial pattern of soil respiration in Tibetan alpine grasslands using Landsat TM images and MODIS data. Ecological Indicators 26: 117–125.
27. Davidson EA, Richardson AD, Savage KE, Hollinger DY (2006) A distinct seasonal pattern of the ratio of soil respiration to total ecosystem respiration in a spruce-dominated forest. Global Change Biology 12, 230–239.
28. Vargas R, Baldocchi DD, Allen MF, Bahn M, Black TA, et al. (2010) Looking deeper into the soil: biophysical controls and seasonal lags of soil CO_2 production and efflux. Ecological Applications 20: 1569–1582.
29. Bondeau A, Smith PC, Zaehle S, Schaphoff S, Lucht W, et al. (2007) Modelling the role of agriculture for the 20th century global terrestrial carbon balance. Global Change Biology 13: 679–706.
30. Foley JA, DeFries RS, Asner GP, Barford C, Bonan G, et al. (2005) Global consequences of land use. Science 309: 570–574.
31. Krugh B, Bickham L, Miles D (1994) The solid-state chlorophyll meter: a novel instrument for rapidly and accurately determining the chlorophyll concentrations in seedling leaves. Maize Genetics Cooperation Newsletter 68: 25–27.
32. Wu CY, Wang L, Niu Z, Gao S, Wu MQ (2010) Nondestructive estimation of canopy chlorophyll content using Hyperion and Landsat/TM images. International Journal of Remote Sensing 31: 2159–2167.
33. Gao Y, Xie Y, Jiang H, Wu B, Niu J (2014) Soil water status and root distribution across the rooting zone in maize with plastic film mulching. Field Crops Research 156: 40–47.
34. Kou TJ, Zhu P, Huang S, Peng XX, Song ZW, et al. (2012) Effects of long-term cropping regimes on soil carbon sequestration and aggregate composition in rainfed farmland of Northeast China. Soil and Tillage Research 118: 132–138.
35. Nelson DW, Sommers LE (1982) Total carbon, organic carbon, and organic matter. In: Page AL, Miller RH, Keeney DR. (Eds.), Methods of Soil Analysis. American Society of Agronomy and Soil Science Society of American, Madison. 101–129.
36. Gallaher RN, Weldon CO, Boswell FC (1976) A semiautomated procedure for total nitrogen in plant and soil samples. Soil Science Society of America Journal 40: 887–889.
37. Wardlow BD, Egbert SL, Kastens JH (2007) Analysis of time-series MODIS 250 m vegetation index data for crop classification in the US Central Great Plains. Remote Sensing of Environment 108: 290–310.
38. Wilson EH, Sader SA (2002) Detection of forest harvest type using multiple dates of Landsat TM imagery. Remote Sensing of Environment 80: 385–396.
39. Zhong B, Ma P, Nie A, Yang A, Yao Y, et al. (2014) Land cover mapping using time series HJ-1/CCD data. Science China: Earth Sciences doi: 10.1007/s11430-014-4877-5.
40. Bian JH, Li AN, Jin HA, Lei GB, Huang CQ, et al. (2013) Auto-registration and orthorecification algorithm for the time series HJ-1A/B CCD images. Journal of Mountain Science 10: 754–767.
41. Liu Y, Li M, Mao L, Cheng L, Chen K (2013) Seasonal pattern of tidal-flat topography along the Jiangsu middle coast, China, using HJ-1 optical images. Wetlands 33: 871–886.
42. Wang SD, Miao LL, Peng GX (2012) An Improved Algorithm for Forest Fire Detection Using HJ Data. Procedia Environmental Sciences 13: 140–150.
43. Gamon JA, Field CB, Goulden ML, Griffin KL, Hartley AE, et al. (1995) Relationship between NDVI, canopy structure and photosynthesis in three Californian vegetation types. Ecological Applications 5: 28–41.
44. Hansen PM, Schjoerring JK (2003) Reflectance measurement of canopy biomass and nitrogen status in wheat crops using normalized difference vegetation indices and partial least squares regression. Remote Sensing of Environment 86: 542–553.
45. Yu X, Yan Q, Liu Z (2010) Atmospheric correction of HJ-1A multi-spectral and hyper-spectral images. Image and Signal Processing (CISP), 2010 3rd International Congress on. IEEE 5: 2125–2129.
46. Li P, Jiang L, Feng Z (2013) Cross-Comparison of Vegetation Indices Derived from Landsat-7 Enhanced Thematic Mapper Plus (ETM+) and Landsat-8 Operational Land Imager (OLI) Sensors. Remote Sensing 6: 310–329.
47. Rouse JW, Haas RH, Schell JA, Deering DW, Harlan JC (1974) Monitoring the vernal advancements and retrogradation of natural vegetation, In: NASA/GSFC, Final Report, Greenbelt, MD, USA, 1–137.
48. Gamon JA, Field CB, Goulden ML, Griffin KL, Hartley AE, et al. (1995) Relationship between NDVI, canopy structure and photosynthesis in three Californian vegetation types. Ecological Applications 5: 28–41.
49. Qi J, Chehbouni A, Huete AR, Kerr YH, Sorooshian S (1994) A modified soil adjusted vegetation index (MSAVI). Remote Sensing of Environment 48: 119–126.
50. Huete A, Didan K, Miura T, Rodriguez EP, Gao X, et al. (2002) Overview of the radiometric and biophysical performance of the MODIS vegetation indices. Remote Sensing of Environment 83: 195–213.
51. Broge NH, Leblanc E (2001) Comparing prediction power and stability of broadband and hyperspectral vegetation indices for estimation of green leaf area index and canopy chlorophyll density. Remote Sensing of Environment 76: 156–172.
52. Haboudane D, Miller JR, Tremblay N, Zarco-Tejada PJ, Dextraze L (2002) Integrated narrow-band vegetation indices for prediction of crop chlorophyll content for application to precision agriculture. Remote Sensing of Environment 81: 416–426.
53. Gitelson AA, Vina A, Ciganda V, Rundquist DC, Arkebauer TJ (2005) Remote estimation of canopy chlorophyll content in crops. Geophysical Research Letters 32: L08403. doi: 10.1029/2005GL022688.
54. Wu C, Niu Z, Tang Q, Huang W (2008) Estimating chlorophyll content from hyperspectral vegetation indices: Modeling and validation. Agricultural and Forest Meteorology 148: 1230–1241.
55. Hirose T, Ackerly DD, Traw MB, Ramseier D, Bazzaz FA (1997) CO_2 elevation, canopy photosynthesis, and optimal leaf area index. Ecology 78: 2339–2350.
56. Gitelson AA, Vina A, Verma SB, Rundquist DC, Arkebauer TJ, et al. (2006) Relationship between gross primary production and chlorophyll content in crops: Implications for the synoptic monitoring of vegetation productivity. Journal of Geophysical Research-Atmospheres 111: D08S11. doi: 10.1029/2005JD006017.
57. Glenn EP, Huete AR, Nagler PL, Nelson SG (2008) Relationship between remotely-sensed vegetation indices, canopy attributes and plant physiological

processes: what vegetation indices can and cannot tell us about the landscape. Sensors 8: 2136–2160.
58. Huang N, Niu Z (2013) Estimating soil respiration using spectral vegetation indices and abiotic factors in irrigated and rainfed agroecosystems. Plant and Soil 367: 535–550.
59. Chevallier T, Voltz M, Blanchart E, Chotte JL, Eschenbrenner V, et al. (2000) Spatial and temporal changes of soil C after establishment of a pasture on a long-term cultivated vertisol (Martinique). Geoderma 94: 43–58.
60. McGrath D, Zhang C (2003) Spatial distribution of soil organic carbon concentrations in grassland of Ireland. Applied Geochemistry 18: 1629–1639.
61. Liu D, Wang Z, Zhang B, Song K, Li X, et al. (2006) Spatial distribution of soil organic carbon and analysis of related factors in croplands of the black soil region, Northeast China. Agriculture, Ecosystems & Environment 113: 73–81.
62. Matheron G (1963) Principles of geostatistics. Economic geology 58: 1246–1266.
63. Webster R, Oliver MA (2007) Geostatistics for environmental scientists. John Wiley & Sons.
64. Raich JW, Tufekciogul A (2000) Vegetation and soil respiration: correlations and controls. Biogeochemistry 48: 71–90.
65. Schaefer DA, Feng W, Zou X (2009) Plant carbon inputs and environmental factors strongly affect soil respiration in a subtropical forest of southwestern China. Soil Biology and Biochemistry 41: 1000–1007.
66. Curiel Yuste J, Baldocchi DD, Gershenson A, Goldstein A, Misson L, et al. (2007) Microbial soil respiration and its dependency on carbon inputs, soil temperature and moisture. Global Change Biology 13: 2018–2035.
67. Pugesek BH, Tomer A, Von Eye A (Eds.) (2003) Structural equation modeling: applications in ecological and evolutionary biology. Cambridge University Press.
68. Iriondo JM, Albert MJ, Escudero A (2003) Structural equation modelling: an alternative for assessing causal relationships in threatened plant populations. Biological Conservation 113: 367–377.
69. Jonsson M, Wardle DA (2010) Structural equation modelling reveals plant-community drivers of carbon storage in boreal forest ecosystems. Biology Letters 6: 116–119.
70. Kim GS (2010) AMOS 18.0: Structural Equation Modeling. Seoul: Hannarae Publishing Co.
71. National Bureau of statistics of China (2006) China social-economic statistical yearbooks for China's counties and cities. China Statistics Press, Beijing.
72. Wang S, Tian H, Liu J, Pan S (2003) Pattern and change of soil organic carbon storage in China: 1960s–1980s. Tellus B 55: 416–427.
73. West TO, Post WM (2002) Soil organic carbon sequestration rates by tillage and crop rotation. Soil Science Society of America Journal 66: 1930–1946.
74. Wilhelm WW, Johnson JM, Karlen DL, Lightle DT (2007) Corn stover to sustain soil organic carbon further constrains biomass supply. Agronomy journal 99: 1665–1667.
75. Foley JA (1995) An equilibrium model of the terrestrial carbon budget. Tellus B 47: 310–319.
76. Lal R (1999) Soil management and restoration for C sequestration to mitigate the accelerated greenhouse effect. Progress in Environmental Science 1: 307–326.
77. Epron D, Bosc A, Bonal D, Freycon V (2006) Spatial variation of soil respiration across a topographic gradient in a tropical rain forest in French Guiana. Journal of Tropical Ecology 22: 565–574.
78. Suyker AE, Verma SB, Burba GG, Arkebauer TJ (2005) Gross primary production and ecosystem respiration of irrigated maize and irrigated soybean during a growing season. Agricultural and Forest Meteorology 131: 180–190.
79. Huang N, Gu L, Niu Z (2014) Estimating soil respiration using spatial data products: A case study in a deciduous broadleaf forest in the Midwest USA. Journal of Geophysical Research: Atmospheres 119. doi:10.1002/2013JD020515.
80. Hudak AT, Lefsky MA, Cohen WB, Berterretche M (2002) Integration of lidar and Landsat ETM+ data for estimating and mapping forest canopy height. Remote Sensing of Environment 82: 397–416.
81. Meng Q, Cieszewski C, Madden M (2009) Large area forest inventory using Landsat ETM+: a geostatistical approach. ISPRS Journal of Photogrammetry and Remote Sensing 64: 27–36.
82. Tsui OW, Coops NC, Wulder MA, Marshall PL (2013) Integrating airborne LiDAR and space-borne radar via multivariate kriging to estimate above-ground biomass. Remote Sensing of Environment 139: 340–352.
83. Verstraeten WW, Veroustraete F, Wagner W, Van Roey T, Heyns W, et al. (2010) Remotely sensed soil moisture integration in an ecosystem carbon flux model-The spatial implication. Climatic Change 103:117–136.
84. Rayner PJ, Scholze M, Knorr W, Kaminski T, Giering R, et al. (2005) Two decades of terrestrial carbon fluxes from a carbon cycle data assimilation system (CCDAS). Global Biogeochemical Cycles 19.
85. Chevallier F, Bréon FM, Rayner PJ (2007) Contribution of the Orbiting Carbon Observatory to the estimation of CO_2 sources and sinks: Theoretical study in a variational data assimilation framework. Journal of Geophysical Research: Atmospheres 112: D09307. doi: 10.1029/2006JD007375.
86. Knorr W, Kaminski T, Scholze M, Gobron N, Pinty B, et al. (2010) Carbon cycle data assimilation with a generic phenology model. Journal of Geophysical Research: Biogeosciences 115: G04017. doi: 10.1029/2009JG001119.

4

Assessing Spatial Accessibility to Maternity Units in Shenzhen, China

Peige Song[1], Yajie Zhu[2], Xi Mao[3], Qi Li[2], Lin An[1]*

1 Department of Child, Adolescent and Women's Health, School of Public Health, Peking University Health Science Center, Beijing, China, 2 Institute of Remote Sensing and GIS, School of Earth and Space Science, Peking University, Beijing, China, 3 Institute of Cartography and GIS, Chinese Academy of Surveying and Mapping, Beijing, China

Abstract

Background: With the rapid development of urbanization, pregnant population is growing rapidly in Shenzhen, and it has been a difficulty to serve more and more pregnant women and reduce spatial access disparities to maternity units (MUs). Understanding of the current status of accessibility to MUs is valuable for supporting the rational allocation of MUs in the future.

Methods: Based on pregnant population data and MUs data, this study uses a two-step floating catchment area (2SFCA) method based on Geographic Information System (GIS) to analyze the current spatial accessibility to MUs, and then make a comparison between that to public MUs and private MUs.

Results: Our analysis of the accessibility to all MUs within a distance of 20 km shows that the accessibilities of the areas alongside the traditional border management line are acceptable, meanwhile highlights some critical areas, such as the west part of Nanshan district and the vast east part of Longgang district. The comparison between spatial accessibility to public MUs and private MUs shows statistically significant difference.

Discussion: Results of this study suggest a great effort should be made to improve the equity of spatial accessibility to MUs in Shenzhen. For policy-making, strategy for the siting and allocation of future MUs, no matter public or private, should guarantee the greatest spatial accessibility for every pregnant woman.

Editor: Josef Neu, University of Florida, United States of America

Funding: The authors have no support or funding to report.

Competing Interests: The authors have declared that no competing interests exist.

* E-mail: anlinbjmu@163.com

Background

Shenzhen is a major city in Guangdong Province of China, situated immediately north of Hong Kong (Figure 1). In 1980, Shenzhen established the first Special Economic Zone (SEZ) in China and became the first experimental city of China's reform and "opening up". As a pioneer of developing socialist market economy, Shenzhen SEZ could use flexible governmental measures to develop economy and get strong economic policy support from central government of China. Over the past three decades, a lot of Chinese and foreign nationals have invested enormous amounts of money to develop manufacturing and service industries in Shenzhen SEZ, such as private real estates, private hospitals, etc. In the process of this experiment, the lessons from SEZ's experience could be used in the rest of Chinese Mainland [1,2]. With the rapid development of urbanization, Shenzhen became the first city without rural areas in China in 2010, which makes this study to be the first typical research of spatial accessibility to health resources in pure urban ground of China. From 2000 to 2010, Shenzhen's urban population density increased from 3,596/km^2 (in 2000, total population 7,008,428) to 8,588/km^2 (in 2010, total population 10,357,938) [3,4], and Shenzhen became the most crowded area in China and fifth in the World [5]. Its health system is already over-burdened, where hospital beds number (including long-term hospital beds, maternity beds and paediatric beds, but not delivery beds) per 1,000 population (a classic indicator to indicate the availability of inpatient services in terms of resources per population and reflect the status of an area's health system) [6,7] has decreased from 2.38 (in 2000) to 2.20 (in 2010) [3,4], falling below the national average (latest in 2009, 4.20 beds per 1,000 population) [7]. Under this circumstance, it is important to know the current situation of health resources allocation in Shenzhen, and make reference for future urban construction and health resource planning.

As populations grow, the annual number of newborns per square kilometer (approximately 94/km^2) in Shenzhen also continually ranks first in China recent years [8]. "Women and children's health is the precondition and foundation of sustained human development" [9]. To ensure safe motherhood, every pregnant woman should accept professional antenatal examinations and skilled attendance at childbirth [10]. In obstetrics, longer travel distances may lead to accidental deliveries outside hospital [11] and may be associated with higher stillbirth rates [12]. In general, the WHO-recommended indicator-"Maternity beds density per 1,000 pregnant women" can be used to evaluate the gross access to maternity

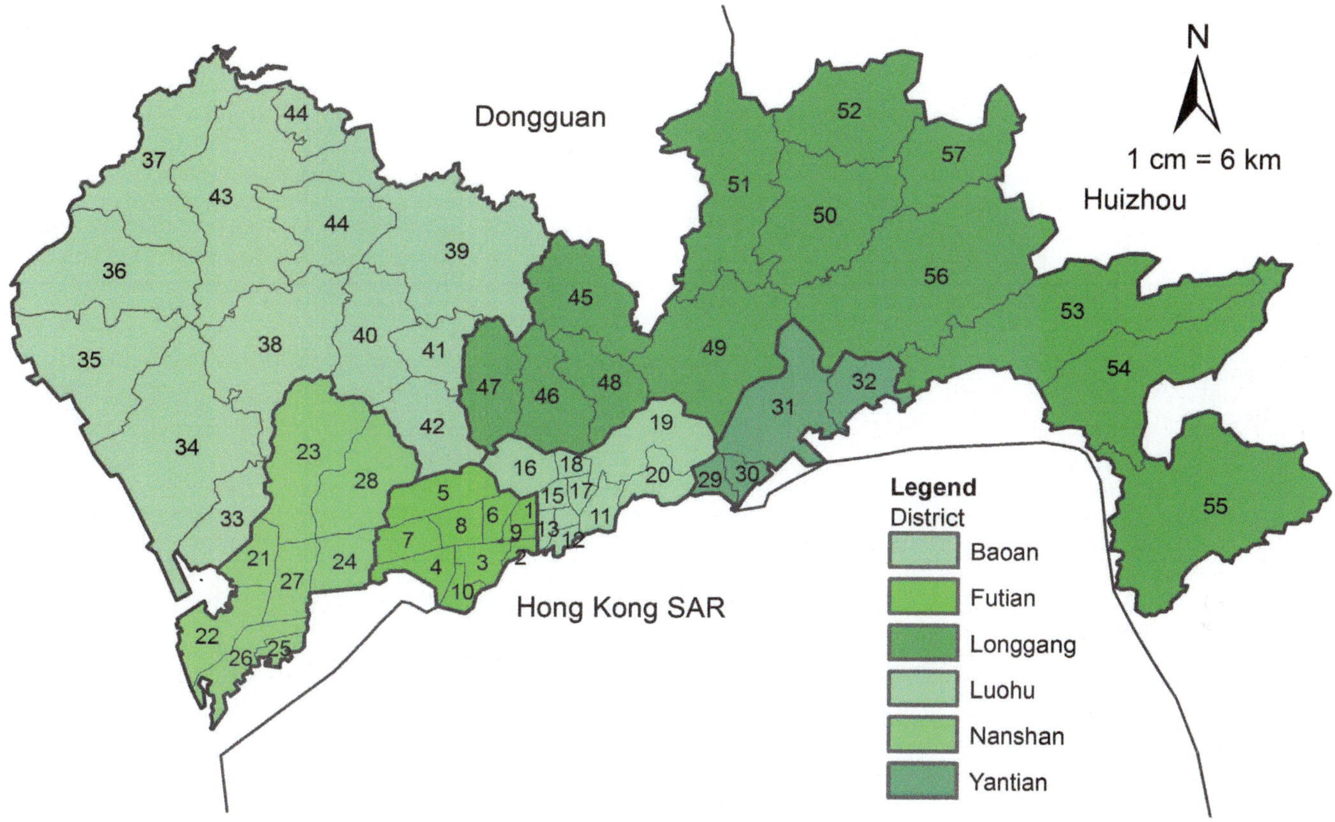

Figure 1. Administrative map of Shenzhen, China.

health care in an area [13]. Besides, a long waiting list or crowded waiting rooms may also reflect poor accessibility to delivery services, especially in urban areas [14]. According to Shenzhen Health and Population and Family Planning Commission, it is a great difficulty to find maternity beds in Shenzhen and pregnant women have to make an appointment of maternity beds in nearby maternity units (MUs) as soon as they are informed of pregnancy [15]. In order to make sure that the number of maternity beds in Shenzhen is sufficient and rationally located for all pregnant women and every woman can access skilled attendance when they deliver, the Shenzhen government is planning to encourage and guide the development of the private gynecological hospitals and make them to be an important complementary resource to public MUs. According to the latest statistic data from Shenzhen health statistics yearbook 2011 (2010 data), there are totally 81 hospitals with qualified MUs in Shenzhen, in which 52 (accounting for 64.2%) are public and 29 (accounting for 35.8%) are private [16]. In order to provide a more rational and efficient planning of MUs allocation, and guarantee the general equity of accessibility to MUs, policy-makers need to have a good understanding of the current distribution pattern of supply and demand, and accessibility to MUs for pregnancy population.

Accessibility is a multidimensional concept, and researchers have defined it from various aspects. Bissonnette et al. emphasized the two components of access: potential and realized access. The former is the supply of health care resources, whereas the latter refers to the actual use of health services [17]. In addition, Peters et al.'s (2008) conceptual framework divided accessibility into four dimensions: availability of services, acceptability, financial accessibility and geographic accessibility [18]. These definitions above demonstrate multidimensionality of accessibility and emphasize the interaction of many factors in the process of access to health resources, including geographical, social, cultural, etc. This study focuses on physical/geographical accessibility, which can measure the spatial distribution of current health resources and assess equality of their distribution [17]. We aim to identify the current underserved areas and compare the different spatial accessibilities to public and private MUs.

There are many methods to measure spatial accessibility, such as regional availability [19], shortest path approach [20], the gravity-based method [14] and the two-step floating catchment area (2SFCA) method [21,22]. The regional availability method is simply the ratio of health capacity to population within an area (generally referring to an administrative area), such as maternity beds per 1,000 pregnant population in one country (or province), and it is often used for comparison among different countries or regions [13]. It is a rough estimate of spatial access and cannot reveal the spatial variation within the target area. The shortest path approach calculates proximity between population and health service locations, not taking availability (supply) into account [20]. The gravity-based method, which concerns distance and demand relative to the supply, is a more realistic model than the shortest path approach because it takes both health service locations (supply) and population (demand) into consideration. It assumes that the attractiveness of a service diminishes with distance and associated increasing travel impedance [14,20], and a continuous coefficient (β), which is also called travel friction coefficient, is used to represent distance impedance between demand and supply. One important limitation of this method is that it tends to conceal health professional shortage areas, which are precisely what we want to designate in study [21]. The 2SFCA method is a recently developed special case of gravity-based method which was first proposed by Radke and Mu [22], and then modified and improved by Luo and Wang [21]. It uses dichotomous distance impedance. First, it calculates the ratio of supply to demand within a service area centered at a supplier's location and then sums up these ratios for demanders living in one area where different provider's services overlap [23]. Through the computation, we can get an accessibility score for every predefined area (e.g., census tract), and this score can be straightforward to interpret because it essentially represents the total services shared by patients living in one area (a special form of beds-to-population ratio) [14,21]. Comparing with the gravity-based method, the 2SFCA method was recommended by Luo and Wang because it is easier to interpret and has an advantage of identifying the poorly spatial access areas which are always the interest of planning and policy making to guarantee society welfare and equity of health resources [21].

This paper explores the distribution of MUs in Shenzhen by using the 2SFCA method, and constructs the appropriate research-scale for exploring the city MUs. Through the comparison of accessibilities to public and private MUs, we provide some advice on MUs planning in the urban area. The specific questions we attempt to answer are:

(1) What is the current spatial distribution of MUs in Shenzhen?
(2) Which regions are underserved and can be appropriate candidate areas for new MUs?
(3) What's the difference between the accessibilities to public and private MUs?

Data and Methods

By the 2SFCA method, the accessibility of a particular census tract can be calculated depending on three main variables: the demand, the supply and travel distance.

The Demand: Pregnant Population

In the first thirty development years (1980–2010) of Shenzhen, the border management line divided this city into two parts in terms of economic policy and development, the special economic zone (SEZ) part and the outside special economic zone (outside SEZ) part. The districts of SEZ were Nanshan, Futian, Luohu, Yantian and those outside SEZ were Baoan and Longgang. In 1 July 2010, the border management line was canceled, the SEZ was expanded to include all districts, and Shenzhen became an integrated SEZ from then on. Now, Shenzhen has six districts and each is further divided into sub-districts. As census tract in Shenzhen, sub-district is the lowest areal unit used for administrative management, and is thus chosen as the analysis unit in this study. There are totally 57 sub-districts in Shenzhen now (see Figure 1 and the corresponding name of each sub-district can be found in Table S1). In our study, the pregnant population refers to women who get pregnant and eventually give birth in Shenzhen MUs, and they are the actual users of the MUs. In Shenzhen, pregnant women are recommended to take antenatal examinations and give birth at a nearby MU by the government.

There is a WHO-recommended method for estimating the number of pregnant women from CBR (crude birth rate) [13]:

a = Estimated number of live births = (CBR per 1,000*total population);

b = Estimated number of pregnancies ending in stillbirths or miscarriages = (a * 0.15);

Estimated pregnancies expected in the year = (a+b);

In our study, we used an actual number of live births and stillbirths to represent pregnant population, which is much closer to the actual situation than the WHO-estimation method. We obtained the latest 2010 data of live births and stillbirths of each district to represent the amount of pregnant population from Shenzhen health statistics yearbook 2011 (2010 data) [16], and by assuming that the distribution of pregnant women is the same as that of total population, we got the pregnant population in each sub-district by multiplying the percentage of the sub-district population by each district population. In 2010, 95,726 pregnant women utilized maternity services in Shenzhen – about 0.9% of total population (10,357,938).

The Supply: MUs

There are a total of 81 qualified MUs in Shenzhen, in which 52 (accounting for 64.2%) are public and 29 (accounting for 35.8%) are private (Figure 2). These MUs have maternity health services technology licenses, meeting the basic conditions of midwifery services in Shenzhen, and have passed the 2008 accreditation (once every three years) held by the administrative department in charge of health. These MUs can provide professional antenatal health examinations during pregnancy and skilled attendance at childbirth for pregnant women. There are many kinds of indicators to reflect a MU's service capacity, such as maternity beds number, health staff number etc. [13,24]. In this study, we chose maternity beds number to reflect the service capacity of a MU's.

Travel Distance

The travel distance between supply and demand is a key component to compute accessibility in the 2SFCA method. There are many indicators to represent travel distance, such as straight-line distance and travel time, etc. [25,26,27]. Travel time is a complicated indicator, which depends on road infrastructure, mode of transport and topography [28]. In this study, the detailed data of Shenzhen road network is unreachable for us, so we chose straight-line distance as a proxy for travel distance between the locations of supply and demand, and used it to delineate the search radius which represents the acceptable cut-off point. So far, there is no regulated or recommended straight-line distance threshold from a pregnant woman's home to the locations of MUs, but there are two relevant time thresholds: a) according to the criteria for health professional shortage areas recommended by US Department of Health and Human Services, the specified threshold travel time for primary medical care should be 30 minutes in rural areas [29,30]; b) according to a Netherlands research [31], the travel time from home to hospital should be within 20 minutes in order to avoid adverse outcomes in women at term. Considering that Shenzhen is an urban area and the regular antenatal health examinations during pregnancy are not emergency, travel time in those situations can largely exceed 20 minutes, so we set a time threshold as 30minutes In order to get the straight-line distance threshold which we actually need, we multiplied 30minutes and average road speed which was around 40 km/h in 2011 reported by Shenzhen Traffic Police Bureau, and get 20 km as the reference threshold straight-line distance value in this study.

Methods

In this study, we used simple geographic centroids in each sub-district to represent pregnant population locations, and the information of MUs (including names, addresses and the number of maternity beds) was obtained from the website of Shenzhen Health and Population and Family Planning Commission [32]. We used the Baidu Map geocoding API (Version v1.0, Baidu Inc., Haidian, BJ) (http://developer.baidu.com/map/geocoding-api.

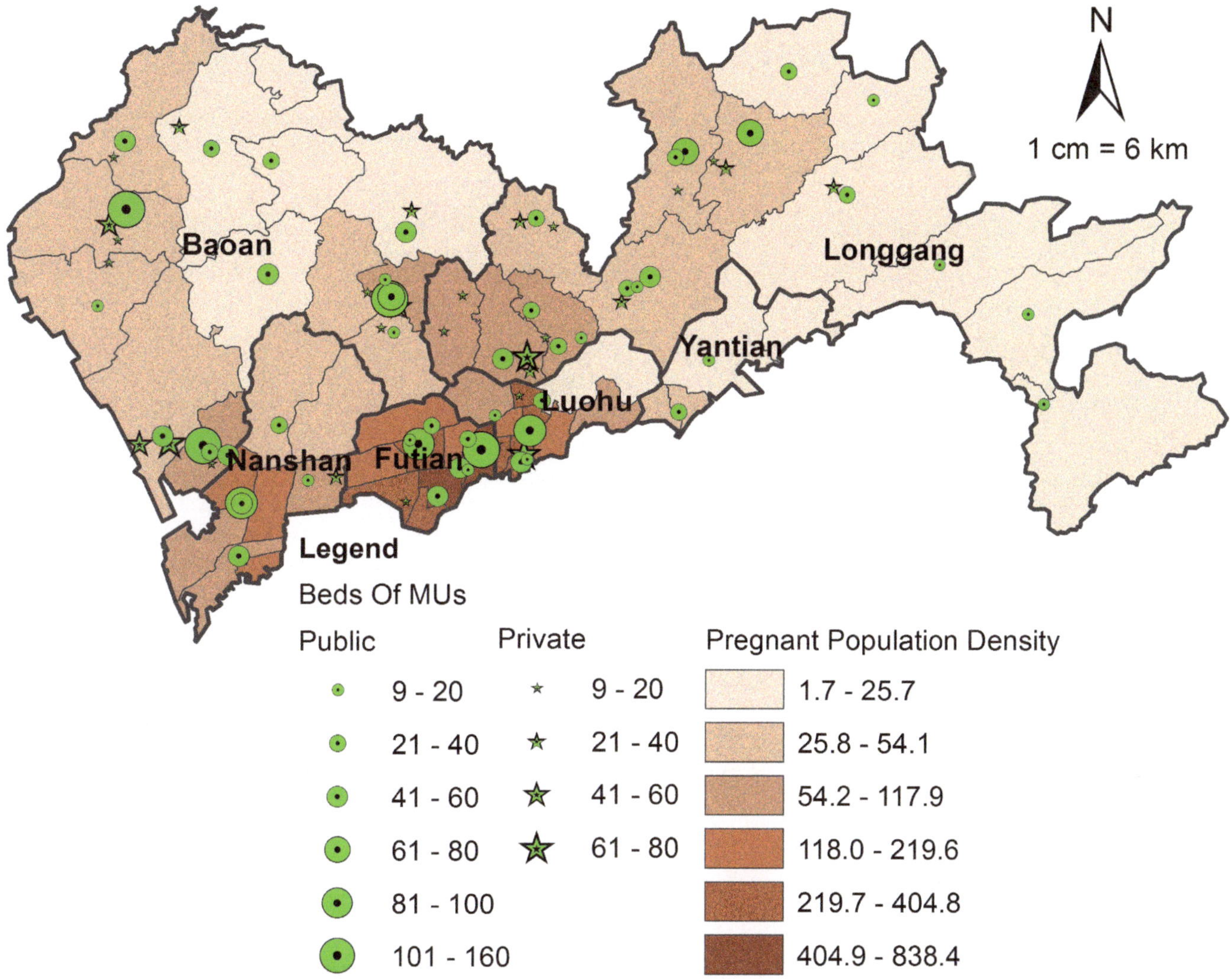

Figure 2. The spatial distribution of pregnant population and MUs in Shenzhen.

htm) to obtain latitudes and longitudes of these MUs according to their street addresses, and then created a point layer (named "locations of MUs") based on their coordinates using ArcGIS (Version 10.1, ESRI Inc., Redlands, CA). Afterwards, we transformed the coordinates from WGS-84 coordinate system to BJ-54 coordinate system.

The 2SFCA method was used in this study. There were two steps in calculating each census tract's (sub-district's) accessibility to MUs. First step, for each supply (MU) location j, search all demand (pregnant population) locations k that are within the threshold distance d_0 (20 km) from location j, which forms the catchment area of supply (MU) location j, and then compute the maternity beds-to-pregnant population ratio R_j within this catchment area (equation 1):

$$R_j = \frac{S_j}{\sum_{k\in(d_{kj}\le d_0)} \frac{P_k}{1000}} \qquad (1)$$

where d_{kj} is the distance between locations k and j, S_j is the number of maternity beds of each MU (supply) at location j, and P_k is the amount of pregnant population (demand) whose location falls within the catchment area ($d_{kj} \le d_0$).

Second step, for each demand (pregnant population) location i, search all supply (MU) locations j that are within the threshold distance d_0 (20 km) from location i, and then sum up all the maternity beds-to-pregnant population ratios R_j (derived in step 1) at those locations j to get the full accessibility A_i^F at each demand location i (equation 2):

$$A_i^F = \sum_{j\in\{d_{ij}\le d_0\}} R_j = \sum_{j\in\{d_{ij}\le d_0\}} \left(\frac{S_j}{\sum_{k\in\{d_{kj}\le d_0\}} \frac{P_k}{1000}} \right) \qquad (2)$$

where A_i^F represents the full accessibility of each census tract to MUs, d_{ij} is the distance between i and j, R_j is the beds number of each MU to pregnant population ratio whose location j falls within the catchment area ($d_{ij} \le d_0$). The larger the value A_i^F, the better accessibility to MUs for one census tract's pregnant population, which can also be interpreted that the more maternity beds are available for per 1,000 pregnant women under the threshold distance 20 km.

At the first step, the 2SFCA method computes the ratio of supply to demand within a threshold distance centered at supply's location, at the second step, it sums up the initial ratios for demand in the overlapped service areas. With this method, we calculated each census tract's accessibilities to all MUs, public MUs and private MUs respectively, and then compared the difference between accessibilities to public and private MUs using paired samples T-test. In order to set a basic level of accessibility and evaluate the conditions of these census tracts, we referred a WHO-recommended method [13]: under the assumption that there should be sufficient beds for all pregnant women within the threshold distance 20 km, the maternity beds occupancy rate was 80% (to account for the uneven spread of demand over time), and the mean duration of stay was 5.1 days (in 2010) [16], the target number should be (1000/.8) * (5.1/365) = 17.5 maternity beds per 1,000 pregnant women. Because accessibility score is a special pattern of beds per 1,000 population, we set 17.5 as the basic level of accessibility score. The sub-districts whose accessibility scores under 17.5 were the poorly served areas and those equal and above 17.5 were acceptable.

The spatial analysis with the 2SFCA method was carried out in ArcGIS (Version 10.1, ESRI Inc., Redlands, CA), and the statistical analysis of the accessibilities to public and private MUs was implemented by SPSS (Version 18.0, SPSS Inc., Chicago, IL).

Results

Figure 2 depicts the spatial distribution of pregnant population at sub-district (census tract) level. The pregnant population density is relatively high in Nanshan, Futian, Luohu and the juncture area of four districts (Futian, Luohu, Baoan and Longgang). A declining tendency can be seen from the central part to the traditional outside SEZ. That is, it is more crowed in the central part of Shenzhen. The highest pregnant population density is about 838.4 pregnant women per km^2 in a sub-district (Nanyuan) of Futian district. It is approximately 493 times higher than the least densely populated sub-district (Nanao) of Longgang district, whose pregnant population density is only about 1.7 women per km^2. The huge variance is one of the most important factors which should be considered when making a reasonable allocation of new MUs.

Figure 2 also shows the spatial distribution of all MUs, including public MUs and private MUs. Consistent with the distribution of pregnant population, the MUs are centered on the most populated regions, such as Futian and Luohu. It also presents a declining tendency to the relatively vast east area. Except Longgang district, each of the other districts has more than one intermediate-to-large capacity (≥60 beds) MUs, most of which are public. Different from the public MUs, the private MUs generally are low-capacity, and tend to be centered in the juncture area of four districts (Futian, Luohu, Baoan and Longgang).

Figure 3 depicts the spatial accessibilities to all MUs, public MUs and private MUs. 1) The accessibility to all MUs shows significant variation, and the scores range from 6.5 to 57.0 with a mean value of 30.93 and a standard deviation of 8.54, which means that within a threshold distance of 20 km, there are averagely 30.9 maternity beds accessible for 1,000 pregnant women. The most accessible areas are located alongside a section of Baoan district's edge which borders Nanshan and Futian. The others are centered on both sides of the traditional border management line and tend to decrease in the city boundary, especially in relatively vast but the least densely populated east sub-districts (Nanao, Dapeng and Kuiyong) of Longgang district, and west sub-districts (Nanshan and Shekou) of Nanshan district. 2) The accessibility scores to public MUs range from 6.5 to 39.4 with a mean value of 22.84 and a standard deviation of 6.16. 3) The accessibility scores to private MUs range from 1.8 to 17.6 with a mean value of 8.09 and a standard deviation of 2.91, and Nanao sub-district of Longgang district has no accessibility to private MUs (score = 0) within the distance of 20 km.

MUs consists of Public and private MUs. They are additional resources. For each sub-district, the sum of its accessibility scores to public and private MUs just equals to the score to all MUs. Through paired t test (Table 1), the difference between accessibility scores to public and private MUs is statistically significant ($P<0.01$).

Discussion

In our study, the distribution of MUs was found to be consistent with the distribution of pregnant population, and clustered in the central part of Shenzhen, especially Futian and Luohu. Though the traditional border management line was cancelled in 2010, border effect still exists. People prefer living in the traditional SEZ, where market-oriented economy has been developed for a longer time, and therefore, economy, traffic, education, etc. are better. However, with the inburst of more and more migrant population these years, the migrant population size has already exceeded the resident population size. Because the house-renting and house-purchasing prices in traditional SEZ are higher than traditional outside-SEZ, many migrant people choose to live in the traditional outside-SEZ fringe zone alongside the traditional border management line [33,34]. This concords with some previous studies [35,36]. These factors may also affect the distribution of the self-financing private MUs, which are also centered in the juncture area of four districts (Futian, Luohu, Baoan and Longgang).

This study highlights critical areas in Shenzhen where new MUs or maternity beds may need to be added. This study shows that the accessibility scores (to all MUs) of sub-districts alongside the traditional border management line is acceptable, while the west part of Nanshan and the vast east part of Longgang have low accessibility scores (<17.5). These areas should be the areas of concern for further knowledge-based policy making. The comparison between accessibility scores to public MUs and to private MUs indicates that private MUs are important additional resources for public MUs. Attributing to the existing of private MUs, the mean accessibility score has increased from 22.84 (only public) to 30.93 (all). It also improves some census tracts' accessibility scores and turns them to acceptable areas, such as the north parts of Baoan and Longgang. Whilst, the results also show that the private MUs are just additional but not complementary resources to public MUs, they own similar distributions. Thus, the function of private MUs is just strengthening the function of public MUs, not complementing the public MUs shortage areas.

Several limitations of this study should be addressed here. 1) The sub-district level pregnant population data is computed based on the percentage of the sub-district population by each district population. It is not accurate and real data to reflect the real world. 2) We used a single point as a proxy to represent the locations of pregnant population within each sub-district, and then calculated the distance from each pregnant population point to each MU, which may result in some aggregation-error [37]. 3) In this study, the bed number of each MU is obtained in 2012 from the Shenzhen Health and Population and Family Planning Commission website, it's not the same year as the population data being used (2010). This may result in a little bias for the real

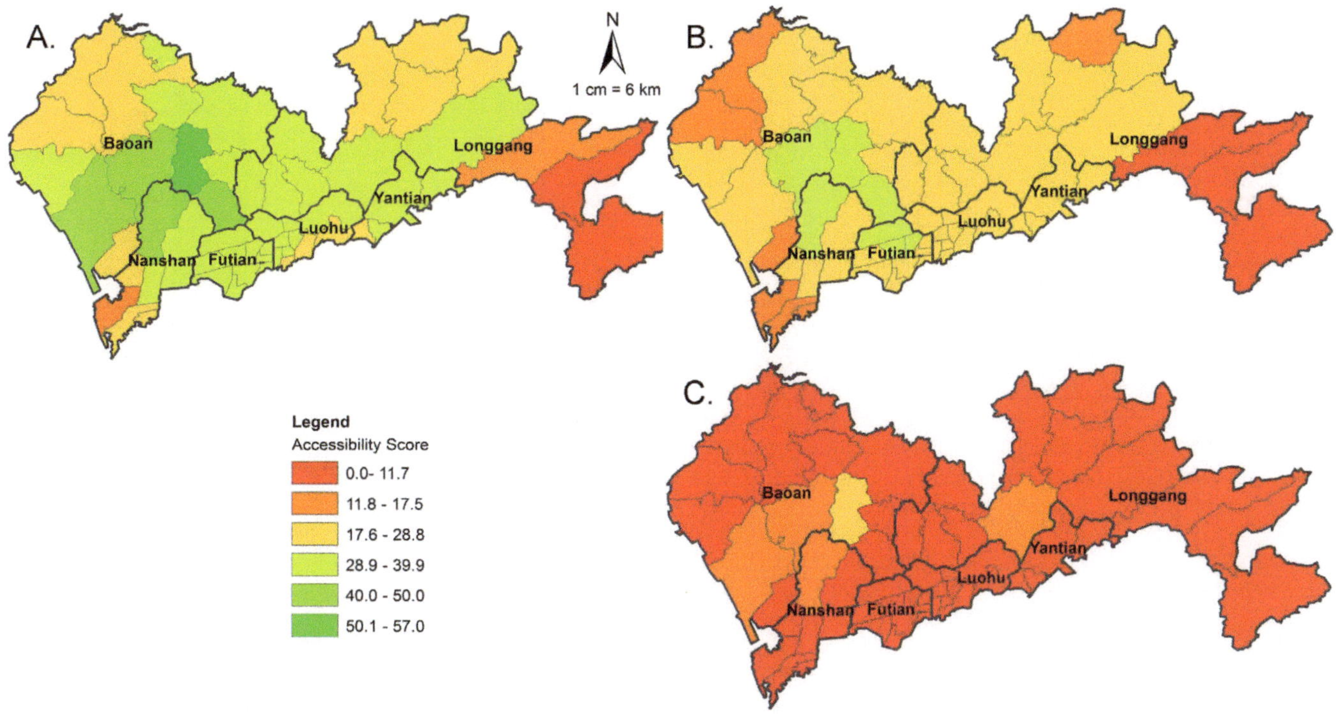

Figure 3. Accessibility to MUs in Shenzhen using the 2SFCA method. A. Accessibility to all MUs. B. Accessibility to public MUs. C. Accessibility to private MUs.

accessibility conditions in 2010. 4) We set a fixed travel distance 20 km as the threshold distance. It means that all travel distance within 20 km are deemed equally accessible, while all travel distance beyond 20 km are not accessible. The choice of 20 km is applied to both steps 1 and 2, meaning that both the service catchment (breadth of population that a service is providing access to) and population catchment (breadth of services that a population has access to) are 20 km. The set of a signal buffer may result in under and overestimation of accessibility across census tracts, especially in the vast size sub-districts [20,27]. Besides, in reality, though most pregnant women give birth in a nearby MU, sometimes, considering the seriousness of the condition, pregnant women or their families may accept a farther distance. As in other studies on spatial accessibility [23,38], this study only took potential spatial accessibility into account, not revealed accessibility (or actual utilization of MUs). Thus, in the process of decision making of MUs allocation, the results need to be interpreted with caution. The accessibility to MUs is complicated, and many other aspatial aspects should be considered in addition to spatialityin order to reflect the revealed access or actual utilization of MUs, such as reputation, policy, etc.

Table 1. Comparison of accessibilities to public and private MUs.

	Accessibility		
	To Public MUs	**To Private MUs**	**P**
Sub-district (n = 57)	22.84±6.16	8.09±2.91	<0.001*

Abbreviation: MU, maternity unit. Data are shown as mean±standard deviation.* indicates a statistically significant difference.

Of course, our study has many advantages. 1) These years, the spatial accessibility methodology has become a popular tool in public health and city planning areas [39,40], but in China, this study is the first to use a spatial method to assess accessibility to MUs. As an application of spatial methodology, this study illustrated the importance of incorporating the locations of MUs into the equitable and efficient allocation of health resources and city planning by providing a visualized and easy-understanding evidence for future planning of MUs. The result of this study will be useful for both health resource planners and other researchers in the field of public health. 2) The 2SFCA method based on GIS instead of the traditional rough indicators, such as service to population ratio or simple shortest distances [38], was chosen to determine areas that are outside of the pre-set catchment, and spatial accessibility was calculated at a much finer spatial resolution, thus the measurement of spatial accessibility was significantly advanced [23]. This is important and with big practical application value for Shenzhen, because taking antenatal examinations and giving birth at a nearby MU is strongly recommended by government [15]. Besides, the capabilities of geographic information systems (GIS) to handle large amounts of data over large geographic areas at fine levels of geographic detail make them suitable to measure geographical accessibility to medical clinics and other healthcare services [41,42]. This study identified three east sub-districts (Nanao, Dapeng and Kuiyong) of Longgang district, and one west sub-district (Nanshan) of Nanshan district as the underserved regions, where health administrators and municipal planners should consider to be the potential areas for further MUs. 3) Through the comparison between accessibility scores to public MUs and private MUs, the evidence of how private MUs benefit the pregnant women of Shenzhen can be found. As the first city establishing socialist market economic system in China, Shenzhen has the most open and active private medical system, the market demand is the most important vane for private medical service planning. However, in this study, the complementary function of private MUs can't be confirmed, their spatial distribution is similar to public MUs, and there are still some potential areas in great needs of maternity services. In recent years, the demand for maternity medical services is growing rapidly [43], and Shenzhen government encourages the development of private MUs. But how to allocate future public MUs and give reasonable guidance for the siting of future private MUs, this study is the first attempt in solving this problem, and our results will provide some advice on planning for both public MUs and private MUs. Furthermore, the research design and methods (including spatial access method, pregnant population calculation, the set of threshold distance and target accessibility score, etc.) in our study can also be applied to the other cities in China in the future.

Acknowledgments

All the authors gratefully thank the reviewers and editor for their insightful and constructive comments, and the Shenzhen Health and Population and Family Planning Commission for providing the data.

Author Contributions

Conceived and designed the experiments: PGS YJZ. Performed the experiments: YJZ XM. Analyzed the data: PGS YJZ XM. Contributed reagents/materials/analysis tools: LA QL. Wrote the paper: PGS LA.

References

1. Ng MK (2005) Planning cultures in two Chinese transitional cities: Hong Kong and Shenzhen. Comparative Planning Cultures, 113–143.
2. Hao P (2010) Spatial evolution of urban villages in Shenzhen, PhD Dissertation, Utrecht Univeristy.
3. Shenzhen Municipal Statistic Bureau (2000) Shenzhen 2000 fifth national population census main data bulletin. Shenzhen, China.
4. Shenzhen Municipal Statistic Bureau (2010) Shenzhen 2010 sixth national population census main data bulletin. Shenzhen, China.
5. Citymayors (2011) City Mayors: Largest cities in the world by population density (1 to 125). [online] Available at: http://www.citymayors.com/statistics/largest-cities-density-125.html [Accessed: 14 May 2013].
6. World Health Organization (2012) World Health Statistics 2012. Indicator compendium. [Report] Geneva: World Health Organization.
7. The United States Agency for International Development (2008) Profiles of Health Facility Assessment Methods: Measure Evaluation. Arlington, VA.
8. Shenzhen News (2012) Shenzhen newborns density ranks first in the country. [online] Available at: http://news.sznews.com/content/2012-02/15/content_6463293.htm [Accessed: 12 Sep 2012].
9. Ministry of Health, People's Republic of China (2011) Report on Women and Children's Health Development in China. China.
10. Dayaratna V, Winfrey W, McGreevey W, Hardee K, Smith J, et al. (2000) Reproductive Health Interventions: Which Ones Work and What Do They Cost? POLICY Project.
11. Viisainen K, Gissler M, Hartikainen AL, Hemminki E (1999) Accidental out-of-hospital births in Finland: incidence and geographical distribution 1963–1995. Acta Obstetricia et Gynecologica Scandinavica 78(5), 372–378.
12. Robson S, Cameron CA, Roberts CL (2006) Birth outcomes for teenage women in New South Wales, 1998–2003. Aust. N. Z. J. Obstet. Gynaecol 46 (4): 305–310.
13. World Health Organization (2012) Measuring service availability and readiness (SARA). Service availability indicators. [Report].
14. Yang DH, Goerge R, Mullner R (2006) Comparing GIS-Based methods of measuring spatial accessibility to health services. J Med Sys 30(1): 23–32.
15. Nanfang Daily (2012) Shenzhen birth rate per square kilometer rank first nationally. [online] Available at: http://gd.nfdaily.cn/content/2012-02/15/content_37950421.htm [Accessed: 12 Sep 2012].
16. Shenzhen Health and Population and Family Planning Commission (2011) Shenzhen health statistics yearbook 2011. Shenzhen, China.
17. Bissonnette L, Wilson K, Bell S, Shah TI (2012) Neighbourhoods and potential access to health care: The role of spatial and aspatial factors. Health Place 18(4): 841–853.
18. Peters DH, Garg A, Bloom G, Walker DG, Brieger WR, et al. (2008) Poverty and access to health care in developing countries. Annals of the New York Academy of Sciences 1136(1): 161–171.
19. Khan AA (1992). An integrated approach to measuring potential spatial access to health care services. Socio-Economic Planning Sciences 26(4): 275–287.
20. McGrail MR, Humphreys JS (2009) Measuring spatial accessibility to primary care in rural areas: improving the effectiveness of the two-step floating catchment area method. Applied Geography 29(4): 533–541.
21. Luo W, Wang F (2003) Measures of spatial accessibility to health care in a GIS environment: synthesis and a case study in Chicago region. Environment and Planning B 30(6): 865–884.
22. Radke J, Mu L (2000) Spatial decompositions, modeling and mapping service regions to predict access to social programs. Geographic Information Sciences 6(2): 105–112.
23. Ngamini Ngui A, Vanasse A (2012) Assessing spatial accessibility to mental health facilities in an urban environment. Spat Spatio-temporal Epidemiol 3(3): 195–203.
24. World Health Organization (2008) SERVICE DELIVERY: Toolkit on monitoring health systems strengthening. [Report].
25. Lasser KE, Himmelstein DU, Woolhandler S (2006) Access to Care, Health Status, and Health Disparities in the United States and Canada: results of a cross-national population-based survey. Am J Public Health 96(7): 1300–1307.
26. Apparicio P, Abdelmajid M, Riva M, Shearmur R (2008) Comparing alternative approaches to measuring the geographical accessibility of urban health services: Distance types and aggregation-error issues. Int J Health Geogr 7(1): 7.
27. Luo W, Qi Y (2009) An enhanced two-step floating catchment area (E2SFCA) method for measuring spatial accessibility to primary care physicians. Health Place 15(4): 1100–1107.
28. Noor AM, Amin AA, Gething PW, Atkinson PM, Hay SI, et al. (2006) Modelling distances travelled to government health services in Kenya. Tropical Medicine & International Health 11(2): 188–196.
29. Fotheringham SA, Rogerson PA (2009) The SAGE Handbook of Spatial Analysis. London: SAGE Publications: 443–64.
30. DHHS (Department of Health and Human Services) (1980) Criteria for designation of health manpower shortage areas: final regulations. Federal Register 45.
31. Ravelli ACJ, Jager KJ, de Groot MH, Erwich JJHM, Rijninks-van Driel GC, et al. (2011) Travel time from home to hospital and adverse perinatal outcomes in women at term in the Netherlands. BJOG: An International Journal of Obstetrics & Gynaecology 118(4): 457–465.
32. Shenzhen Health and Population and Family Planning Commission (2012) the using condition of obstetric beds of Shenzhen healthcare institutions [online] Available at: http://www.szhealth.gov.cn/wsj/view?id = 84 [Accessed: 12 Mar 2012].
33. Dai XM (2010) Shenzhen: new dual structure of housing market. China Real Estate Market 6: 50–51.
34. Lin YC, Feng J (2011) The bounder effect and space-time evolution in the integration progress of Shenzhen special economic zone and outside economic zone. Tropical Geography 31 (6): 580–585.
35. Lowell-Smith EG (1993) Patterns of dentist location: Lansing, Michigan 1950–1991. Soc Sci Med 36(4): 569–573.
36. Steele MT, Schwab RA, McNamara RM, Watson WA (1998) Emergency medicine resident choice of practice location. Ann Emerg Med 31(3): 351–357.
37. Hewko J, Smoyer-Tomic KE, Hodgson MJ (2002) Measuring neighborhood spatial accessibility to urban amenities: does aggregation error matter? Environment and Planning 34(7): 1185–1206.
38. Cheng Y, Wang J, Rosenberg MW (2012) Spatial access to residential care resources in Beijing, China. Int J Health Geogr 11(1): 32.
39. Ngui AN, Apparicio P (2011) Optimizing the two-step floating catchment area method for measuring spatial accessibility to medical clinics in Montreal. BMC health services research 11(1): 166.
40. Kwan MP, Murray AT, O'Kelly ME, Tiefelsdorf M (2003) Recent advances in accessibility research: Representation, methodology and applications. Journal of Geographical Systems 5(1): 129–138.
41. Benigeri M (2007) Geographic information systems (GIS) in the health field: an opportunity to bridge the gap between researchers and administrators. Can J Public Health 98.
42. Phillips RL, Kinman EL, Schnitzer PG, Lindbloom EJ, Ewigman B (2000) Using geographic information systems to understand health care access. Arch Fam Med. 9(10): 971–978.
43. China health network (2011) Shenzhen: maternal population increasing and maternity beds saturated. [online] Available at: http://www.69jk.cn/yuer/hy/yqzs/80076.html [Accessed: 24 Mar 2012].

5

marmap: A Package for Importing, Plotting and Analyzing Bathymetric and Topographic Data in R

Eric Pante*[⁹], Benoit Simon-Bouhet[⁹]

Littoral, Environnement et Sociétés Joint Research Unit 7266 Centre National de la Recherche Scientifique, Université de La Rochelle, La Rochelle, France

Abstract

In this communication we introduce marmap, a package designed for downloading, plotting and manipulating bathymetric and topographic data in R. marmap can query the ETOPO1 bathymetry and topography database hosted by the NOAA, use simple latitude-longitude-depth data in ascii format, and take advantage of the advanced plotting tools available in R to build publication-quality bathymetric maps. Functions to query data (bathymetry, sampling information...) are available interactively by clicking on marmap maps. Bathymetric and topographic data can also be used to calculate projected surface areas within specified depth/altitude intervals, and constrain the calculation of realistic shortest path distances. Such information can be used in molecular ecology, for example, to evaluate genetic isolation by distance in a spatially-explicit framework.

Editor: Guy J.-P. Schumann, NASA Jet Propulsion Laboratory, United States of America

Funding: No current external funding sources for this study.

Competing Interests: The authors have declared that no competing interests exist.

* E-mail: pante.eric@gmail.com

[⁹] These authors contributed equally to this work.

Introduction

Marine ecologists still lack simple, highly-customizable tools for building and using marine charts, open-source tools being particularly lacking. Currently, apart from commercial solutions, marine ecologists can use GIS tools such as GRASS GIS [1], or the graphical user interface provided by GeoMapApp [2] and Google Earth [3] to prepare publication-quality bathymetric charts, but these tools either require knowledge in the field of GIS, or are limited in the number of functions, datasets, and analyses available. The R environment [4] is a convenient platform for generating maps, thanks to its built-in functions for data matrix manipulation and advanced graphics. The package marmap [5], introduced herein, takes advantage of these built-in functions, as well as packages for spatial analysis developed by others, to provide a simple and flexible tool for importing, manipulating, plotting and exporting bathymetric data. It provides a simple platform for bathymetric data analysis and plotting even with limited experience with R, and allows production of publication-quality maps. The entirety of the marmap package was coded in the R language, providing advanced R users with the opportunity to fine-tune or expand any function. Also, because all bathymetric data are imported as R objects, the arsenal of R tools can be utilized for analyses not implemented in the marmap package.

Description of the marmap Package

Portability

Binaries and source code for the marmap package are freely available on the Comprehensive R Archive Network (CRAN [6]), along with a tutorial (also called "R vignette"). marmap contains R code exclusively, which maximizes its portability across platforms. CRAN makes the source code and binaries available for Unix-like and Windows operating systems [5].

Importing and Plotting Bathymetric Data

marmap offers different ways to upload bathymetric information, which should provide flexible data preparation, depending on need and access to online resources. First, bathymetric data can be downloaded directly from within R by querying the ETOPO1 database [7] hosted by the NOAA. Second, three-column data frames (with longitude, latitude, and depth data) can be imported from different sources, such as the US National Ocean and Atmospheric Administration (NOAA) [8]. Third, marmap has functions to prepare and use a local SQL database, stored on the user's hard drive. Use of a personalized SQL bathymetry database is convenient for querying subsets of very large bathymetric datasets such as the 5Go ETOPO1. Finally, data can be imported like any other dataset into R, and converted using marmap functions. While marmap was originally designed for marine ecologists, it is fully compatible with topographical data, which can be treated along with bathymetric data, or analysed on their own.

Once loaded into R, bathymetric data are re-organized as a matrix that is used for manipulation, plotting and exporting. These data are represented as an R class called "bathy", valid within a marmap session. The creation of a custom R class allows the use of generic functions such as plot and summary. Bathymetric data of class "bathy" can be plotted with the streamlined marmap plotting tools, or used with other geographic analysis packages (e.g. "maps" [9]). Bathymetric data can be plotted as simple contour plots, with control over the range, density and look of isobaths, or with automatic isobath choice and placement. Contour plots can be coupled with heat maps with built-in and customizable color ramps (Figure 1A,E). Sampling data (GPS points, tracks, polygons,

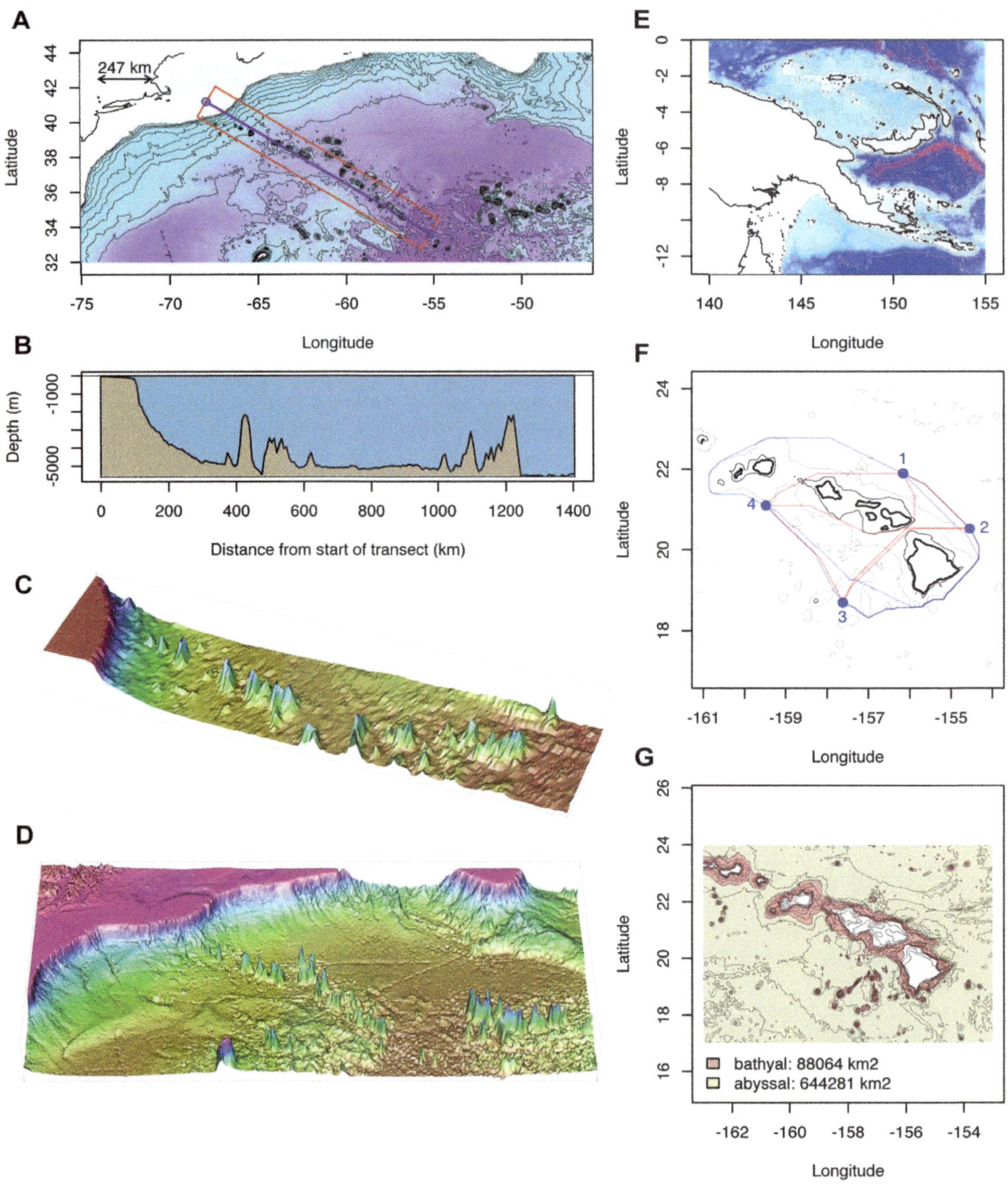

Figure 1. Examples of two and three dimensional plots created using marmap tools. Left panel (**A**): data from the NW Atlantic Ocean, showing the NE coast of the USA and the New England and Corner Rise seamounts chains. The blue line represents the position of two- (**B**) and three- (**C**) dimensional cross sections, the red rectangle delimiting the area covered by the belt transect. The bottom left figure (**D**) represents the NW Atlantic data as plotted with the wireframe function from package lattice [21], based on data imported with marmap. Right panel: map of Papua New Guinea and satellite islands (**E**; see text). The middle figure (**F**) represents the results of a least-cost path analysis around the Hawaiian islands (coastline in thick black, dark grey: 1000 m isobath, light grey 4000 m isobath; red line: path avoiding waters shallower than 1000 m, blue line: path avoiding waters shallower than 4000 m). The bottom figure (**G**) represents the results of projected surface area calculations for the bathyal and abyssal areas around the Hawaiian islands. R code is available in File S1.

etc…) and text (legends, labels, scales, etc…) can simply be added to marmap maps using R low-level functions.

For example, creating a map of Papua New Guinea can be done in three lines of code, in a new R session:

```
> library(marmap)
> getNOAA.bathy(lon1 = 140,lon2 = 155,lat1 = −13,lat2 = 0, resolution = 1) ->papoue
>plot(papoue, image = TRUE)
```

The first line loads marmap, the second queries bathymetric data (resolution of 1 minute) and stores the result in a variable of class "bathy," and the third line creates a map with automatic isobath choice and built-in color ramp. The upper right map on Figure 1, plotted with isobaths of different widths and colors as well as a custom color ramp was produced with these two lines (the first creating and storing a custom palette, and the second creating the map; details on commands and their arguments are available in the marmap documentation on CRAN):

```
> colorRampPalette(c("red","purple","blue","cadetblue1",
  "white")) -> blues
> plot(papoue, image = T, bpal = blues(100),
  deep = c(−9000, −3000,0), shallow = c(−3000, −10, 0),
  step = c(1000, 1000, 0),
  col = c("lightgrey","darkgrey","black"), lwd = c(0.3,0.3,0.6),
  lty = c(1,1,1),
  drawlabel = c(F,F,F))
```

Using Bathymetric for Further Ecological Analysis

By interactively clicking on a map, the user can retrieve bathymetric data (either from a point or an area), sampling information (e.g. list of specimens sampled within a particular geographic area), or two- and three-dimensional depth cross-sections (Figure 1B–C). Projected surface areas can also be estimated for specified depth ranges, such as the bathyal and abyssal zones around Hawaii (Figure 1G).

Based on functions developed by Jacob van Etten in the raster and gdistance packages [10,11], marmap facilitates analysis of geographically-explicit ecological data, by allowing least-cost path calculation constrained by bathymetry and/or topography. Shortest great-circle (haversine) distance between pairs of sites, avoiding land masses or depth layers, are calculated. Outputs of least-cost path analysis can be in the form of great-circle distance, taking into account the curvature of the Earth, and in the form of geographical paths that can be plotted on marmap maps, and used to compute two- and three-dimensional depth cross-sections. Distance data can be easily exported for direct use in landscape genetics software such as TESS [12]. Data can also be used directly from within R to perform Mantel tests or other analyses [13–15].

Interactions with other Packages

marmap interacts with multiple existing R packages for visualization and analysis, such as lattice for building three-dimensional plots (Figure 1C–D), and gdistance for least-cost path calculations (see above). marmap also contains functions to ease interactions with other packages dedicated to the analysis of spatial data. Data from class bathy can be transformed into raster objets for use in the raster package [10] (functions trans.mat and as.raster). Bathy objects can also be converted into the class SpatialGridDataFrame for use in the sp package [16,17] (function as.SpatialGridDataFrame). These transformed objects can be converted back into objects of class bathy using the as.bathy function. In the example below, the package raster is used to change the projection of a dataset extracted from ETOPO1. The first line creates a raster object from a bathy object, the second defines the target projection (PROJ.4 format), the third creates a new projected raster object, and the last converts the new raster projection into a bathy object for plotting with marmap. Figure 2 provides an example of such transformed data.

```
> r1<- as.raster(original.marmap.data)
> new.projection <- "+proj = ortho"
> r2<- projectRaster(r1,crs = new.projection)
>projected.marmap.data <- as.bathy(r2)
```

Conclusions

marmap was designed to provide easy-to-use tools for importing and using bathymetric and topographic data in R. While marmap has been primarily designed for research, and the rapid production

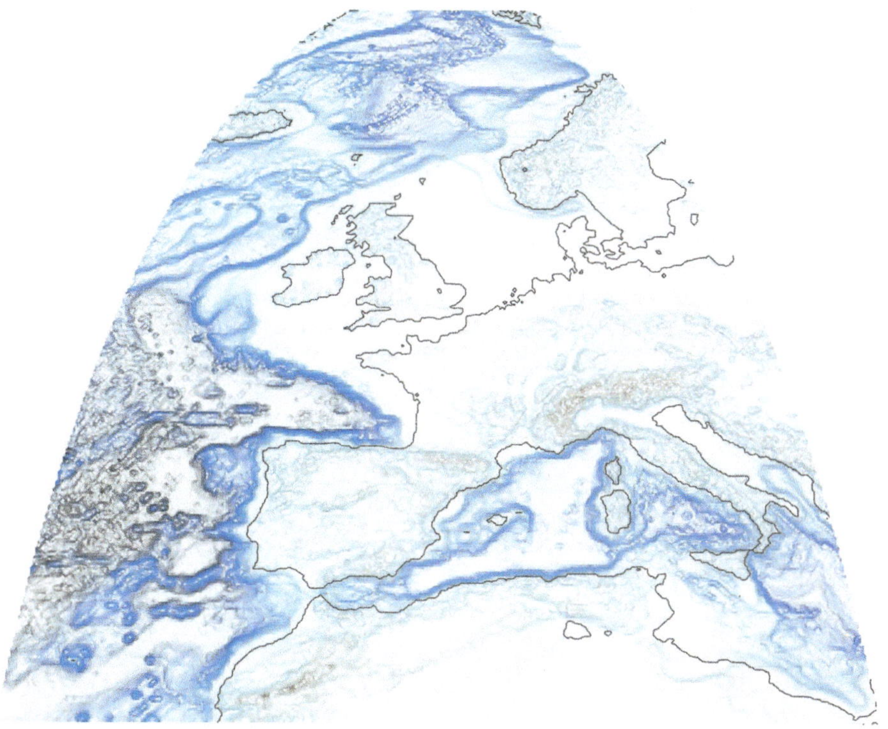

Figure 2. Exemple of ETOPO1 data downloaded with the marmap package, transformed with the raster package, and plotted with marmap. Orthographic projection, resolution of 10 minutes. R code is available in File S1.

of publication-quality maps (examples: [18–20]), its simplicity of use should make it interesting as a teaching tool as well.

Acknowledgments

The authors thank Mikael Guichard for his help with remote data accession, as well as Amélia Viricel and two anonymous reviewers for constructive criticism of the manuscript.

Author Contributions

Conceived and designed the experiments: EGP BSB. Performed the experiments: EGP BSB. Analyzed the data: EGP BSB. Contributed reagents/materials/analysis tools: EGP BSB. Wrote the paper: EGP BSB.

References

1. GRASS Development Team (2012) Geographic Resources Analysis Support System (GRASS GIS) Software. Open Source Geospatial Foundation, USA. Available: http://grass.osgeo.org. Accessed: 2013 Apr 5.
2. Ryan W, Carbotte S, Coplan J, O'Hara S, Melkonian A, et al. (2009) Global multi-resolution topography synthesis. Geochemistry, Geophysics, Geosystems 10: Q03014.
3. Google (2013) Google Earth. Available: http://www.google.fr/earth/. Accessed: 2013 Mar 27.
4. R Development Core Team (2012) R: A Language and Environment for Statistical Computing R Foundation for Statistical Computing, Vienna, Austria. ISBN 3-900051-07-0. Available: http://www.R-project.org. Accessed: 2013 Apr 5.
5. Pante E, Simon-Bouhet B (2013) marmap: Import, plot and analyze bathymetric and topographic data. R package version 0.5 Available: http://CRAN.R-project.org/package = marmap. Accessed: 2013 Apr 10.
6. CRAN (2013) The Comprehensive R Archive Network. Available: http://cran.r-project.org/. Accessed: 2013 Apr 5.
7. Amante C, Eakins BW (2009) Etopo1 1 arc-minute global relief model: Procedures, data sources and analysis. NOAA Technical Memorandum NESDIS NGDC-24 : 1–19.
8. NOAA National Geophysical Data Center (2013) GEODAS Grid Translator - Design a grid. Available: http://www.ngdc.noaa.gov/mgg/gdas/gd_ designa-grid.html. Accessed: 2013 Apr 5.
9. Original S code by Richard A Becker and Allan R Wilks R version by Ray Brown-rigg Enhancements by Thomas P Minka (2013) maps: Draw Geographical Maps. Available: http://CRAN.R-project.org/package = maps. R package version 2.3–2. Accessed: 2013 Apr 10.
10. van Etten RJHJ (2012) raster: Geographic data analysis and modeling. Available: http://CRAN.R-project.org/package = raster. R package version 2.0–41. Accessed: 2013 Apr 10.
11. van Etten J (2012) gdistance: Distances and routes on geographical grids. R package version 1.1–4. Available: http://CRAN.R-project.org/package = gdistance. Accessed: 2013 Apr 10.
12. Durand E, Jay F, Gaggiotti O, François O (2009) Spatial inference of admixture proportions and secondary contact zones. Molecular Biology and Evolution 26: 1963–1973.
13. Dray S, Dufour A, Chessel D (2007) The ade4 package-II: Two-table and K-table methods. R News 7: 47–52.
14. Dray S, Dufour A (2007) The ade4 package: implementing the duality diagram for ecologists. Journal of Statistical Software 22: 1–20.
15. Chessel D, Dufour A, Thioulouse J (2004) The ade4 package -I- One-table methods. R News 4: 5–10.
16. Pebesma EJ, Bivand RS (2005) Classes and methods for spatial data in R. R News 5: 9–13.
17. Bivand RS, Pebesma EJ, Gomez-Rubio V (2008) Applied spatial data analysis with R. Springer, NY. Available: http://www.asdar-book.org/.
18. Thoma JN, Pante E, Brugler MR, France SC (2009) Deep-sea octocorals and antipatharians show no evidence of seamount-scale endemism in the NW Atlantic. Marine Ecology Progress Series 397: 25–35.
19. Pante E, Corbari L, Thubaut J, Chan TY, Mana R, et al. (2012) Exploration of the deep-sea fauna of Papua New Guinea. Oceanography 25: 214–225.
20. Pante E, Watling L (2012) *Chrysogorgia* from the New England and Corner Seamounts: Atlantic – Pacific connections. Journal of the Marine Biological Association of the United Kingdom 92: 911–927.
21. Sarkar D (2008) Lattice: Multivariate Data Visualization with R. New York: Springer. ISBN 978-0-387-75968-5. Available: http://lmdvr.r-forge.r-project.org. Accessed: 2013 Apr 10.

6

Assessing Associations between Socio-Economic Environment and Self-Reported Health in Amsterdam Using Bespoke Environments

Eleonore M. Veldhuizen[1]*, **Karien Stronks**[2], **Anton E. Kunst**[2]

1 Department of Human Geography, Planning & International Development Studies, Faculty of Social and Behavioural Sciences, University of Amsterdam, Amsterdam, The Netherlands, **2** Department of Public Health, Academic Medical Center, University of Amsterdam, Amsterdam, The Netherlands

Abstract

Background: The study of the relationship between residential environment and health at micro area level has a long time been hampered by a lack of micro-scale data. Nowadays data is registered at a much more detailed scale. In combination with Geographic Information System (GIS)-techniques this creates opportunities to look at the relationship at different scales, including very local ones. The study illustrates the use of a 'bespoke environment' approach to assess the relationship between health and socio-economic environment.

Methods: We created these environments by buffer-operations and used micro-scale data on 6-digit postcode level to describe these individually tailored areas around survey respondents in an accurate way. To capture the full extent of area effects we maximized variation in socio-economic characteristics between areas. The area effect was assessed using logistic regression analysis.

Results: Although the contribution of the socio-economic environment in the explanation of health was not strong it tended to be stronger at a very local level. A positive association was observed only when these factors were measured in buffers smaller than 200 meters. Stronger associations were observed when restricting the analysis to socioeconomically homogeneous buffers. Scale effects proved to be highly important but potential boundary effects seemed not to play an important role. Administrative areas and buffers of comparable sizes came up with comparable area effects.

Conclusions: This study shows that socio-economic area effects reveal only on a very micro-scale. It underlines the importance of the availability of micro-scale data. Through scaling, bespoke environments add a new dimension to study environment and health.

Editor: Harry Zhang, Old Dominion University, United States of America

Funding: The authors have no support or funding to report.

Competing Interests: The authors have declared that no competing interests exist.

* E-mail: e.m.veldhuizen@uva.nl

Introduction

Since the mid-1990s, a great deal of research has been conducted with the aim to assess area effects on health (for an overview, see Smyth) [1]. A key aim in this research has been to demonstrate the independent effect, if any, that area-level socio-economic factors have on health. Most studies have concluded that living in a socio-economically disadvantaged neighbourhood is associated with only relatively small effects on health outcomes. Furthermore, the health effects shown in observational studies often disappear after extensive adjustment for individual socio-economic characteristics (see for example, Robert, Pickett et al., Reijneveld, Yen et al.) [2–5].

An explanation for this lack of strong association may be that area effects are difficult to measure. It is widely recognized that the selection of the spatial unit is an important consideration in accurately detecting area effects. However, the definitions of 'neighbourhood' used in most studies are not based on theoretical considerations but instead on data availability [6]. As a result, in many cases, administratively defined areas have been used to define the spatial units for analysis.

Using administratively defined areas poses two types of problems related to, respectively, scale effects and boundary effects. **Scale effects** refer to the influence of the spatial scale used on the measurement of area effects. It is commonly agreed upon that the existence and strength of area effects on health are scale dependent [7–12]. Generally, stronger effects may be found if a smaller spatial scale is used [13]. **Boundary effects** occur especially when administrative boundaries do not accurately reflect appropriate neighbourhood boundaries. Administrative borders may not be relevant in the daily lives of residents. Residents living near the border of administrative areas may relate more to neighbouring administrative zones [14]. Due to such effects, the use of administratively defined areas may underestimate or skew geographical effects that would otherwise be observed within more relevantly defined areas.

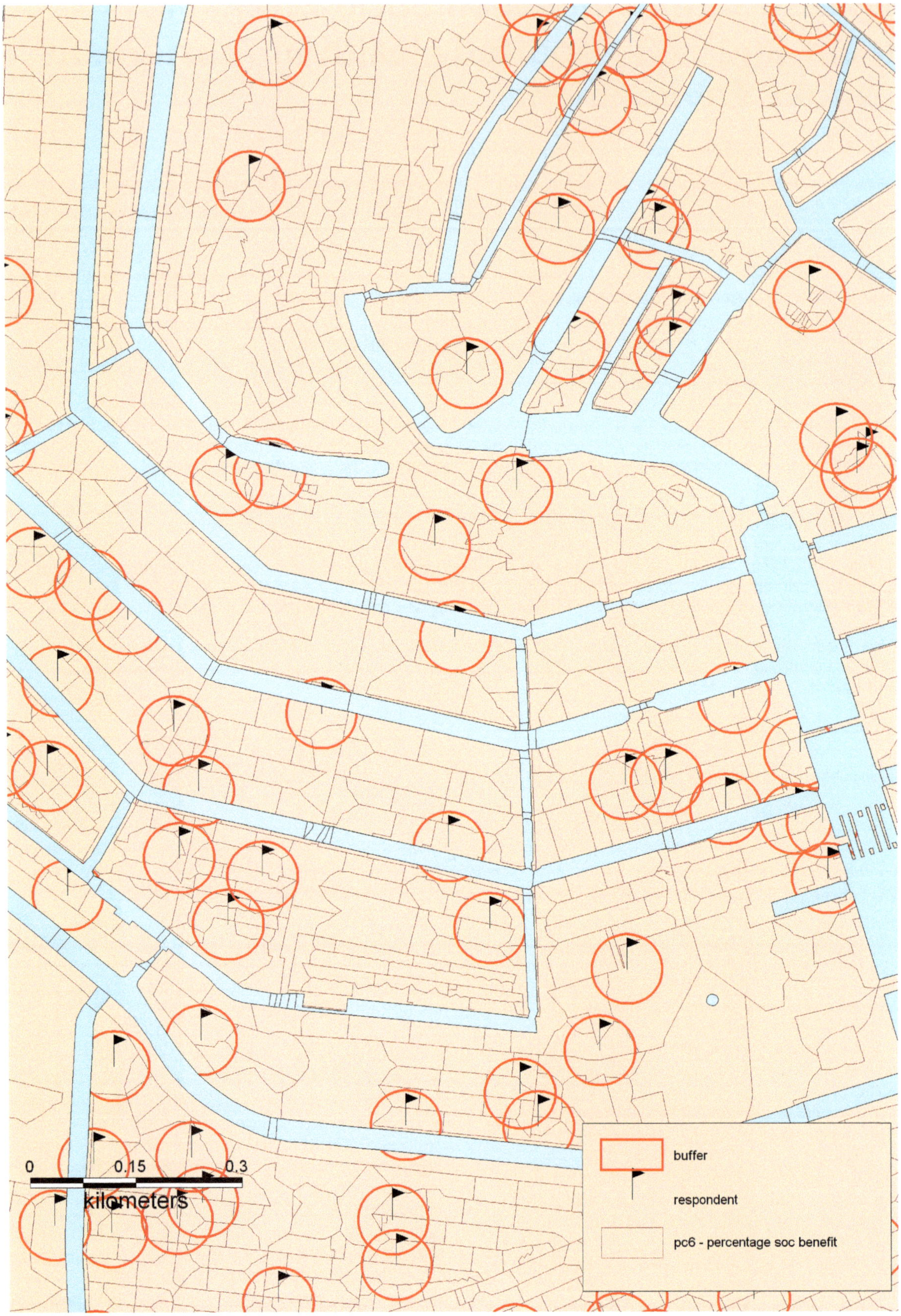

Figure 1. Data aggregation from 6-digit postcode areas to 50 meters radius bespoke environments.

Scale and boundary effects could in principle be avoided when so-called "bespoke environments" are used. In this approach, separate neighbourhoods are created for each individual resident. These neighbourhoods are centered on each respondent's home and are independent of administrative boundaries. The size of these neighbourhoods can be determined flexibly in terms of distance (the radius of the buffer) or counts (e.g. the number of residents).

This methodology has been applied in several research fields. Bespoke environments were introduced in the 1990s in studies of voting [15] and of social exclusion [16]. Studies on voting behaviour observed clear links between the characteristics of local milieus and voting behaviour [15], [17]. Anderson et al. [11] used, aside from administrative units, bespoke environments of 100 meters around each individual's home to study area effects on income. They found area effects to be strong at this very local level while non-existent or weak at the municipal level. Bolster et al. [18], investigated the effect of neighbourhood disadvantage on income dynamics using bespoke environments of different scales. They too found that the local level had a stronger association with individual outcomes.

The concept of bespoke environments has been applied in epidemiology at only a limited scale. Frank et al. [19] used this approach to assess the effect of the neighbourhood environment on walking behaviour and obesity. Each household was designated an area of one kilometer around the home. They found that the greater the diversity in land use within the bespoke environment, the lower the risk of obesity. Propper et al. [20] also used bespoke environments in their study on local neighbourhood conditions and mental health. The bespoke environment consisted of the area around each individual that contained the nearest 500–800 people. They found that the prevalence of common mental disorders was related to the socio-economic composition of the surrounding population, although the impact was limited. Maas et al. [21] used bespoke environments to measure the amount of green space in people's direct residential environment. A weak positive relationship was found with levels of physical activity.

To our knowledge, this study is the first in using bespoke environments to assess the association between socio-economic environment and self-reported health. The main aim of this study is to take into account scale effects and boundary effects when assessing the relationship between socio-economic environment and self-reported health in Amsterdam. The analysis consisted of three steps. First, we compared bespoke environments defined at eight different scales, with a radius ranging from 50 meters to 1500 meters, and assessed whether the association between socio-economic factors and self-reported health was strongest at smaller scales. Next, we distinguished between areas that were socio-economically homogeneous and heterogeneous and assessed whether the association between socio-economic factors and self-reported health was stronger among homogeneous areas. Finally, we compared the results with analysis based on administrative areas and assessed whether the bespoke approach showed a stronger association between socio-economic area and self-reported health.

Materials and Methods

Ethics Statement

The interview survey data were obtained and analysed anonymously. As the Dutch Act on Medical Research Involving Human Subjects (WMO) does not apply to this study, an official approval of this study by the Medical Ethics Review Committee was not required.

Data

The data was obtained from the 2009 "State of the City" survey conducted by the Municipality of Amsterdam's Department of Research and Statistics. The State of the City surveyed 4351 inhabitants of Amsterdam. Stratified sampling was used to ensure that residents of all districts and ethnic groups within Amsterdam were represented, and respondents from five socially deprived neighbourhoods were oversampled. Data was collected by telephone (29 percent of all respondents), face-to face interviews (16 percent) and postal questionnaires (56 percent), with response rates of 34, 30 and 14 percent respectively (because the documentation of the source data mentions only rounded percentages the sum is not equal to 100). In the analysis we excluded respondents living in buffers with less than 10 inhabitants and/or less than 10 houses, because for these areas we could not obtain valid measures of the socio-economic environment. In the final analysis 4131 respondents were included.

Table 1. Mean and variation of contextual variables.

Spatial unit	Area (km^2)	Percentage receiving social benefit		Percentage of social housing		Average property value	
		mean	standard deviation	mean	standard deviation	mean	standard deviation
Buffer 50	0.0078	15.37	9.05	56.55	37.28	23.35	9.47
Buffer 100	0.0311	15.49	7.81	56.70	33.52	23.96	9.05
Buffer 150	0.0699	15.49	7.15	56.82	30.59	24.51	8.75
Buffer 200	0.1244	15.53	6.71	56.94	28.24	24.99	8.53
Buffer 300	0.2798	15.61	6.08	57.04	24.92	25.77	8.22
Buffer 600	1.1191	15.52	5.09	56.85	20.45	27.66	7.39
Buffer 1000	3.1087	15.27	4.25	56.14	17.34	29.28	6.43
Buffer 1500	6.9945	14.77	3.57	54.48	15.42	30.35	5.43
Ward	0.4826	15.45	6.41	56.36	26.83	25.58	8.66
District	1.8573	15.58	5.53	56.89	23.11	26.89	7.92
Postcode 4	2.5104	15.25	5.12	56.34	20.99	29.40	8.03

Table 2. Number of respondents and percentage reporting fair/poor health by individual characteristics.

Individual variable	N	% reporting poor health
Sex		
Male	1821	23.7
Female	2310	25.9
Age	4131	
Household composition		
Single-parent family	322	33.9
Two adults with child	1244	21.8
Two adults without child	1210	22.5
Single	1172	28.7
Other	183	23.0
Ethnicity		
Natives	2209	18.9
Surinamese	308	35.1
Atillean	63	31.7
Turks	343	38.5
Moroccans	461	33.6
Other non-western immigrants	192	30.2
Western immigrants	458	22.1
Rest of Asia	97	39.2
Education		
No education	496	55.0
Low	791	33.5
Medium	971	20.9
High	1578	10.6
Other	295	41.4
Income net (Euros)		
700	192	44.8
700–1000	442	50.0
1000–1350	503	38.2
1350–2050	831	20.6
2050–3200	757	14.8
3200 and more	590	6.3
Missing	816	25.9
Receiving social benefit		
Yes	267	62.9
No	3864	22.3
Home ownership		
No owner	2558	29.7
Owner	1518	16.7
Living on household income		
Very difficult	209	61.2
Quite difficult	660	44.8
Difficult	536	33.0
Quite easy	784	20.0
Easy	1327	14.7
Very easy	467	8.8
Missing	148	34.5

The survey asked respondents about their living situation, such as housing and neighbourhood conditions, socioeconomic position and health. Perceived health status was measured by the response to the question, 'All in all, would you say your health is excellent, good, fair or poor?' The answers were classified into two categories: excellent/good and fair/poor. From the same survey, we obtained data on characteristics of the respondents that were used as control variables at the individual level. These include age, sex, ethnicity, household composition, educational level, income level, receipt of social benefits, home ownership and a measure of general wealth (whether the respondent experienced difficulties living on his or her current household income).

To measure the socio-economic characteristics of each respondent's environment, we used integral socio-economic registries maintained by the Municipality of Amsterdam. The registries were obtained by aggregating information from individual residents, households or houses to the level of 6-digit postcodes. A 6-digit postcode area, originally used for postal delivery, is the smallest geographical unit available. In urban areas, these units are sized approximately 50×50 meters and include 10 to 20 households. For each postcode area, we constructed three socio-economic variables: the percentage of residents receiving a social benefit (unemployment or welfare), the percentage of social housing, and the average property value of houses.

We constructed 'bespoke environments' or 'buffers' for each respondent using Geographic Information Systems (GIS) based on the central point location of the respondent's six-digit postal code. Buffers of eight sizes, with a radius ranging from 50 to 1500 meters, were created around each respondent. Finally, the socioeconomic characteristics of each of these buffers were estimated based on the data aggregated by postcode. Postcode areas that were only partially located within the buffer were weighted based on the percentage of the area contained. For this process, we performed an overlay operation, which joins data layers based on common geographical location. This approach is illustrated in figure 1.

In addition, we optimized the geographic delimitation of the buffers, and the measurement of their socioeconomic characteristics, by taking into account natural barriers. These barriers included the Amstel River, the IJ River and the Ring Road (highway). The resulting, more strictly delimited areas were expected to correspond more closely to the mental map and the immediate living environment of the respondents.

For further analyses, we classified the buffers based on whether they were socioeconomically homogeneous or heterogeneous. This determination was made by calculating the standard deviation of each of the socioeconomic variables for the postcode areas within the buffer. The buffer was considered relatively homogeneous if the standard deviation was smaller than average for at least two socioeconomic variables. All other buffers were considered to be heterogeneous.

We also analysed respondents' administratively defined areas for comparison with the bespoke environment. We used three types of administrative areas: the 4-digit postcode area (on average 2.5 km^2, or approximately 1.6 km by 1.6 km); districts (referred to as 'wijken' in the Netherlands, on average 1.8 km^2); and wards ('buurten', on average 0.4 km^2). Wards and districts in Amsterdam are considered to be socioeconomically homogeneous. The boundaries for wards are primarily determined by physical boundaries and often correspond to specific periods of construction. Wards are a common unit of geographical analysis by statistical bureaus and municipal offices.

Table 1 shows the extent of geographical variation in the three socioeconomic variables based on the spatial unit (bespoke and

Table 3. Changes in R^2 and AIC by inclusion of neighbourhood-SES variables compared to a model without neighbourhood-SES*.

Buffer size	Percentage receiving social benefit		Percentage of social housing		Average property value		Together	
	R^2 (%)	Increase	R^2	Increase	R^2	Increase	R^2	Increase
50	18.5	1.3	18.4	1.2	18.0	0.8	18.8	1.6
100	18.3	1.1	18.2	1	17.8	0.6	18.5	1.3
150	18.1	0.9	18.1	0.9	17.7	0.5	18.2	1
200	18.1	0.9	18.0	0.8	17.6	0.4	18.2	1
300	17.9	0.7	17.8	0.6	17.6	0.4	17.9	0.7
600	17.7	0.5	17.7	0.5	17.5	0.3	17.8	0.6
1000	17.6	0.4	17.5	0.3	17.3	0.1	17.6	0.4
1500	17.5	0.3	17.5	0.3	17.3	0.1	17.5	0.3
	AIC	Decrease	AIC	Decrease	AIC	Decrease	AIC	Decrease
50	4120	38	4122	36	4134	24	4114	44
100	4124	34	4128	30	4140	18	4124	34
150	4130	28	4133	25	4143	15	4131	27
200	4131	27	4135	23	4146	12	4134	24
300	4139	19	4142	16	4148	10	4141	17
600	4143	15	4146	12	4152	6	4146	12
1000	4148	10	4150	8	4156	2	4152	6
1500	4151	7	4152	6	4158	0	4154	4

*base model without neighbourhood-ses (age, sex, household composition, ethnicity: Nagelkerke R^2=.172; AIC=4158).

administrative). If the size of the buffer increases, the standard deviation of socioeconomic variables decreases. The standard deviations for the percentage of the population receiving a social benefit decreased from 9.1 to about 3.6; the standard deviation for average property value decreased from 9.5 to 5.4. The standard deviations for the percentage of the population living in social housing are high at a small buffer size, but quickly decrease with increasing buffer size.

Statistical Analysis

The relationship between socio-economic characteristics of areas and self-reported health was assessed using logistic regression analysis, with fair/poor health as the dependent variable. We controlled for age, sex, ethnicity and household composition (model 1), as well as for education, income, receipt of social benefit, home ownership and the proxy for wealth (model 2). The results of these models are expressed in terms of odds ratios, which are derived from the regression coefficient for the socioeconomic characteristics. The 95 percent confidence intervals are derived from the standard errors of the regression coefficients.

To enable comparison between the different buffer sizes, we also present the odds ratios corresponding to standardized regression coefficients (which is equivalent to transforming the socioeconomic variables into z-scores before performing a logistic regression). These odds ratios can be interpreted as the increase in the odds of fair/poor perceived health if the socioeconomic level of a neighbourhood changes with one standard deviation. This measure takes into account the large differences in standard deviation according to buffer size (table 1).

In order to quantify the explanatory power of socioeconomic characteristics of areas, we also applied a regression strategy involving two steps: first we included only the individual-level characteristics, and next we added the socioeconomic characteristics of areas. Using Nagelkerke R^2 and AIC, we quantified the increase in explained variance by adding the latter terms.

Results

Table 2 illustrates the percentage of respondents reporting fair/poor health, broken out by the respondents' individual characteristics. Fair/poor health is more often reported by single parent families (33.9 percent), non-western migrants (on average 33.5), respondents with no education (55) or a low educational-level (33.5), lower income groups (about 50), those receiving social benefit (62.9) and those having difficulties in making ends meet (61.2).

Table 3 quantifies the explanatory power of models including socioeconomic characteristics of areas, in terms of increase in percentage of variance explained and decrease of AIC. The explanatory power is strongest for small buffers, and it declines with increasing buffer size. For the percentage of residents living on social benefit, the percentage explained declines from 1.3 percent for 50 meter buffer size to 0.3 percent for 1500 meter buffer size. Notably, the percentage explained when the three socioeconomic variables are combined hardly exceeds the percentage that could already be explained by variable on residents living on social benefit.

Table 4 presents the effect of controlling for individual-level socioeconomic variables. We pay particular attention to the standardized odds ratio of columns 4 and 5. For example, the odds ratio in column 4 is 1.30 for the share of people living on social benefit within 50-meter buffers. This implies that if the share of people living on social benefit increases by 1 standard deviation (in this case 9 percent; see table 1), the odds of fair/poor health increases by 30 percent. After controlling for individual-level socioeconomic variables, this odds ratio declines to 1.10.

Table 4. Comparison of the effects of the contextual variables between model 1 (base model) and model 2 (extensive model) at different scales.

Buffer size	Odds ratio's (95% CI)		Standardized OR	
	Model 1	Model 2	Model 1	Model 2
receiving social benefit (%)				
50	1.029 (1.018;1.039)	1.011 (1.001;1.022)	1.30	1.10
100	1.032 (1.022;1.043)	1.009 (0.997;1.021)	1.28	1.07
150	1.032 (1.020;1.045)	1.007 (0.993;1.021)	1.25	1.05
200	1.033 (1.021;1.046)	1.007 (0.993;1.021)	1.24	1.05
300	1.032 (1.017;1.046)	1.005 (0.989;1.021)	1.21	1.03
600	1.033 (1.017;1.050)	1.007 (0.989;1.023)	1.18	1.04
1000	1.032 (1.012;1.053)	1.008 (0.988;1.028)	1.14	1.03
1500	1.034 (1.011;1.056)	1.014 (0.991;1.038)	1.13	1.05
social housing (%)				
50	1.007 (1.005;1.009)	1.002 (1.000;1.004)	1.30	1.08
100	1.007 (1.005;1.009)	1.002 (1.000;1.004)	1.26	1.07
150	1.007 (1.005;1.009)	1.001 (0.997;1.005)	1.24	1.03
200	1.007 (1.005;1.009)	1.001 (0.997;1.005)	1.22	1.03
300	1.007 (1.003;1.011)	1.001 (0.997;1.005)	1.19	1.03
600	1.007 (1.003;1.011)	1.001 (0.997;1.005)	1.15	1.02
1000	1.007 (1.003;1.011)	1.001 (0.995;1.007)	1.13	1.02
1500	1.007 (1.001;1.013)	1.001 (0.995;1.007)	1.11	1.02
average property value (in 10.000 Euros)*				
50	1.023 (1.013;1.033)	1.006 (0.996;1.016)	1.25	1.05
100	1.020 (1.011;1.031)	1.004 (0.994;1.014)	1.20	1.04
150	1.019 (1.007;1.030)	1.003 (0.993;1.013)	1.18	1.03
200	1.017 (1.008;1.029)	1.002 (0.992;1.012)	1.16	1.02
300	1.017 (1.007;1.028)	1.003 (0.991;1.015)	1.15	1.02
600	1.015 (1.005;1.026)	1.006 (0.994;1.018)	1.12	1.04
1000	1.010 (0.999;1.022)	1.006 (0.992;1.020)	1.06	1.04
1500	1.009 (0.995;1.024)	1.006 (0.990;1.021)	1.05	1.03

*For average property value, the OR is inverted to make it more directly comparable to the other SES indicators. The OR represents the increase in odds of poor health if property value decreases with 10,000 Euro's.

Generally, after controlling for all individual-level variables, the association between health and socioeconomic factors is strongest at small buffer sizes. Statistically significant associations are found only for 50-meter buffers and 100-meter buffers (for the percentage of social housing). For buffer sizes larger than about 200 meters, the associations are not statistically significant. Moreover, above 200 meters, the odds ratios in columns 3 and 5 do not provide indications of a consistent relationship with buffer size.

In table 5, the results are compared across buffers that are relatively homogeneous in terms of the percentage of people receiving social benefits. In this sub-set of buffers, the association with health is stronger. Standardised odds ratios are highest for homogeneous buffers of 300-meters or less, up to an odds ratio of 1.15 for homogeneous buffers of 50-meters. No associations were observed in the larger buffers, irrespective of their degree of homogeneity. For the other two socioeconomic variables (percentage of social housing; property values), we also found that associations were evident only in relatively homogeneous buffers of 300 meters and smaller (results not shown).

Table 6 explores whether analysis using administratively defined areas yields different results compared to analysis using bespoke environments. The results turn out to be similar: when socioeconomic factors are measured at the level of the smallest administrative unit, the ward, they can explain most of the variance in fair/poor health. The percentage explained at the ward level is about as large as when socioeconomic factors are measured at the level of buffers of 200 meters or smaller (cf. table 3). The AIC results indicate the same: the model improves if neighbourhood-SES variables are included and the effect decreases as administrative scales increase.

Figure 2 shows the standardised odds ratios, as estimated for different buffers. The odds ratios are plotted against the average size of the surface of the buffers. In general, the odds ratios decrease with increasing area surface of buffers. This implies that the association between health and socioeconomic factors is weaker when the latter are measured to larger buffers. For average property value, this trend is less consistent as odds ratios sharply increase for buffers smaller than 600 meter buffers. For the other two area characteristics, the association becomes consistently weaker with increasing area surface.

In addition, in Figure 2, a comparison can be made between buffers and administratively defined areas, while taking area surface into account. Analyses at the 4 digit postcode yield smaller effect estimates as compared to analyses using buffers of about similar size. However, when socioeconomic factors are measured at the level of districts, they perform equally well as socioeconomic variables measured at the level of buffers of a comparable size.

Discussion

Previous studies may have underestimated the association between health and socioeconomic characteristics of areas due to scale and boundary effects. We aimed to address these effects by using "bespoke environments" or "buffers" to study the relationship between health and the surrounding socio-economic environment. By comparing buffers of different sizes, we observed that the association between socio-economic environment and self-reported health could be demonstrated only for small buffers with a radius of 50 or 100 meters. Stronger associations were observed in analyses that only compared relatively homogeneous areas. When socioeconomic factors were measured to small administrative units (wards), they performed equally well as socioeconomic variables measured at the level of buffers of comparable size.

Evaluation of Methodology

Our method and results should be considered in the light of the modifiable areal unit problem (MAUP). The MAUP states that area-level effects are dependent on the form, size and location of the sub-areas used. This dependency is particularly important when using administratively defined areas. Administrative zones have a form, size and location that are often quite arbitrary. In studies using administrative areas, the results therefore could be strongly sensitive to the precise delineation of these areas [22].

Theoretically, bespoke environments should solve some aspects of this problem. By using bespoke environments, all areas have the same form (distances are equal in all directions) and location (each area is based around the center point of individual respondents), thus avoiding potential boundary effects. In addition, the size aspect can be addressed by using bespoke environments of different sizes.

Table 5. Comparison of neighbourhood effects between relatively homogeneous buffers and all buffers together.

Percentage of people receiving social benefit

Buffer size	Group	Odds ratio's (95% CI)		Standardized OR	
		Model 1	Model 2	Model 1	Model 2
50	homogeneous	1.035 (1.021;1.050)	1.018 (1.004;1.032)	1.37	1.15
	all	1.029 (1.018;1.039)	1.011 (1.001;1.002)	1.30	1.10
100	homogeneous	1.038 (1.021;1.054)	1.017 (0.999;1.035)	1.34	1.12
	all	1.032 (1.022;1.043)	1.009 (0.997;1.021)	1.28	1.07
150	homogeneous	1.027 (1.008;1.045)	1.009 (0.989;1.029)	1.21	1.06
	all	1.032 (1.020;1.045)	1.007 (0.993;1.021)	1.25	1.05
300	homogeneous	1.019 (0.999;1.040)	0.998 (0.978;1.020)	1.12	0.99
	all	1.032 (1.017;1.046)	1.005 (0.989;1.021)	1.21	1.03
600	homogeneous	1.026 (1.011;1.063)	1.016 (0.989;1.044)	1.13	1.08
	all	1.033 (1.017;1.050)	1.007 (0.989;1.023)	1.18	1.04

The use of bespoke environments as a geographic method might however bring new challenges as well. Because buffers overlap, especially the larger ones, observations for individual respondents are not entirely independent. Failure to take into account this dependency may result in overestimation of the precision and statistical significance of the area-effects. The use of multi-level models, using a restricted number of environments, would address this problem. However, when applying bespoke environments, such models cannot be easily integrated as respondents do not share identical environments and thus cannot be aggregated into the same high-order level category. We would like to note that, in our analyses, the strongest effects were observed at a smaller scale (50 meters) where buffers rarely overlapped.

We might have failed to control for potentially important confounders at the area level, such as land use mix, or noise nuisance caused by Schiphol Airport. We checked for area-level confounding by mapping the residuals of the regression analyses, with full control for individual-level variables. However, we did not observe spatial clusters of residuals, suggesting that there are no area-level confounders that could have biased our results to a significant extent.

In the analysis, we aimed to control for individual-level demographic and socioeconomic characteristics that could be considered to be potential confounders to the association between health and the surrounding socioeconomic environment. As controlling for these characteristics had an important effect on our effect estimates, we cannot exclude the possibility that more detailed control would remove even more of the area-level effect. At the same time, we would like to stress that we already had controlled both for education, income and wealth (by proxy), and that the potential for residual confounding by SES thus seems limited. However, we cannot exclude potential confounding by other factors and capabilities that may determine where people can choose to live [23].

The overall response rate to the survey was only 23 percent. It is documented that, in general, non responders are often young, of non-Western origin and have a low income [24]. These characteristics were strongly related to self perception of health. Given these relationships, we cannot exclude the possibility that selective non-response may have biased our estimates of the association between health and the socioeconomic environment. Most likely, we think that this association may have been underestimated to some extent.

Studies comparing administrative areas and alternative definitions of a neighbourhood found similar associations with health outcomes irrespective of the way in which the neighbourhood

Table 6. Changes in R^2 and AIC for three neighbourhood-SES variables at different administrative scales compared to a model without neighbourhood-SES*.

	% social benefit		% social housing		Average property value		Together	
administrative-zone	**R^2 (%)**	**Increase**	**R^2**	**Increase**	**R^2**	**Increase**	**R^2**	**Increase**
ward (0.4 km^2)	17.9	0.7	17.9	0.7	17.7	0.5	18.1	0.9
combination (1.8 km^2)	17.5	0.3	17.4	0.2	17.5	0.3	17.7	0.5
postcode 4 (2.5 km^2)	17.5	0.3	17.5	0.3	17.4	0.2	17.6	0.4
	AIC	**Decrease**	**AIC**	**Decrease**	**AIC**	**Decrease**	**AIC**	**Decrease**
ward (0.4 km^2)	4135	23	4137	21	4143	15	4131	27
combination (1.8 km^2)	4137	21	4138	20	4152	6	4132	25
postcode 4 (2.5 km^2)	4150	8	4151	7	4154	4	4148	10

*base model without neighbourhood-ses (age, sex, household composition, ethnicity: Nagelkerke R^2 = .172; AIC = 4158).

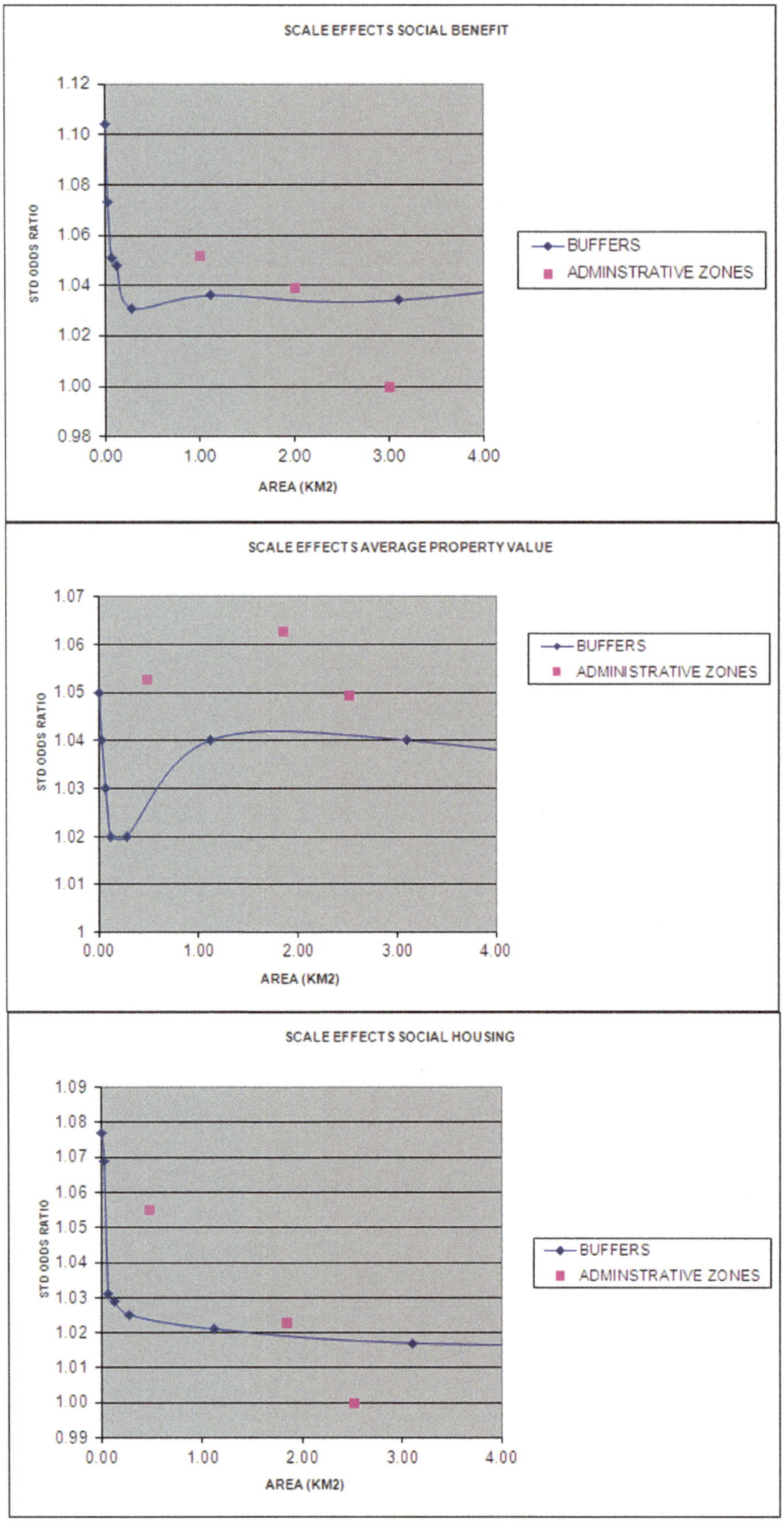

Figure 2. The strength of socioeconomic area effects according to scale.

boundaries were defined [25], [26]. This corresponds to our finding that the analysis of wards yielded similar results as the analysis with similarly-sized buffers. However, we might have expected associations to be stronger with the buffers, as buffers may be a better representation of one's immediate living environment and activity space. Our results however suggest that administrative areas that are defined with regards to socioeconomic and geographic criteria, such as wards (in the case of Amsterdam), may function equally well.

By using GIS techniques we had the opportunity to construct residential areas on a very local scale. We observed this to be an important advantage, as the association between socioeconomic variables and health was found to be the strongest, and only demonstrable with statistical significance, at the level of very small buffers (50- or 100-meter). In addition, GIS techniques make it possible to perform additional geographic operations such as measuring the degree of homogeneity of areas. This offered the opportunity to restrict the analysis to a subset of areas with greater contrast in socioeconomic conditions.

Interpretation and Comparison to Previous Studies

Other studies have also observed that the association between health and area-level socioeconomic characteristics was stronger in smaller areas. For example, one Dutch study assessed the effects of area-level socio-economic factors on mortality within postcode areas, districts, and wards. That study showed that differences in mortality chance of men were most pronounced at the lowest scale level of postcode areas [27].

We observed that the effect of area-level socioeconomic factors was small in comparison to the effects of individual-level socioeconomic variables on health (cf. table 2 and 4). A relatively small effect was also found in other Dutch studies [4], [28] and should possibly be considered in a national context. We postulate that effects of socio-economic conditions of areas may be small in a welfare state such as the Netherlands due to, among other factors, social housing policies and urban renewal schemes that that have limited sharp differences in living conditions amongst its population.

The fact that effects are observed only at the level of small (50–100 meter) buffers is suggestive of an effect of factors with a highly local reach. Among these, social networks might play an important role. In the case of voting behaviour, Johnston [17] and McAllistar [15] found clear links between local milieus and how people behave. Those who live in relative close proximity are more likely to think and act in similar ways. Other localized factors may include neighbourhood-level psycho-social stressors (e.g., nuisance from neighbours, feeling unsafe, drug abuse, etc.), many of which have been found to be related to self-rated health, including in Amsterdam [29]. Generally, these stressors may produce health effects on local scales, especially in socio-economically deprived areas [30].

Conclusions

To conclude, this study observed scale effects to be highly important when studying socio-economic area effects on health. The measurement of socioeconomic factors for large areas might result in a substantial underestimation, or even a negligence, of the effects of socioeconomic environment on health. The results stress the importance of using micro-scale data on the environment as well as health outcomes in order to study the relationship between these two. When such data are available, the methodology of bespoke environments could be applied to many environmental features and health-related outcomes. An important advantage of this methodology is that the buffer width can be tuned to the scale at which processes are expected to operate – whether a few meters or a few kilometres. The most relevant scale is likely to vary based on the health outcome and population group (e.g. children vs. middle-aged men) being measured. Through scaling, bespoke environments add a new dimension to study environment and health.

Acknowledgments

The authors gratefully acknowledge the Municipality of Amsterdam's Department of Research and Statistics for making their State of the City Survey data available.

Author Contributions

Conceived and designed the experiments: EV AK. Performed the experiments: EV AK. Analyzed the data: EV. Wrote the paper: EV AK KS.

References

1. Smyth F (2008) Medical geography: understanding health inequalities. Prog Hum Geogr 32: 119–127.
2. Robert SA (1999) Socioeconomic Position and Health: the Independent Contribution of Community Socioeconomic Context. Annu Rev Sociol 25: 489–516.
3. Pickett KE, Pearl M (2001) Multilevel analyses of neighbourhood socioeconomic context and health outcomes: a critical review. J Epidemiol Community Health 55: 111–122.
4. Reijneveld SA (1998) The impact of individual and area characteristics on urban socioeconomic differences in health and smoking. Int J Epidemiol 27: 33–40.
5. Yen IH, Kaplan GA (1999) Poverty area residence and changes in depression and perceived health status: evidence from the Alameda County Study. Int J Epidemiol 28: 90–94.
6. Dietz RD (2002) The estimation of neighbourhood effects in the social sciences: An interdisciplinary approach. Soc Sci Res 31: 539–575.
7. Macintyre S, Ellaway A, Cummins S (2002) Place effects on health: how can we conceptualise, operationalise and measure them? Soc Sci Med 55: 125–139.
8. Diez Roux AV (2001) Investigating Neighbourhood and Area Effects on Health. Am J Public Health 11: 1783–1789.
9. Diez Roux AV (2007) Neighbourhoods and health: Where are we and where do we go from here? Rev Epidemiol Sante Publique 55: 13–21.
10. Lupton R (2003) Neighbourhood Effects: Can We Measure Them and Does It Matter? CASE paper 73, London School of Economics: 1–24.
11. Anderson A, Musterd S (2010) What scale matters? Exploring the relationships between individuals' social position, neighbourhood context and the scale of neighbourhood. Geogr Ann Ser B 92: 1–21.
12. Spielman SE, Yoo E (2009) The spatial dimensions of neighbourhood effects. Soc Sci Med 68: 1098–1105.
13. Reijneveld SA, Verheij RA, de Bakker DH (2000) The impact of area deprivation on differences in health: does the choice of the geographical classification matter? J Epidemiol Community Health 54: 306–313.
14. Flowerdew R, Manley DJ, Sabel CE (2008) Neighbourhood effects on health: Does it matter where you draw the boundaries? Soc Sci Med 66: 1241–1255.
15. MacAllistar I, Johnston RJ, Pattie CJ, Tunstall H, Dorling DFL, et al. (2001) Class dealignment and the neighbourhood effect: Miller revisited. Br J Polit Sci 31: 41–60.
16. Buck N (2001) Identifying neighbourhood effects on social exclusion. Urban Stud 38: 2251–2275.
17. Johnston RJ, Pattie CJ, Dorling DFL, MacAllistar I, Tunstall H, et al. (2000) The neighbourhood effect and voting in England and Wales: real or imagined? British Elections and Parties Review 10: 47–63.
18. Bolster A, Burgess S, Johnston R, Jones K, Propper C, et al. (2007) Neighbourhoods, households and income dynamics: a semi-parametric investigation of neighbourhood effects. Journal of Economic Geography 7: 1–38.
19. Frank LD, Andresen MA, Schmid TL (2004) Obesity Relationships with Community Design, Physical Activity and Time Spent in Cars. Am J Prev Med 27: 87–96.
20. Propper C, Jones K, Bolster A, Burgess S, Johnston R, et al. (2005) Local neighbourhood and mental health: Evidence from the UK. Soc Sci Med 61: 2065–2083.
21. Maas J, Verheij RA, Spreeuwenberg P, Groenenwegen PP (2008) Physical activity as a possible mechanism behind the relationship between green space and health: A multi-level analysis. BMC Public Health 8: 1–13.

22. Manley D, Flowerdew R, Steel D (2006) Scale, levels and processes: Studying spatial patterns of British census variables. Comput Environ Urban Syst 30: 143–160.
23. Jongeneel-Grimen B, Droomers M, Stronks K, Kunst AE (2011) Migration does not enlarge inequalities in health between rich and poor neighbourhoods in the Netherlands. Health Place 17: 988–995.
24. City of Amsterdam Department of Research and Statistics (2009) De Staat van de Stad Amsterdam V [The State of the City of Amsterdam V] Amsterdam: Department of Research and Statistics. 202 p.
25. Stafford M, Duke-Williams O, Shelton N (2008) Small area inequalities in health: Are we underestimating them? Soc Sci Med 67: 891–899.
26. Ross NA, Tremblay S, Graham K (2004) Neighbourhood influence on health in Montreal, Cananda. Soc Sci Med 59: 1485–1494.
27. Smits J, Keij I, Westert G (2002) Sociaal-economische status en sterfte. Economisch Statistische Berichten 87: 419.
28. Reijneveld SA, Schene AH (1998) Higher prevalence of mental disorders in socioeconomically deprived urban areas in the Netherlands: community or personal disadvantage? J Epidemiol Community Health 52: 2–7.
29. Agyemang C, van Hooijdonk C, Wendel-Vos W, Lindeman E, Stronks K, et al. (2007) The association of neighbourhood psychosocial stressors and self-rated health in Amsterdam, The Netherlands. J Epidemiol Community Health 61: 1042–1049.
30. Hill T, Ross CE, Angel RJ (2005) Neighbourhood Disorder, Psychophysiological Distress and Health. J Health Soc Behav 46: 170–186.

7

Spatial Distribution of Soil Organic Carbon and Total Nitrogen Based on GIS and Geostatistics in a Small Watershed in a Hilly Area of Northern China

Gao Peng[1]*, Wang Bing[2]*, Geng Guangpo[1], Zhang Guangcan[1]

1 Shandong Agricultural University, College of Forestry/Taishan Mountain Forest Ecosystem Research Station/Shandong Provincial Key Laboratory of Soil Erosion and Ecological Restoration, Tai'an, Shandong, P.R. China, **2** Research Institute of Forest Ecology Environmental Protection, Chinese Academy of Forestry, Beijing, P.R. China

Abstract

The spatial variability of soil organic carbon (SOC) and total nitrogen (STN) levels is important in both global carbon-nitrogen cycle and climate change research. There has been little research on the spatial distribution of SOC and STN at the watershed scale based on geographic information systems (GIS) and geostatistics. Ninety-seven soil samples taken at depths of 0–20 cm were collected during October 2010 and 2011 from the Matiyu small watershed (4.2 km^2) of a hilly area in Shandong Province, northern China. The impacts of different land use types, elevation, vegetation coverage and other factors on SOC and STN spatial distributions were examined using GIS and a geostatistical method, regression-kriging. The results show that the concentration variations of SOC and STN in the Matiyu small watershed were moderate variation based on the mean, median, minimum and maximum, and the coefficients of variation (CV). Residual values of SOC and STN had moderate spatial autocorrelations, and the Nugget/Sill were 0.2% and 0.1%, respectively. Distribution maps of regression-kriging revealed that both SOC and STN concentrations in the Matiyu watershed decreased from southeast to northwest. This result was similar to the watershed DEM trend and significantly correlated with land use type, elevation and aspect. SOC and STN predictions with the regression-kriging method were more accurate than those obtained using ordinary kriging. This research indicates that geostatistical characteristics of SOC and STN concentrations in the watershed were closely related to both land-use type and spatial topographic structure and that regression-kriging is suitable for investigating the spatial distributions of SOC and STN in the complex topography of the watershed.

Editor: Ben Bond-Lamberty, DOE Pacific Northwest National Laboratory, United States of America

Funding: This research was funded by the Special Fund for Forestry Scientific Research in the Public Interest (No. 201204101), the National Basic Research and Development Program of China (973 Program, Nos. 2011CB403201 and 2012CB416904), and was subsidized by CFERN&GENE Award Funds on Ecological Paper. Funders had a role in the data collection and decision to publish the manuscript.

Competing Interests: The authors have declared that no competing interests exist.

* E-mail: gaopengy@163.com (GP); wangbing@caf.ac.cn (WB)

Introduction

As important factors in global biogeochemical cycling of the terrestrial ecosystem, soil organic carbon (SOC) and total nitrogen (STN) are important in alleviating global warming, mitigating land degradation, improving food security, and enhancing crop production [1,2,3]. They are also at the heart of the global carbon-nitrogen cycle and climate change research. SOC and STN are carbon and nitrogen sources for plant growth. They also affect soil biodiversity and the structure and physical stability of soil that enables it to resist erosion [4]. SOC and STN have strong spatial heterogeneity, with internal changes in the vertical and horizontal directions and external exchanges with the atmosphere and biosphere [5]. Many factors, such as topography, land-use type, field management and vegetation, can control SOC and STN spatial variability at various scales. Understanding and incorporating such heterogeneity and spatial distribution characteristics can improve the precision of carbon-nitrogen budgets and assist in implementation of effective measures toward vegetation recovery.

There is considerable research into SOC and STN spatial distributions on different scales [6,7,8,9,10], and results show that SOC and STN have a changing continuum with a non-uniform spatial distribution. Several studies have shown variability in topography, vegetation, cultivation, land use and parent material [11,12,13,14].

However, current research mainly relies on the ordinary kriging method, which produces large uncertainty in the prediction of soil spatial distribution because of the impacts of land use, topographic characteristics and other factors [15]. In recent years, some methods have been proposed to solve this problem, such as the regression-kriging method, the geographically weighted regression method (GWR) and the geographically weighted regression kriging method (GWRK) [16,17]. However, among these methods, only the regression-kriging method can incorporate topographic factors, vegetation coverage and other elements and thereby improve the accuracy of spatial prediction [18,19,20].

Some researchers have studied SOC and STN spatial variability in different regions of China, including the black-soil region in the northeast and the Loess Plateau in the northwest [21,22,23]. However, there is generally little quantitative information on this spatial variability on small-watershed scales in the rocky mountain area of the northern part of the country. This is especially true regarding the few spatial variability studies using regression-kriging

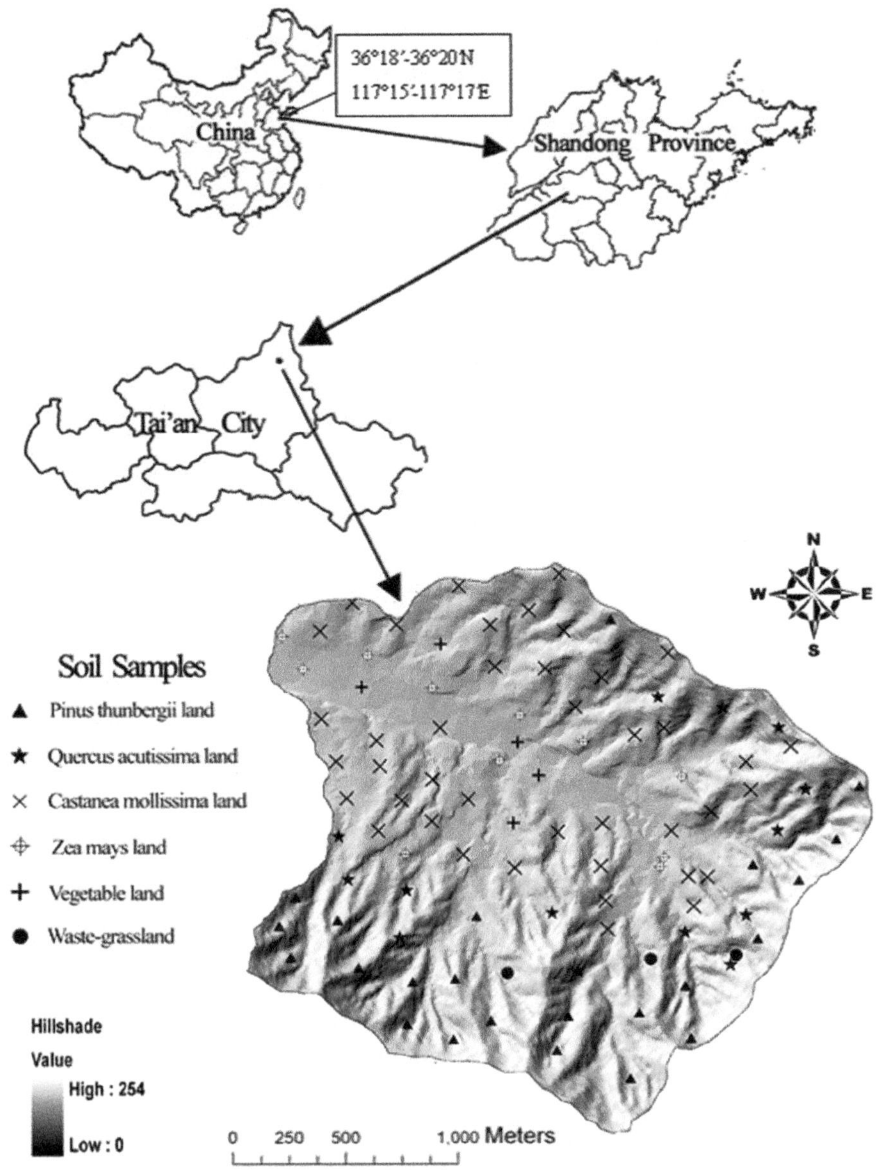

Figure 1. Distribution map of soil sample points (0–20 cm) in Matiyu small watershed.

and geographic information system (GIS) technology, limiting the capability of evaluating the carbon budget and predicting the ecosystem response to environmental and climate change.

In this paper, we selected the Matiyu small watershed, which is typical of the rocky mountain areas of northern China, as a research site. We used regression-kriging and GIS to achieve the following objectives: 1) to reveal and analyze spatial SOC and STN distributions at the scale of a small watershed; 2) to address the impacts of different land-use types, elevation, vegetation coverage, and other factors on those spatial distributions; and 3) to compare SOC and STN prediction accuracy using regression-kriging and ordinary kriging methods.

Materials and Methods

Ethics Statement

The research station for this study is managed by Shandong Agricultural University. The study was approved by the Mountain Tai Forest Ecosystem Research Station of the State Forestry Administration.

Study area condition

The Matiyu watershed is in Tai'an city, Shandong Province. It has an area of 4.2 km^2 and lies between 36°18′–36°20′N and 117°15′–117°17′E (Figure 1), with elevations ranging from 219 m to 595 m. The climate is semiarid, with an average of 765 mm of annual precipitation; 70% of this precipitation falls from June through September. Annual average air temperature is 12.8°C, the accumulated temperature greater than 10°C is 4120°C, and there are 200 frost-free days.

The dominant soil type is cinnamonsoil. The study area includes the following major land-use types: forest (including *Pinus thunbergii*, *Quercus acutissima*, and *Castanea mollissima*), farmland (including *Zea mays* and vegetables), and waste-grassland.

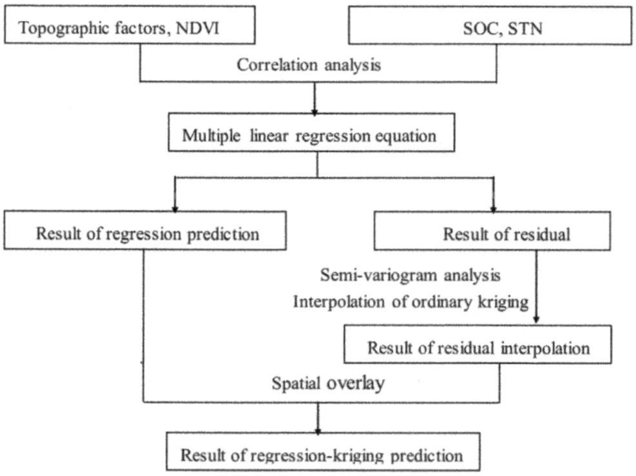

Figure 2. Flowchart of regression-kriging.

Sampling and measurement methods

The soil sampling design was based off of a digital topographic map of the watershed at a scale of 1:10000, which was used to generate a raster digital elevation model (DEM) map with a resolution of 10 m×10 m. A grid square crossed by the study boundary was regarded as a distinct unit if more than half of it lay inside the study area; otherwise, it was merged into a neighboring grid. A Real Time Kinetic – Global Positioning System (RTK-GPS) receiver was used to locate each grid center in the field and to record latitudes, longitudes and elevations of sampling points. The terrain information was then recorded in detail. Within a radius of approximately 20 cm at each site, four separate sub-samples were taken from the top 20-cm of soil using a 5.0-cm diameter soil auger. These samples were then mixed and were considered representative of the soil at the site.

In total, 97 soil samples (0–20 cm) were collected in Matiyu watershed during October 2010 and October 2011 (Figure 1) [24,25]. Soil samples were air-dried for approximately 7 days and then passed through a 0.25-mm sieve. SOC and STN contents were determined in duplicate for each sample, using the potassium dichromate oxidation and micro-Kjeldahl methods [26].

Acquisition of topographic factor attributes and NDVI

Topographic factors were computed using a spatial analysis model and digital topography analysis, and included elevation (H), slope (β), sinα and cosα of aspect, compound topographic index (CTI), stream power index (SPI), and sediment transport index (STI). The normalized difference vegetation index (NDVI) was acquired from a Landsat 5 image from August 30, 2010.

Interpolation of regression-kriging

Figure 2 shows a flowchart of the regression-kriging method. Taking into account the spatial distribution law and geographic influence factors, the method simulates both spatial distribution trends and uncertainty. Equations were established between the assistant variable and target variables, and then ordinary kriging was used to separate the trend item and interpolate the residual. Finally, the spatial overlay was carried out between the trend item of regression prediction and the residual value of ordinary kriging to obtain the predicted value of the target variable [27,28].

Precision evaluation of prediction results

Twenty-six of the 97 soil samples were randomly extracted from the data to test the model's predictive accuracy based on data from the remaining 71 soil samples. Prediction accuracy was evaluated by comparing observed and predicted values of SOC and STN from validation point locations. Mean prediction error (*MPE*) and root mean square prediction error (*RMSE*) were selected as evaluation indexes, and R_I was used to analyze improvement of the prediction accuracy by comparing regression-kriging with ordinary kriging [29,30].

$$MPE=\frac{1}{n}\sum_{j=1}^{n}(Z_{oi}-Z_{pi})$$

$$RMSE=\sqrt{\frac{1}{n}\sum_{j=1}^{n}(Z_{oi}-Z_{pi})^2}$$

$$R_I=\frac{RMSE_R-RMSE_{RK}}{RMSE_R}\times 100\%$$

where *MPE* is the mean prediction error; *RMSE* is the root mean square prediction error; *n* represents sampling points of the test dataset; Z_{oi} and Z_{pi} are observed and predicted values of the sampling points, respectively; and R_I is the improvement of prediction accuracy from comparing regression-kriging with ordinary kriging. If R_I is positive, it means that prediction accuracy of regression-kriging was higher than that of ordinary kriging, and vice-versa for a negative R_I value. $RMSE_R$ is the root mean square prediction error of ordinary kriging, and $RMSE_{RK}$ is that of regression-kriging.

Software platform. SPSS17.0 software was used for classical statistical and regression analyses of SOC and STN and for testing their differences under different land use types via ANOVA.

Table 1. Statistical results of mean concentrations of SOC and STN in Matiyu small watershed†.

Soil nutrients‡	Mean (g·kg^{-1})	Median (g·kg^{-1})	Min. (g·kg^{-1})	Max. (g·kg^{-1})	Range (g·kg^{-1})	Std. Deviation	CV (%)
SOC	11.91	11.73	6.20	21.67	15.47	2.87	24.10
STN	1.18	1.16	0.54	2.57	2.03	0.28	23.73

"†" Numbers represent the mean of 2 years (2010–2011a). The following is the same.
"‡" SOC = soil organic carbon; STN = soil total nitrogen.

Table 2. Correlation analysis of SOC and STN with the environmental factors.

Soil nutrients‡	Elevation	slope	cosα	sinα	CTI	SPI	STI	NDVI
SOC	0.495**	0.339**	0.298*	−0.052	−0.094	0.030	0.117	0.199*
STN	0.415**	0.249*	0.400**	−0.020	−0.083	0.016	0.042	0.215*

*and ** indicate significance at $P<0.05$ and $P<0.01$, respectively.
"‡" SOC = soil organic carbon; STN = soil total nitrogen; cosα = cosine of aspect; sinα = sine of aspect; CTI = compound topographic index; SPI = stream power index; STI = sediment transport index; NDVI = Normalized difference vegetation index.

ArcGIS 9.2 was used to establish the small-watershed DEM and extract its topographic factors. GS+7.0 was used for semi-variogram analysis of regression prediction and residual values. ArcGIS9.2 was then used to perform the spatial plus operation between the trend item of regression prediction and the residual value of ordinary kriging [31]. Finally, the spatial prediction distribution map of SOC and STN in the Matiyu watershed was produced.

Results

Descriptive statistics of SOC and STN in Matiyu watershed

Mean SOC and STN concentrations in the watershed were 11.91 g·kg^{-1} and 1.18 g·kg^{-1}, respectively. Respective coefficients of variation (CVs) were 24.10% and 23.73%, respectively, and both CVs were moderate (Table 1).

Correlations of SOC and STN with environmental factors

Table 2 shows that both SOC and STN had highly significant positive correlations with elevation (correlation coefficients of 0.495 and 0.415, respectively). This indicates that the higher the elevation, the larger the SOC and STN concentrations. SOC and STN also had highly significant positive and significant positive correlations with slope, respectively. This reveals that the greater the slope, the larger the two concentrations. The increasing trend of SOC was much more significant than that of STN. Both concentrations had little correlation with the sine of aspect. However, SOC and STN had highly significant positive and significant positive correlations with the cosine of aspect, respectively. This cosine measures the degree of northward orientation; thus, the greater the northward orientation, the larger the SOC and STN concentrations.

Both SOC and STN had significant positive correlations with NDVI, which indicates that the greater the vegetation cover, the higher the two concentrations. Both SOC and STN had little correlation with CTI, SPI or STI.

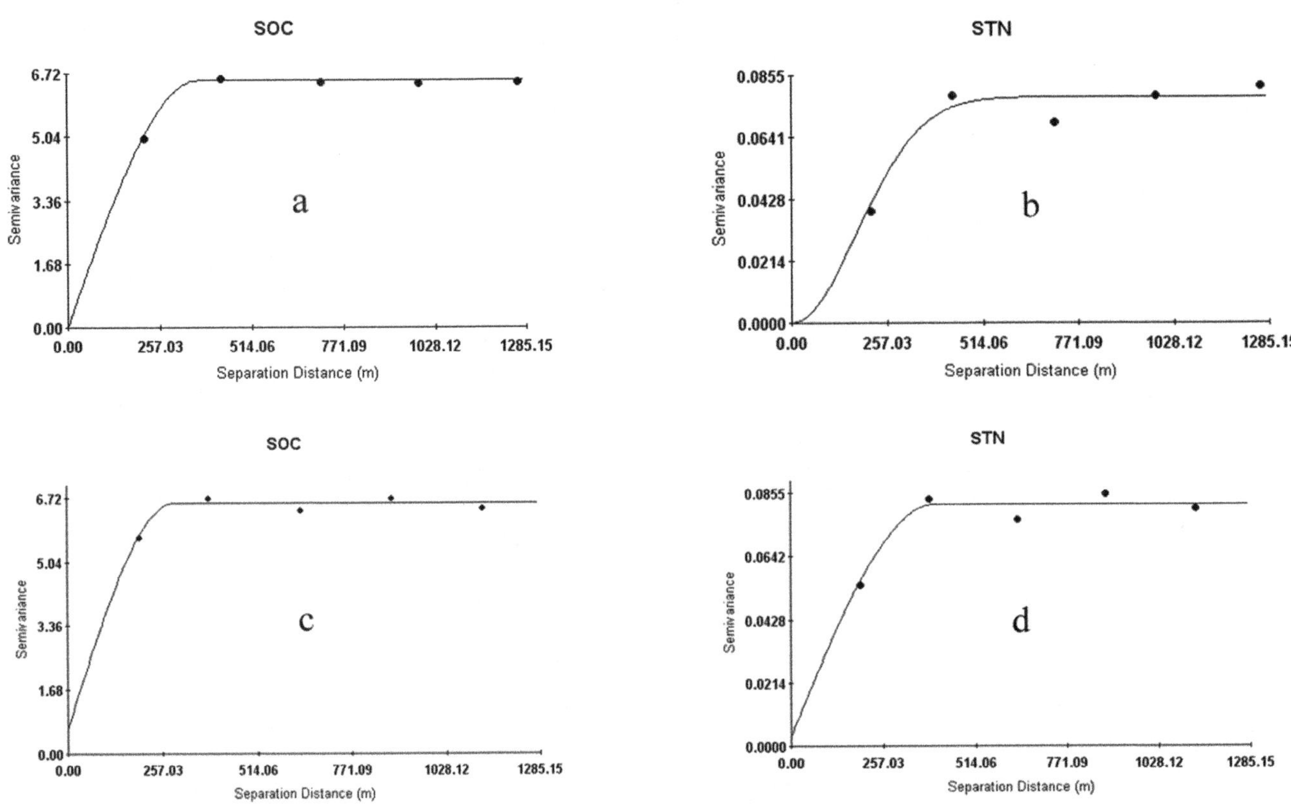

Figure 3. Semi-variogram analysis of SOC and STN for regression-kriging (a, b) and ordinary kriging (c, d).

Table 3. Semi-variogram analysis of SOC and STN and its parameters.

Soil nutrients‡	Data Item	Active lag distance (m)	Lag interval (m)	Model	Nugget C_0	Sill C_0+C	Nugget/Sill C_0/(C_0+C) (%)	Range (m)	R^2
SOC	residual error(RK)	1285.15	280	Spherical	0.0100	6.5230	0.2	365	0.996
STN	residual error(RK)	1285.15	280	Gaussian	0.0001	0.0779	0.1	424	0.915
SOC	raw data(OK)	1285.15	250	Spherical	0.6500	6.5740	9.9	288	0.880
STN	raw data(OK)	1285.15	250	Spherical	0.0031	0.0817	3.8	399	0.908

"‡" SOC = soil organic carbon; STN = soil total nitrogen; RK = regression-kriging; OK = ordinary kriging.

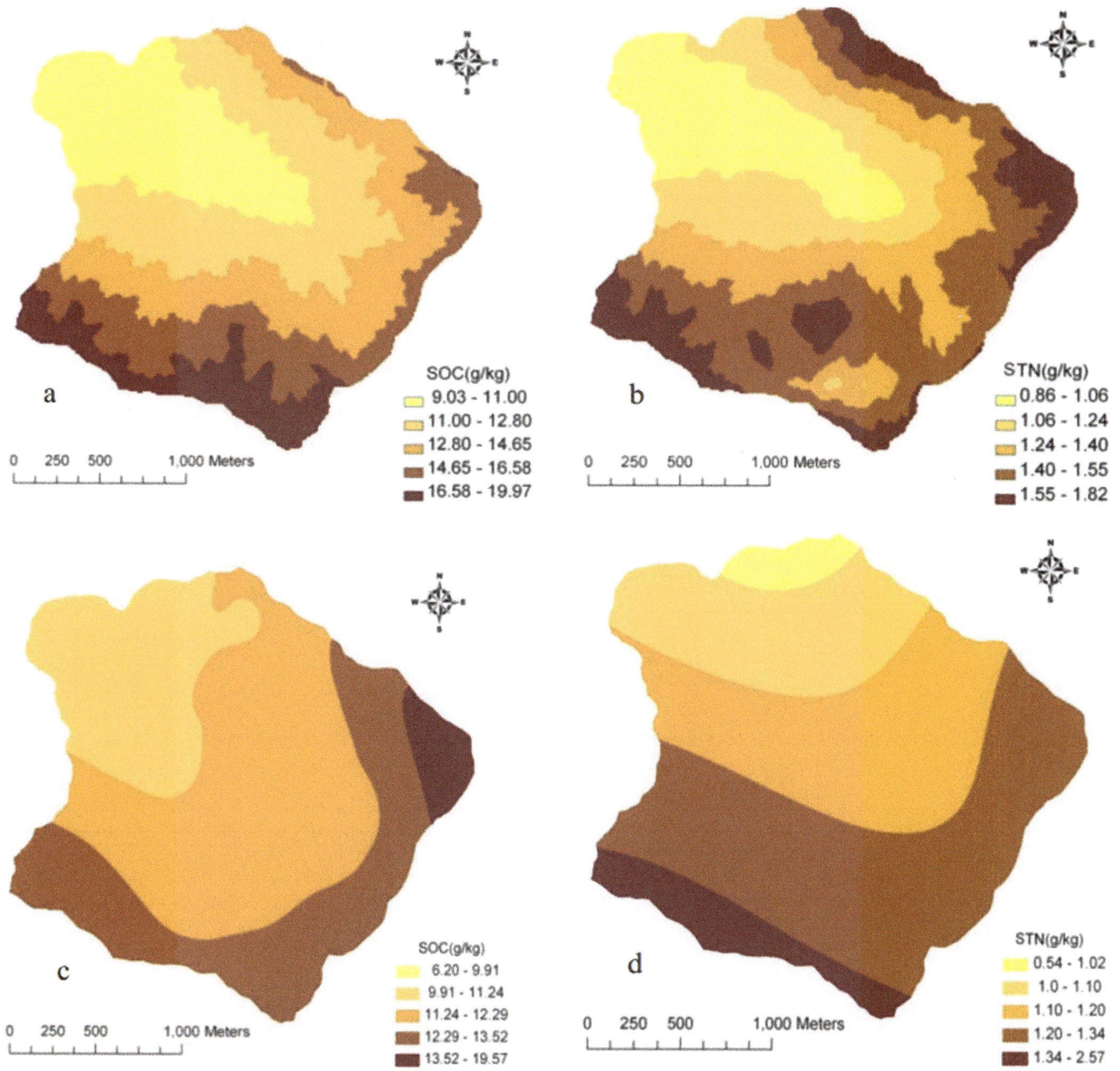

Figure 4. Distribution map of SOC and STN concentrations by regression-kriging (a, b) and ordinary kriging (c, d) in Matiyu small watershed.

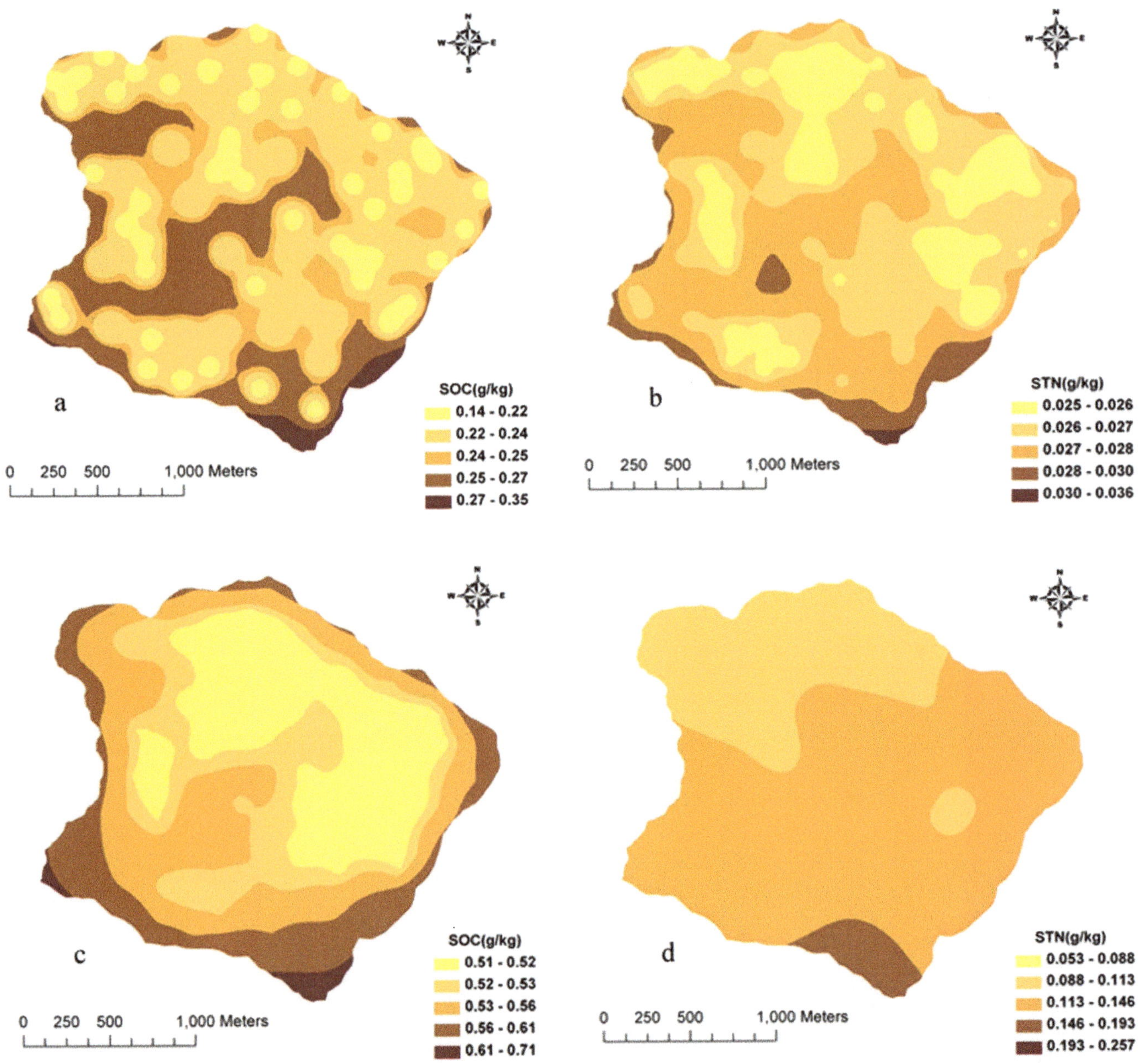

Figure 5. Maps of Prediction Standard Error of SOC and STN by regression-kriging (a, b) and ordinary kriging (c, d) in Matiyu small watershed.

Table 4. Comparison of prediction accuracy between regression-kriging and ordinary kriging.

Soil nutrients‡	Ordinary kriging		Regression-kriging		R_I(%)
	MPE	*RMSE*	*MPE*	*RMSE*	
SOC	−0.5	3.63	0.39	3.02	16.8
STN	−0.06	0.29	−0.01	0.24	17.24

"‡" SOC = soil organic carbon; STN = soil total nitrogen; MPE = mean prediction error; RMSE = root mean square prediction error; RI = improvement of prediction accuracy.

Regression-kriging models and semi-variogram analysis

Using the multiple linear stepwise regression method, all topographic factors and NDVI were used to explain the variability of SOC and STN concentrations. Equation fitting results were as follows:

$r^2 = 0.330$ ($P<0.05$)

$r^2 = 0.328$ ($P<0.01$)

The elevation and cosine of aspect that joined the prediction of SOC and STN were the optimal factors from all selected environmental factors. The determination coefficients of SOC and STN (r^2) were 0.330 and 0.328, and the equation fits were good.

Spatial autocorrelations of SOC and STN from regression-kriging (RK) and ordinary kriging (OK) were both low; however,

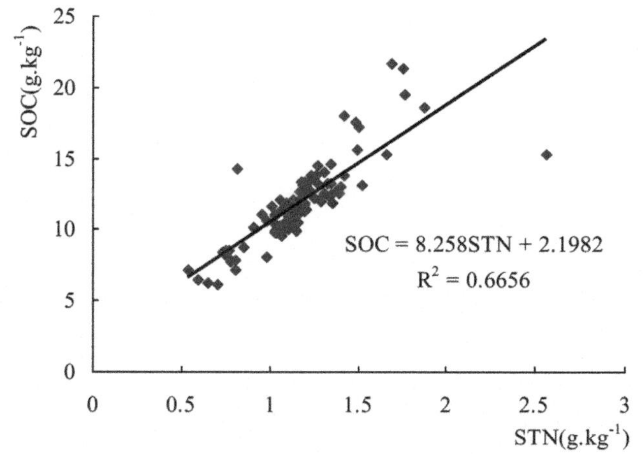

Figure 6. Regression equations of SOC and STN in Matiyu small watershed.

the effect of regression-kriging was better than that of ordinary kriging, and the Nugget/Sill were 0.2% and 0.1%, respectively (Table 3, Figure 3). The semi-variogram test showed that structural factors such as topography, vegetation and climate were the primary causes of SOC and STN spatial variation; random factors had little influence.

Evaluation of prediction accuracy using different prediction methods

In accordance with the flow chart of regression-kriging, linear combinations of environmental factors were considered as the external drift trend term to separate residual error and eliminate non-stationary properties to increase prediction accuracy. Seventy-one samples were randomly selected to conduct ordinary kriging interpolation for regression residual error of SOC and STN in the Matiyu small watershed. Meanwhile, as a control, those 71 samples were used to conduct ordinary kriging interpolation. From the prediction errors and distribution maps of regression-kriging and ordinary kriging (Figure 4, Figure 5), the result showed that the regression-kriging was better than that of ordinary kriging. The regression-kriging predictions were much more detailed concerning the partly variation and topographical relationships and much closer to the observed spatial distribution of SOC and STN. The other 26 samples were used to compare the accuracy of the two prediction methods (Table 4). The regression-kriging predicted values were very close to the measured ones, and its prediction accuracy was superior to that of ordinary kriging. The improvements of prediction accuracy (R_I) of SOC and STN were 16.80% and 17.24%, respectively (Table 4).

SOC and STN spatial distribution characteristics and their correlations in small watershed

The distribution maps of regression-kriging showed that both SOC and STN concentrations decreased from southeast to northwest in the watershed, similar to the DEM results (Figure 4). SOC and STN were strongly correlated, with a correlation coefficient of 0.6656 ($P<0.01$). Regression analysis revealed that SOC and STN had a highly significant linear relationship (Figure 6).

Table 5. Changes of SOC and STN concentrations from different land use types*.

Land use types	Samples	SOC(g·kg^{-1}) ‡	TN(g·kg^{-1})
Pinus thunbergii land	14	14.07±0.57c	1.41±0.07c
Quercus acutissima land	23	13.05±0.93bc	1.20±0.07bc
Castanea mollissima land	41	10.65±0.33ab	1.03±0.03ab
Zea mays land	11	10.70±0.41ab	1.16±0.04abc
Vegetable land	5	13.12±1.18bc	1.37±0.08c
Waste-grassland	3	9.65±1.52a	0.93±0.15a

*means within a column that contain different letters are significantly different at p<0.01;.
means within a column that contain the same letters are not significantly different at p<0.05.
"‡" SOC = soil organic carbon; STN = soil total nitrogen.

Table 6. Changes of SOC and STN concentrations at different elevations*.

Elevation(m)	samples	SOC(g·kg^{-1})‡	STN(g·kg^{-1})
220–290	35	10.70±0.36a	1.11±0.04a
290–370	29	11.32±0.42a	1.08±0.04a
370–500	33	13.71±0.55b	1.32±0.06b

*Means within a column that share different letters are significantly different at p<0.05; the same letters are not significantly different at p<0.05.
"‡" SOC = soil organic carbon; STN = soil total nitrogen.

Table 7. Changes of SOC and STN concentrations from different aspects*.

Aspect	samples	SOC(g·kg^{-1})‡	STN(g·kg^{-}1)
shady slope	36	12.06±0.50a	1.25±0.06b
half-sunny slope	44	12.03±0.45a	1.16±0.03ab
sunny slope	17	11.26±0.53a	1.08±0.05a

*Means within a column that share different letters are significantly different at p<0.05; the same letters are not significantly different at p<0.05.
"‡" SOC = soil organic carbon; STN = soil total nitrogen.

Discussion

Influences on SOC and STN spatial distributions under different land use types

A change of land-use type can generate corresponding changes of microenvironment for every land type and subsequently alter soil nutrients [32,33]. Table 5 reveals that SOC concentrations in varying land use types were ranked as *P. thunbergii* > *Q. acutissima* > vegetable > *Z. mays* > *C. mollissima* > waste-grassland. STN concentrations were ranked as *P. thunbergii* > vegetable > *Q. acutissima* > *Z. mays* > *C. mollissima* > waste-grassland. This suggests that land-use type had a significant impact on spatial SOC and STN patterns.

The variance analysis showed that differences of SOC and STN concentrations for varying land use types were identical ($P<0.01$), whereas differences among vegetable, *Z. mays* and *C. mollissima* lands were not significant ($P>0.05$). We speculate that these results may be attributed to the following four factors: 1) Forest land (such as *P. thunbergii* and *Q. acutissima*) could fix plentiful SOC and STN because of flourishing soil plant roots and a thicker forest litter layer, which were easily resolved and beneficial to SOC and STN accumulation; 2) Litterfall on *C. mollissima* land was often collected by humans, which reduced land cover and increased soil erosion with rainfall, and therefore, SOC and STN concentrations were lower than on other land use types (except waste-grassland); 3) Farmland, especially vegetable land, was close to residential areas and had increased application of chemical fertilizers, so SOC and STN concentrations were higher than in waste-grassland; and 4) Soil and water loss was very serious in the waste-grassland, so its SOC and STN concentrations were the lowest of all land use types.

Influences on SOC and STN spatial distributions at different elevations and aspects

The spatial distributions of soil nutrients were correlated with many environmental factors. Some research indicates that SOC and STN concentrations had significant positive correlations with elevation, cosine of aspect, and other topographic factors [34]. However, this research was mainly on pure forest land or farmland and used the ordinary kriging method. Little work has been conducted for a small watershed with varying elevations and aspects. Table 6 again shows that SOC and STN concentrations had significant positive correlations with elevation, which is consistent with the distribution map of those concentrations (Figure 4).

The variance analysis showed that the effects of elevation on SOC and STN concentrations were identical. For both nutrients, the 370–500 m elevation band was significantly different from 220–290 m and 290–370 m ($P<0.05$), but 220–290 m was not significantly different from 290–370 m ($P>0.05$). We suggest that these results may be due to the following factors: 1) The sample area at high elevation (370–500 m) supported SOC and STN accumulation because it had a low average soil temperature, slow nutrient decomposition and fewer human impacts; and 2) The sample area at low elevation (220–290 m) was not beneficial for accumulation due to rapid nutrient decomposition and greater human influence.

Aspect can also modify the spatial distribution of soil nutrients [35,36]. In the study, the aspects were divided into three classes at 90° intervals from due north; 0°–45° and 315°–360° were shady slopes, 45°–135° and 225°–315° were half-sunny slopes, and 135°–225° were sunny slopes. Table 7 shows that SOC and STN concentrations with different aspects were ranked as follows, shady slopes > half-sunny slopes > sunny slopes. These results are consistent with the distribution map of SOC and STN concentrations (Figure 1, Figure 4).

The variance analysis revealed that the differences of SOC concentrations with aspect were not significant ($P>0.05$). STN concentrations on half-sunny slopes were significantly different from sunny slopes ($P<0.05$), and shady slopes were significantly different from half-sunny and sunny slopes ($P>0.05$). These findings are attributed to the following factors: 1) On sunny slopes, SOC decomposes rapidly and STN concentrations are low, the opposite of conditions on shady and half-sunny slopes; and 2) fewer samples, which caused slight changes among shady, half-sunny and sunny slopes, and little significant difference of SOC and STN. Therefore, to improve the accuracy of SOC and STN spatial distributions in the study area, more attention should be given to the influences of soil temperature and the total number of samples.

Conclusion

1) Compared with the ordinary kriging method, improvement of SOC and STN prediction accuracies by regression-kriging was 16.80% and 17.24%, respectively. This demonstrates that regression-kriging is suitable for studying SOC and STN spatial distributions in a watershed with a complex topography.
2) The semi-variogram test of the regression-kriging models revealed intermediate spatial autocorrelations of both SOC and STN, and the Nugget/Sill were 0.2% and 0.1%, respectively. The test also showed that structural factors such as topography, vegetation and climate were the main factors causing SOC and STN spatial variation in the small Matiyu watershed.
3) The distribution maps of regression-kriging revealed that both SOC and STN concentrations decreased from southeast to northwest in the watershed, similar to the DEM trend.
4) Both land use types and elevation factors had significant influences on the spatial distributions of SOC and STN concentrations in the watershed. Aspect had a smaller but still appreciable effect.

Acknowledgments

The corresponding author sincerely thanks Prof. Chen Fusheng for reviewing the manuscript.

Author Contributions

Conceived and designed the experiments: GP WB. Performed the experiments: GP GG. Analyzed the data: WB GG GP. Contributed reagents/materials/analysis tools: WB GG. Wrote the paper: GP GG ZG.

References

1. Elliott (1986) Aggregate Structure and Carbon, Nitrogen, and Phosphorus in Native and Cultivated Soils. Soil Science Society of America Journal, 50: 627–633.
2. Kucharik CJ, Brye KR, Norman JM, Foley JA, Gower ST, et al. (2001) Measurements and modeling of carbon and nitrogen cycling in agroecosystems of Southern Wisconsin: potential for SOC sequestration during the next 50 years. Ecosystems, 4: 237–258.
3. Lal R (2004) Soil carbon sequestration impacts on global climate change and food security. Science, 304: 1623–1627.
4. Cao LH, Zhao SW (2007) Progress of study on factors of affecting the SOC pool and measures for its regulating and controlling. Journal of Northwest Agriculture and Forestry University (Nat. Sci. Ed.), 35 (3): 177–182.
5. Zhao Z, Li P, Wang NJ (2000) Distribution patterns of root systems of main planting tree species in Weibei Loess Plateau. Chinese Journal of Applied Ecology, 11 (1): 37–39.
6. Wang HJ, Shi XZ, Yu DS, David C, Huang B (2009) Factors determining soil nutrient distribution in a small-scaled watershed in the purple soil region of Sichuan Province, China. Soil & Tillage Research, 105: 300–306.
7. Dou FG, Yu X, Chen LP, Gary M, Guo LG, et al. (2010) Spatial variation of tundra soil organic carbon along the coastline of northern Alaska. Geoderma, 154: 328–335.
8. Gebreyesus BT, Lulseged T, Paul LG (2011) Catchment-scale spatial variability of soil properties and implications on site-specific soil management in northern Ethiopia. Soil & Tillage Research, 117: 124–139.

9. Mohammadi J, Motaghian MH (2011) Spatial Prediction of Soil Aggregate Stability and Aggregate-Associated Organic Carbon Content at the Catchment Scale Using Geostatistical Techniques. Pedosphere, 21(3): 389–399.
10. Fantappiè M, L'Abate G, Costantini EAC (2011) The influence of climate change on the soil organic carbon content in Italy from 1961 to 2008. Geomorphology, 135 (3–4): 343–352.
11. Wang HQ, Hall CAS, Cornell JD, Hall MHP (2002) Spatial dependence and the relationship of soil organic carbon and soil moisture in the Luquillo, Experimental Forest, Puerto Rico. Landscape Ecol., 17: 671–684.
12. Tan ZX, Lal R (2005) Carbon sequestration potential estimates with changes in land use and tillage practice in Ohio, USA. Agr. Ecosyst. Environ., 111: 140–152.
13. Liu DW, Wang ZM, Zhang B, Song KS, Li XY, et al. (2006) Spatial distribution of soil organic carbon and analysis of related factors in croplands of the black soil region, Northeast China. Agr. Ecosyst. Environ., 113: 73–81.
14. Su ZY, Xiong YM, Zhu JY (2006) Soil organic carbon content and distribution in a small landscape of Dongguan, South China. Pedosphere, 16: 10–17.
15. Stacey KF, Lark RM, Whitmore AP, Milne AE (2006) Using a process model and regression kriging to improve predictions of nitrous oxide emissions from soil. Geoderma, 135: 107–117.
16. Harris P, Fotheringham AS, Crespo R, Charlton M (2010) The use of geographically weighted regression for spatial prediction: An evaluation of models using simulated data sets. Mathematical Geosciences, 42: 657–680.
17. Kumar S, Lal R, Liu DS (2012) A geographically weighted regression kriging approach for mapping soil organic carbon stock. Geoderma, 189: 627–634.
18. Hengl T, Heuvelink GBM, Stein A (2004) A generic framework for spatial prediction of soil variables based on regression-kriging. Geoderma, 120: 75–93.
19. Sumfleth K, Duttmann R (2008) Prediction of soil property distribution in paddy soil landscapes using terrain data and satellite information as indicators. Ecological Indicators, 8(5): 485–501.
20. Simbahan GC, Dobermann A, Goovaerts P, Jianli P, Haddix ML (2006). Fine-resolution mapping of soil organic carbon based on multivariate secondary data. Geoderma, 132 (3–4):471–489.
21. Wang YQ, Zhang XC, Zhang JL, Li SJ (2009) Spatial Variability of Soil Organic Carbon in a Watershed on the Loess Plateau. Pedosphere, 19 (4): 486–495.
22. Li Y (2010) Can the spatial prediction of soil organic matter contents at various sampling scales be improved by using regression kriging with auxiliary information? Geoderma, 159, 63–75.
23. Sun W, Minasny B, McBratney A (2012) Analysis and prediction of soil properties using local regression-kriging. Geoderma, 171–172, 16–23.
24. Wang K (2013) Application of Regression Kriging on the Spatial Prediction of Total Soil Nitrogen. Chinese Agricultural Science Bulletin, 29(20): 142–147.
25. Liu S, Ren HY, Shi XZ, Pan JJ (2010) Different methods for prediction of spatial patterns of paddy soil organic carbon density in Changxing County of Zhejiang Province. Journal of Geo-information Science, 12(2): 180–185.
26. Nelson DW, Sommers LE (1982) Total carbon, organic carbon and organic matter. Agronomy Monograph No. 9. ASA and SSSA, Madison, WI., 534–580.
27. Lu WD (2005) SPSS 12.0 for windows statistics analysis. Beijing: Electron Industry Press, 1–135.
28. Zhao YC, Shi XZ, Yu DS (2005) Different methods for prediction of spatial patterns of soil organic carbon density in Hebei Province, China. Acta Pedol. Sin., 42: 379–385.
29. Batjes NH, Sombroek WG (1997) Possibilities for carbon sequestration in tropical and subtropical soils. Global Change Biology, 3 (2): 161–173.
30. Guo LB, Gifford RM (2002) Soil carbon stocks and land use change: a meta analysis. Global Change Biology, 8 (4): 345–360.
31. Fang X, Xue Z, Li B, An S (2012) Soil organic carbon distribution in relation to land use and its storage in a small watershed of the Loess Plateau, China. Catena 88: 6–13.
32. McGrath DA, Smith CK, Gholz HL, Oliveira FA (2001) Effects of land-use change on soil nutrient dynamics in Amazonia. Ecosystems, 4: 625–645.
33. Odeh IOA, McBratney AB, Chittleborough DJ (1995) Further results on prediction of soil properties from terrain attributes: heterotopic cokriging and regression-kriging. Geoderma, 67, 215–225.
34. Hicks RR, Frank PS (1984) Relationship of aspect to soil nutrients, species importance and biomass in a forested watershed in West Virginia. Forest Ecology and Management, 8 (3–4): 281–291.
35. Wang J, Fu B, Qiu Y, Chen LD (2003) Analysis on soil nutrient characteristics for sustainable land use in Danangou catchment of the Loess Plateau, China. CATENA, 54 (1–2): 17–29.
36. Zhang S, Huang Y, Shen C, Ye H, Du Y (2012) Spatial prediction of soil organic matter using terrain indices and categorical variables as auxiliary information. Geoderma, 171–172, 35–43.

8

Uncontacted Waorani in the Yasuní Biosphere Reserve: Geographical Validation of the *Zona Intangible Tagaeri Taromenane (ZITT)*

Salvatore Eugenio Pappalardo[1]*, Massimo De Marchi[2,3], Francesco Ferrarese[4]

1 Department of Agronomy, Food, Natural Resources, Animals and the Environment, University of Padova, Padova, Italy, **2** Department of Civil, Environmental Architectural Engineering, University of Padova, Padova, Italy, **3** Prometeo Project, Universidad Nacional de Chimborazo, Riobamba, Ecuador, **4** GIS Laboratory, Department of Historical, Geographical and Antiquity Sciences, University of Padova, Padova, Italy

Abstract

The Tagaeri Taromenane People are two indigenous groups belonging to the Waorani first nation living in voluntary isolation within the Napo region of the western Amazon rainforest. To protect their territory the Ecuadorean State has declared and geographically defined, by Decrees, the *Zona Intangible Tagaeri Taromenane* (ZITT). This zone is located within the UNESCO Yasuní Biosphere Reserve (1989), one of the most biodiverse areas in the world. Due to several hydrocarbon reserve exploitation projects running in the area and the advancing of a large-scale deforestation front, the survival of these groups is presently at risk. The general aim was to validate the ZITT boundary using the geographical references included in the Decree 2187 (2007) by analyzing the geomorphological characteristics of the area. Remote sensing data such as Digital Elevation Models (DEM), Landsat imagery, topographic cartography of IGM-Ecuador, and fieldwork geographical data have been integrated and processed by Geographical Information System (GIS). The ZITT presents two levels of geographic inconsistencies. The first dimension is about the serious cartographical weaknesses in the perimeter delimitation related to the impossibility of linking two rivers belonging to different basins while the second deals with the perimeter line not respecting the hydrographic network. The GIS analysis results clearly show that ZITT boundary is cartographically nonsense due to the impossibility of mapping out the perimeter. Furthermore, GIS analysis of anthropological data shows presence of Tagaeri Taromenane clans outside the ZITT perimeter, within oil production areas and in nearby farmer settlements, reflecting the limits of protection policies for non-contacted indigenous territory. The delimitation of the ZITT followed a traditional pattern of geometric boundary not taking into account the nomadic characteristic of Tagaeri Taromenane: it is necessary to adopt geographical approaches to recognize the indigenous right to their liveable territories in the complex territorialities enacted by different stakeholders.

Editor: Dennis O'Rourke, University of Utah, United States of America

Funding: This work was supported by the University of Padova, Veneto Region (ESF). The funder had no role in study design, data collection and analysis, decision to publish, or preparation of the manuscript.

Competing Interests: The authors have declared that no competing interests exist.

* E-mail: eugenio.pappalardo@geoyasuni.org

Introduction

The Tagaeri Taromenane are two indigenous groups living in "Voluntary Isolation" in the Napo Moist terrestrial ecoregion, located in the western Amazon rainforest [1–5]. This ecoregion is one of the most biologically and culturally diverse areas on the planet [6,7] representing an extraordinary richness across several taxa (amphibian, mammal, bird and plants), a high level of regional endemism [8–10], and is home to several indigenous ethnic groups including some of the world's last uncontacted peoples [11].

The Tagaeri Taromenane are settled in the Ecuadorean Amazon Region between the Yasuní and Curaray rivers and within the ancestral territory of the Waorani (or Huaorani) indigenous first nation [12,13].

To protect the territory of these uncontacted indigenous people in the Yasuní Biosphere Reserve, the national government declared (1999) and mapped (2007), through two Presidential Decrees, the "No-Go-Zone" called Intangible Zone or *Zona Intangible Tagaeri Taromenane* (ZITT) [14,15].

Decree 2187 of 2007 represents the first official text that uniquely gives, using a geographical language, spatially explicit information to define the perimeter of ZITT.

The boundary of the ZITT is clamped onto 17 univocal points geographically expressed by pairs of metrical coordinates declared in the Decree 2187 (Figure 1A). To map the borders of the ZITT it is required to join the pairs of coordinates by rectilinear segments and by sections of river courses according to the geomorphological descriptions included in the Decree 2187.

Due to different interpretations of the geographical references expressed in the Decree 2187 and to the complex geomorphological characteristics of the Amazon sub-basins, different map representations of the ZITT are demarcated both in the official and unofficial cartography of the Yasuní Biosphere Reserve [16–20].

Several studies about the ZITT and the issue on the uncontacted indigenous groups have been produced using mainly social and anthropological approaches [2,3,12,13,21–25]. No

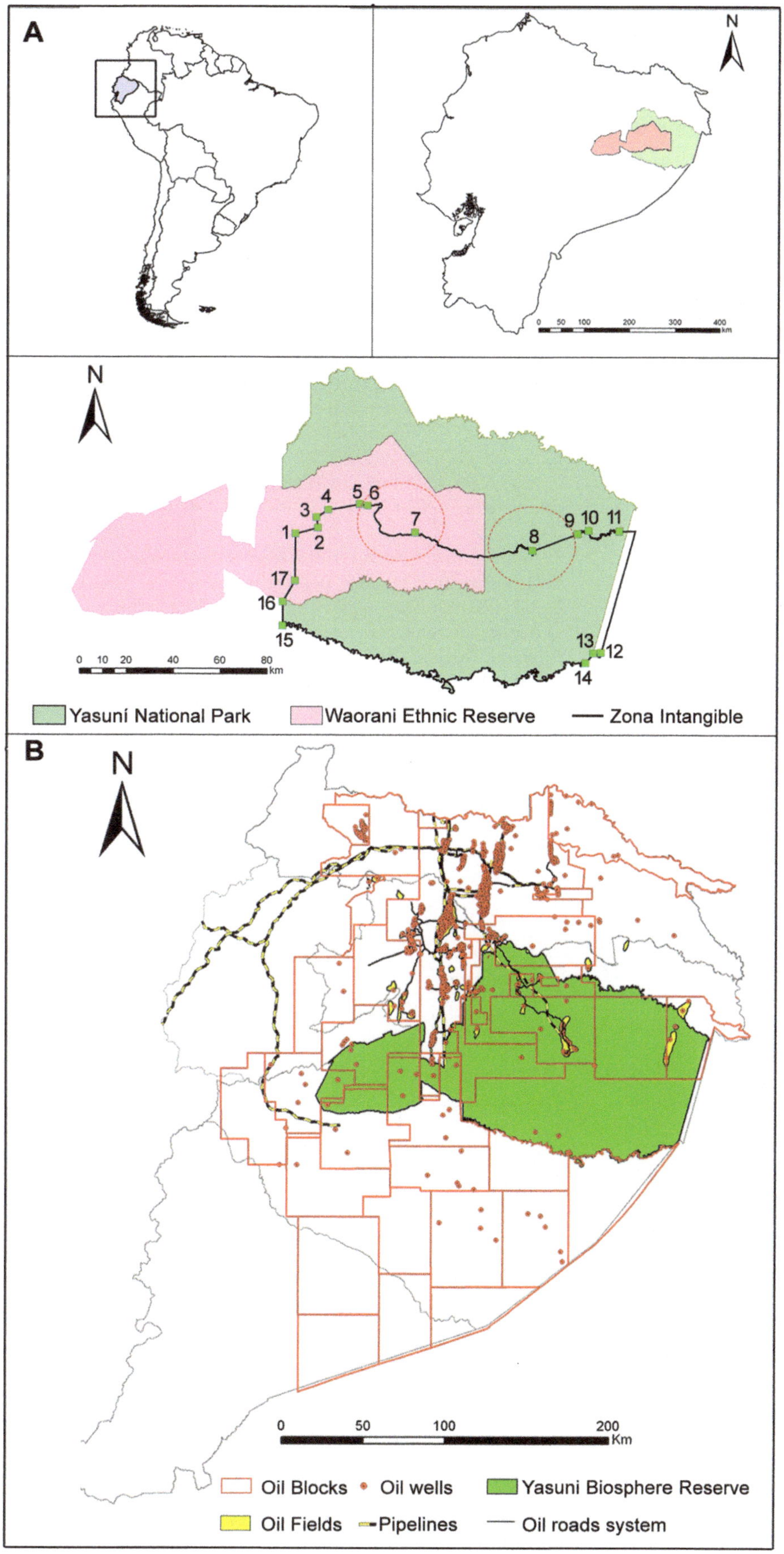
A
N
0 50 100 200 300 400 km
N
1 2 3 4 5 6 7 8 9 10 11 12 13 14 15 16 17
0 10 20 40 60 80 km
Yasuní National Park
Waorani Ethnic Reserve
Zona Intangible
B
N
0 50 100 200 Km
Oil Blocks
Oil wells
Yasuni Biosphere Reserve
Oil Fields
Pipelines
Oil roads system

Figure 1. Yasuní Biosphere Reserve and *Zona Intangible Tagaeri Taromenane* (ZITT): geographical framework. A) Delimitation of the *Zona Intangible* by given geographical coordinates and critical hotspots (red circles); B) Oil production in the Ecuadorian Amazon Region (9th concession licensing round, 2001).

geographical studies and GIS spatial analysis on the ZITT have been produced in the scientific literature.

Fieldwork activities in the Yasuní Biosphere Reserve, geographical analysis of the Decree 2187, and studies on geomorphological characteristics of the area suggested the hypothesis that the official boundary of the ZITT is cartographically inaccurate with serious geographical inconsistencies along the perimeter.

Crucial geographical inconsistencies seem localized between points No. 6 and No. 7 and between No.7 and No. 8.

The general aim of this paper is to validate the boundary using the geographical references expressed in the Decree 2187 of 2007 by analyzing the geomorphological characteristics of the area in order to re-map the perimeter of the ZITT through spatial and remote sensing analysis.

The GIS analysis results clearly show geographical incoherencies in two hotspots of the Yasuní and unequivocally demonstrate the impossibility to map the perimeter of the ZITT.

Analyses of both anthropological geospatial data about the Tagaeri Taromenane and their spatial relationships with extractive activities and roads confirm the existence of uncontacted groups in a wider area crossing oil fields and farmer settlements outside the same boundary of the *Zona Intangible*.

Furthermore, violent contacts between Tagaeri Taromenane group and external *mestizo* farmers within the territory district of Dayuma (close to the Hormiguero extractive platform and the Armadillo oil field in 2009) [26–28] and maybe with Waorani relative (in 2013 within oil block 16, near the Maxus Road) happened in the area of Yasuní. Between the 4th and 5th of March 2013 two old Waorani were speared to death. The attack is under investigation by Ecuadorian Authorities to confirm responsibility of Tagaeri Taromenane groups [29]. These facts confirmed once more the presence of the Tagaeri Taromenane also outside the ZITT perimeter (Figures S1, S2), fuelling a hot debate inside national society questioning the relationship among the majority and minorities and asking for a different way of conceiving territories: from the Westphalian rigid geometry to a geographic network of cultural and environmental relationships.

The overlapping systems of territorial complexities

Through the Man and Biosphere Program, the United Nation's Agency UNESCO included this geographical area into the Biosphere Reserve network (1989), declaring the Yasuní National Park and the land-titled of Waorani territory (known as Waorani Ethnic Reserve) as Yasuní Biosphere Reserve. It was added in order to prioritize and to conciliate biodiversity conservation with sustainable development in the territory planning [30,31].

However, due to huge non-renewable energy reserves and to the crucial role they play in the national economy [32–34], the Ecuadorean State zoned specific geographical areas in the Amazon Region for hydrocarbon industrial activities – the so called "oil blocks" – so that, according to the 9th licensing round established in 2001, roughly the 79% of the YBR [35–37] was overlapped by concessions for oil extraction and production (Figure 1B) [38].

Oil development in the Ecuadorian Amazon (Figure 1B) has already directly and indirectly caused major social and environ-

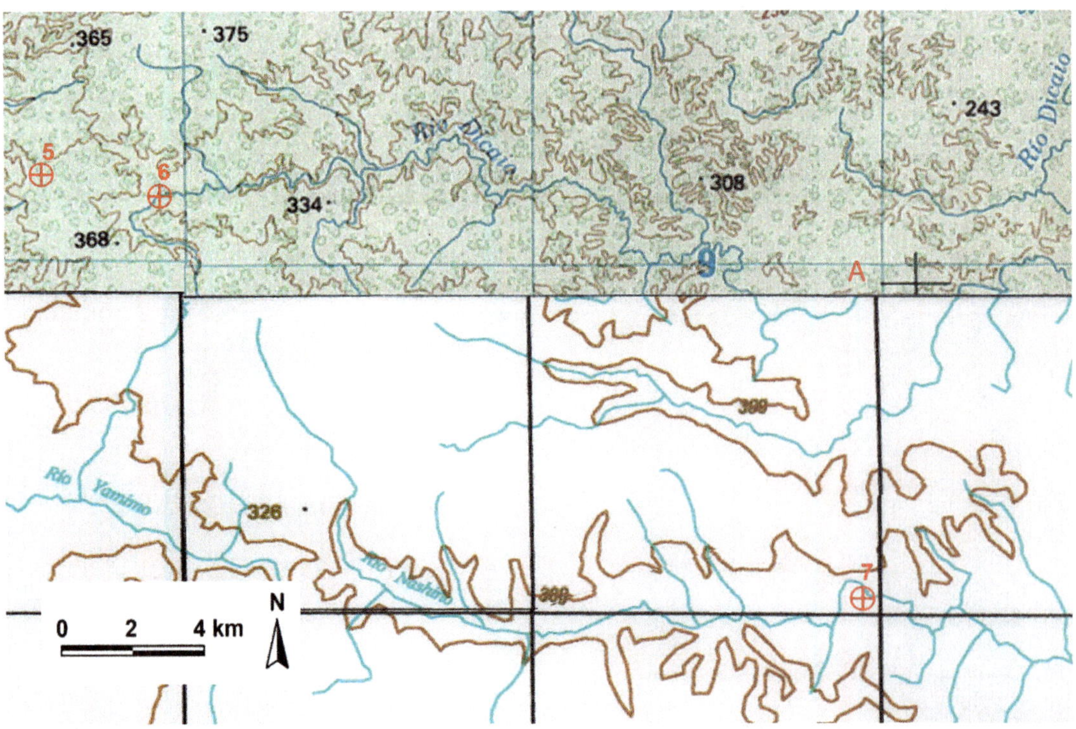

Figure 2. Perimeter of *Zona Intangible Tagaeri Taromenane* (ZITT) including points No. 5, No. 6, No. 7. (IGM Ecuador, scale 1:250,000). Points No. 6 and No. 7 belong to different rivers: the former to the Rio Dicaro, the latter to the Rio Nashiño.

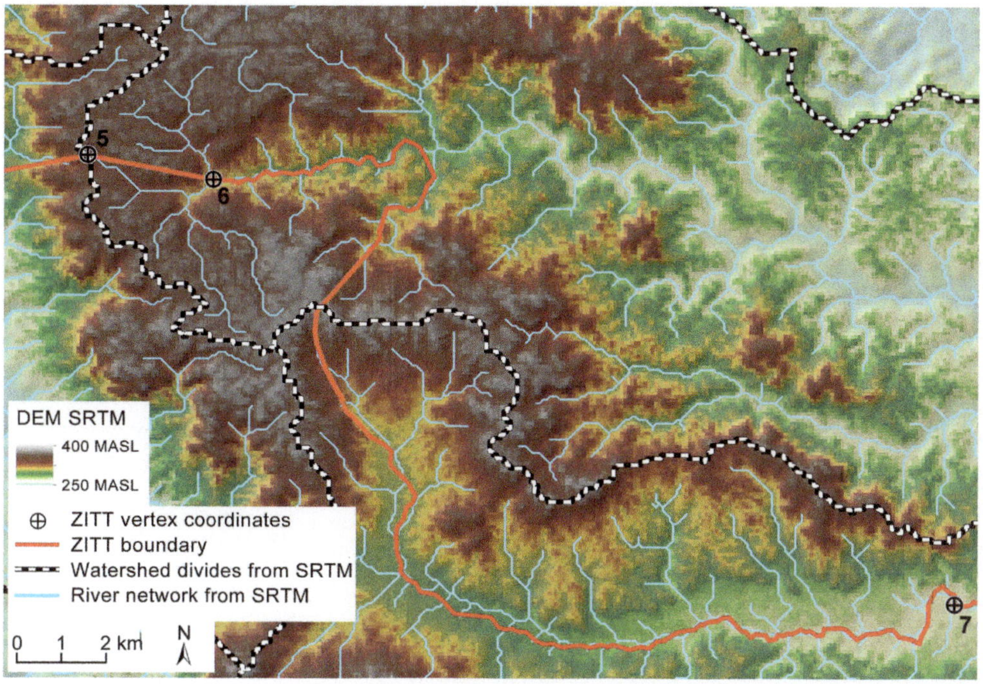

Figure. 3. Digital Elevation Model (SRTM) in the area of points No. 5 and 6. DEM analysis of river basins and catchment's divides: Rio Dicaro and Rio Nashiño belong to different basins: the first one flows into the Yasuní River, the second one into the Curaray River. Therefore "following downstream the Rio Dicaro till point No. 7 by the Rio Nashiño" (Art. 1, indent 3) [15] becomes a "geographical nonsense" in mapping out the perimeter section from points No. 6 to that No. 7.

mental change to the Amazon people and ecosystems. Direct effects include deforestation for drilling platforms, pipelines, access roads, seismic prospecting activities, and chemical contamination of water bodies from wastewater discharges, oil spills and gas

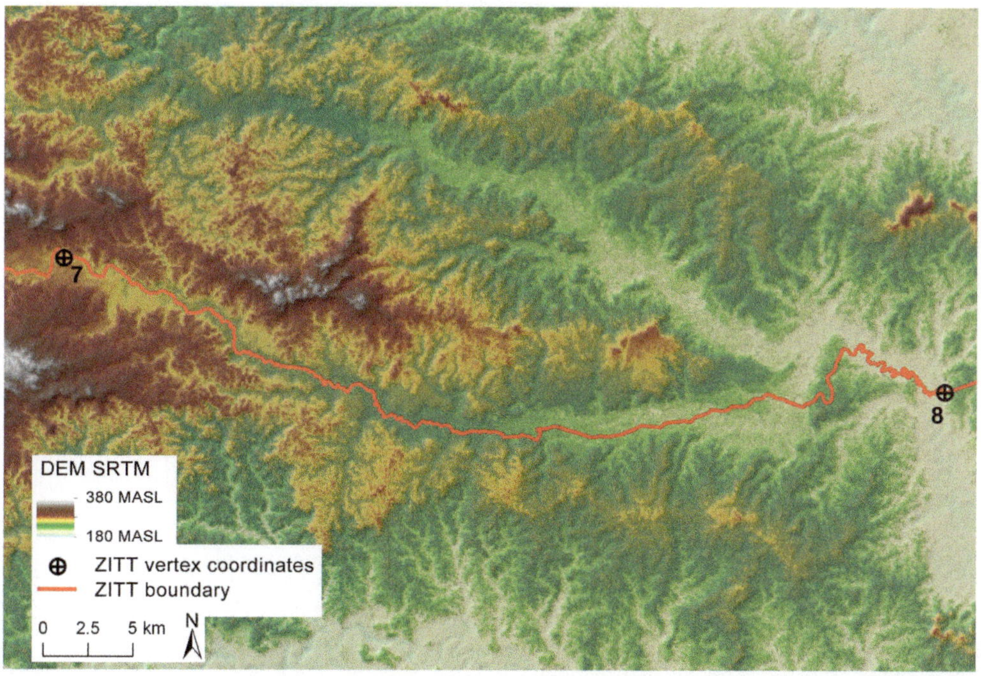

Figure 4. Digital Elevation Model (SRTM) analysis between points No. 7 and 8: a critical mapping out of the *Zona Intangible Tagaeri Taromenane* (ZITT). The DEM analysis of river dynamic clearly shows that the boundary section, rather than overlapping the Rio Nashiño river bed, drifts away from the river course running over a hills line upper slope. This mapping out violates both the terrain morphology and the same official text (Decree 2187, 2007).

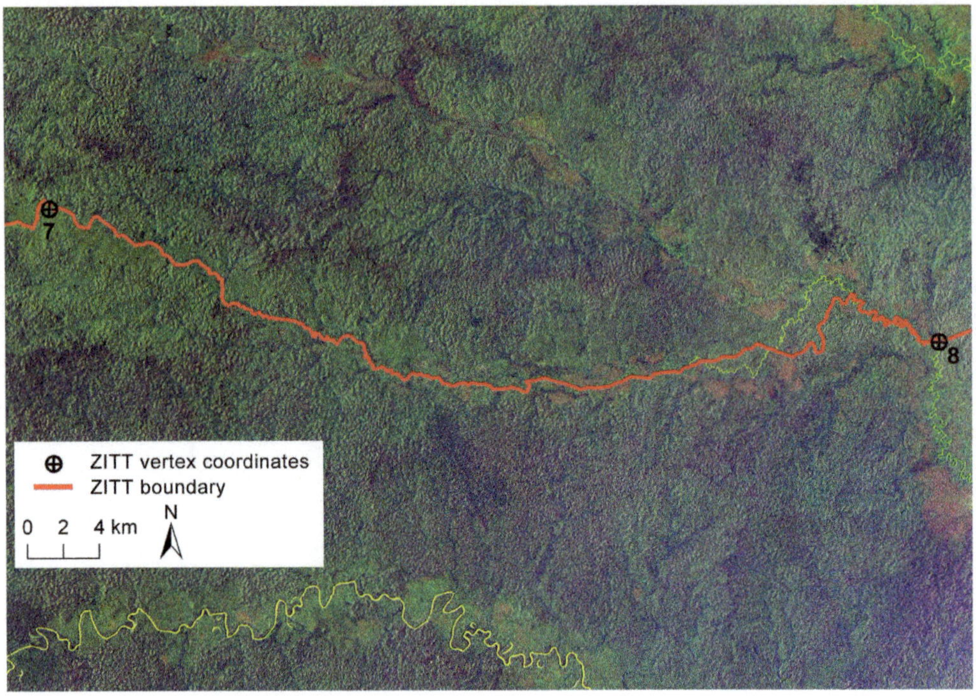

Figure 5. Spatial analysis of perimeter section between points No. 7 and No. 8 processing satellite scenes from Landsat TM 5. Landsat Thematic Mapper (TM) false color composite (4 blue, 5 green, 7 red) of the territory encompassed between points No. 7 and No. 8 in the southern sector of the Oil Block 31 (PetroAmazonas). Boundary section (red color line) does not match the Rio Nashiño river bed (yellow color line).

emissions [39–43]. Indirect effects are related to the opening of roads for oil exploration and transportation which turns terrestrial communications infrastructures into the main vector for colonization of primary forest [44–46], represented by the Moist Tropical Forest. Colonization of Moist Tropical Forest is causing increased deforestation as well as logging and hunting from human settlements [47–50]. As it is interpreted by the satellite image of the Via Auca Road (Landsat ETM+, 2002; Figure S1) – the most important oil terrestrial infrastructure at the western sector of the Yasuní Biosphere Reserve – colonization processes of the forests are usually driven by the main road axis [51]. They represent the main driver of extensive deforestation activity traced out by orthogonal and parallel processes resulting in land cover/land use change dynamics in the typical deforestation spatial pattern called fish-bone (Figure S1) [52–56]. Moreover, rapid expansion of Via Auca road system to the west underlies direct anthropogenic pressure to Tagaeri Taromenane mobility space [38]. Previous GIS analysis by processing GPS data of the Via Auca road network showed, between 2009 and 2012, a spatial evolution to the west of 100 linear km (Figure S2) [38,57]. Land use and land cover change dynamics result in agricultural activities both at intensive cultivation level (i.e. African Palm, *Elaeis guineensis* bot. sp.) and non-traditional farming (Figures S1, S2).

Therefore, hydrocarbon reserves exploitation, by its direct and indirect impacts on tropical forest ecosystems, played a pivotal role in turning the Napo ecoregion into one of the 14 mayor deforestation fronts in the world [58–60].

This set of socio-environmental processes taking place in the Ecuadorian Amazon (especially in the Auca territory system) is also affecting the social reproduction and the same survival of the Tagaeri Taromenane, clans of uncontacted hunters, and collectors living by semi-nomadic lifestyle in a wide territory of the Yasuní Biosphere Reserve's western and southern sector [3,13,61].

Additionally, in the last decades, several violent contacts between uncontacted indigenous people and external stakeholders induced the Inter-American Commission on Human Rights (IACHR, 2006) to grant precautionary measures in favor of the Tagaeri Taromenane demanding that "the Ecuadorian State adopt measures necessary to protect the territory inhabited by beneficiaries from third parties" [62].

The institutional history of the Intangible Zones began in the 1999 by the Decree 552, declaring the *Zona Intangible* to protect the territory of the uncontacted Tagaeri Taromenane indigenous groups. This Decree stated to identify and geographically define the ZITT within a time range of 120 days from the 2nd of February 1999 [14]. After 4 years, by the Inter-ministerial Agreement 092, a Technical Commission was saddled with establishing the perimeter of the ZITT (Quito, 2004) along with the task of identifying criteria including the integration of monitoring and controlling systems of the area [63]. The delimitation of the ZITT has been a long and controversial process [2,64] complicated by the fact that this area is "geographically embedded" within the industrial concession areas for oil exploitation [65]. On the north and west sides, there are oil companies operating in industrial production for the past three decades while on the south side there are the exploratory blocks which represent, by the 11th licensing round opened in November 2012 [66], a spatial shift to the Southern Amazonian sector of the extractive frontier (Figure 1B) [67–70].

Finally, a geographic delimitation of the ZITT was issued on the 16th of January 2007 through the Decree 2187 signed by the President of the Republic Alfredo Palacio [15]. To mitigate external anthropic influences, a buffer zone of 10 km has been defined around its perimeter. Within this additional protected area, wood extraction activities and new oil concessions are forbidden, while traditional activities such as hunting, fishing, and

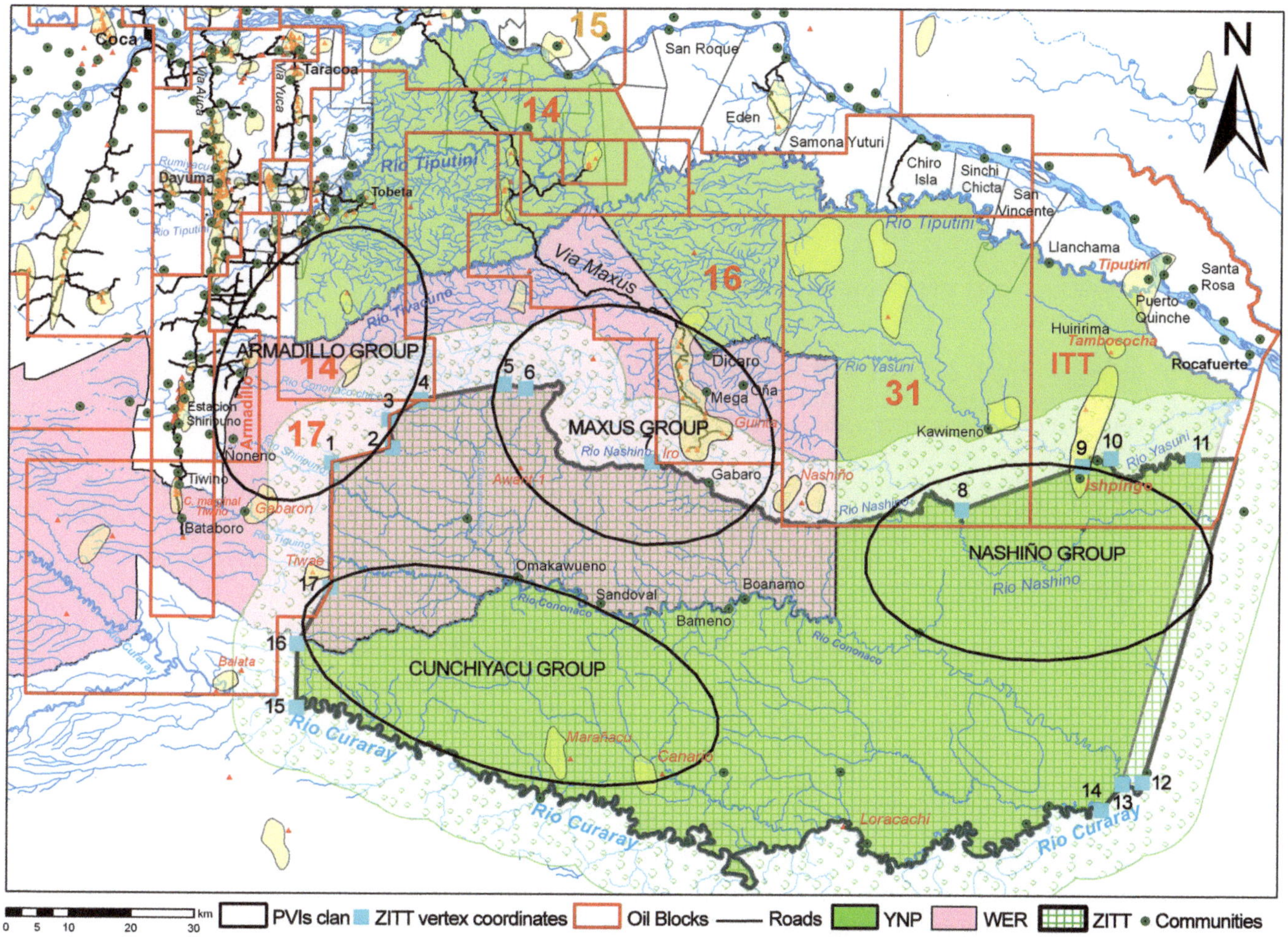

Figure 6. Synthesis map: Yasuní National Park (YNP), Waorani Ethnic Reserve (WER), *Zona Intangible Tagaeri Taromenane* (ZITT) and oil blocks. The overlapping systems of territorial complexity in and around the Yasuní Biosphere Reserve: the Zona Intangible (ZITT) is geographically adjoined with productive oil concessions area at the north-west side; on the contrary the north-east sector overlaps both with oil blocks No. 31 (operated by the Petroamazonas national company) and that ITT block (Ishipingo-Tiputini-Tambococha block, Secreteria de Hidrocarburos of Ecuador). The spatial distribution of the Tagaeri Taromenane clans is partially outside the ZITT perimeter in the cases of the Nashiño and the Maxus groups; the Armadillo group is completely outside the ZITT boundary overlapping oil blocks No. 14 and No. 17. Their home range overlaps several titled and untitled land of mestizo (colonos) farmers and non-autochthonous indigenous settlements in the Dayuma and Ines Arango districts, implicating a sensible proximity with roads system related to the Via Auca main axis. The presence of the Armadillo group in this area is also confirmed by recent violent contacts between the Hormiguero oil platform and the Armadillo oil field.

traditional use of biodiversity is allowed to the ancestral indigenous communities (the Waorani first nation) (Art. 2, Decree 2187) [15].

According to the Ecuadorean Decree the Intangible Zones are defined as "protected areas of high cultural and biological importance in which whatever extractive activities are forbidden due to its high value for the Amazon rainforest, Ecuador, the World and the present and future generations" (1999, Presidencia de la Repubblica) [71]. It is therefore an "untouchable zone" in which oil, gas, and other external anthropic activities must be off limits.

The ZITT is also a crucial issue because it is related to the "Yasuní-ITT initiative" undertaken by the Ecuadorean Government (2007) [72–74], an international initiative oriented to protect biodiversity, recognize indigenous peoples' territory, and contrast climate change. The "Yasuní-ITT Initiative" proposal aims to keep locked underground, in perpetuity, at least 850 million barrels of heavy crude oil. This will prevent the emission of CO_2 associated with burning fossil fuels of about 410 million metric tons of out of the atmosphere, in exchange of financial compensation from the developed countries concerned about tropical deforestation and climate change [75,76].

Results and Discussion

Cartographic analysis of the 17 vertex coordinates

The boundary of the ZITT is cartographically defined by Art. 1 of Decree 2187. The perimeter is spatially determined by 17 given points expressed by metric coordinates (UTM projection, zone 18, PSAD1956 geographic projection system) which have been cartographically validated. They are all geographically plausible and do not have any transcription errors. On the one side, all of the 17 vertex coordinates are consistent both with the geomorphological description and the measurements of straight lines included in the text of Decree 2187; however, on the other side sections of river courses linking vertex coordinates are in some segments ambiguous. In the latter case, mapping out the boundary becomes more complicated.

Marking out river segments of the ZITT boundary imply the use large scale maps (1:50,000 or 1:25,000), including a detailed representation of hydrographic networks, the relief system, and toponyms. Due to a complex and vast territory characterized by the typical Moist Tropical Forest land cover and by a hydrographic system evolving into a very dense network, geographic interpretation could be difficult for cartographers since it is easily prone to cartographic errors.

Cartographic delimitation of the ZITT starts from point No. 1 which reaches by a straight line point No. 2. From here the boundary overlaps the Cononaco Chico River to the north until point No. 3. This segment of the limit is short (4,550 meters) and it does not create any problems of interpretation due to the importance of this river which is well represented in the cartography production of IGM. From point No. 3 to No. 4, the boundary is marked out by a straight line of 6,140 meters according to the given coordinates. The border continues from point No. 4 to No. 5 again by a straight line. From point No. 5 to No. 6 the perimeter line follows once again a straight line "till reaching the confluence of two tributaries of the Rio Bahameno in point No. 6, having 9892355 North and 339450 Este" (2nd indent, Art. 1, Decree. 2187) (Document S1). We noticed that, although the linear distance between these points is consistent to the values of the given coordinates (Table S1) and point No. 6 exactly lies on a river confluence, the two tributaries do not belong to the Rio Bahameno but to the Rio Dicaro (Figure 2, 3). As a matter of fact, it is the same geomorphological description in the official text stating that from point No. 6 the boundary follows the Rio Dicaro downstream in the east direction.

The most controversial part has been identified in the perimeter section between point No. 6 and No. 7: "(the boundary) follows to East direction by the Rio Dicaro course, then it runs downstream by the Rio Nashiño till reaching point No. 7" (3rd indent, Decree 2187) (Document S1) [15]. As mentioned above, the ZITT limit runs downstream along the Rio Dicaro in the east direction. Once proved that point No. 6 matches up with the riverbed of the Rio Dicaro (not the Rio Bahameno) and point No. 7 properly overlies the Rio Bahameno, it is important to understand how it has been operated the leap of river basins. It is clear that the Rio Dicaro and Rio Nashiño belong to two different river basins. The first one flows into the Rio Yasuní and the second into the Rio Curaray.

Assuming the linear distance between point No. 6 and No. 7 is about 23,300 meters (Table S1), this particular step of the Decree 2187 technically gives infinite solutions. Therefore, if there is no better explanation, this perimeter section becomes cartographically unresolved. In any case, it is not possible to follow downstream the Rio Dicaro until point No. 7. This discrepancy results in geomorphological inconsistencies if it is not made explicit where to leave the Rio Dicaro course.

From point No. 7 to No. 8 the boundary follows the river course of Rio Nashiño. The distance corresponds to more than 50 km (Table S1). From this point the limit continues until point No. 9 by a straight line approximately 20.6 km to the north-east direction, according to the values of coordinates declared in the Decree 2187.

From point No. 9 to No. 10 the ZITT perimeter follows the river course of Rio Yasuní and from that point by a straight line of 6,850 meters reaches point No. 11. This point coincides to the same state border between Ecuador and Peru.

From point No. 11 to that No. 12 the boundary is traced out by a straight line of 53,978 meters oriented in the south-west direction. The given coordinates of point No. 12 matches the river course of Rio Cononaco which has to be followed upstream to point No. 13. From here, by a straight line of 5,438 meters, the ZITT limit reach point No. 14 which overlaps the course of Rio Curaray. From this point the perimeter section follows the Curaray River upstream in the west direction until point No. 15. From point No. 15 to No. 16 the boundary is trace out by a straight line 10,190 meters long orientated in the north direction. It then heads north-east by another straight segment of 10,500 meters to reach point No. 17. From this point to that No. 1 the limit line close the perimeter by another straight line headed north.

According to geographic interpretation of spatially explicit data contained in Art. 1 of the Decree 2187, the ZITT perimeter from point No. 7 to No. 17 does not show cartographic inconsistencies nor geomorphological ones.

Rio Dicaro and Rio Nashiño: the river basins analysis

Basins and river dynamics analyses have been carried out to clarify the most critical perimeter section of the ZITT expressed between points No. 6 and No. 7.

The GIS analysis is based on comparing three different Digital Elevation Models (DEM) that clearly represent the terrain morphology: an IGM DEM map [77], an Aster GDEM (second version) map [78], and a SRTM map [79].

The DEM analysis of river basins spatial relationships clearly show the catchment's divides of the Rio Dicaro and the Rio Nashiño river basins. Using the results from three different DEM, maps demonstrate that it is not possible to operate a leap of basin to link out points No. 6 to that No. 7 (Figures 3, S3); however, according to the Decree 2187 text, it is clear that somewhere this leap has to be done.

Therefore, "following downstream the Rio Dicaro till point No. 7 by the Rio Nashiño" (Art. 1, indent 3, Decree 2187) (Document S1) [15] becomes "geographical nonsense" in mapping out the perimeter section from points No. 6 to that No. 7. Due to this geographic ambiguity, different representations of the ZITT boundary in this area have been mapped out in official and unofficial cartography production on the Yasuní Biosphere Reserve [16–20].

Perimeter section between points No. 7 and No. 8: a critical mapping out

According to the Decree 2187 the ZITT boundary section between points No. 7 and No. 8 has to run exactly along the riverbed of the Rio Nashiño and consequently following the bottom valley line. On the contrary, the official cartographic representation of the ZITT boundary [17,71,80] seems to have mapped out a different limit line on the section close to point No. 7.

To validate this section of the limit by remote sensing technologies, a geospatial analysis has been developed using both DEM cartography (SRTM and GDEM maps) and satellite scenes (Landsat 5 TM) in order to deeply investigate the terrain morphology and the river dynamics in this area.

To deploy a crossed spatial analysis we have examined four Landsat 5 TM satellites scenes in the years 1990, 2004, 2005, and 2008. The DEM analysis of river dynamics clearly shows that the boundary section drifts away from the Rio Nashiño riverbed running over a hill's ridge (Figures 4, S4).

The Rio Nashiño river course is consistent with all the analyzed time series satellite scenes. Despite the previous cartographic base used to trace out the ZITT limits, this boundary section does not match with results obtained by remote sensing analysis of Landsat 5 TM satellite imagery (Figures 5, S5), SRTM DEM, nor Aster

GDEM. Following its own trace violates both the terrain morphology and the same official text (Decree 2187).

The Zona Intagible Tagaeri Taromenane and the Yasuní Biosphere Reserve: a synthesis map

The nine thematic variables assembled in the synthesis map give rise to the multiple territories setting up the Yasuní Biosphere Reserve space: the hydrocarbon dimension represented by concession, oil fields and wells for extraction and re-injection, protected areas recognized as IUCN category II for conservation of biodiversity by Yasuní National Park [65] and the Waorani indigenous territory by the titled-land (Waorani Ethnic Reserve), and finally the *Zona Intangible* for the Tagaeri Taromenane. Moreover, geodata about the historic incidents between Tagaeri Taromenane and external actors has been included in the map.

The hydrographic network (unfortunately the map presents a spatial gap in the south-east sector of the Yasuní Biosphere Reserve due to the unavailability of data) has been included in the map background because sections of river are often used as a "natural boundary" to delimitate protected areas (Yasuní National Park, Waorani Ethnic Reserve and ZITT) and they also play a crucial role in the indigenous territory building [56,81–83]. Furthermore, according to anthropological studies, we considered Tagaeri Taromenane as indigenous groups who traditionally establish their territory along "inter-fluvial" spaces [3,65,84].

The road network is represented by the complex of the Via Auca (at the west sector, built in 1972), Via Maxus (north-west sector, built in 1994), Via "Oxy" (built in 2003 by the Occidental Petroleum company), and the Via "Petrobras" (built in 2004 by the homonymous Brazilian oil company). All the roads in the Yasuní Biosphere Reserve are important terrestrial infrastructures for hydrocarbon reserves exploitation [38,85,86].

As it is shown in the synthesis map (Figure 6), presently the ZITT is geographically adjoined with productive oil concessions confirming the overlapping territorial policies. GIS analysis of spatial relationships between oil block areas and the *Zona Intangible* demonstrates that – except with blocks No. 16, No. 14 and No. 17 which seem to cut out the same ZITT perimeter – hydrocarbon concessions No. 31 (Petroamazonas) and the ITT block (managed by Secretaria de Hidrocarburos of Ecuador) result in, at the present time, an overlap of 8,542 and 9,952 ha (Figure 6; Table 1). It is relevant to clarify that the actual spatial dimension of oil blocks reflects the most recent changes operated by SHE in 2011 [87]. Before these relevant changes, previous GIS analysis had calculated a wider overlap of 84,789 ha by the overall oil blocks, corresponding to 10% of the ZITT. Areal overlap becomes greater if we consider the 10 km buffer zone of ZITT – 190,929 hectares, corresponding to 22,7% (Table 1) [20].

GIS analysis of the spatial distribution of the four Tagaeri Taromenane clans [88] shows geographical weaknesses of the ZITT perimeter and the Buffer Zone in protecting the PVIs territory (Figure 6). As it is disclosed in the synthesis map, the ZITT area does not completely fit with the home range of Tagaeri Taromenane. The Cunchiyacu group is settled within the Curaray and Cononaco rivers within the ZITT. In the case of the Maxus and Nashino groups, the home range of these two clans is partially out of the ZITT area. On the contrary, the Tagaeri Taromenane Armadillo clan is definitely out of the same perimeter of the ZITT, completely overlapping oil blocks No. 14 and No. 17 (Figure 6) which are confirmed hydrocarbon reserve areas to be exploited by the Petrooriental oil company [89,90].

Furthermore, the Armadillo clan home range overlaps several titled and untitled lands of *mestizo* (colonos) farmers and non-autochthonous indigenous settlements in the Dayuma and Ines Arango districts, implicating a sensible proximity with roads system (Figures S1, S2) related to the Via Auca main axis [38].

The fact is a crucial zone in which different actors and different territory projects are at stake is also confirmed by the dramatic contacts between *colonos* and uncontacted people in 2009. Only 400 meters from the Hormiguero extraction platform (oil blocks No.17), a Tagaeri Taromenane group speared to death three local farmers [3,27,28,91].

The Nashiño and the Maxus groups are partially outside the ZITT perimeter overlapping three oil blocks (No. 14, No. 17, No. 16, No. 31 and the ITT). In the case of the Maxus group most of their territory, having a home range of roughly 30 km, is outside the ZITT within the oil blocks No. 16 and No. 14. Within block No. 16 the news documented a possible contact in March 5th (2013) (under verification by Ecuadorian Authorities) confirming it is a critical area [29,92,93]. This spatial distribution of these clans is also confirmed by previous anthropological analysis on the Tagaeri Taromenane issue [2,3,28,84,94], in which several witnesses by the Waorani communities, settled within the same oil blocks No. 16, report the presence of uncontacted people in this precise area. Moreover, qualitative field based research by Colleoni and Proaño refer to signs of presence by footprints and typical symbolic territory demarcations (i.e. crossed broken branches) in the area of the Rio Dicaro sources, between oil blocks No. 14 and No. 17., identifying a mobility space of uncontacted people that are used to cross from Rio Dicaro to Rio Yasuní (Figure 6). This behaviour is also confirmed by the database of Ministerio de Justicia y derechos Humanos related to the operation derived from direct survey and patrols of the ZITT and surrounding areas.

A synthesis map of the critical zone defined between points No. 5, 6, and 7 of the ZITT has been produced. Geospatial data on

Table 1. Oil Blocks overlaps on the Zona Intangible Tagaeri Taromenane (ZITT) and the Buffer Zone (10 km).

Oil Blocks[a]	Total area Ha	ZITT Ha	Buffer Zone Ha	ZITT Buffer Zone Ha	ZITT %	Buffer Zone %	ZITT and Buffer Zone %
ITT (Secretaria de Hidrocarburos de Ecuador)	192,255	29,818	37,621	67,439	15,5	19,5	35
31 (Petroamazonas)	201,556	9,954	43,477	53,431	5	21,5	26,5
17 (Petrooriental)	186,186	0	50,720	50,720	0	27,2	27,2
14 (Petrooriental)	201,090	0	12,030	12,030	0	6	6
16 (Repsol-YPF)	130,642	0	13,715	13,715	0	10,5	10,5

Areal measures (hectares) and corresponding percentage. Data on oil blocks refers to the last updates of the 10th oil concession licensing round.

hydrocarbon concessions (10th licensing round), oil fields and wells, Waorani communities, and the home range of Tagaeri Taromenane have been assembled using a SRTM DEM map to also show geomorphological characteristics. This map shows spatial relationships between oil production areas, the ZITT perimeter, and the Maxus uncontacted group (Figure 7).

Drawing a new map: recognizing the spatial patterns of Tagaeri Taromenane

From the perspective of the geography discipline, the territory is not just a polygon closed by a perimeter linking the adjoining areas. Territories are based on junctions, corridors, multiple and non-continuous polygons, centres and peripheries, intensity and directions of functions, multiple uses of natural resources, and coexistence of actors [56,81,82,95–97]. As it as shown by the geographical analysis of the ZITT, every geometrical space may include multiple territories, often conflicting each other.

In a geographical perspective, territories are produced not only by physical intervention, but also (and before physical intervention) by images and particularly maps since they are the strongest way to produce imaginary geographies [98–102]. Normally maps presenting the ZITT are produced in the framework of the Westphalian territorial idea based on a rigid geometry of surface and perimeter which is the classical and most common representation of the ZITT. Figure 8 tries to produce a map representation closer to human right policies and the spatial patterns of Tagaeri Taromenane. First of all limits are abolished. Second, there is a representation of the buffer zone (normally not communicated). Third, a territorial continuity between the nuclei of the four groups and the official ZITT has been represented. This representation shows that the traditional geometric ZITT seems unable to contain the territory reconstructed from the traces of practical territorialization implemented by Tagaeri Taromenane. The decision to use a texture instead of a strong tick colour or an empty area with a marked perimeter is a typical choice from a wide array of tools available to cartographic science. Its usage stresses the communication message of a cultural geography of network relationships among societies and ecosystems [103–106].

This map is only a first attempt to highlight how the recognition of indigenous territory in the Amazon basin needs to implement

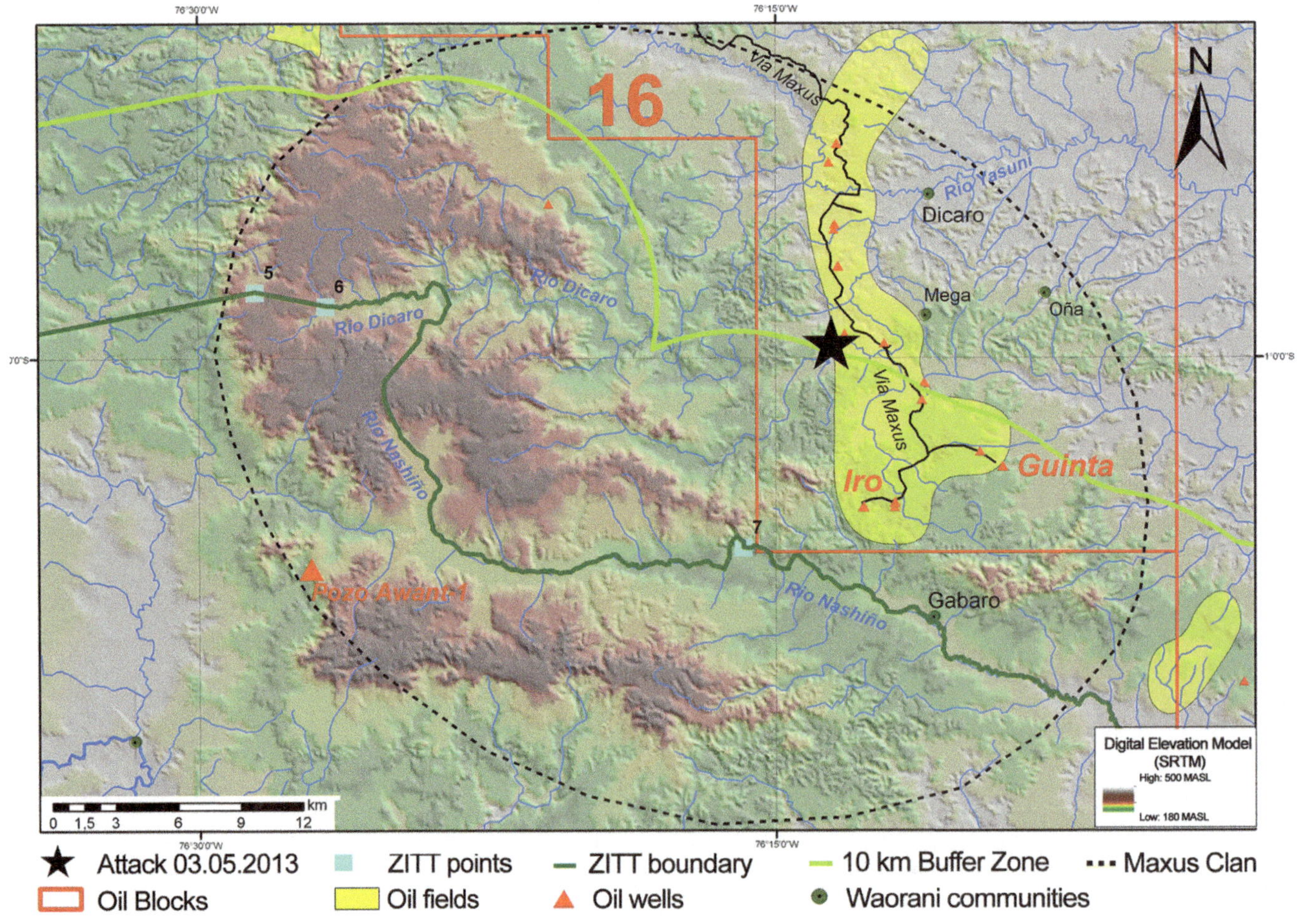

Figure 7. Synthesis map of the critical zone around points No. 5, 6 and 7 of the perimeter section: spatial relationships between oil production (block 16), home range of uncontacted Maxus group, and the spot of the Tagaeri Taromenane attack occurred on the 5th March 2013. This synthesis map of the critical zone defined by points No. 5, 6, 7 of the *Zona Intangible* show sensible spatial relationships between oil production (blocks, fields and wells), the home range of uncontacted Maxus clan, the ZITT perimeter and the Waorani communities settlements. Spatial relationships between oil production areas and the *Zona Intangible* perimeter section are clearly shown in the map. The attack has major significance in terms of its location since it occurred near the geographic inconsistencies highlighted by this paper: about 100 meters from the Wipsi-1 oil platform, 10 km from point 7, almost 20 km from point 7, along the border of the Buffer Zone. Background raster map has been assembled using a SRTM DEM model to also show geomorphological characteristics of the area.

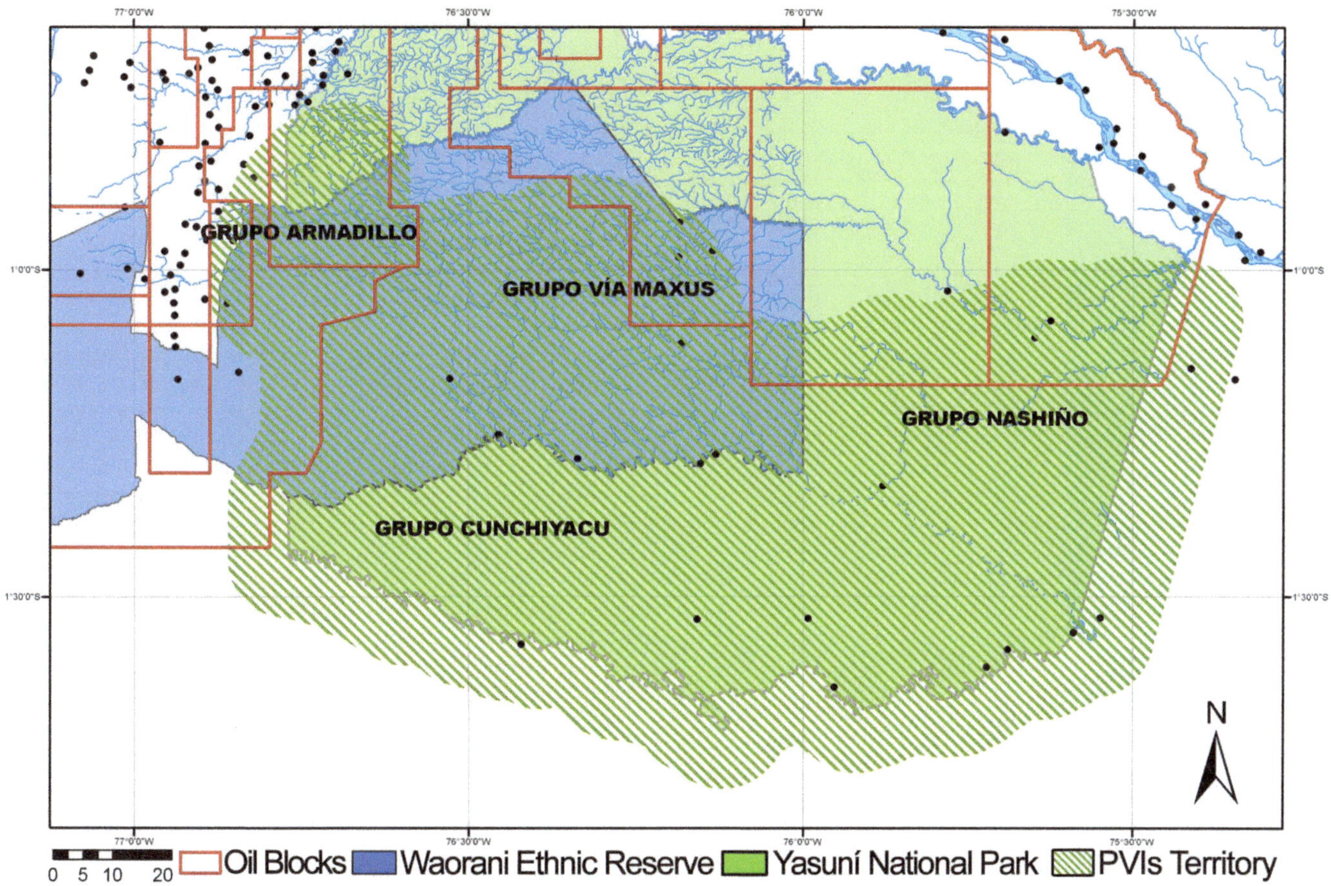

Figure 8. Synthesis map of the *Zona Intangible*. Recognizing the complex territory of uncontacted Tagaeri Taromenane: from geometry to geography. The map results from the combination of human right policies *(Zona Intangible Tagaeri Taromenane)* representing also the Buffer Zone (normally avoided in official maps), the oil concessions for hydrocarbon production and the spatial patterns of Tagaeri Taromenane (combining the results of data about paths and presences). Considering the power of images in shaping territorial representation, cartography science has the responsibility in formulating more complex discourses about controversial territories.

geographical approaches in order to deal with the complex dynamics of territories and territorialities in relation to actors and conflicts.

Materials and Methods

To develop geographical analysis of the ZITT limits we primarily examined the Decrees No. 552 (1999) and No. 2187 (2007). The first one officially declares the No-Go-Zone as "an Intangible Zone for conservation of the homeland of the Waorani groups known as Tagaeri Taromenane in which it is forbidden in perpetuity whatever sort of extractive activities" (see the Spanish original text in supporting materials) while the second one determines the boundary of the ZITT providing geospatial information and detailed geomorphological description of the area [14,15]. In the second Decree (2007) the ZITT is cartographically defined by 17 pairs of metrical coordinates conjoined to each other by straight lines or sections of river courses (Figure 1A). It also declares the Geographical System adopted to mark off the ZITT: Provisional South America Datum 1956 (PSAD-1956), 18 South Zone. The total area declared is 758,051 ha.

In order to validate the perimeter of the ZITT, we analyzed topographic sheets by the Military Geographic Institute of Ecuador (Instituto Geografico Militar, IGM) at different scales (1:50,000, 1:250,000). The IGM cartography at a 1:50,000 scale consists of four sheets named as Rio Bahameno [107], Rio Yasuní Este [108], Rio Nashiño [109], and Rio Yamino [110]. The sheets adjoin each other to form four quadrants including, in their common corner, the perimeter between points No. 5, No. 6, and No. 7. The sheets are provided of geographical metadata defining as projected into a UTM 18 South, Datum WGS 1984 except the Rio Nashiño map that is projected in the Zone 17 South. All four sheets have been assembled using aerial photography mainly taken in the year 1986 while the "Yasuní" sheet is derived by photographic coverage from 1963 and the "Rio Bahameno" sheet from 1976. Contour lines show an equidistance mainly of 20 meters and occasionally of 10 meters. These values are obtained by subtracting a constant of 30 meters of altitude assuming an average canopy of the MTF of 30 meters high.

Altimetry points show a distribution average of 0.6 values per km^2, calculated by three information aliquots of 66 km^2. Geodetic and trigonometric points are rare most likely due difficulty of accessing to the Amazon territory.

To perform a comparative cartographic analysis we also used the 1:250,000 IGM sheets named Shushufindi [111] and Curaray [112]. Unfortunately, the Military Geographic Institute does not cover the eastern territory of the ZITT by the other two adjoining quadrants. However, the "Shushufindi" sheet represents a more detailed cartographic landscape compared to the ones at 1:50,000 scale. On the contrary, the Curaray sheet is definitely poor, without any basic metadata. These topographic sheets were produced in the year 1998 and are basically derived from the 1:50,000 scale sheets. Therefore, we do not expect different information according to the larger scale. In any case, both the 1:250,000 scale sheets represent a poor and incomplete hydrographic network. For instance, there are some river sources ending at the border of the sheet and not continuing onto the adjoining one.

In order to analyze the spatial relationships both of catchment's divides and hydrographic network related the ZITT perimeter we acquired raster thematic cartography (Table S2). Three types of Digital Elevation Models (DEM) maps have been used to deeply investigate both the terrain morphology and river dynamics [113–116]: an IGM-DEM map [117], a Shuttle Radar Topographic Mission (SRTM) map [79], and a GDEM derived by the Aster satellite [78].

The IGM-DEM map presents a geometric resolution of 30 meters (one spot elevation every 900 m^2), the SRTM map of 92 meters (one spot elevation every 8,400 m^2), and the GDM Aster satellite of 30 meters. The SRTM elevation spot derives directly by radar interferometry technique. On the contrary, the IGM-DEM map is obtained not by remote sensing technology, but by cartographic data included in maps at 1:50,000 and 1:250,000 scale.

The remote sensing data from the Aster satellite (GDEM) presents a higher geometric resolution to the ground – 30 instead of 90 meters – but they show several false records due to the intense land cover of the MTF. Therefore, these satellite derived maps present a repetitive error more extensive in the representation, but homogeneous and, consequently, tolerable.

Moreover, Landsat satellite imagery has been used to validate some uncertain segments of the perimeter represented by river course sections. Three satellite time-series of Landsat Thematic Mapper (TM) false colour composite (4 blue, 5 green, 7 red) [118] of the territory encompassed between points No. 7 and No. 8 in the southern sector of the Oil Block 31 (PetroAmazonas) have been acquired. The Landsat satellite scenes acquired are from the years 1991 and 2005 (2nd of July and 28th of August respectively). For a complete overview of the raster cartography used in the spatial analysis see Table S1.

Other geodata has been used to perform the spatial analysis of the ZITT perimeter together with other key components shaping the Yasuní territory in order to have a systemic understanding of the complex territorial processes taking place in the area [20,96].

Geospatial information on protected areas (Yasuní National Park, Waorani Ethnic Reserve), hydrographic network, and urban centres are from institutional web sources and were publicly available at the time of the submission. These include the World Database on Protected Areas (UNEP-WCMC, IUCN World Commissions on Protected Areas, 2012) [119] and the EcuadorIan IGM Geoportal. The official ZITT perimeter and geographical data on Tagaeri Taromenane clans distribution have been acquired from the Ministerio de Justicia, Derechos Humanos y Cultos of Ecuador (2012), who from October 2010 holds the Plan of Precautionary Measures for the Protection of Isolated Indigenous Peoples of Ecuador (Plan de Medidas Cautelares, PMC) [88].

Road network of Ecuadorian Amazon was taken by the Instituto Nacional para el Eco-Desarrollo de la Region Amazonica Ecuatoriana (ECORAE) in 2008 [120].

Oil fields and oil wells have been acquired by the ECORAE institution (2008). Oil blocks with the latest changes (2011) of the 10th hydrocarbon licensing round have been derived by digitizing oil maps released in 2012 by the *Secreteria de Hidrocarburos Ecuatorianos* (SHE) [87]. An overview of characteristics of all vector geodata used in the spatial analysis are shown in Table S3.

The main tools to conduct cartographic and spatial analysis of the ZITT have been Geographical Information Systems (GIS) and remote sensing technologies. GIScience approach has been used to collect and elaborate different types of data, as long as spatially explicit, referring both to the biophysical components and to the anthropic one, so that cartographic outputs and synthesis maps make visible spatial relationships between relevant components of environmental systems and that of the anthropic one [121–124].

All the anthropic, physical, and geographical data collected have been integrated in a GIS database in order to manage spatially explicit, temporally referenced, and thematically differentiated information.

Spatial analyses are based both on a critical examination and integration of the physical and thematic cartography previously acquired and on GIS elaboration in order to produce analytical and synthesis maps showing the following data: hydrographic network, river basins, protected areas (Yasuní National Park, Waorani Ethnic Reserve, ZITT) and buffer zones, urban centres and local communities, road network, and the spatial dimension of the hydrocarbon exploitation (blocks, oil fields and wells). Moreover, we have also acquired spatial anthropological data concerning the Tagaeri Taromenane as the family clans spatial distribution [88].

ArcGIS™ 10.1 software has been used for GIS analysis mainly conducted by spatial functions such as overlay, geometrical interception, linear, and area measurements. Raster data analysis has been carried out using IDRISI-GIS software.

To validate important data such as spatial information of key rivers, local communities and roads the GPS ground-truth methodology has been adopted during fieldwork activities (2005, 2006, 2010, 2011) [38] [125].

Conclusions

Results of GIS and remote sensing analysis, fieldwork data, and examination of all the available sources show different critical issues in the Decree 2187 when it comes to defining the perimeter of the ZITT. They belong both at an interpretation level and, therefore, in the same cartographic representations. On the other side, the policies implemented in the ZITT and in the Buffer Zone show conflicting territorial objectives.

During the revision of this article, in the night between the 4th and 5th of March 2013, the old Waorani man Ompore and his wife Buganey were speared to death in proximity of Iro oil field, in Block 16, which is operated by Repsol-YPF (Figure 6, 7). The attack should be confirmed, and at the moment investigation into the incident has been initiated by police and Ecuadorian authorities to ascertain the responsibility of Tagaeri Taromenane groups. Information collected by Waorani relatives of Ompore Omeway and Buganey Cayga refer to repetitive requests by Tagaeri Taromenane directed to Ompore to act with the oil company to stop oil extraction, pollution of water and air, and noise [92,93,126]. Meanwhile, a group of Waorani from the Yarentaro community organized an expedition to avenge their dead relatives, entered the ZITT, and killed approximately dozens of Tagaeri Taromenane as well as abducted two uncontacted youths. This fact, confirmed by an official twitter account of Ministerio de Justicia on the 3rd of April 2013, is also under investigation.

These two dramatic events should be deeply analyzed in order to understand the complex relationship among uncontacted Waorani groups and the wide group of Waorani, the discontinued presence of the State, the role of Waorani organizations, and the deterministic approach of some NGO advisors.

The incident of 5th of March also demonstrates the long and complex process to enact human right policies in a controversial and complex territory. Moreover, the attack has major significance in terms of its location since it occurred near the geographic inconsistencies highlighted by this paper: about 100 meters from the Wipsi-1 oil platform, 10 km from point 7, almost 20 km from point 7, along the border of the Buffer Zone (Figure 7).

So it is interesting to analyze how the limit line of the ZITT runs when it moves – close to point No. 6, from west to east direction – within this zone inward the oil block 16 (Repsol-YPF): after a sudden curve upper slope it drastically runs to south leaving the oil blocks by the Nashiño river, setting up a kind of cartographic enclave by a "bird beak" shape (Figure 7). After considering previous anthropological studies on Tagaeri and Taromenane in this area, one can logically ask the question: is the ZITT boundary line between the Dicaro and Nashiño rivers shaping an ethnical enclave as well?

Using a systemic approach to better define this issue, it is relevant to also take into account the geographical position and the same history of the ZITT (Figure 6).

During three years of the delimitation process (2004–2007), the Technical Commissions established by the Environment Ministry of Ecuador had to deal with blocks for oil exploitation which extensively overlap the northwest sector of the ZITT (Figure 6). This is the case of oil block No. 14 and No. 17 (Andes-Petroleum Company, China) which presented an overlap of 2,719 and 24,534 ha, respectively, in the same area of the ZITT, namely 1.4% and the 49.6% of each oil concession (Table 1) [64].

Due to a possible and relevant reduction of oil block areas during the ZITT delimitation process, the Andes-Petroleum company suggested to the Ministry of Energy (Ministerio de Energia y Minas) and its Environmental Department (DINAPA) to modify the perimeter in order to facilitate oil exploitation in the area maintaining direct access to the proved oil field called Awant-1. Previous GIS analysis [20] showed the "cartographic suggestion" of Andes Petroleum to create – according to its productive requirements – a special corridor to exploit the Awant-1 oil field (Fig. S6) which is localized 12,8 km to south point No. 6, inside the same ZITT area (Figure 7).

All of these geographical elements display the complexity of the Yasuní Biosphere Reserve in which different projects of territory – and, therefore, different modalities to use natural resources – are overlapped and often in conflict each other. A geography of energetic resources marked in maps by productive zoning (oil blocks) is superposed to that of protected areas (Yasuní National Park), to that of farmers settlements (Agrarian Reform), to that of indigenous territory (Waorani Ethnic Reserve), and finally to that of the ZITT [38,51,56].

The same delimitation of the ZITT induced a national and international debate on uncontacted people and their territorial rights. Aguirre (2007) stresses that since first steps of delimitation (2004) the process has been sinuous and thorny, made by consultation and negotiations, *in primis* with oil companies operating in the area of the ZITT. Later with officers of United Nation and IUCN, with the Waorani indigenous Organization (NAWE) and the National Indigenous Confederation of Ecuador (CONAIE), and finally national and international NGOs, without avoiding the weak but existing pressures from the local *colonos* (settled farmers of agrarian reform).

In general, beyond the technical issue to map out a perimeter for the ZITT, anthropological studies argue that a lack of understanding of territory and the same territoriality of the Tagaeri Taromenane is one of the main factors of the weakness in delimitating an *Zona Intangible* [2,3,6,25,41].

It is not just a technical issue considering the geographical and the biophysical characteristics of an area placed in the core of the Amazon region. It is a remote zone in which the Moist Tropical Forest as well as the dense and homogeneous hydrographic network, both set up a territory that is difficult to study, understand, and control. Therefore, mapping out the boundaries by "remote" analyses, using unverified geospatial data and not considering the complex arena in which territorial dynamics and the same territoriality of the Tagaeri Taromenane take place, is very difficult without holistic approaches and knowledge based on fieldwork.

Moreover, if we consider that the Decree 2187 uses "natural boundaries", such as section of rivers, to delimit the ZITT perimeter, cartographic operations become more complicated, especially if the geospatial data is not precise, they present data gaps, or relevant geographic errors.

On the other side, a ZITT perimeter only based on straight lines could be easily traced out. However, in tropical forest territory, whatever border marked out that does not have clear natural spatial references, such as rivers or hills, results in a concealed boundary. A ZITT cartographically perfect will be produced, but at the same time it is invisible on the ground to both residents and to the people "from outside".

A scientific contribution in the ZITT debate is necessary to make visible this territory (physically and graphically) to different components of Ecuadorian society because the ultimate objective of this policy is the right to liveable place of Tagaeri Taromenane uncontacted people.

In a structural context where media debate about the Amazon as "space of exception" [100], and ZITT territoriality is disputed, it is important to remember what "*The protection guidelines for indigenous peoples in isolation*" issued in May 2012 by the Office of the High Commissioner for Human Rights (OHCHR, 2012) states: "*While there is no consensus on the term to be used to refer to these people, internationally the most widely used is the concept of "isolated peoples", in some countries they are known as, inter alia, free people, uncontacted, hidden, invisible, in voluntary isolation. Despite the different formulations, all refer to the same concept "isolation was not a voluntary option but a survival strategy (par. 9)".*

Supporting Information

Figure S1 Land cover map (Landsat ETM+, 2002) of the Via Auca territory and anthropic pressures on the *Zona Intangible* and the uncontacted indigenous group. Deforestation processes developed by road systems around the Via Auca main axis, African Palm cultivations (yellow areas at the north sector), *Zona Intangible Tagaeri Taromenane,* Armadillo uncontacted clan and oil blocks.

Figure S2 Expansion of the Via Auca road network towards the Yasuní National Park and the Waorani Ethnic Reserve. Oil blocks (yellow lines), historic incidents (red stars) between uncontacted Tagaeri Taromenane groups and external actors, and protected areas. Road network in 2009 by black lines (MAE, 2009) and spatial evolution to the east in 2011 developed by GPS survey (red lines). The map also show the uncontacted Armadillo home range overlapping several oil blocks and the mestizo farmer settlements.

Figure S3 Comparative GIS analysis of catchment's divides by different Digital Elevation Models. Catchment's divides are obtained by three different DEM maps (IGM Ecuador, SRTM, GDEM). Even if they present some spatial differences, all of them confirm that points No. 5 and No. 6 belong to different river basins. The first one belong to the Rio Yasuni, the second one to the Rio Curaray.

Figure S4 Details of the boundary section nearby point No. 8 which drifts away from the Rio Nashiño river bed.

Figure S5 Detail of the Landsat TM 5 satellite scene and the boundary section close to point No. 8. The boundary section drifts away from the Rio Nashiño riverbed runs over a hill's ridge in two spots, violating both the terrain morphology and the same official text (Decree 2187, 2007).

Figure S6 Andes-Petroleum company and the cartographic suggestion to modify the perimeter of the *Zona Intangible*. Andes-Petroleum company (China) suggested to the Ministry of Energy (Ministerio de Energia y Minas) and its Environmental Department (DINAPA) to modify the perimeter in order to facilitate oil exploitation in the area maintaining direct access to the proved oil field called Awant-1. This map shows the "cartographic suggestion" of Andes Petroleum to create – according to its productive requirements – a special corridor to exploit the Awant-1 oil field which is localized 12,8 km to south point No. 6, inside the same ZITT area.

Document S1 Presidential Decree 2187 signed by President Alfredo Palacio in 2007. This is the official document which establishes and maps the *Zona Intangible Tagaeri Taromenane* by given vertex coordinates and geomorphological description of the area.

Table S1 Metric coordinates of the 17 given points by the Presidential Decree 2187 (2007) defining the Intangible Zone Tagaeri Taromenane.

Table S2 Geospatial raster data used in GIS analyses: coverage, typology, resolution and sources.

Table S3 Characteristics and sources vector geodata.

Acknowledgments

We firstly thank all the mestizos and indigenous local communities such as Nueva Esperanza, La Perla de la Amazonia, Tobeta, Gabaro, el Eden and Chiro Isla which facilitated and supported our fieldwork activities in the Yasuní Biosphere Reserve during all these years.

We also wish to thank the local Institution such as Province of Orellana (Gobierno Autonomo Provincial de Orellana, GAPO), the Gobierno Autonomo Descentralizado Francisco de Orellana (El Coca), the Parroquia de Dayuma and Eden Yuturi.

We also thank the local ngo's and organizations involved since many years in the struggle for the socio-environmental rights of the Yasuní communities: the Red Indigena Angel Shingre (Ciudad Del Coca, Orellana), Fundación Alejandro Labaka, Vicariato Apostolico de Aguarico and the Solidaridad Internacional de Catalunya.

This paper would not have been possible without the geographic information and the logistic support of the Ministerio de Justicia, Derechos Humanos y Cultos team holding the Plan of Precautionary Measures for the Protection of Isolated Indigenous Peoples (Plan de Medidas Cautelares, PMC) that have dedicated resources to facilitate and/or undertake research on the *Zona Intangible Tagaeri Taromenane.*

We also thank all the Waorani staff working at the Estación de Monitoreo de la *Zona Intangible* in Shiripuno (Orellana).

We finally thank Chavon Grande for assisting and supporting our scientific production.

Author Contributions

Conceived and designed the experiments: SEP MDM. Performed the experiments: SEP MDM. Analyzed the data: SEP MDM FF. Wrote the paper: SEP MDM.

References

1. United Nations, Economic and Social Council, Official Records SN23 (2009) Permanent Forum on Indigenous Issues, Report on the eight session. Available: http://www.un.org/esa/socdev/unpfii/documents/E_C_19_2009_14_en.pdf.
2. Cabodevilla MA (2008) Zona intangible: peligro de muerte! Quito, Ecuador: CICAME. p. 206.
3. Colleoni P, Proaño J (2010) Camminantes en la selva. Los Pueblos en Aislamiento de la Amazonia Ecuatoriana. Informe 7 IWGIA. Grupo Internacional de Trabajo sobre Asuntos Indígenas (IWGIA) – 2010. p. 48. Available: http://www.iwgia.org/publicaciones/buscar-publicaciones?publication_id=275.
4. Beckerman S, Erickson PI, Yost J, Regalado J, Jaramillo L, et al. (2009) Life histories, blood revenge, and reproductive success among the Waorani of Ecuador. Proceedings of the National Academy of Sciences of the United States of America 106: 8134–8139. doi: 10.1073/pnas.0901431106.
5. Napolitano DA, Ryan ASS (2007) The dilemma of contact: Voluntary isolation and the impacts of gas exploitation on health and rights in the Kugapakori Nahua Reserve, Peruvian Amazon. Environmental Research Letters 2: 045005. doi: 10.1088/1748-9326/2/4/045005.
6. Olson DM, Dinerstein E (2002) The Global 200: Priority Ecoregions for Global Conservation. Annals of the Missouri Botanical Garden 89: 199–224.
7. Maffi L, Woodley E (2010) Biocultural Diversity Conservation: A Global Sourcebook. London, UK: Earthscan Ltd. p. 313.
8. Steege HT, Pitman N, Sabatier D, Castellanos H, Hout PVD, et al. (2003) A spatial model of tree α-diversity and -density for the Amazon. Biodiversity and Conservation: 2255 – 2277.
9. Erwin TL, Pimienta MC, Murillo OE, Aschero V (2005) Mapping Patterns of Diversity for Beetles Across the Western Amazon Basin: A Preliminary Case for Improving Inventory Methods and Conservation Strategies Mapping Patterns of Diversity for Beetles Across the Western Amazon Basin: A Preliminary Case. Proceedings of the California Academy of Sciences 56.
10. Pimm BYSL, Jenkins C (2005) Sustaining the Variety of Life. Scientific American 293: 66–73. Available: http://www.eebweb.arizona.edu/courses/Ecol206/PimmEA_Sept2005_SciAm_Biodiversity.pdf.
11. Gamboa Balbin CL (2007) Régimen especial transectorial de protección a favor de pueblos indígenas en aislamiento y en contacto inicial. Lima, Peru: Bellido Ediciones EIRL. p. 254.
12. Cabodevilla MA (1996) Coca: la región y sus historias. Quito, Ecuador: CICAME. p. 203.
13. Cabodevilla MA (1994) Los Huaorani en la historia de los pueblos del Oriente. Quito, Ecuador: Cicame. p. 488.
14. Presidencia de la Republica de Ecuador (1999) Decreto Ejecutivo No. 552. Quito, Ecuador.
15. Presidencia de la Republica (2007) Decreto Ejecutivo No. 2187. Quito, Ecuador. Available: http://saveamericasforests.org/Yasuni/News/ZonaIntangible/DECRETO ZONA INTANGIBLE.pdf.
16. Quintero ME (2011) Manejo sostenible del patrimonio natural y cultural de la Reserva de la Biosfera Yasuní. Quito. p. 73. Available: http://www.fao.org/climatechange/34137-047a043520d4271e782221326960e8858.pdf.
17. Ecociencia (2007) Bloques Petroleros (ITT y 31), Reserva de Biosfera Yasuní. Quito. Available: http://ecuaflorida.wordpress.com/2010/11/05/la-iniciative-yasuni-itt-nos-necesita/. Accessed 2012 Sep 9.
18. Duerto G, David S-D, Tatiana C, Susan P (2011) Propuesta de control y vigilancia para la Reserva de Biosfera Yasuní. Quito: Corporación Grupo Randi Randi. p. 94.
19. Aguirre M (2010) ITT, Pueblos Ocultos, Zona Intangible del Yasuní. Available: http://alejandrolabaka.blogspot.it/2010/02/blog-post_09.html. Accessed 2012 Jun 12.
20. Pappalardo SE (2011) Biodiversity conservation strategies and water conflicts in and around the Yasuní Biosphere Reserve of the Ecuadorean Amazon Region, in Czellecz Buttercup, Stephen Lazar, Zoltan Paul (eds.), Geography of Water. Cluj Napoca, Romania: Collegium Geographicum, Abel Publishing House.

21. Alex Rivas Toledo Rommel LP (2001) Conservacion y petroleo en la Amazonia Ecuatoriana. Un acercamiento al caso huaorani. Quito, Ecuador: Ecociencia, Abya Yala. p. 139.
22. Rival L (2008) Resistance in an Amazonian Community. Huaorani Organizing against the Global Economy. Journal of Latin American Anthropology 10: 257–259. doi: 10.1525/jlca.2005.10.1.257.
23. Holt FL (2005) The Catch-22 of Conservation: Indigenous Peoples, Biologists, and Cultural Change. Human Ecology 33: 199–215.
24. Belaunde LE (2008) Trekking through history: the Huaorani of Amazonian Ecuador – By Laura M Rival. Journal of the Royal Anthropological Institute 14: 460–461. doi: 10.1111/j.1467-9655.2008.00511_27.x.
25. Chávez D (2010) La situación de los pueblos indígenas aislados en el Ecuador. Quito, Ecuador. p. 37. Available: observatorio.cdes.org.ec/documentos/doc_download/982-la-situacion-de-los-pueblos-indigenas-aislados-en-el-ecuador.
26. Cabodevilla MA (2003) Reflexiones sobre un ataque Huao. Ikonos 17. Quito, Ecuador: FLACSO. Available: http://www.flacsoandes.org/dspace/handle/10469/2181.
27. Aguirre M (2011) Armadillo: la otra cara del ITT. Letras Verdes 8. Quito, Ecuador. Available: http://flacsoandes.org/dspace/bitstream/10469/3166/1/RFLACSO-L8-07-Aguirre.pdf.
28. Aguirre M, Cabodevilla MA, Andueza JC (2009) Otra Historia de Violencia y Desorden, Lanzas y Muerte en Los Reyes. Quito, Ecuador: CICAME, Fundación A. Alabaka.
29. Plaza J (2013) Dos huaorani murieron atravesados por lanzas en su territorio. El Comercio website. Available: http://www.elcomercio.com/pais/Amazonia-Ecuador-huaorani-ataque-tagaeri_0_877112605.html. Accessed 2012 Mar 12.
30. UNESCO (1989) MAB Biosphere Reserves Directory. Available: http://www.unesco.org/mabdb/br/brdir/directory/biores.asp?code = ECU+02&mode = all. Accessed 2012 Sep 8.
31. Villaverde X, Ormaza F, Marcial M, Jogenson J (2005) Parque Nacional y Reserva de Biosfera Yasuní: historia, problemas y perspectivas .Quito, Ecuador: FEPP, WCS. p. 283.
32. Fontaine G, Narvaez I (2008) Yasuní en el siglo XXI. El Estado Ecuatoriano y la Conservacion de la Amazonia. Ikonos 30: 121–123. Quito, Ecuador: FLACSO. Available: http://www.flacsoandes.org/dspace/handle/10469/4057.
33. Fontaine G (2003) Petróleo y desarrollo sostenible en Ecuador. Las reglas de juego. Quito, Ecuador: FLACSO, Petroecuador. p. 208.
34. Vallejo Espinosa A (2003) Modernizando la naturaleza: desarrollo sostenible y conservación de la naturaleza en la Amazonía ecuatoriana Quito, Ecuador: SIMBIOE. p. 100.
35. Petroecuador (2010) El Petroleo en Ecuador. 1: 1–133. Available: http://www.eppetroecuador.ec/idc/groups/public/documents/archivo/001138.pdf. Accessed 2012 Sep 14.
36. El Universo website (2002) Petroecuador prepara nueva licitación petrolera. Available: http://www.eluniverso.com/2002/05/27/0001/9/3F69A7216A3D426AA0EC7E6EE4E92D2E.html. Accessed 2012 Sep 14.
37. Hoy website (2002) La Novena Ronda petrolera despierta poco interés. Available: http://www.explored.com.ec/noticias-ecuador/la-novena-ronda-petrolera-despierta-poco-interes-133565.html. Accessed 2012 Sep 15.
38. Pappalardo SE (2010) Espansione della frontiera estrattiva e conflitti ambientali nell'area della Riserva della Biosfera Yasuní: la Via Auca. Padova, Italy: Cluep, Quaderni di Dottorato 5: 1–14.
39. Wunder S (2003) Oil Wealth and the Fate of the Forest: A Comparative Study of Eight Tropical Countries. Routledge Explorations in Environmental Economics. Routledge. p. 456.
40. Initiative E (2003) The Energy & Biodiversity Initiative. Integrating Biodiversity Conservation into Oil and Gas Development Washington DC: Conservation International. p. 65.
41. Finer M, Jenkins CN, Pimm SL, Keane B, Ross C (2008) Oil and gas projects in the Western Amazon: threats to wilderness, biodiversity, and indigenous peoples. PloS One 3: e2932. Available: http://www.plosone.org/article/info%3Adoi%2F10.1371%2Fjournal.pone.0002932. Accessed 2012 Jul 18.
42. Thomsen J, Mitchell C, Piland R, Donnaway J (2001) Monitoring Impacts of Hydrocarbon Exploration in Sensitive Terrestrial Ecosystems. Footprints in the Jungle 1: 90 – 113.
43. Alberto A (2009) La maldición de la abundancia. Quito, Ecuador: Abya-Yala. p. 240.
44. Galo Z-R, Esteban SR, Víctor UB, Javier VOA (2006) Evaluation of Anthropogenic Threats in Yasuní National Park and its Implications for Wild Mammal Conservation. Lyonia 10: 31–41.
45. Messina JP, Walsh SJ (2001) 2001 Open Meeting of the Human Dimensions of Global Environmental Change Research Community in Rio de Janeiro. Simulating Land Use and Land Cover Dynamics in the Ecuadorian Amazon through Cellular Automata Approaches and an Integrated GIS. Rio de Janeiro. 1–13.
46. Narvaez I (2009) Petróleo y poder: el colapso de un lugar singular. Yasuní. Quito, Ecuador: Flacso. p. 584.
47. Greenberg JA, Kefauver SC, Stimson HC, Yeaton CJ, Ustin SL (2005) Survival analysis of a neotropical rainforest using multitemporal satellite imagery. Remote Sensing of Environment 96: 202–211. doi: 10.1016/j.rse.2005.02.010.
48. Southgate D, Sierra R, Brown L (1991) The causes of tropical deforestation in Ecuador: A statistical analysis. World Development 19: 1145–1151. doi: 10.1016/0305-750X(91)90063-N.
49. Rodrigo Sierra JS (1998) The Dynamics and Social Organization of Tropical Deforestation in Northwest Ecuador' 1983-19951. Human Ecology 26.
50. Bilsborrow RE, Barbieri AF, Pan W (2004) Changes in population and land use over time in the Ecuadorian Amazon. Acta Amazonica 34: 635–647. doi: 10.1590/S0044-59672004000400015.
51. Pappalardo SE (2012) Amazzonia ecuadoriana: la Riserva della Biosfera Yasuní tra progetti per la conservazione della natura e conflitti ambientali. In: Donatella S, Spagna F (eds.) Etnografie collaborative e questioni ambientali Ricerche nell'America indigena contemporanea. Padova, Italy: Cleup. 113–138.
52. Rudel TK, Horowitz B (1996) La deforestación tropical pequeños agricultores y desmonte agrícola en la Amazonía Ecuatoriana. Quito, Ecuador: Abya-Yala. p. 233.
53. Laurance WF, Albernaz AKMLM, Schroth GG, Fearnside PM, Bergen S, EM Venticinque, Da Costa C (2002) Predictors of Deforestation in the Brazilian Amazon. Journal of Biogeography 29: 737–748. Available: http://hdl.handle.net/10088/7036. Accessed 2012 Sep 13.
54. Metzger JP (2001) Effects of deforestation pattern and private nature reserves on the forest conservation in settlement areas of the Brazilian Amazon. Revista Biota Neotropica 1.
55. Imbernon J, Branthomme A (2001) Characterization of landscape patterns of deforestation in tropical rain forests. International Journal of Remote Sensing 22: 1753–1765. doi: 10.1080/01431160118426.
56. De Marchi M (2004) I conflitti ambientali come ambienti di apprendimento. Trasformazioni territoriali e cittadinanza in movimento in Amazzonia. Padova, Italy: Cleup. p. 216.
57. Pappalardo SE (2013) Expansion de la frontera extractiva y conflictos ambientales en la Region Amazonia Ecuatoriana: el caso Yasuni. PhD Thesis. Padova, Italy: unpublished.
58. Myers N (1988) Threatened Biotas: "Hot Spots" in Tropical Forest. Environmentalist 8: 187–208.
59. Rieckmann M, Adomßent M, Härdtle W, Aguirre P (2011) Sustainable Development and Conservation of Biodiversity Hotspots in Latin America: The Case of Ecuador. in Zachos FE, Habel J. (eds.), Biodiversity Hotspots: Distribution and Protection of Conservation Priority Areas: 435–452. Berlin, Heidelberg: Springer.
60. Myers N (1993) Tropical Forests – The Main Deforestation Fronts. Environmental Conservation 20: 9–16.
61. Cabodevilla MA (2004) El exterminio de los pueblos ocultos. Quito, Ecuador: CICAME. p. 211.
62. Inter-American Commission on Human MC-91/6 (2006) Precautionary Measures, The Tagaeri and Taromenane Indigenous Peoples, Ecuador. Available: http://www.oas.org/en/iachr/mandate/what.asp. Accessed 2012 Sep 12.
63. Ministerio del Ambiente, Ministerio de Energia y Minas de Ecuador (2004) Acuerdo Interministerial No 092, Registro Oficial 440: 2–3. Available: http://www.derechoecuador.com/index.php?option = com_content&view = article&id = 1476:registro-oficial-12-de-octubre-del-2004&catid = 195:octubre&Itemid = 332#anchor560178.
64. Aguirre M (2007) ¡A quién le importan esas vidas!: Un reportaje sobre la tala ilegal en el Parque Nacional Yasuní. Quito, Ecuador: Cicame. p. 232.
65. Bass MS, Finer M, Jenkins CN, Kreft H, Cisneros-Heredia DF, et al. (2010) Global conservation significance of Ecuador's Yasuní National Park. PloS one 5: e8767. Available: http://www.plosone.org/article/info:doi/10.1371/journal.pone.0008767. Accessed 2012 Jul 13.
66. Xinhua (2013) Ecuador abre licitación de XI Ronda Petrolera para campos del sur de la Amazonía. America Economia. Available:http://www.americaeconomia.com/negocios-industrias/ecuador-abre-licitacion-de-xi-ronda-petrolera-para-campos-del-sur-de-la-amazonia. Accessed 2012 Sep 2.
67. Oilwatch (2012) ECUADOR – La firma de contratos petroleros dio fin a la décima ronda. Available: http://www.oilwatchsudamerica.org/petroleo-en-sudamerica/ecuador/3995-ecuador-la-firma-de-contratos-petroleros-dio-fin-a-la-decima-ronda.html. Accessed 2012 Sep 9.
68. Petroecuador website (2012) Mapa Bloques Región Oriente. Available: http://www.eppetroecuador.ec/idc/groups/public/documents/archivo/001374.pdf. Accessed 2012 Aug 12.
69. Hoy.com.ec website (2011) Ecuador no ha explorado crudo durante 12 años. Available: http://www.hoy.com.ec/noticias-ecuador/ecuador-no-ha-explorado-crudo-durante-12-anos-559713.html. Accessed 2012 Sep 18.
70. El Universo website (2012) Se quiere evitar la explotación petrolera. Available: http://www.eluniverso.com/2012/04/22/1/1447/se-quiere-evitar-explotacion-petrolera.html. Accessed 2012 Sep 18.
71. Jeffrey P. Jorgenson MCR (2001) Conservación y desarrollo sostenible del Parque Nacional Yasuní y su área de influencia. In: Ministerio del Ambiente, WCS (eds.). Quito, Ecuador: Editorial Simbioe.
72. Finer M, Moncel R, Program E, Jenkins CN (2010) Leaving the Oil Under the Amazon: Ecuador's Yasuní-ITT Initiative. Biotropica 42: 63–66.
73. Rival L (2010) Ecuador's Yasuní-ITT Initiative: The old and new values of petroleum. Ecological Economics 70: 358–365. doi: 10.1016/j.ecolecon.2010.09.007.

74. Larrea C, Warnars L (2009) Ecuador's Yasuni-ITT Initiative: Avoiding emissions by keeping petroleum underground. Energy for Sustainable Development 13: 219–223. doi: 10.1016/j.esd.2009.08.003.
75. Larrea C, Greene N, Rival L, Sevilla E, Warnars L (2009) Yasuni–ITT Initiative: A Big Idea from a Small Country. Quito, Ecuador: Ministerio del Ambiente de Ecuador, UNDP Ecuador, MdGs.
76. Vallejo MC (2010) Biophysical structure of the Ecuadorian economy, foreign trade, and policy implications. Ecological Economics 70: 159–169. doi: 10.1016/j.ecolecon.2010.03.006.
77. Instituto Geografico Militar (2010) IGM Digital Elevation Model, raster file.
78. NASA website (2012) Advanced Spaceborne Thermal Emission and Reflection Radiometer (ASTER), EOS Program. Available: http://gdem.aster.ersdac.or.jp. Accessed 2012 Aug 20.
79. NASA website (2012) Shuttle Radar Topographic Mission (SRTM). Available: http://srtm.csi.cgiar.org. Accessed 2012 Aug 12.
80. Ministerio del Ambiente de Ecuador (2010) Zona Intangible Tagaeri Taromenane (ZITT), geospatial data shapefile.
81. Faggi P, Turco A (2001) Conflitti Ambientali: Genesi, Sviluppo, Gestione. Milano, Italy: Unicopli. p. 372.
82. Turco A (1988) Verso una teoria geografica della complessità. Milano, Italy: Unicopli. p. 184.
83. Bertoncin M (2004) Logiche di terre e acque. Le geografie incerte del delta del Po. Roma, Italy: Cierre Edizioni. p. 496.
84. Cabodevilla MA, Smith R, Rivas Toledo A (2004) Tiempos de guerra: Waorani contra Taromenane. Quito, Ecuador: Abya-Yala. p. 164.
85. Pappalardo SE (2009) Conservazione della biodiversità, aree protette e conflitti ambientali nell'Amazzonia ecuadoriana: la Riserva di Biosfera Yasuní. Master thesis, Padova, Italy: Available: http://www.slideshare.net/biorebel/pappalardo-tesis-yasuni. Accessed 2013 Mar 12.
86. Finer M, Vijay V, Ponce F, Jenkins CN, Kahn TR (2009) Ecuador's Yasuní Biosphere Reserve: a brief modern history and conservation challenges. Environmental Research Letters 4: 034005. Available: http://stacks.iop.org/1748-9326/4/i=3/a=034005?key=crossref.87a7df3368c4de2aa7971dbff8589f86. Accessed 2012 Aug 19.
87. Secretaria de Hidrocarburos Ecuador Ministerio de Recursos No Renovables website (2012) Mapa de Bloques Petroleros del Ecuador. Available: http://www.she.gob.ec/portal/es/web/hidrocarburos/mapa-de-bloques-petroleros-del-ecuador. Accessed 2012 Sep 8.
88. Ministerio del Ambiente (2010) Plan de Medidas Cautelares. Distribución espacial de los clanes Tagaeri Taromenane. Geospatial data, shapefile.
89. Secretaria de Hidrocarburos de Ecuador Ministerio de Recursos No Renovables website (2011) Firma de Tres Contratos Petroleros de la X Ronda Petrolera. Available: http://www.she.gob.ec/portal/es/web/hidrocarburos/firma-de-tres-contratos-petroleros-de-la-x-ronda-petrolera. Accessed 2012 Sep 16.
90. El Telégrafo website (2012) Tres campos marginales entregarán una renta petrolera de 500 millones de dólares al país. Available: http://www.eltelegrafo.com.ec/index.php?option=com_zoo&task=item&item_id=37901&Itemid=2. Accessed 2012 Sep 16.
91. El Universo website (2009) Matanza en Oriente sería por petroleras. Available: http://www.eluniverso.com/2009/08/16/1/1447/matanza-oriente-seria-petroleras.html. Accessed 2012 Sep 9.
92. Aguirre M (2013) De luto. El Comercio website. Available: http://www.elcomercio.com/milagros_aguirre/luto_0_882511782.html.
93. ONWO (2013) Comunicado Oficial Organización de la Nacionalidad Waorani de Orellana. Observatorio de Derechos Colectivos de Ecuador. Available: http://observatorio.cdes.org.ec/113-noticias/ultimo-minuto/387-comunicado-oficial-de-onwo-por-incidente-con-aislados. Accessed 2013 Mar 13.
94. Proaño J, Colleoni P (2008) Taromenane Warani Nani. Quito, Ecuador: Abya-Yala. p. 120.
95. Raffestin C (1980) Pour une géographie du pouvoir. Paris, France: Librairies techniques. p. 249.
96. De Marchi M, Natalicchio M, Ruffato M (2010) I territori dei cittadini: il lavoro dellOLCA (Observatorio Latinoamericano de Conflictos Ambientales). Padova, Italy: Cleup. p. 204. Available: http://iterritorideicittadini.files.wordpress.com/2010/06/demarchi-iterritorideicittadini.pdf.
97. De Marchi M (2011) Conflitti socio-ambientali e cittadinanza in movimento. in Dansero E, Bagliani M (eds.), Politiche per l'ambiente dalla natura al territorio. Torino, Italy: UTET. 317–348.
98. Boulding KE (1956) The image: knowledge in life and society. University of Michigan Press. p. 184.
99. Dorling D, Fairbairn D (1997) Mapping: ways of representing the world. London: Addison Wesley Longman. p. 184.
100. Derek G (2005) Geographies, publics, and politics. Progress in Human Geography: 367–378.
101. Wood D (2002) The map as a kind of talk: Brian Harley and the confabulation of the inner and outer voice. Visual Communication 1: 139–161. Available: http://vcj.sagepub.com/cgi/doi/10.1177/147035720200100201. Accessed 2012 Oct 6.
102. Harley JB (2002) The New Nature of Maps: Essays in the History of Cartography. In: Laxton P, editor. Baltimore, Maryland: The Johns Hopkins University Press. p. 331.
103. Brewer CA (1997) Spectral Schemes: Controversial Color Use on Maps. Cartography and Geographic Information Science 24: 203–220. doi: 10.1559/152304097782439231.
104. Monmonier M (1991) How to lie with maps. Chicago, USA: The University of Chicago Press. p. 207.
105. Robinson AH (1977) The uniqueness of the map. Cartography 10: 73–74. Available: http://www.tandfonline.com/doi/abs/10.1080/00690805.1977.10437983. Accessed 2012 Oct 7.
106. MacEachren A (2004) How Maps Work: Representation, Visualization, and Design. New York: The Guilford Press. p. 526.
107. Instituto Geográfico Militar de Ecuador (IGM) (1996) Hoja Rio Bahameno, serie cartográfica J721, edición 1, PIII-F3-4391-III, escala 1:50,000. Quito, Ecuador: IGM impresión.
108. Instituto Geográfico Militar de Ecuador (IGM) (1997) Hoja Rió Yasuní Este, serie cartográfica J721, edición I, PIII-F4-4391-II, escala 1:50,000. Quito, Ecuador: IGM impresión.
109. Instituto Geográfico Militar de Ecuador (IGM) (1993) Hoja Rió Nashiño, serie cartográfica J721, PIV-B2-4390-I, escala 1:50,000. Quito, Ecuador: IGM impresión.
110. Instituto Geográfico Militar de Ecuador (IGM) (1997) Hoja Yamino, serie cartográfica J721, edición 1, PIV-B1-4390-IV, escala 1:50,000. Quito, Ecuador: IGM impresión.
111. Instituto Geográfico Militar de Ecuador (IGM) (1998) Hoja Shushufindi, edición 1, P-III-SA 18-2 Norte, escala 1:250,000. Quito, Ecuador: IGM impresión.
112. Instituto Geográfico Militar de Ecuador (IGM) (1986) Hoja Curaray, PIV-SA 18-6, escala 1:250,000. Quito, Ecuador: IGM impresión.
113. Rodrigues TG, Paradella WR, Oliveira CG (2011) Evaluation of the altimetry from SRTM-3 and planimetry from high-resolution PALSAR FBD data for semi-detailed topographic mapping in the Amazon Region. Anais da Academia Brasileira de Ciências 83: 953–966.
114. Birkett CM (2002) Surface water dynamics in the Amazon Basin: Application of satellite radar altimetry. Journal of Geophysical Research 107. doi: 10.1029/2001JD000609.
115. Falorni G (2005) Analysis and characterization of the vertical accuracy of digital elevation models from the Shuttle Radar Topography Mission. Journal of Geophysical Research 110: 1–20. doi: 10.1029/2003JF000113.
116. Gesch D., J-P. Muller TGF (2006) The Shuttle Radar Topography Mission-Data validation and applications. Photogramm Eng Rem Sens: 233–234.
117. Instituto Geográfico de Ecuador (IGM) (2010) Modelo Digital del Terreno (DEM), Región Amazonica Ecuatoriana. Geodata raster file.
118. NASA website (2012) Landsat Thematic Mapper (TM). Available: http://glovis.usgs.gov. Accessed 2012 Aug 12.
119. NASA Earth Observing System Data and Information System (EOSDIS) (2012) LBA-ECO LC-01 National, Provincial and Park Boundaries, Ecuador doi:10.3334/ORNLDAAC/1057. Available: http://daac.ornl.gov/cgi-bin/dsviewer.pl?ds_id=1057. Accessed 2012 Sep 10.
120. Instituto para el Ecodesarrollo de la Region Amazonica Ecuatoriana (ECORAE) (2008) Road Network in the Amazon Region. Geospatial data shapefile.
121. Walsh SJ, Crews-Meyer, A K (2002) Linking People, Place, and Policy: A Giscience Approach. Springer. p. 360.
122. Wilson JP (2012) GIScience Research at the Thirty-second Annual Esri International User Conference. Transactions in GIS 16: 267–269. doi: 10.1111/j.1467-9671.2012.01345.x.
123. Duckham M, Goodchild MF, Worboys M (2003) Foundations of Geographic Information Science CRC Press. p. 272.
124. Moran EF (2010) Environmental Social Science: Human – Environment interactions and Sustainability. Oxford: Wiley-Blackwell. p. 232.
125. Brandon R, Bottomley B. (2000) Mapping Rural Land Use and Land Cover Change in Carroll County, Arkansas Utilizing Multi-Temporal Landsat Thematic Mapper Satellite Imagery. University of Arkansas. p. Available: http://www.cast.uark.edu/home/research/environmental-studies/lulc-change-in-carroll-county.html.
126. Servindi website (2013) Ecuador: Waoranis exigen proteger a sus hermanos, los aislados Tagaeri y Taromenane. Servindi. Available: http://servindi.org/actualidad/84705. Accessed 2013 Apr 2.

9

The Loss of Efficiency Caused by Agents' Uncoordinated Routing in Transport Networks

Zhongzhi Xu[☯], **Li Sun**[☯], **Junjie Wang, Pu Wang***

School of Traffic and Transportation Engineering, Central South University, Changsha, Hunan, P.R. China

Abstract

Large-scale daily commuting data were combined with detailed geographical information system (GIS) data to analyze the loss of transport efficiency caused by drivers' uncoordinated routing in urban road networks. We used Price of Anarchy (POA) to quantify the loss of transport efficiency and found that both volume and distribution of human mobility demand determine the POA. In order to reduce POA, a small number of highways require considerable decreases in traffic, and their neighboring arterial roads need to attract more traffic. The magnitude of the adjustment in traffic flow can be estimated using the fundamental measure traffic flow only, which is widely available and easy to collect. Surprisingly, the most congested roads or the roads with largest traffic flow were not those requiring the most reduction of traffic. This study can offer guidance for the optimal control of urban traffic and facilitate improvements in the efficiency of transport networks.

Editor: Wen-Bo Du, Beihang University, China

Funding: The work is supported by the National Natural Science Foundation of China (No. 51208520), the FokYing Tong Education Foundation (No. 141075), and the Shenghua Scholar Program of Central South University. The funders had no role in study design, data collection and analysis, decision to publish, or preparation of the manuscript.

Competing Interests: The authors have declared that no competing interests exist.

* Email: wangpu@csu.edu.cn

☯ These authors contributed equally to this work.

Introduction

In this era of unprecedented global urbanization, the fast growth of human mobility has put immense pressure on urban roads [1–3], which has manifested in the form of severe traffic congestion and traffic-related air pollution [4–7]. Improving the efficiency of transport networks has become an urgent problem to solve and it has recently attracted widespread attention from scientific and engineering fields [8–14]. The transport efficiency of a road network is primarily determined by its network topology [15–21], by the volume and distribution of travel demand [9,18–20], and by drivers' routing behavior (the manner in which the road network is used) [11–14]. Studies have shown that increasing the capacity of the important backbone of transportation networks [10] or removing specific segments [11] can make these networks more efficient. It has also been discovered that traffic congestion can be mitigated efficiently by intelligently reducing a small fraction of travel demand [9].

Despite the intensive investigations on the effects of network topology and travel demand on the efficiency of transport networks, the effect of agents' routing behavior has only been studied on theoretical networks or simplified road networks, without considering actual travel demand [11–15]. Using three urban road networks, one type of transport networks, as an example, we show a comprehensive image of the loss of transport efficiency caused by agents' (drivers') uncoordinated routing. Furthermore, we explored the way that can lead a road network to its optimal state. We also believe our findings can shed light on improving other types of transport networks experiencing a lack of coordination among agents, such as the Internet [22].

Data and Methods

The road networks in San Francisco, Santa Clara, and Alameda were extracted from the Bay Area road network (Figure 1), which was provided by NAVTEQ, a commercial provider of geographical information systems data [23]. The road networks are composed of highways and arterial roads. There are 2,816 road segments in San Francisco, 7,269 in Santa Clara, and 5,805 in Alameda. There are 1,144 intersections in San Francisco, 3,420 in Santa Clara, and 2,744 in Alameda. More detailed information regarding the three road networks was provided in Table S1 and Figure S1.

In the following, the Bay Area commuting OD (origin-destination matrix) was first estimated, and consequently the commuting OD was calculated for each county. The Bay Area daily home-work commuting data were provided by the U.S. census bureau [24]. The numbers of trips from residents' home locations to work locations at a street-block level were recorded. The street blocks were grouped into census tracts where they were located (1,398 in total) for generating the census tract based OD.

People use various modes of transportation in their daily commutes, these include cars (driving alone), carpooling, public transportation, bicycling, and walking. Based on the mode split data, the vehicle use rate was calculated for each census tract [25]:

$$VUR(i) = P_{driver}(i) + P_{carpool}(i)/S \quad (1)$$

Here, $P_{driver}(i)$ and $P_{carpool}(i)$ are the fractions of residents in census tract i who drive alone or share a car. The average carpool size in California ($S = 2.25$) was used for these calculations [26]. Next, a mode of transportation (vehicle or non-vehicle) was randomly assigned to each of the residents of each census tract according to the estimated vehicle usage rate VUR (Figure S2). Then the trips not completed by vehicles were filtered out, generating the vehicle-based commuting OD.

The average number of daily trips per person is about 4 in the U.S., this generates about 22 million trips in the Bay Area [27]. Based on the daily distribution of traffic volume, average hourly trip production W during the morning commute (6:00 a.m.–10:00 a.m.) was estimated [28]. The vehicle-based daily commuting OD was rescaled using W to estimate the morning peak hourly commuting OD.

To assign trips to the road networks, the census tract based OD was mapped to the intersection-based OD. For each trip in the census tract based OD, the road intersections within the origin census tract and destination census tract were identified (Figure S3). One intersection in the origin census tract and one intersection in the destination census tract were randomly selected as the origin and destination of the trip in the intersection-based OD.

Finally, the commuting ODs for the three road networks were extracted. Using San Francisco as an example, trips start and end within the county formalize the internal-internal OD, trips that start within the county and end outside it and trips start outside the county and end within it formalize the internal-external OD and the external-internal OD. Trips that start and end outside the county formalize the external-external OD. The origins and destinations of each internal-external trip, external-internal trip, and external-external trip were mapped to the San Francisco road network (Figure 2a). For each internal-external trip, the Dijsktra algorithm was used to find the shortest path (measured in travel time) within the Bay Area road network [29]. The destination of the trip was replaced with the last road intersection that the driver passed before leaving San Francisco. For each external-internal trip, the origin of the trip was mapped to the first road intersection

Figure 1. Bay Area road network and locations of three counties.

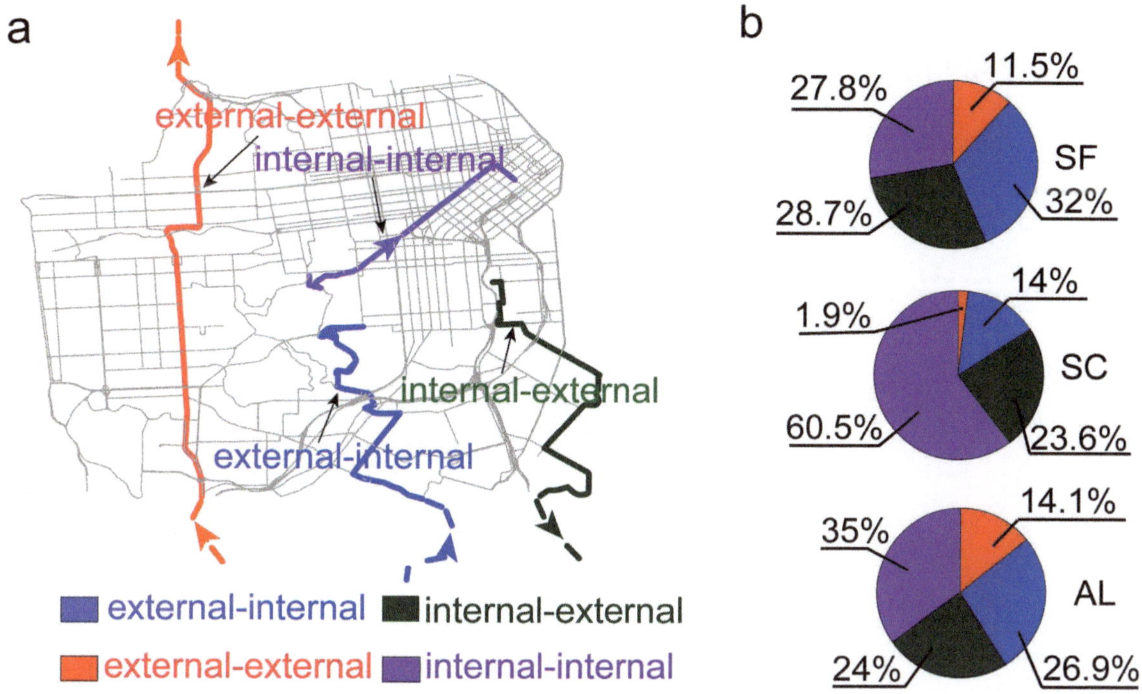

Figure 2. Illustration and statistics of the four types of trips. (*a*) Four types of trips defined by the locations of the origin and destination. Purple, green, blue, and red lines represent the path of an internal-internal trip, an internal-external trip, an external-internal trip, and an external-external trip respectively. (*b*) The statistics of the four types of trips in San Francisco (SF), Santa Clara (SC) and Alameda (AL) (also see Table S2).

that the driver passed upon entering San Francisco. For each external-external trip, the new origin and destination were mapped to the first road intersection that the driver passed upon entering San Francisco and the last intersection passed before leaving San Francisco. The four types of ODs were also generated for Santa Clara and Alameda.

Due to different geographic locations and patterns of land use, the four types of ODs showed different combinations in the three counties (Figure 2b). In San Francisco, there were similar numbers of internal-internal trips, internal-external trips, and external-internal trips. There were slightly more external-internal trips than internal-external trips, indicating that more people enter San Francisco during the morning peak. In Santa Clara, the majority of trips (60.5%) were within the county. There were many more people leaving than entering the county during the morning. People rarely drove across the county. In Alameda, although the majority of trips began and ended within the county, the number of internal-internal trips (35%) was only slightly higher than that of internal-external trips (24%) and external-internal trips (26.9%). Alameda had the most cross-county trips.

The number of trips between a pair of origin and destination N can be approximated by a power-law distribution $P(N) = 509.2N^{-3.32}$ for all the three counties $R^2 > 0.99$, showing that travel demand between most pairs of locations was small, but there was high volume between a few origins and destinations (Figure 3d). The random OD, which had the same number of OD pairs and the same number of trips as the San Francisco OD, was generated and used for conducting comparative studies. The trips in the random OD were randomly assigned to pairs of origins and destinations.

User Equilibrium and Social Optimum

Based on the ODs generated for the three road networks, traffic flow along each road segment was estimated under two scenarios. In the first scenario, all drivers were assumed to know all information regarding the road network and put their own interests first. In this scenario, the whole system reached the user equilibrium (UE) such that no driver could reduce travel time any further by switching paths. This is also known as the Nash equilibrium [12]. The second scenario was the social optimum (SO), the state that was most beneficial overall, meaning that it minimized the total travel time across the whole system [11,30]. The price of anarchy POA was then defined as the ratio of the total travel time of the Nash equilibrium and the total travel time of the social optimum:

$$\mathrm{POA} = T_{\mathrm{UE}}/T_{\mathrm{SO}} \tag{2}$$

The price of anarchy POA quantifies the loss of transportation efficiency caused by drivers' selfish routing. Given the huge traffic volume in a big city, a small POA can still mean a big loss of efficiency. Understanding the method to reduce POA has important consequences for the optimal design and control of transportation systems.

To calculate the equilibrium flow f_{UE} under UE scenario, Wardrop's principle, as described by the Beckmann model, was used [31]. In this case, the objective function in Eq. (3) was minimized:

$$T_{\mathrm{UE}} = \sum \int_0^{f_{ij}} t_{ij}(f)\mathrm{d}f \tag{3}$$

Here, f_{ij} and t_{ij} represent the traffic flow and travel time between a

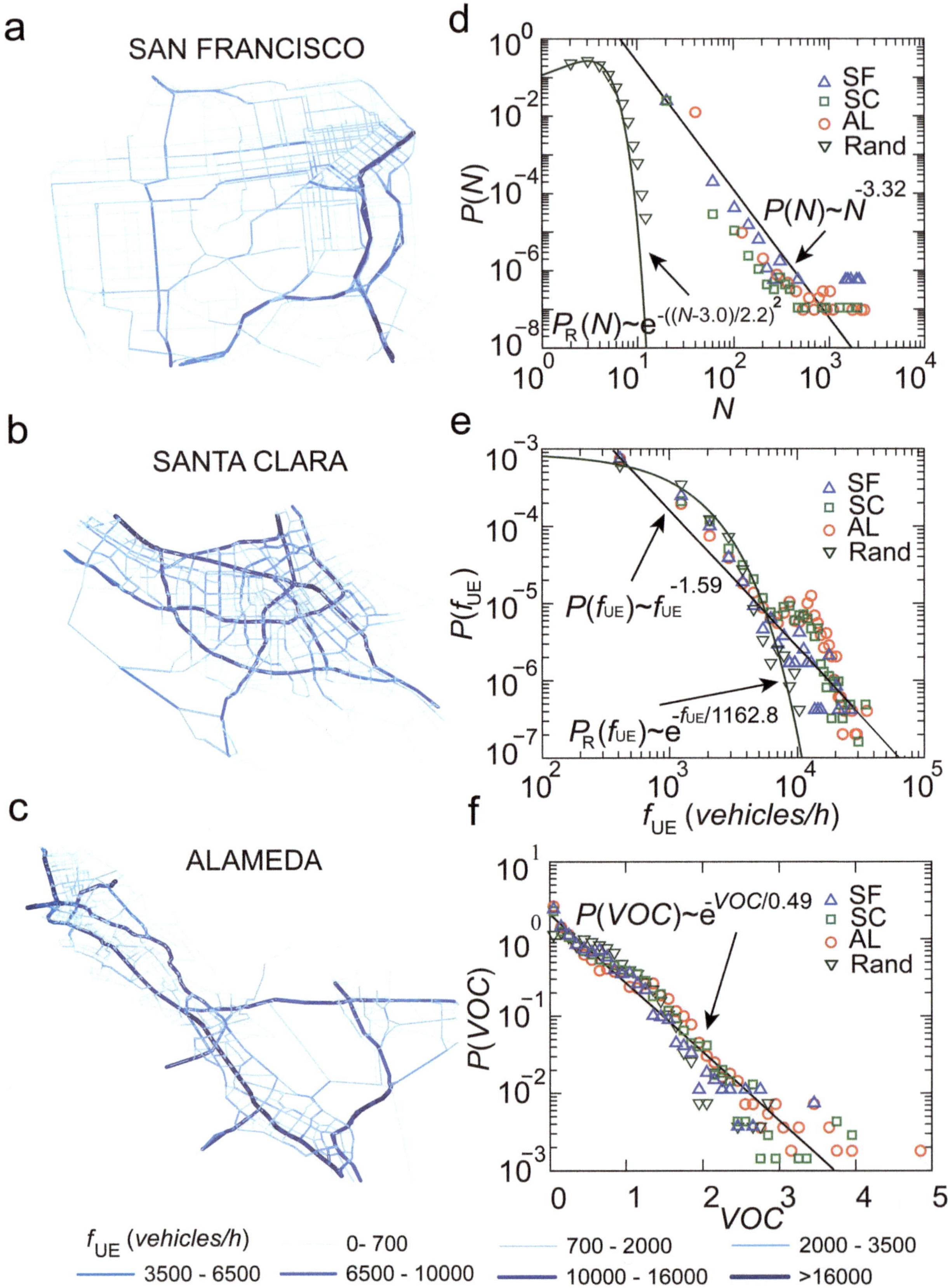

Figure 3. Statistical analysis of traffic flow and volume over capacity (VOC). (*a*), (*b*), (*c*) Spatial distributions of traffic flow are shown for San Francisco (SF), Santa Clara (SC), and Alameda (AL). (*d*) The number of trips between a pair of origin and destination. (*e*) Traffic flow f_{UE} was estimated using the actual ODs and the random OD. (*f*) VOC was estimated using the four ODs.

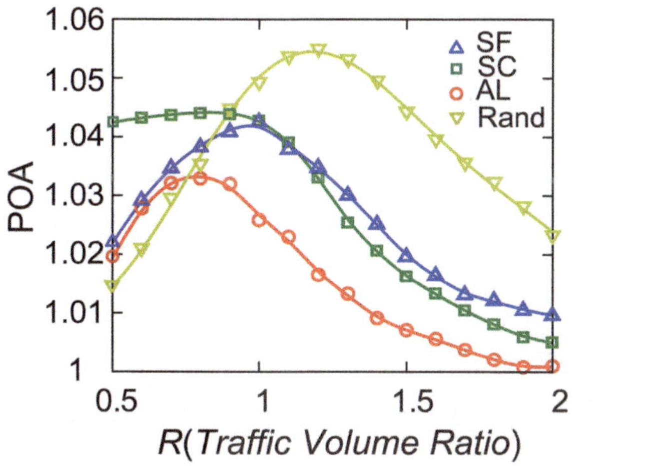

Figure 4. Price of anarchy (POA) versus traffic volume ratio *R*. Using actual travel demand distributions, the maximum POA = 1.043 and POA = 1.033 were observed at $R = 0.8$ for Santa Clara and Alameda. In San Francisco, the maximum POA estimated using actual travel demand distributions and random travel demand distributions were observed at $R = 1.0$ (POA = 1.041) and $R = 1.2$ (POA = 1.056) respectively.

pair of road intersections i and j. Travel time t_{ij} was estimated using the Bureau of Public Roads (BPR) function, which is widely used in civil engineering:

$$t_{ij}(f_{ij}) = t_f(1 + \alpha(f_{ij}/c_{ij})^{\beta}) \tag{4}$$

Here, c_{ij} is the capacity of the road segment from road intersection i to road intersection j. Commonly used values $\alpha = 0.15$ and $\beta = 4$ were selected [25]. The equilibrium flows f_{UE} were then numerically calculated using the Frank-Wolfe algorithm (see Method S1) toolkit provided by TransCAD 5.0, a transportation planning software [25,32].

Under the SO scenario, the socially optimal traffic flows f_{SO} were estimated by minimizing total travel time in Eq. (5):

$$T_{\mathrm{SO}} = \sum f_{ij} t_{ij}(f_{ij}) \tag{5}$$

To calculate the socially optimal flow f_{SO}, the function of travel time was converted to the following:

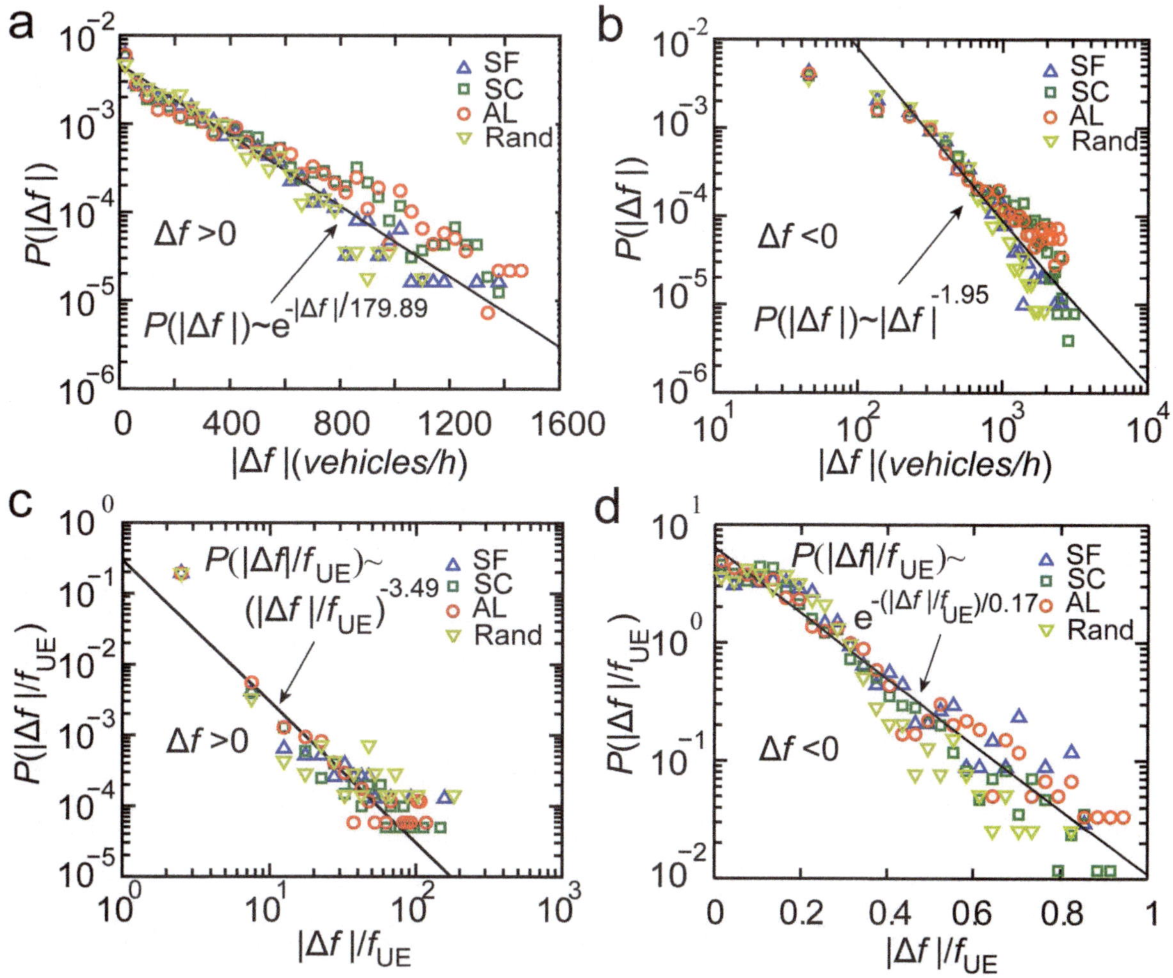

Figure 5. Difference between equilibrium flow f_{UE} and socially optimal flow f_{SO}. (*a*), (*b*) The difference between equilibrium flow and socially optimal flow $|\Delta f|$ in the cases of $\Delta f > 0$ and $\Delta f < 0$. (*c*), (*d*) The rescaled difference between equilibrium flow and socially optimal flow $|\Delta f|/f_{\mathrm{UE}}$ in the cases of $\Delta f > 0$ and $\Delta f < 0$.

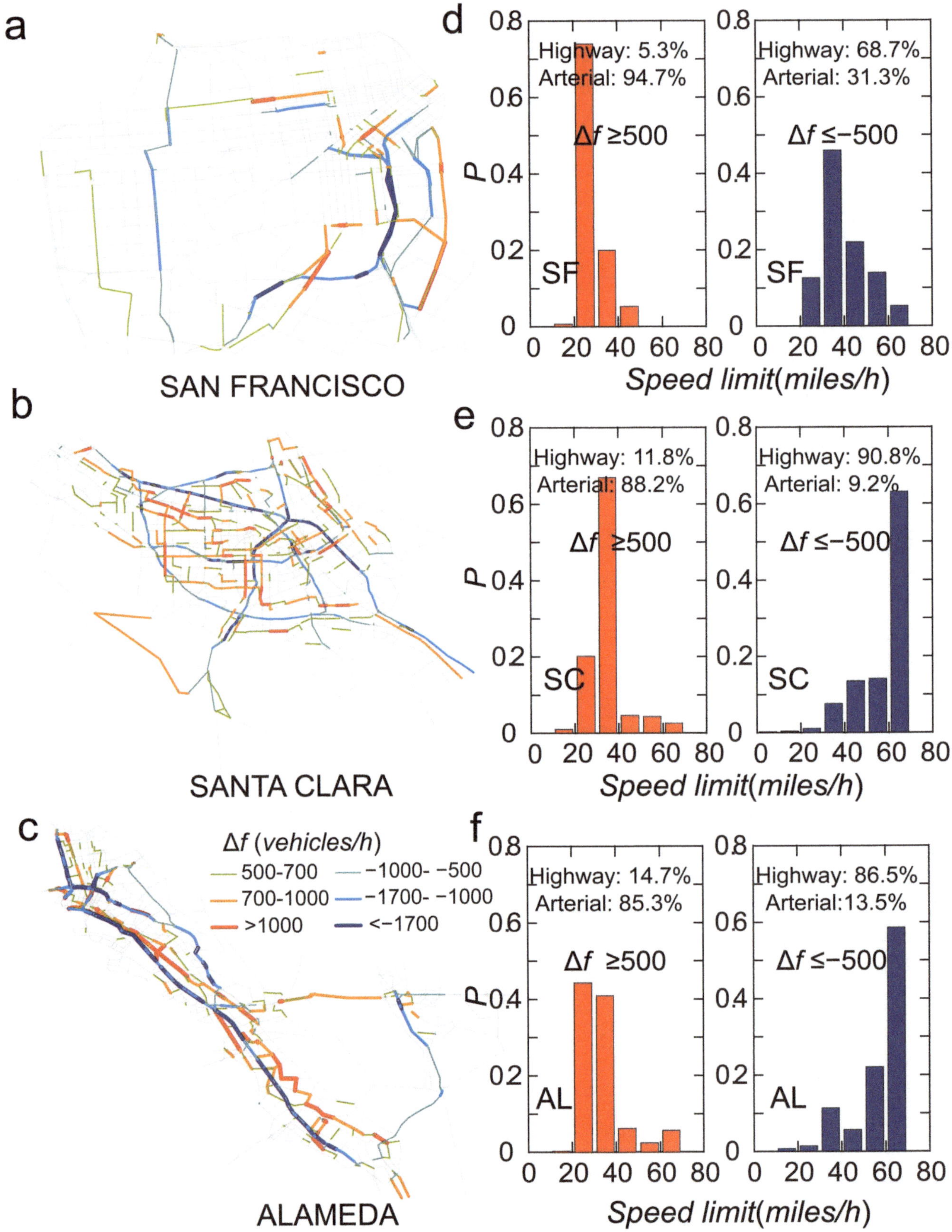

Figure 6. Illustration of road segments with large $|\Delta f|$. *(a), (b), (c)* Road segments with $\Delta f \geq 500$ *(vehicles/hour)* and road segments with $\Delta f \leq -500$ *(vehicles/hour)* in San Francisco, Santa Clara, and Alameda. *(d), (e), (f)* Speed limits of road segments with $\Delta f \geq 500$ *(vehicles/hour)* and $\Delta f \leq -500$ *(vehicles/hour)* in San Francisco, Santa Clara, and Alameda.

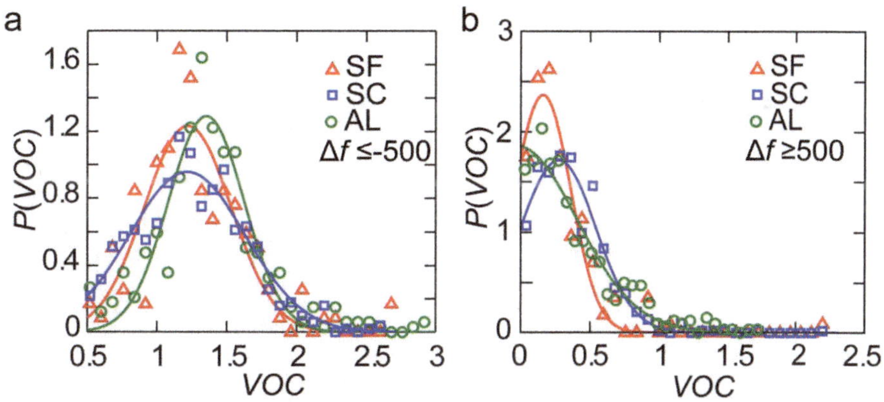

Figure 7. Volume over capacity (VOC) of road segments with large $|\Delta f|$. (*a*) VOC of road segments with $\Delta f \le -500$ (*vehicles/hour*) can be approximated by Gaussian distributions $P(VOC)=ae^{-((VOC-b)/c)^2}$ with $a=1.24$ (0.96, 1.29), $b=1.22$ (1.22, 1.35), $c=0.44$ (0.59, 0.38), and $R^2=0.80$ (0.92, 0.89) for San Francisco (Santa Clara, Alameda). (*b*) VOC of road segments with $\Delta f \ge 500$ (*vehicles/hour*) can be approximated by Gaussian distributions $P(VOC)=ae^{-((VOC-b)/c)^2}$ with $a=2.37$ (1.72, 1.86), $b=0.16$ (0.27, -0.04), $c=0.28$ (0.39, 0.62), and $R^2=0.95$ (0.97, 0.95) for San Francisco (Santa Clara, Alameda).

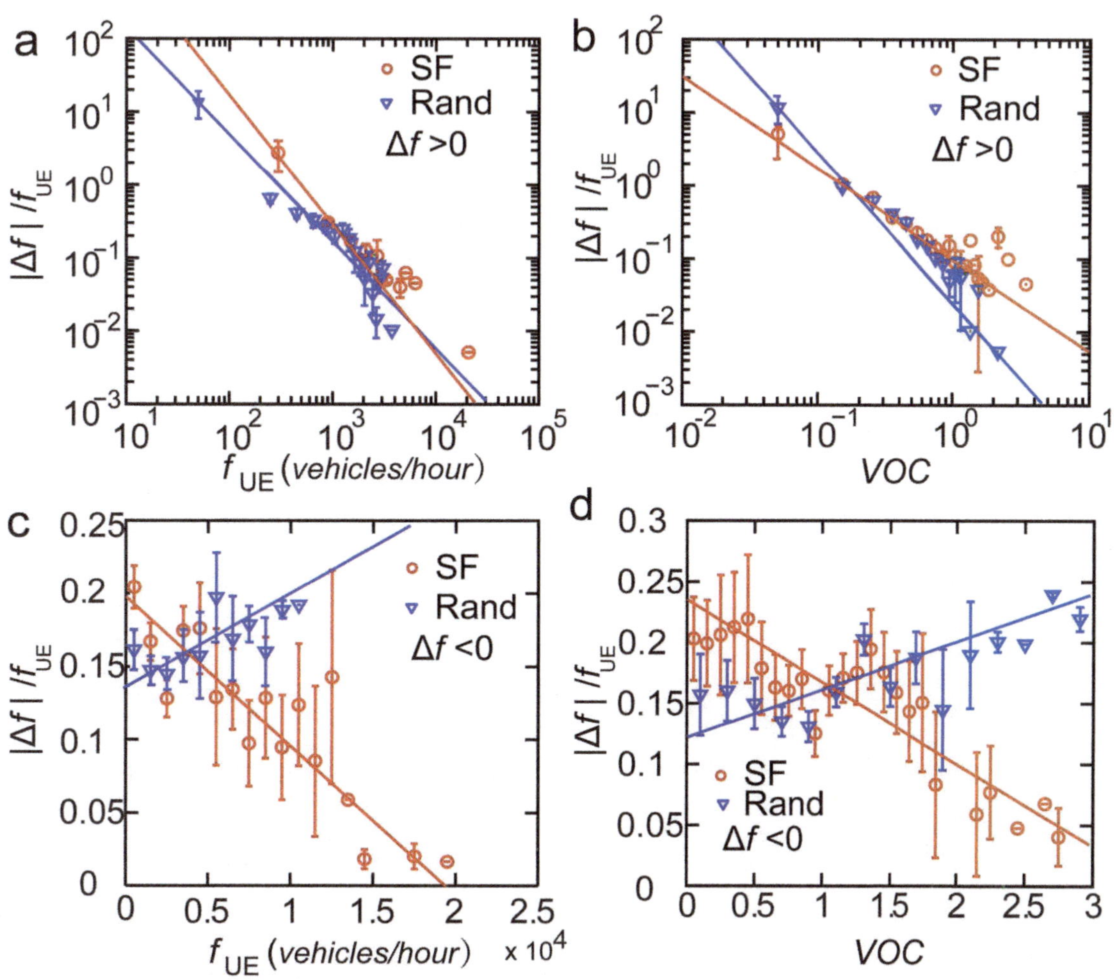

Figure 8. The magnitude of traffic flow adjustment for reducing POA. (*a*), (*b*) When increasing traffic flow of a road segment to reduce POA ($\Delta f > 0$), $|\Delta f|/f_{UE}$ can be approximated by two power-law functions $|\Delta f|/f_{UE}=8.18\times 10^4 f_{UE}^{-1.81}$ ($R^2>0.99$) or $|\Delta f|/f_{UE}=0.09VOC^{-1.34}$ ($R^2>0.99$). (*c*), (*d*) When decreasing traffic flow of a road segment to reduce POA ($\Delta f < 0$), $|\Delta f|/f_{UE}$ can be approximated by $|\Delta f|/f_{UE}=-1.03\times 10^{-5} f_{UE}+0.20$ ($R^2=0.83$) and $|\Delta f|/f_{UE}=-0.06VOC+0.23$ ($R^2=0.80$). Error bar here represents a 95% confidence interval.

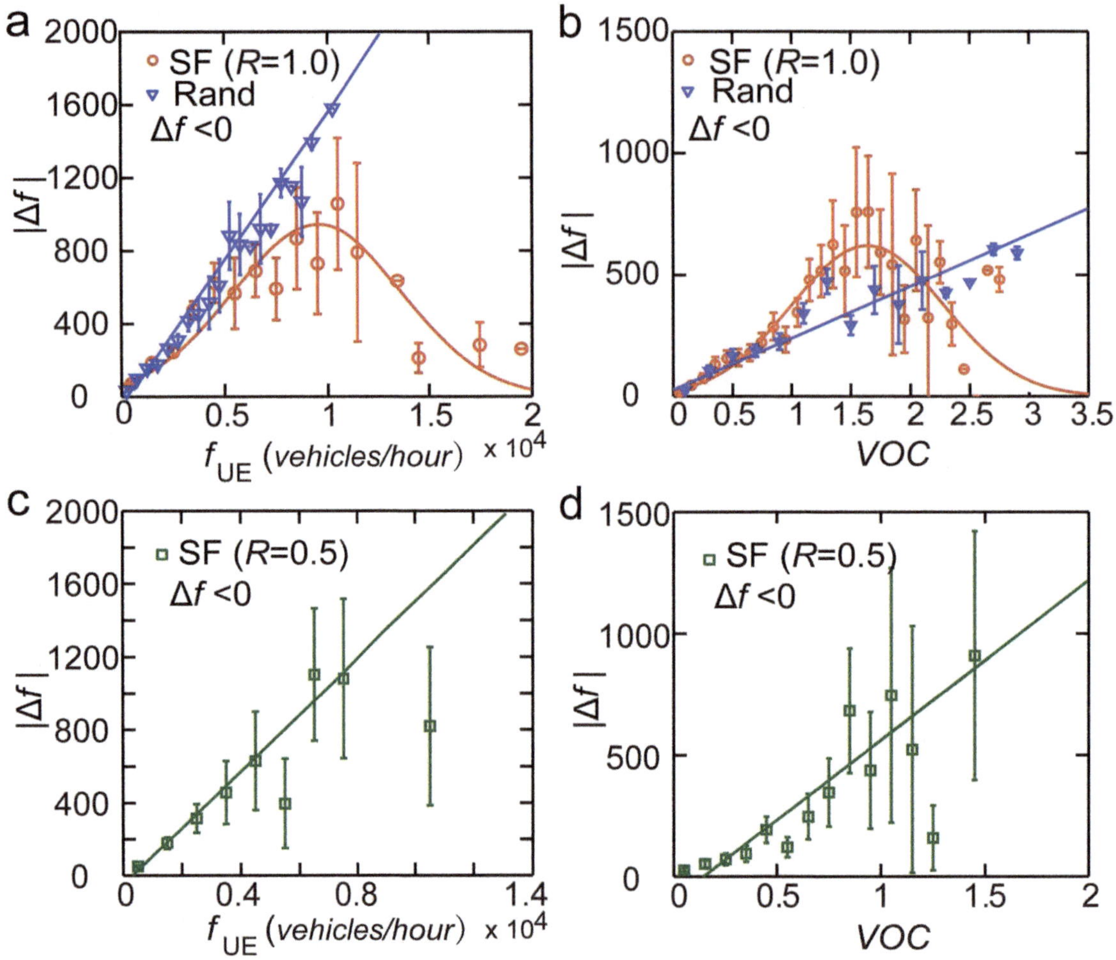

Figure 9. The distribution and volume of travel demand determines the adjustment of traffic flow. (*a*), (*b*) Δf estimated using the San Francisco random OD and actual OD show different correlations with f_{UE} and VOC. Statistical fits $|\Delta f| = 0.15 f_{UE} - 7.9$ ($R^2 > 0.99$), $|\Delta f| = 923e^{-((f_{UE}-9805)/5683)^2}$ ($R^2 = 0.91$), $|\Delta f| = 214.3VOC + 23.0$ ($R^2 = 0.87$) and $|\Delta f| = 630.7e^{-((VOC-1.7)/0.8)^2}$ ($R^2 = 0.94$) were plotted to guide the eyes. (*c*), (*d*) $|\Delta f|$ estimated using the re-scaled ($R = 0.5$) San Francisco actual OD. Statistical fits $|\Delta f| = 0.16 f_{UE} - 49.6$ ($R^2 = 0.89$) and $|\Delta f| = 655.2VOC - 93.4$ ($R^2 = 0.91$) were plotted to guide the eyes. Error bar here represents a 95% confidence interval.

$$\tilde{t}_{ij}(f_{ij}) = t_{ij}(f_{ij}) + f_{ij}\frac{dt_{ij}(f_{ij})}{df_{ij}} \tag{6}$$

Computing the integral of $\tilde{t}_{ij}(f_{ij})$, the following was determined:

$$\begin{aligned}\int_0^{f_{ij}} \tilde{t}_{ij}(\omega)d\omega &= \int_0^{f_{ij}} [t_{ij}(\omega) + \omega\frac{dt_{ij}(\omega)}{d\omega}]d\omega \\ &= \int_0^{f_{ij}} d[t_{ij}(\omega)\omega] = f_{ij}t_{ij}(f_{ij})\end{aligned} \tag{7}$$

$$T_{SO} = \sum \int_0^{f_{ij}} \tilde{t}_{ij}(f)df \tag{8}$$

In this way, equation (7) possesses the same form as equation (3). Traffic flow f_{SO} can also be numerically calculated using the Frank-Wolfe algorithm toolkit with the travel time function $\tilde{t}_{ij}(f_{ij})$ (Eq. (6)).

Results

Experimental tests show that humans find paths within networks in the perspective of minimizing their own travel costs [33,34]. For this reason, the equilibrium flow f_{UE} was used to analyze road usage patterns (Figure 3a–c). As Figure 3e shows that the equilibrium flow f_{UE} can be well approximated by a power-law distribution $P(f_{UE}) = 10.5 f_{UE}^{-1.59}$ for San Francisco, Santa Clara, and Alameda, indicating that road usages are similar and heterogeneous in the three counties ($R^2 = 0.99$). Traffic flow for more than 80% of the road segments was below 2,000 (*vehicles/hour*), and there were 1.0%, 3.4%, and 5.6% of the road segments in San Francisco, Santa Clara, and Alameda having their $f_{UE} > 10{,}000$ (*vehicles/hour*). The equilibrium flow f_{UE} was also measured using the random OD of San Francisco and observed to follow an exponential distribution $P(f_{UE}) \sim e^{-f_{UE}/1{,}162.8}$ (Figure 3e). This indicates a much faster decay.

The volume over capacity $VOC = f_{UE}/C$ was also measured (C is the capacity of a road segment). As shown in Figure 3f, VOC

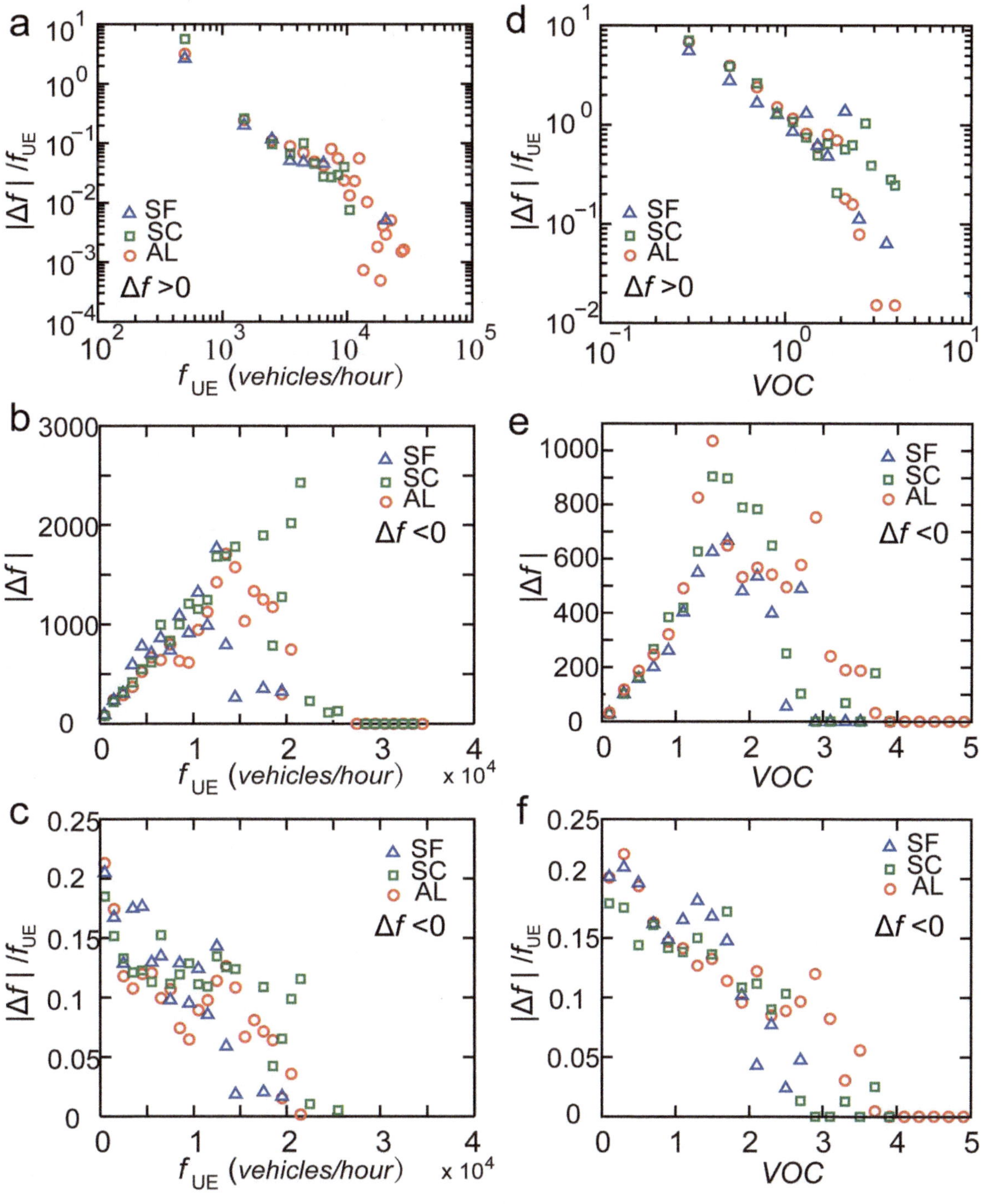

Figure 10. The magnitude of traffic flow adjustment for reducing POA in the three counties. (*a*) $|\Delta f|/f_{UE}$ versus f_{UE} when $\Delta f > 0$. (*b*) $|\Delta f|$ versus f_{UE} when $\Delta f < 0$. (*c*) $|\Delta f|/f_{UE}$ versus f_{UE} when $\Delta f < 0$. (*d*) $|\Delta f|/f_{UE}$ versus VOC when $\Delta f > 0$. (*e*) $|\Delta f|$ versus VOC when $\Delta f < 0$. (*f*) $|\Delta f|/f_{UE}$ versus VOC when $\Delta f < 0$.

estimated using either the actual ODs or the random OD can be approximated using an exponential distribution $P(VOC) = 2.04e^{-2.04VOC}$ ($R^2 = 0.98$), showing traffic flow in most road segments to be well within the roads' capacities, but there was a small number of congested road segments.

To determine how varying traffic volumes (total number of trips) affect POA, traffic volume ratio R was defined as the potential traffic volume over the current traffic volume. Traffic volumes of the actual ODs and the random OD were scaled up or down using parameter R, at the same time the original distributions of travel demands were kept. As shown in Figure 4, the maximum POA = 1.043 and POA = 1.033 were both observed at $R = 0.8$ in Santa Clara and Alameda, and the maximum POA estimated using actual distribution of travel demand was 1.041 in San Francisco when $R = 1.0$ (Table S3). All of the POA first increased with R and then decreased with R, suggesting that the traffic flow patterns are similar under UE and SO scenarios when traffic volume is very small or very large. This validates the generality of the findings obtained in [11]. Because when employing more-detailed travel demand information and road network information, similar pattern of POA versus traffic volume was still observed.

In Figure 4, the random OD of San Francisco was used to show how distribution of travel demand determines the POA. Unlike the maximum POA, which was predicted using the actual OD, a larger maximum POA = 1.056 was observed at a larger traffic volume ratio $R = 1.2$. POA was estimated using the actual OD and the random OD, which were 1.021 and 1.015, respectively, for small values of R, such as $R = 0.5$. They were 1.01 and 1.023, respectively, for large values of R, such as $R = 2.0$. The different patterns observed for POA versus R in these three counties also confirm that the distribution of travel demand needs to be considered in estimating the price of anarchy.

Given the huge traffic volumes in urban areas, a small POA can still mean a big loss of transport efficiency. Taking San Francisco as an example (POA = 1.041), drivers' selfish routing produced 1073.3 hours more travel time during only one hour of the morning rush. To offer guidance to reduce POA in urban road networks, the differences in equilibrium flow and socially optimal flow $\Delta f = f_{SO} - f_{UE}$ were measured for each road segment. Results showed $|\Delta f|$ to be heterogeneously distributed in the road networks ($|\Delta f| = \Delta f$ when $\Delta f > 0$ and $|\Delta f| = -\Delta f$ when $\Delta f < 0$). Most road segments have a small $|\Delta f|$ and a few road segments, the targets of urban traffic controls, have $|\Delta f|$ values as large as 3,000 (*vehicles/hour*). In the three counties, distributions of $|\Delta f|$ can be approximated by an exponential distribution $P(|\Delta f|) \sim e^{-|\Delta f|/179.9}$ ($R^2 = 0.88$) when $\Delta f > 0$ and by a power-law distribution $P(|\Delta f|) \sim |\Delta f|^{-1.95}$ ($R^2 = 0.98$) when $\Delta f < 0$ (Figure 5a and b). The rescaled traffic flow difference $|\Delta f|/f_{UE}$, which quantifies the relative difference between f_{SO} and f_{UE}, can be approximated by a power-law distribution $P(|\Delta f|/f_{UE}) \sim (|\Delta f|/f_{UE})^{-3.49}$ ($R^2 > 0.99$) when $\Delta f > 0$ and by an exponential distribution $P(|\Delta f|/f_{UE}) \sim e^{-(|\Delta f|/f_{UE})/0.17}$ ($R^2 = 0.88$) when $\Delta f < 0$ (Figure 5c and d).

The heterogeneously distributed Δf suggests that traffic flow in only a small number of roads requires considerable adjustment. We found only 5.3% of the road segments in San Francisco, 9.3% of those in Santa Clara, and 9.8% of those in Alameda had $\Delta f \geq 500$ (*vehicles/hour*) or $\Delta f \leq -500$ (*vehicles/hour*) (Figure 6a–c, Table S3). In San Francisco, Santa Clara, and Alameda, 94.7%, 88.2%, and 85.3% of the road segments with $\Delta f \geq 500$ (*vehicles/hour*) are arterial roads, and 68.7%, 90.8%, and 86.5% of the road segments with $\Delta f \leq -500$ (*vehicles/hour*) are highways (Figure 6d–f). Targeted highways can noticeably reduce travel time, so they attracted large amounts of traffic and were overly used (most of them had $VOC > 1.0$) (Figure 7a). Targeted arterial roads near the targeted highways may be suitable as alternative paths, but they do not see as much use as they could (most of them had $VOC < 1.0$) (Figure 7b). The present findings held true for all the three counties, indicating a general explanation for the price of anarchy in road networks. These results can provide insight that can be used to identify roads in need of traffic control in practical situations.

We next explored the magnitude of traffic flow adjustment to reduce POA. Taking San Francisco as an example, when increasing traffic flow of a road segment ($\Delta f > 0$), the rescaled difference of traffic flows $|\Delta f|/f_{UE}$ can be approximated by power-law functions $|\Delta f|/f_{UE} \sim f_{UE}{}^{-1.81}$ and $|\Delta f|/f_{UE} \sim VOC^{-1.34}$. This offers a guidance for increasing traffic flow in the targeted road segments when their f_{UE} or VOC is known (Figure 8a and b). The increase of traffic flow in congested roads or roads with large traffic would be very tiny. Obvious increases in traffic flow were only observed in roads with low f_{UE} and VOC. This law also holds for the case of the random OD.

When decreasing traffic flow of a road segment to reduce POA ($\Delta f < 0$), the rescaled difference of traffic flows $|\Delta f|/f_{UE}$ can be approximated by linear functions $|\Delta f|/f_{UE} = -1.03 \times 10^{-5} f_{UE} + 0.20$ ($R^2 = 0.83$) and $|\Delta f|/f_{UE} = -0.06\,VOC + 0.23$ ($R^2 = 0.80$) (Figure 8c and d). However, for the case of the random OD, different functional relationship was found, $|\Delta f|/f_{UE}$ was observed to increase linearly with f_{UE} and VOC (Figure 8c and d).

To understand the observed different relationships between $|\Delta f|/f_{UE}$ and f_{UE} (VOC), we analyzed the traffic flow difference $|\Delta f|$ for the case of the random OD and the case of the actual OD. We found that $|\Delta f|$ estimated using the random OD increased linearly with f_{UE} and VOC. However, counter intuitively, $|\Delta f|$ estimated using the actual OD first increased and then decreased with f_{UE} and VOC. The roads with the largest flow or VOC were not those requiring the highest reduction of traffic. The highest reduction of traffic flow was observed at intermediate values of $f_{UE} \sim 9{,}805$ (*vehicles/h*) and $VOC \sim 1.7$ (Figure 9a and b).

The counter intuitive result could be resulted from the different distributions of traffic flow estimated by the random OD and the actual OD. Comparing with the case of the random OD, traffic flow estimated using the actual OD was more heterogeneously distributed. There are much more road segments with large traffic flow (Figure 3e), preventing drivers from selfishly switching routes. This phenomenon is similar to that when traffic volume is very large, traffic flow patterns under UE scenario and SO scenario are similar (POA~1.0). To further test this explanation, the actual ODs were scaled down using parameter $R = 0.5$, the observed maximum traffic flow is largely reduced. As Figure 9c and d shows, the difference in traffic flow $|\Delta f|$ shows similar pattern with the case of the random OD rather than the case of the actual OD.

We further analyzed the magnitude of traffic flow adjustment for reducing POA for Santa Clara and Alameda. Similar functional relationships between $|\Delta f|/f_{UE}$ and f_{UE}, VOC were observed (Figure 10). The significance of these findings is that it offered useful guidance that could reduce POA for urban road networks, using traffic flow f_{UE} only. This was widely available and recorded by many devices.

Conclusions

To conclude, we generate morning-peak commute ODs for three Bay Area counties and study the price of anarchy using actual travel demand in large-scale road networks. The different patterns observed for POA versus R in the three counties showed that both volume and distribution of travel demand determine the POA in a road network. The difference of equilibrium flow and socially optimal flow $|\Delta f|$ was measured for each road segment. It was found that $|\Delta f|$ for most roads is tiny. For this reason, adjusting traffic flow of a small number of roads can push a system toward its social optimum. Here, roads with large $|\Delta f|$ were found to be only a few highways that saw excessive use due to their high speed limits and the neighboring arterial roads, which can offer alternative paths. The rescaled traffic flow difference $|\Delta f|/f_{UE}$ was found to have power-law functional relationships with f_{UE} and VOC when $\Delta f > 0$ and have linear functional relationships with f_{UE} and VOC when $\Delta f < 0$. Surprisingly, the roads with the largest traffic flow and VOC did not have the largest adjustment of traffic flow.

The present work can inform specific intervention strategies on reducing the loss of efficiency caused by agents' selfish routing in urban road networks, a crucial infrastructure that billions of people use every day. The present work uses real-world travel demand data to explore the way to reduce POA. The elucidation of these findings may draw more attention to the use of actual transport demand information in the optimal control of a broad set of networks experiencing a lack of coordination among agents.

Supporting Information

Figure S1 Distributions of lengths and free travel times of road segments. Both lengths and free travel times of road segments follow similar distributions in three counties.

Figure S2 Vehicle usage rates (*VUR*) in census tracts. Different colors represent different vehicle usage rates. Urban areas have lower *VUR* than suburban areas.

Figure S3 Illustration of census tracts and convention from census tract based OD to intersection based OD. The road segments in the vicinity of San Francisco downtown are depicted by gray lines and the small black dots are the road intersections that lie in the census tracts inside.

Table S1 Census data and road networks in San Francisco, Santa Clara, and Alameda.

Table S2 Statistics of four types of trips in San Francisco, Santa Clara, and Alameda.

Table S3 A summarization of the results in figure 4 and figure 6 for San Francisco, Santa Clara, and Alameda.

Method S1 Solving the Beckmann model by the Frank-Wolfe algorithm.

Acknowledgments

The authors thank Marta González and Lijun Sun for valuable discussions.

Author Contributions

Conceived and designed the experiments: PW. Performed the experiments: ZX LS. Analyzed the data: ZX LS JW. Contributed to the writing of the manuscript: LS PW.

References

1. Brockmann D, Hufnagel L, Geisel T (2006) The scaling laws of human travel. Nature 439: 462–465.
2. González MC, Hidalgo CA, Barabási AL (2008) Understanding individual human mobility patterns. Nature 435: 779–782.
3. Song C, Qu Z, Blumm N, Barabási AL (2010) Limits of predictability in human mobility. Science 327: 1018–1021.
4. Helbing D (2003) A section-based queueing-theoretical traffic model for congestion and travel time analysis in networks. J. Phys. A: Math. Gen. 36: 593–598.
5. Schrank D, Lomax T (2009) Annual urban mobility report (Texas Transportation Institute). http://mobility.tamu.edu/ums/.
6. Chin ATH (1996) Containing air pollution and traffic congestion: transport policy and the environment in Singapore. Atmospheric Environment 30(5): 787–801.
7. Rosenlund M. Forastiere F, Stafoggia M, Porta D, Perucci M, et al (2008) Comparison of regression models with land-use and emissions data to predict the spatial distribution of traffic-related air pollution in Rome. Journal of Exposure Science and Environmental Epidemiology 18: 192–199.
8. Du WB, Wu ZX, Cai KQ (2013) Effective usage of shortest paths promotes transportation efficiency on scale-free networks. Physica A 392(17): 3505–3512.
9. Wang P, Hunter T, Bayen AM, Schechtner K, González MC (2012) Understanding the road usage patterns in urban areas. Scientific Reports 2: 1001.
10. Wu Z, Braunstein LA, Havlin S, Stanley HE (2006) Transport in weighted networks: partition into superhighways and roads. Phys. Rev. Lett. 96: 148702.
11. Youn H, Gastner MT, Jeong H (2008) Price of anarchy in transportation networks: efficiency and optimality control. Phys. Rev. Lett. 12(101): 0712–1598.
12. Roughgarden T (2005) Selfish routing and the price of anarchy. Cambridge, USA: MIT Press.
13. Roughgarden T (2006) On the severity of Braess's Paradox: Designing networks for selfish users is hard. Journal of Computer and System Sciences 72: 922–953.
14. Nagurney A (2010) The negation of the Braess's paradox as demand increases: The wisdom of crowds in transportation networks. EPL 91(4): 48002.
15. Koutsoupias E, Papadimitriou C (1999) Worst-case equilibria. STACS 99 Springer Berlin Heidelberg: 404–413.
16. Barthélemy M, Flammini A (2008) Modeling urban streets patterns. Phys. Rev. Lett 100: 138702.
17. Crucitti P, Latora V, Porta S (2006) Centrality measures in spatial networks of urban streets. Phys Rev E 73: 036125.
18. Yan G, Zhou T, Hu B, Fu ZQ, Wang BH (2006) Efficient routing on complex networks. Phys. Rev. E. 73: 046108.
19. Jiang B, Zhao S, Yin J (2008) Self-organized natural roads for predicting traffic flow: a sensitivity study. J Stat Mech: P07008.
20. Kurant M, Thiran P (2006) Layered complex networks. Phys. Rev. Lett. 96: 138701.
21. Liu C, Du WB, Wang WX (2014) Particle Swarm Optimization with Scale-Free Interactions. PloS one 9(5): e97822.
22. Acemoglu D, Johari R, Ozdaglar A (2007) Partially Optimal Routing. IEEE Journal on selected areas in communications 25(6): 1148–1160.
23. Navteq Official website. Available: http://www.navteq.com/. Accessed 2014 Oct 5.
24. US Census Bureau website. Available: http://www.census.gov/geo/www/tiger/tgrshp2010/tgrshp20 10.html. Accessed 2014 Oct 5.
25. Caliper website. Available: http://www.caliper.com/PDFs/TravelDemandModelingBrochure.pdf. Accessed 2014 Oct 5.
26. National TDM and Telework Clearinghouse website. Available: http://www.nctr.usf.edu/clearinghouse/censusavo.htm. Accessed 2014 Oct 5.
27. Bureau of Transportation Statistics website. Available: http://www.rita.dot.gov/bts/sites/rita.dot.gov.bts/files/subject_areas/national_household_travel_survey/index.html. Accessed 2014 Oct 5.
28. Seto YW, Holt A, Rivard T, Bhatia R (2007) Spatial distribution of traffic induced noise exposures in a US city: an analytic tool for assessing the health impacts of urban planning decisions. International Journal of Health Geographics 6: 24.
29. Dijkstra EW (1959) A note on two problems in connexion with graphs. NumerischeMathematik 1: 269–271.
30. Nathan H Gartner (1980) Optimal traffic assignment with elastic demands: a review part ii. algorithmic approaches. Transportation Research 14(2): 192–208.
31. Beckmann MJ, Meguire CR, Winsten CB (1956) Studies in the economics of transportation. New Haven, USA: Yale University Press.
32. LeBlanc LJ, Morlok EK, Pierskalla WP (1975) An efficient approach to solving the road network equilibrium traffic assignment problem. Transportation Research 9(5): 309–318.
33. Selten R, Chmura T, Pitz T, Kube S, Schreckenberg M (2007) Commuters route choice behaviour. Games and Economic Behavior 58(2): 394–406.
34. Stein WE, Rapoport A, Seale DA, Zhang H, Zwick R, et al. (2007) Batch queues with choice of arrivals: Equilibrium analysis and experimental study. Games and Economic Behavior 59(2): 345–363.

10

Analysis of Feature Intervisibility and Cumulative Visibility Using GIS, Bayesian and Spatial Statistics: A Study from the Mandara Mountains, Northern Cameroon

David K. Wright[1]*, Scott MacEachern[2], Jaeyong Lee[3]

1 Department of Archaeology and Art History, Seoul National University, Seoul, Republic of Korea, **2** Department of Anthropology, Bowdoin College, Brunswick, Maine, United States of America, **3** Department of Statistics, Seoul National University, Seoul, Republic of Korea

Abstract

The locations of *diy-geδ-bay* (DGB) sites in the Mandara Mountains, northern Cameroon are hypothesized to occur as a function of their ability to see and be seen from points on the surrounding landscape. A series of geostatistical, two-way and Bayesian logistic regression analyses were performed to test two hypotheses related to the intervisibility of the sites to one another and their visual prominence on the landscape. We determine that the intervisibility of the sites to one another is highly statistically significant when compared to 10 stratified-random permutations of DGB sites. Bayesian logistic regression additionally demonstrates that the visibility of the sites to points on the surrounding landscape is statistically significant. The location of sites appears to have also been selected on the basis of lower slope than random permutations of sites. Using statistical measures, many of which are not commonly employed in archaeological research, to evaluate aspects of visibility on the landscape, we conclude that the placement of DGB sites improved their conspicuousness for enhanced ritual, social cooperation and/or competition purposes.

Editor: Lalit Kumar, University of New England, Australia

Funding: This work was supported by an international faculty research grant from Seoul National University, the National Geographic Society's Committee for Research and Exploration (grant # 9494-13), and the National Research Foundation of Korea Grant #2013S1A5B6043901 funded by the Korean government. The funders had no role in study design, data collection and analysis, decision to publish, or preparation of the manuscript.

Competing Interests: The authors have declared that no competing interests exist.

* Email: msafiri@snu.ac.kr

Introduction

How and why people settle onto landscapes have been focal research questions in archaeology since the advent of the discipline (e.g., [1–3]). Post-processual archaeology challenged the assumption that the motivations guiding human behavior can be analytically reduced to testable sets of assumptions [4]. However, it is now widely accepted that a middle ground exists between generating absolute laws of governing human behavior and human agents behaving independent of systems or repeatable principles [5–7]. The use of scientific mapping tools can be used to pattern human settlements, the construction of archaeological features and provide insight into how human agency expresses itself onto a landscape.

Monumental architecture is commonly seen as a means for solidifying political hegemony [8–10] or promoting social cohesion [11–13] within or between communities. In Africa, the explanations for the construction of walled settlements and monumental architecture are as diverse as the continent itself [14–17]. The diversity of human populations on the African continent gave rise to different forms of political complexity that exhibit unique characteristics worthy of situating in the global discourse of social ascendancy [18].

In the mid-1990s, a series of tell mound sites at the edge of the Mandara Mountains were excavated in order to study the evolution of social and political systems over the last three millennia (Fig. 1) [19]. The culmination of these cultural developments resulted in the rise of predatory, slave-raiding plains states. More recent research at a set of monumental stone sites (the *diy-geδ-bay* or DGB sites) in the northwestern Mandara massif indicates a parallel history of variable political complexity in the mountains over the last 800 years at least, with some evidence for occupation at least a century or two before that [20]. Archaeological and historical data strongly indicate that the relationship between mountains and plains was extremely complex, and that both regions offered resources and challenges that mutually influenced political and cultural developments in the other [20–22].

The motivations behind the construction of DGB sites remain a source of speculation, with one theory being that they were initially constructed as a response to profound droughts in the 15th and 16th centuries CE [19]. This seems to have been a particularly intense period of occupation at the two largest sites, DGB-1 and DGB-2, but excavations on the former has established that initial stages of construction two place at least two centuries before the period of drought [19]. Based on examination of site locations on 1:50,000 topographic maps, Nicolas David [23] has hypothesized

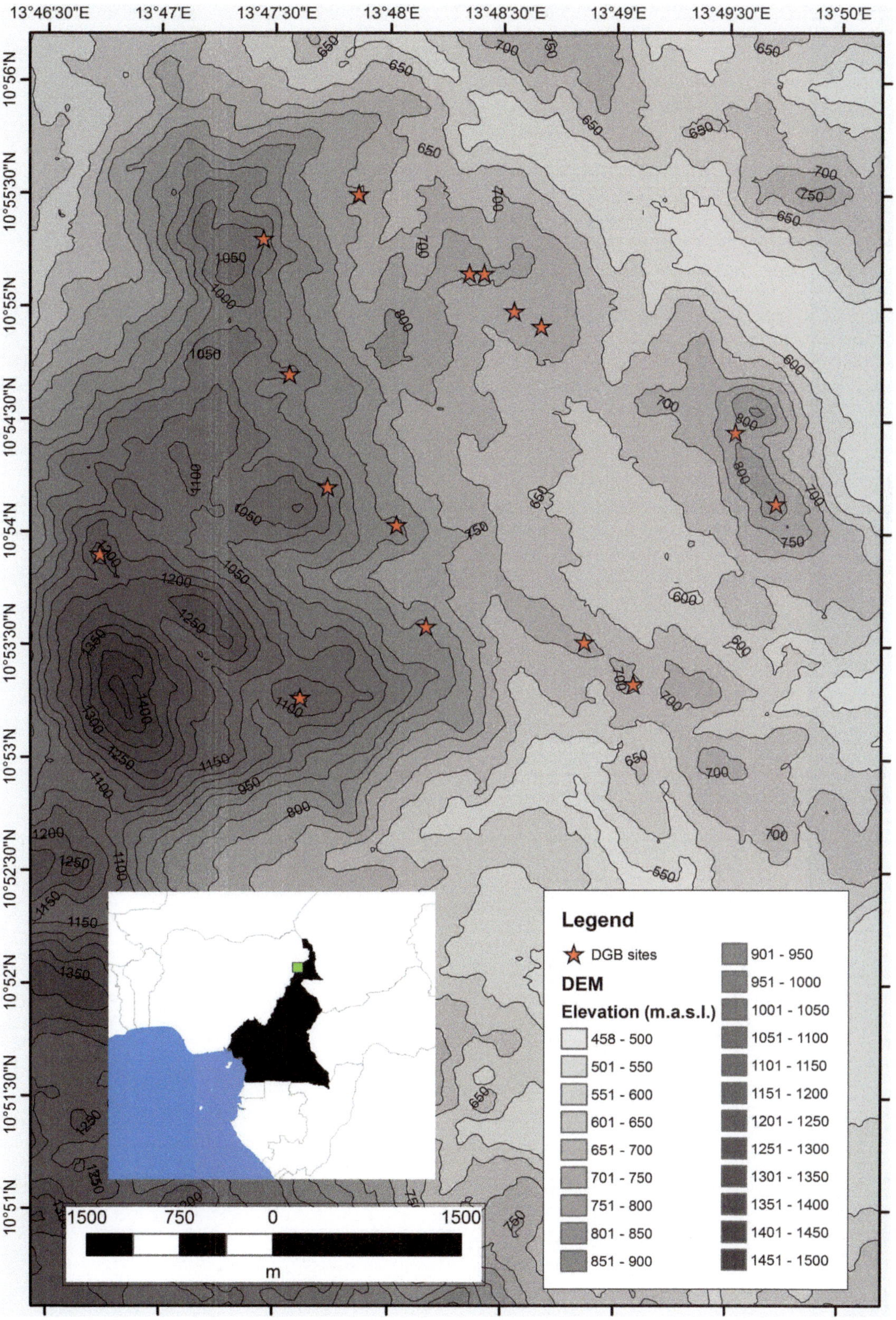

Figure 1. Location of the project area and DGB sites.

that the sites were situated non-randomly on high ground in order to maximize their visibility over surrounding terrain and their intervisibility, possibly in the course of ritual and/or political competition between local communities and their leaders. However, this hypothesis of non-random location in the service of intervisibility has heretofore remained untested using modern computing methods.

In this paper, we describe novel ways of determining relative intervisibility of archaeological sites as well as the cumulative viewshed of the surrounding landscape based on a Geographical Information Systems (GIS) analysis. Using DGB sites as our case study, we statistically analyze whether the locations of sites were randomly located above an elevational threshold of 700 m.a.s.l. or were selected in order to maximize their viewshed. The results of our statistical tests reject the null hypothesis that DGB site locations were randomly selected and we argue that although other causative site selection factors cannot be ruled out, the viewscape of the DGB sites was certainly a factor in where sites were constructed.

Background to the research area: DGB sites

The landscape of the northern Mandara Mountains in northern Cameroon and Nigeria is one of the most densely populated in Central Africa, with densities of up to 220 people/km^2 in a dry Sudanian environment. It is thus by necessity an almost entirely domesticated landscape, with all of the arable land under agriculture and with virtually every tree and shrub exploited for human needs. The region is extremely diverse ethnically, linguistically and culturally, with more than 20 different languages spoken in an area of about 2000 km^2, and with a complex sociopolitical history to match.

The 6.35-×-6.5 km area covered in this study includes 16 DGB sites that have been previously identified by archaeologists [20,23]. These sites are scattered across volcanic inselbergs of the Mandara Mountains, which comprise the northeastern segment of the Cameroon Volcanic Line [24]. The Mandara Mountains are mapped as trachyte and rhyolitic plugs [25]. The topography is highly irregular with steep peaks and dissected valleys. The native vegetation is comprised of mosaics of *Pennistum* sp. grasslands and *Acacia* sp. woodlands and allies [26], but has been heavily anthropogenically modified into an agricultural landscape during the Late Holocene with intensive human settlement [20].

The 16 DGB sites are prehistoric monumental architectural complexes made up of varying combinations of rubble-filled platforms and terraces, and faced with distinctive and carefully arranged dry-stone facades. A number of them, and all so far excavated, contain a variety of interior passages, staircases, courtyards and chambers. The two largest, DGB-1 and DGB-2, are situated just over 100 m apart and are at least in part contemporary; they thus form a single monumental site complex over an area of about 2.5 ha (Fig. 2.). The other DGB sites are significantly smaller. Seventeen dates from DGB-1 establish occupation during the 13th to 17th centuries AD, while the one reliable date from DGB-2 is from the 15th century [19]. Two dates from DGB-1, the only other excavated DGB site, are of the late 14th–15th century AD [23]. It is quite possible that further excavations would increase the period of site usage at these latter sites.

The functions of the DGB sites are not completely understood. They do not appear to have defensive functions, given their

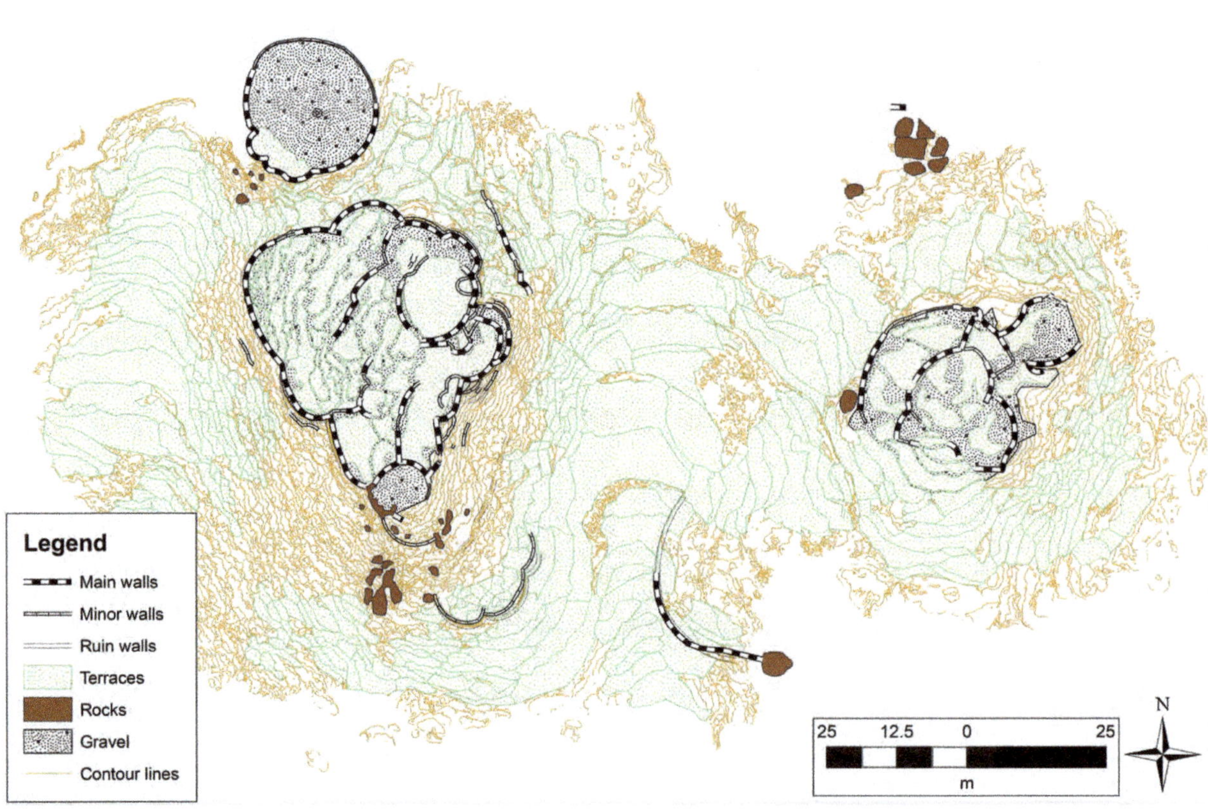

Figure 2. Laser phase scans using a Leica HDS of the DGB-1 and DGB-2 complexes (complements of Heinz Rüther and the Zamani Project).

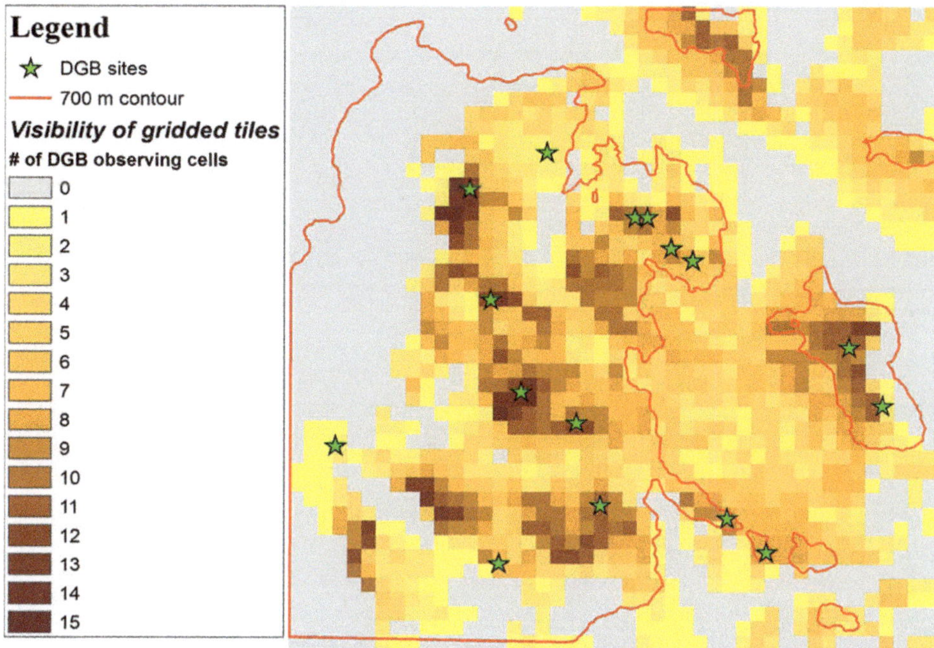

Figure 3. Observation point analysis of 2025 tiles showing the number of Actual DGB sites able to see each cell.

positioning, unfortified wall features and lack of close access to water. There is some evidence for contemporary domestic occupation on DGB-1 and in the area between DGB-1 and DGB-2, but they do not seem to have been primarily habitation sites. They provide a striking impression of being in some sense platforms for display, almost for theater: their raised open platforms would certainly afford easy visibility for public activities, while sunken courtyards, subterranean passages and chambers would accommodate private ceremonialism, with stairways linking the two worlds.

David [23] has located the functioning of the DGB sites in the context of modern Mandara montagnard ritual practice, in ceremonialism involving water and fecundity, while MacEachern [27] emphasizes their potential role in the increasingly complex political world of the southern Lake Chad Basin in the mid-second millennium AD, when predatory states and slave-raiding were becoming important elements in the plains north of the Mandara Mountains. The contemporary capital of one of these states, Wandala, was located only 30 km north of the DGB sites, at Keroua/Kirawa on the modern Cameroon-Nigeria border. It is very likely that the DGB sites fulfilled both of these roles simultaneously. Their size and imposing architectural features, along with the labor that would have been invested in their construction, makes it almost certain that they would have played an important role in sociopolitical relationships in the region during their period of use. Given that fact, it will be extremely useful to know if they were situated to maximize their viewsheds and their intervisibility. This would imply that the 16 sites were in fact of an interacting regional cultural system, and might increase the likelihood that some element of comparison, cooperation and/or competition was involved in their use. There are precedents for such regional ceremonialism in the area [28], particularly with the *maray* ceremony held at varying intervals among mountain communities. The 'passing' of the ritual between communities might be relevant to questions of intervisibility among ceremonial sites.

The locations of the sites have been previously hypothesized to be situated according to principles of intervisibility [23], potentially for reasons of interregional cooperation and/or competition. Herein, we articulate two sets of testable hypotheses:

(1a) H1: The locations of DGB sites are situated to maximize intervisibility of the sites to one another.

(1b) H0(null): The locations of DGB sites are randomly situated above 700 m.a.s.l. and are not situated to maximize intervisibility to one another.

(2a) H2: The locations of DGB sites are situated to maximize the cumulative viewshed of the surrounding valleys.

(2b) H0: The locations of DGB sites are randomly situated above 700 m.a.s.l. and are not situated to maximize the cumulative viewshed of the surrounding valleys.

By comparing the locations of actual DGB sites (hereafter, Actual DGB) to permutations of stratified random "DGB sites" (hereafter, Random DGB-1, 2, 3...), we intend to accept or reject the hypotheses based on statistical measures of randomness. We are not attempting to eliminate other potential factors for site selection (geology, aspect, soil fertility of adjacent land, visibility for astronomical purposes, etc.), merely testing an earlier hypothesis using the best fit of data generated with modern computing techniques.

Visibility studies in archaeology

Landscape visibility and feature intervisibility are related concepts in GIS data processing. The landscape visibility is concerned with determining whether features are located in order to see positions at lower elevations, and potentially vice-versa [29]. Methods analyzing feature intervisibility typically determine whether features at relatively similar elevations might have been constructed in their given locations in order to facilitate their ability to see one another. In both cases, the "viewshed" is considered a tangible cultural asset, worthy of enhancing or restricting for some socially important reason. While determining whether or not sites were constructed in order to enhance visibility

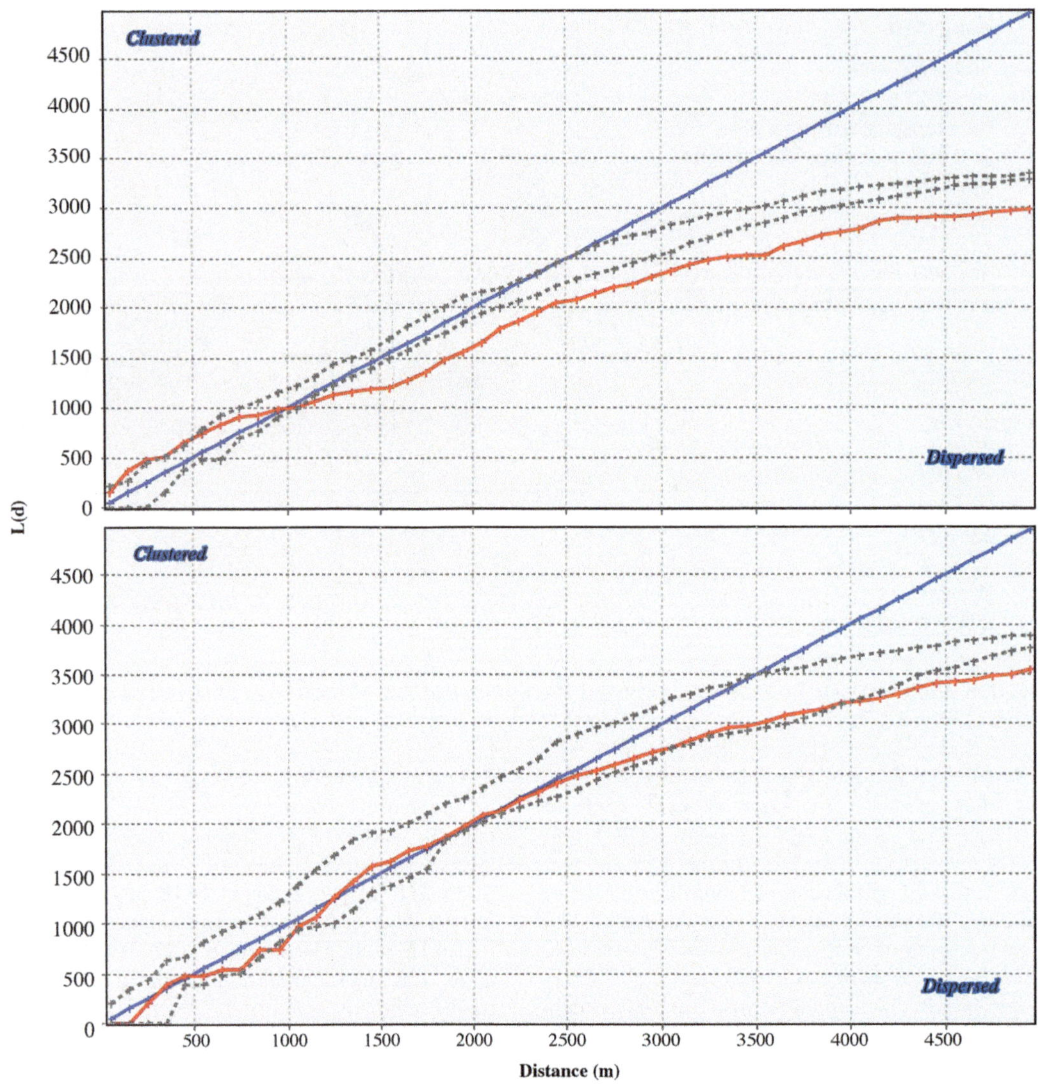

Figure 4. Ripley's K analysis of clustering of Actual DGB sites (top) and Random DGB-1 (bottom). The blue line represents the predicted values, the red line represents the actual values analyzed and the dotted lines represent the 2-σ confidence envelope.

has become relatively simple, getting at the emic cultural dimensions of feature construction has become no less simple than in the days preceding automated spatial computational techniques.

The intervisibility of archaeological sites or features has been a topic of discussion since the advent of GIS. An early pioneer, Kvamme [30], advocates testing samples of sites against a "background standard" in order to identify random vs. non-random patterns of site placement. Wheatley [31] analyzes the placement Neolithic funerary monuments using a Kolmorgov-Smirnoff (K–S) test of cumulative frequency distributions to argue that the patterns show intentional intervisibility of "long barrow" sites in Stonehenge, but the degree to which the sites are visible to the landscape in general is variable according to region (see also [32] for a similar study on rural settlement in Greece). K–S tests are used to determine whether samples exceed the statistical probability of being randomly distributed. In Kvamme's [30] analysis, "the sampling frame of...cell values...serves as the background referent distribution". However, Fisher [33] argues that such studies are most effective if there is a control set of randomized points to compare in order to test random vs. non-random effects. Additionally, in recent years, K–S has been determined to be a less effective test of normality in datasets compared to other non-parametric tests such as Shapiro-Wilk, Lilliefors and Anderson-Darling [34].

A second type of visibility analysis includes the ability of observers to see points on a landscape from archaeological features [33,35,36]. This has been called "cumulative viewshed" analysis and has been employed in numerous archaeological studies [30–33,37,38]. Good visibility of a landscape to and from an archaeological site or feature has been argued to have military applications of site defensibility [39–42], resource acquisition [43–45] or to solidify some aspect of social power/cohesion through a feature's visibility [46–50]. Regardless of the purported reasons for doing so, it is clear that prehistoric people frequently made informed choices about how to situate archaeological sites to either enhance or reduce visibility relative to points on a landscape.

Early pioneering work on landscape visibility was very labor intensive and lacked statistical rigor because of a lack of data processing tools (e.g., [37,51]), but Geographic Information

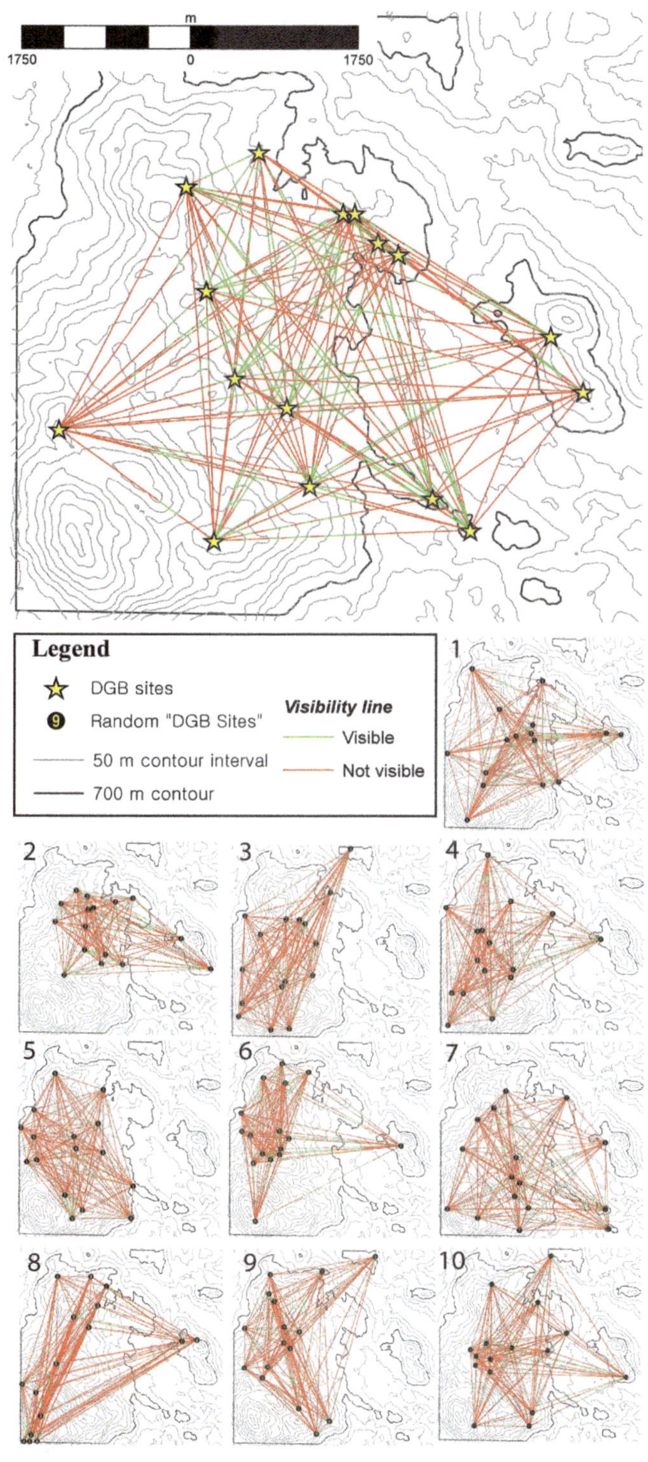

Figure 5. Intervisibility lines generated using the "Line of Sight" tool in ArcGIS10.1 Spatial Analyst (vectors created using v.net.visibility in GRASS6.4) of Actual DGB and 10 permutations of Random DGB sites.

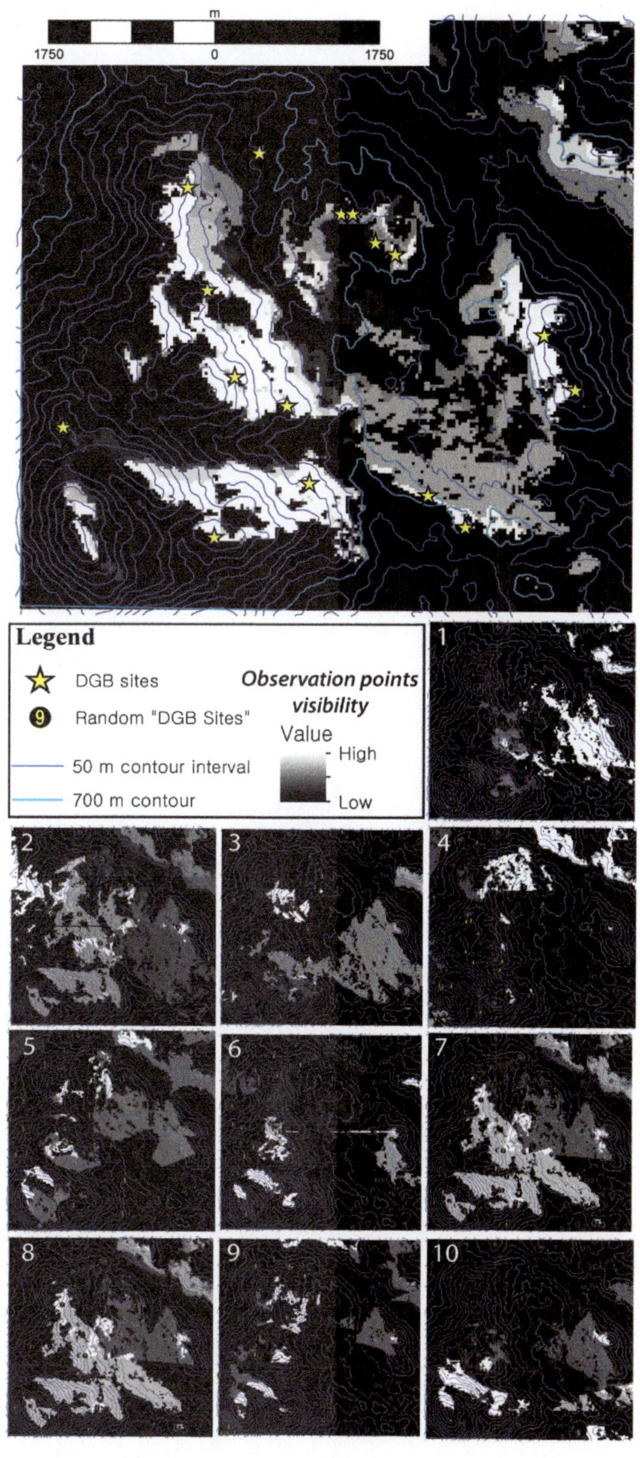

Figure 6. Viewshed analysis in ArcGIS 10.1 of Actual DGB and 10 permutations of Random DGB sites.

Systems (GIS) now provide ample computing power and statistical tools for determining visible and invisible aspects of a landscape from an observer point. The use of Monte-Carlo simulations to create randomized sets of data has become increasingly popular to create control samples of data populations [33,52–54]. The advantage of such simulations is that they remove any possible subjectivity in the analysis, while the disadvantage is that they are difficult to accommodate using stratified random sampling methods.

The fundamental principle of visibility and intervisibility in GIS is to generate lines of sight between given points on a landscape. In the case of two features (Point A, Point B), a line of sight is created as a straight vector between A and B and the topography which

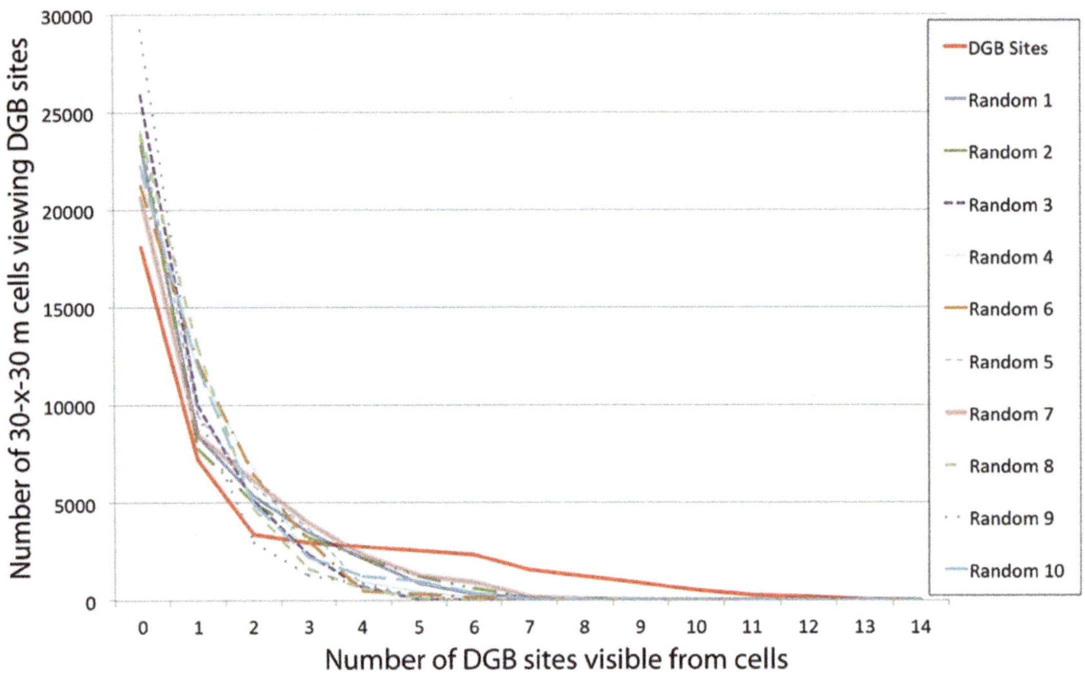

Figure 7. Comparison of the number of DGB sites (Actual, Random-1– Random-10) visible from 30-×-30-m grid cells within the study area.

would prohibit or facilitate intervisibility is generated from a Digital Elevation Model (DEM) of the terrain. Today, DEMs are commonly available to resolutions of up to 10 m, but most portions of Africa (the subject of this study) are covered by 30-×-30 m grids by United States Geological Survey (USGS) satellites (e.g., http://glovis.usgs.gov). Whether or not Point A is visible to Point B is determined on the basis of analyzing the presence/absence of topographic features that might interfere with the vector connecting the two nodes. Although visibility itself is hindered by physical distance relating to the curvature of the earth and recognizability of objects to the human eye over long distances [35], we presume that the DGB sites were theoretically intervisible on the basis that no two intervisible sites were more than 5 km from each other (which is roughly the distance at which the earth curves out of sight to the naked eye on a continually flat surface).

Methods

Spatial organization of the data using GIS

Previous studies of intervisibility have advocated one-sample tests comparing a site sample against a hypothetical "background" sample (the strongest advocate and most commonly cited of these studies is [30]). However, in applying the K–S test, background samples have typically been truly randomized to check distributions of actual vs. hypothetical data to delineate statistical outliers. In the present analysis of DGB sites, we apply a stratified random sampling method for the distribution of hypothetical DGB sites (> 700 m.a.s.l.), because a truly randomized distribution of hypothetical sites across the project area would not reflect the true conditions of site selection.

After a DEM of the project area was generated, a spider diagram of DGB sites was created in ArcGIS10.1 as potential sight lines between the facilities. In order to test whether the distribution of DGB sites was situated for intervisibility, random points were generated in a stratified fashion within a polygon shapefile created from a contour above 700 m.a.s.l. This corresponds to the approximate minimum elevation on which DGB sites were located. Therefore, we make the assumption that for any myriad of reasons (e.g., defense, drainage, visibility), there was a cultural proclivity to select locations above 700 m.a.s.l. Ten sets of 16 random points were generated using the "Create Random Points" tool in ArcGIS10.1 with the constraining feature being the elevationally-derived contour line shapefile. A minimum distance of 121 m was assigned between the random points based on the distance between DGB-1 and DGB-2. Lines of sight vectors were constructed using the"v.net.visibility" tool in GRASS6.4. The result was the construction of 120 vectors, which is the total number of lines that interconnect all 16 points (points do not connect to themselves and connections are not made twice). Intervisibility lines were created using the 3D Analyst toolbox ("Line of Sight" tool) from the lines of sight vectors. The DEM was the input surface, and line features were the sight lines created.

Slope of Actual and Random DGB sites were calculated from the raster DEM, then aggregated and averaged to control whether site construction was a function of factors other than intervisibility. Slope values were generated automatically in ArcGIS10.1 by extracting the elevation of plotted DGB points from the Mandara Mountain DEM compared to the eight surrounding raster cells then calculating a % of elevation change. The data were tabulated to show the variance between the slope of Actual and Random DGB sites.

A separate raster map of the observability of DGB sites was created to assess whether site locations were situated so that they were more visible from different landscape points. Based on the resolution of the DEM, the raster cell value was set to include a 30-×-30 m area. Observer points were set as the Actual DGB sites or Random DGB sites according to the point features created randomly. A viewshed analysis was executed according to a simple yes/no (binary) function to determine whether a raster cell has visibility of at least one observer point factoring in potential

Table 1. Comparative analysis of numbers of DGB sites (Actual, Random-1-10) visible from 30-×-30-m grid cells within the study area.

Number of Sites Visible	DGB Actual	Random-1	Random-2	Random-3	Random-4	Random-5	Random-6	Random-7	Random-8	Random-9	Random-10	mean (ALL)	1-σ	mean (Random-1-10)
0	18132	23372	23846	25894	21873	24090	21274	20721	23996	29189	22240	23147.9	2884.3	23649.5
1	7226	8427	7782	10028	11853	9532	12254	8468	13043	9609	12017	10021.7	1991.1	10301.3
2	3384	5329	5029	5111	6716	5863	6434	6136	4678	2985	4945	5146.4	1170.5	5322.6
3	2961	3474	3218	2363	2135	3691	3069	3985	1612	1253	2209	2724.5	874.6	2700.9
4	2762	2145	2178	635	917	780	488	2359	605	865	1240	1361.3	830.5	1221.2
5	2545	845	1215	47	434	102	312	1272	117	183	991	733.0	755.2	551.8
6	2362	313	606	6	150	24	126	916	30	0	340	443.0	698.6	251.1
7	1568	132	163	0	6	2	87	172	3	0	79	201.1	458.4	64.4
8	1227	38	44	0	0	0	40	46	0	0	21	128.7	364.8	18.9
9	906	9	3	0	0	0	0	9	0	0	2	84.5	272.5	2.3
10	499	0	0	0	0	0	0	0	0	0	0	45.4	150.5	0
11	250	0	0	0	0	0	0	0	0	0	0	22.7	75.4	0
12	185	0	0	0	0	0	0	0	0	0	0	16.8	55.8	0
13	60	0	0	0	0	0	0	0	0	0	0	5.5	18.1	0
14	17	0	0	0	0	0	0	0	0	0	0	1.5	5.1	0
MEDIAN	1568	132	163	0	6	2	87	172	3	0	79	201.1	458.4	64.4
1-σ	4601.7	6173.7	6221.5	6936.5	6211.0	6493.1	6120.0	5561.9	6767.7	7681.3	6218.9	6229.3	807.0	6438.6
invis. km^2	16.3	21.0	21.5	23.3	19.7	21.7	19.1	18.6	21.6	26.3	20.0	20.8	2.6	21.3
visible km^2	23.4	18.6	18.2	16.4	20.0	18.0	20.5	21.0	18.1	13.4	19.7	18.8	2.6	18.4
weighted avg.	**2.3**	**1.0**	**1.1**	**0.7**	**0.9**	**0.8**	**0.9**	**1.3**	**0.7**	**0.5**	**0.9**	*1.0*	*0.5*	0.9

Invisible vs. visible km^2 represent the total area that can see ≥1 DGB site. Weighted average multiplies the number of sites visible by the number of cells visible to and from the sites divided by the total number of cells in the project area. Note: Total cells analyzed are 44,084; each cell is 900 m^2.

Table 2. Spearman's-ρ ranked analysis comparing the average cumulative viewshed of cells from DGB Random-1–10 against individual values.

Average DGB Random 1–10 vs.	Spearman's R	Spearman's t	p-level
Actual DGB	0.9820	18.7350	0.0000
Random1	1.0000	0.0000	1.0000
Random2	1.0000	0.0000	1.0000
Random3	0.9389	9.8330	0.0000
Random4	0.9661	13.4907	0.0000
Random5	0.9661	13.4907	0.0000
Random6	0.9860	21.3307	0.0000
Random7	1.0000	0.0000	1.0000
Random8	0.9661	13.4907	0.0000
Random9	0.9027	7.5631	0.0000
Random10	1.0000	0.0000	1.0000
Average*	**0.9734**	**13.9906**	

*Average of Spearman's t includes only those with positive values.

topographic barriers. The "Viewshed" tool in ArcGIS10.1 determines the number of observer points visible from a given raster cell on the landscape. The values are then classified from high to low with "high" values being a computer-determined value of numerous sites visible from an observer point, and "low" values are those with no visibility to an observer point. The data are then be displayed graphically or exported into table format for statistical analysis.

Cumulative viewshed analysis

Once stratified random sample sets were created and exported into tabular format, statistical means for testing the distribution of actual vs. random sites were generated to test the hypotheses. Weighted means of cumulative viewsheds [33] were generated from the number of DGB sites (Actual DGB, Random DGB) visible from 30-×-30 m cells within the project area ($n = 44084$). Spearman's-ρ correlations were performed on comparing individual DGB sites (Actual and Random) to the computed average of the stratified Random DGB (1–10) sites using R statistical software. Spearman's-ρ is a measure of concordance between two related, but independent sets of variables, taking into account the related degree of monotonicity of the datasets [55]. Spearman's-ρ values vary between +1 and −1 depending on the degree of correlation of ranked data. A perfect positive correlation of variables is a +1, a perfect negative correlation is a −1, and a non-correlation value is 0.

Two-way statistical tests

Additionally, binary sets of intervisible DGB sites were created from lines of sight generated from both Actual DGB and Random DGB points. A Cochran's Q test was performed on the intervisible points of the 16 DGB sites compared to the 10 permutations of stratified random DGB sites located above 700 m.a.s.l. Cochran's Q is designed to analyze statistical significance of differences in binary populations [56]. For the purpose of the present study, intervisible sites were classified as "1" and non-intervisible sites were classified as "0". For example, DGB sites 3 and 4 were intervisible, so they were assigned a 1, while DGB sites 3 and 14 were not intervisible, so they were assigned a 0. McNemar's contingency tests were performed comparing the randomness of Actual DGB vs. Random DGB-1, 2, 3… as well as Random DGB sites to one another. The McNemar's contingency tests determine whether two sets of binary data are normally distributed [57,58]. Such methods have not been applied to spatial data in archaeology, but are common in medical (e.g., [59]) and ecology-related (e.g., [60]) fields.

Bayesian logistic regression

We postulate a Bayesian logistic regression model tests the hypotheses $H_{0(null)}$ and H_1, i.e. whether the locations of DGB sites are situated to maximize intervisibility of the sites to one another. The test model follows several steps beginning with a spatial organization of the data and concluding with a probability-based calculation. First, we retained the same rectangular-shaped region as previous analyses containing 16 DGB sites and subdivided this region by segmenting the west-to-east line and the south-to-north line into 45 equal length intervals, respectively, resulting in 2025 tiles (Fig. 3). The decision was made to analyze ~2000 tiles in order to capture enough geographical resolution to have meaningful insight into apparent visibility or non-visibility of sites, but not too much data that would make the analysis overly complex for the 16 DGB sites. The area captured within each tile approximated nine DEM cells plus perimeter values of an additional 16 cells extracted by a rectangular mask in ArcGIS10.1 to estimate the elevation of each 20,375 m^2 tile.

Among the 2025 tiles within the overall project area, we take $n = 1236$ tiles located ≥700 m.a.s.l. for the logistical regression analysis. For tile $i = 1,2,\ldots,1236$, we created two variables, y_i and x_i. We let $y_i = 1$ if a DGB site was built in tile i and $y_i = 0$ otherwise, and let x_i be the number of DGB sites that are visible from tile i excluding the DGB site on tile i. If the DGB sites are situated to maximize intervisibility of the sites to one another, the chance of $y_i = 1$ would increase, as x_i gets larger.

Second, we postulate the logistic regression model assuming that $i = 1,2,\ldots, n$, y_i follows Bernoulli distribution with probability $p(x_i)$ which is a function of the number of DGB sites intervisible from tile i. The function $p(x)$ is the probability that a DGB site is built in a tile when there are x DGB sites visible from the tile. The functional form that relates to the probability and the number of visible DGB sites is the logit function; in particular, $\mathrm{logit}(p(x)) = \log$

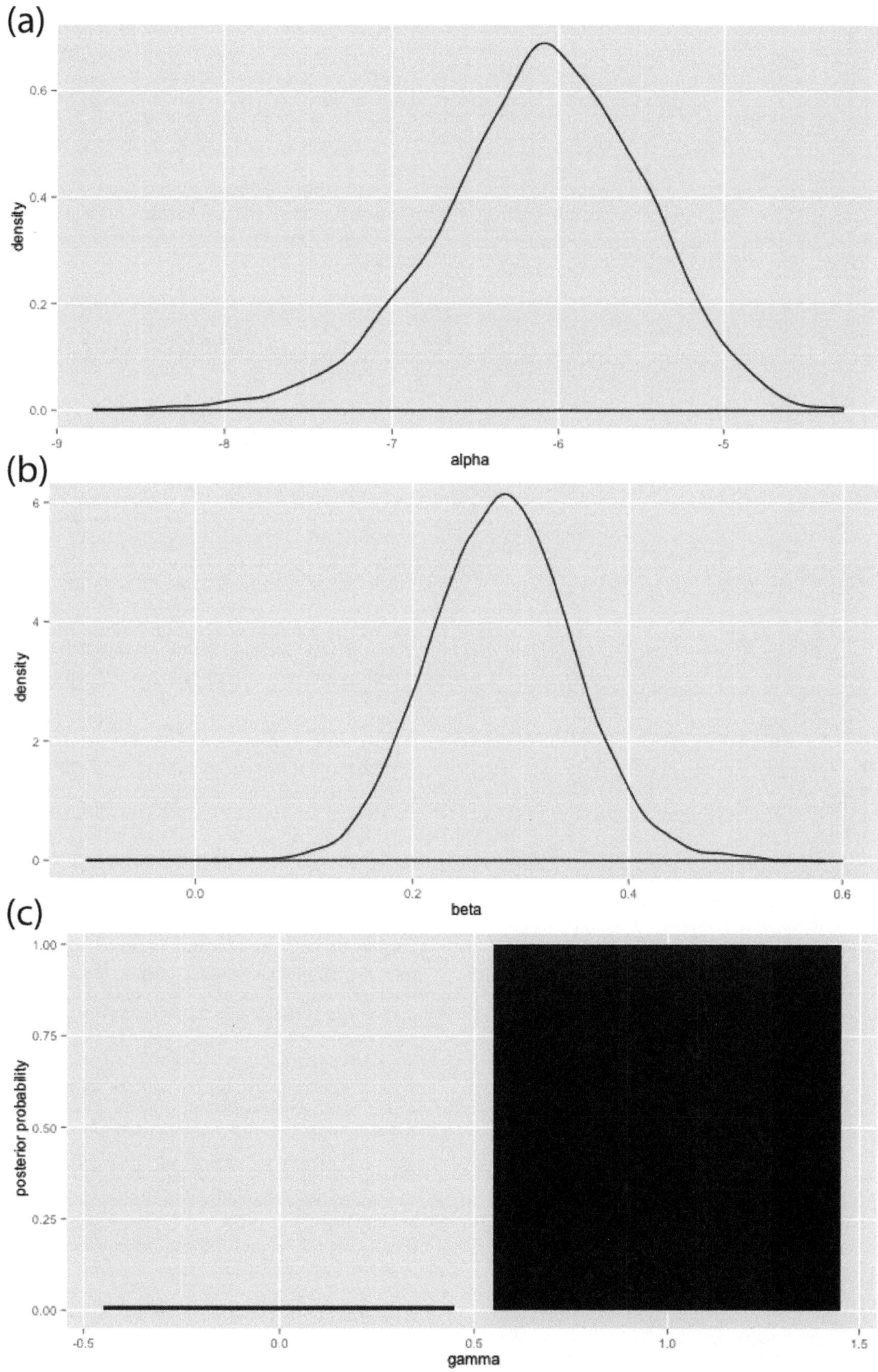

Figure 8. Posterior means and standard deviations of α, β, and γ using a Bayesian logistical regression model to test the intervisibility DGB sites across 2025 tiles, each of which measures 20,375 m^2.

Table 3. Average and 1-σ calculations of elevation (m.a.s.l.) and slope (%) of Actual and Random DGB sites.

	DGB	DGB	DGB	DGB	DGB	DGB	DGB	DGB	DGB	DGB	DGB	mean		mean
	Actual	Random-1	Random-2	Random-3	Random-4	Random-5	Random-6	Random-7	Random-8	Random-9	Random-10	(ALL)	1-σ	(Random-1-10)
Average elevation	866.4	890.1	846.5	952.5	967.3	913.5	946.2	917.1	971.6	911.1	918.6	918.3	39.9	923.4
1-σ elevation	157.3	193.4	107.8	187.3	191.9	172.4	170.1	163.5	215.4	122.3	169.3	168.2	31.2	169.3
Average slope	10.9	18.8	20.3	19.9	18.6	16.4	15.0	17.3	20.5	16.0	14.6	17.1	2.9	17.7
1-σ slope	9.2	7.7	9.0	8.9	6.7	9.2	6.2	11.4	10.7	9.6	7.6	8.7	1.6	8.7

$[p(x)/(1-p(x))] = \alpha+\beta x$. The parameter α is logit (p) when $x = 0$, i.e., there are no visible DGB sites from the tile. The parameter β is amount of increase in probability that a DGB site is built in the logit scale as x increases. If $\beta = 0$, the number of visible DGB sites does not affect the probability of building DGB site; if $\beta > 0$, the probability of building DGB sites gets larger as x gets larger; if $\beta < 0$, the probability of building DGB sites gets smaller as x gets larger.

Since we are interested in whether the number of intervisible DGB sites affect the probability of building DGB sites, we formulate the question in the following statistical hypothesis testing problem:

$$H_{0(null)}: \beta = 0 \text{ vs } H_1 : \beta^{1} 0$$

For the prior of α and β, we take $\alpha \sim N(0, 10000)$ whose variance is very large representing vague prior knowledge on α, and $\beta \sim 0.5\ \delta_0 + 0.5\ N(0, 10000)$, where δ_0 is the probability distribution which has probability 1 at 0. The latter prior represents that β is 0 with probability 0.5 and follows N (0, 10000) with probability 0.5. We are effectively putting equal prior probabilities to H_0 and H_1, respectively. In the actual implementation of the prior of β, we introduce another random variable γ, which follows Bernoulli distribution with probability 0.5. In particular, $\gamma = 1$ if $\beta \neq 0$ and $\gamma = 0$ if $\beta = 0$. The samples from the posterior distribution of α, β, and γ are obtained by Markov chain Monte Carlo; in particular we use the R2jags package of R statistical language to generate the posterior sample from the posterior.

Ripley's K function

Ripley's K function (multidistance spatial cluster analysis) was run on the projected data points to determine the degree of clustering or dispersion at different distances. The Ripley's K analysis determines whether the number of neighbors for a set of points at a given distance is greater or less than that of a hypothetical distribution of points [61]. A total of 50 distance bands were generated with a computed confidence envelope of nine permutations per analytical distance. The outer boundary was simulated using the ArcGIS10.1 "Outer Boundary Simulator," which simulates points outside the study area in order to minimize edge effects. The beginning distance band was 50 m, increasing by 100-m increments in each iteration of the analysis. We applied the "Simulate Outer Boundary Values" using an automatically-generated, minimum-enclosing rectangle for the project area as the feature class of the study area.

The data are not perfectly suited for Ripley's K analysis because the randomized site locations were stratified with the user-defined minimum elevation of site locations being ≥700 m.a.s.l. However, because the distribution of land ≥700 m.a.s.l. is relatively well distributed across the study area, Ripley's K patterns were deemed as potentially instructive in showing a site selection bias in the distribution of Actual DGB sites when compared to their stratified random counterparts [53,54].

Results

Ripley's K

The various measures of spatial analysis demonstrate that the placement of DGB sites on the landscape for intervisibility concerns was not random. Ripley's K analysis can be highly sensitive to the size of the project area being evaluated [62,63], and the results of our analysis bear this out. The Ripley's K analysis demonstrates clustering of Actual DGB sites at the scale of ≤300 m (exceeding the 95% confidence envelope), a moderate degree of clustering ≤1000 m (within the 95% confidence envelope), but dispersal at >1000 m (Fig. 4). In contrast, the randomly placed points simulating the potential DGB sites (Random DGB 1 was selected for the figure, but all results are similarly distributed) show a random distribution ≤2500 m with dispersion thereafter (Fig. 4). Dispersion measured >2500 m is statistically significant, but analytically insignificant as the project area itself is only 39.7 km^2. Therefore, dispersion measured above the 2500-m scale in Random DGB sites reflects the effects of the boundaries in the simulation, as the computations include hypothetical neighbors outside the project area itself. When compared to the 10 permutations of stratified Random DGB sites, the clustering of the Actual DGB sites ≤300 m and moderate clustering between 300 and 1000 m is provisionally interpreted as a function of a necessary principle of intervisibility-the sites need to be within view of each other and dispersion of the sites would hamper observers' abilities to see other sites. The clustered pattern of Actual DGB sites ≤1000-m scale is interpreted to reflect the apparent preference to locate DGB sites in locations in which other DGB sites are visible. However, the 10 permutations of Random DGB sites do not reflect this bias, so, predictably, their distribution is not patterned to accommodate this concern.

Cochran's Q and McNemar's contingency tests

In order to further test this assumption, a strict measure of the degree to which the placement of sites was related to their ability to see one another was generated. The comparison of the intervisibility of Actual DGB sites compared to the 10 permutations of random sites was performed using the Cochran's Q test (Fig. 5; Table S1). The test statistics of the 11 sites ($n = 240$) yielded a Cochran's Q value of 444.502 ($df = 10$, asymptotic significance = 0.000). With only two degrees of freedom (comparing Actual DGB sites with Random DGB-1 and Random DGB-2), the Cochran's Q value was 98.709 (asymp. sig. = 0.000). When removing the Actual DGB sites from the Cochran's Q, the value was 59.621 ($df = 9$, asymp. sig. = 0.000); while with two degrees of freedom (comparing Random DGB-1, −2 and −3), the value was 15.955 (asymp. sig. = 0.000). We interpret the asymptotic significance of Cochran's Q among the Random DGB sites as reflecting the bimodal nature of the intervisibility of the stratified random sample with Random DGB-1, −2, −6, −7 and −8 having 10 or more intervisible sites and Random DGB-3, −4, −5, −9 and −10 having six or less intervisible sites. The stratified random sampling of the test DGB sites skewed intervisibility of sites higher than a truly random distribution, but not as high as the actual distribution of DGB sites.

Therefore, in order to test whether there was a systematic bias in selecting stratified random DGB sites, a McNemar contingency test was performed comparing all stratified random samples (Random DGB-1– Random DGB-10) dichotomously with Actual DGB sites. The results are shown in Table S2 and demonstrate that the differences between the actual distribution of intervisible DGB sites and stratified random samples are statistically significant above the 99% confidence envelope. Furthermore, five permutations of McNemar's test were performed on the samples. The samples selected were non-biased (1 vs. 2, 3 vs. 4, 5 vs. 6, 7 vs. 8, 9 vs. 10) and demonstrate no statistical difference in the intervisibility of sites at the 95% confidence interval (Table S2). The results clearly demonstrate that the stratified random selection of hypothetical DGB sites did not inherently bias the intervisibility of sites against the actual distribution of sites.

Spearman's-ρ test

Analyzing landscape visibility (a.k.a. cumulative viewshed) of Actual DGB sites vs. Random DGB-1– Random DGB-10, shows that Actual DGB sites command a higher visibility of the surrounding landscape compared to the test sites (Fig. 6; Fig. 7; Table 1). The Spearman's-ρ analysis shows almost no differences in the ranks of data between Actual and Random DGB sites and an average of the stratified random DGB sites. A Spearman's-ρ analysis of ranks of the number of sites visible from individual 30-×-30 m grid cells from the Actual and Random DGB sites vs. the average ranks of Random DGB-1– Random DGB-10 (Table 2) show strong positive relationships between the ranks of data. All R values were >0.9 with Random DGB-9 having the weakest correlation coefficient and Random 1, 2, 7 and 10 showing perfect (+1) positive correlations. Actual DGB sites had R-values of 0.98 and τ-values of 18.74, which wasn't significantly different than the average values. We interpret the strong positive relationships between these data as reflecting the relatively good aspect of visibility of sites >700 m.a.s.l. selected in the stratified random sampling method undertaken in this study.

However, a more categorical approach to the data shows that subtle visibility conditions may have played a bigger role than the scaled data analysis suggests. Visible portions of the landscape from Actual DGB sites (23.4 km^2) was almost 2-σ above the mean distribution compared to Random DGB-1– Random DGB-10. Additionally, while there were no instances of portions of the landscape where ≥10 Random DGB sites were visible, there were 1011 cells totaling 0.9 km^2 where ≥10 Actual DGB sites were visible. The weighted average of visible cells on the landscape of Actual DGB sites (2.3 sites per cell) was >2-σ above the next highest permutation of Random DGB-2 (1.1 sites per cell). The results demonstrate that not only were intervisible locations factored into the construction of Actual DGB sites, but the ability of DGB sites to view the surrounding landscape (and probably vice versa) appears to have been a consideration.

Testing with Bayesian logistic regression model

The posterior means and standard deviations of α, β, and γ calculated from the posterior samples are −6.1014, 0.2828, 0.9990 and 0.61727, 0.07143, 0.03161, respectively (Fig. 8). The posterior probability of $\beta \neq 0$ or equivalently $\gamma = 1$ is 0.0001. Since we used the equal prior probability for $H_{0(null)}$ and H_1, the ratio of the posterior probabilities of $H_{0(null)}$ and H_1 is the same as the Bayes factor B_{10} and we obtained $\log B_{10} = 6.90$. According to Kass and Raftery's [64] interpretation of the Bayes factor, this is decisive evidence of H_1; furthermore, the posterior probability of $\beta > 0$ is 0.999. On that basis, we conclude that rejecting $H_{0(null)}$ indicates that the probability of building a DGB site is larger for the tiles with a greater number of intervisible DGB sites.

Secondary factors (elevation and slope)

Secondary factors of site location were also evaluated pertaining to elevation and slope. Actual DGB sites are ≤1.3-σ below the average elevation of all DGB sites (Actual and Random DGB-1– Random DGB-10; Table 3). However, the slope of Actual DGB sites are 2.1-σ less than average DGB sites (Actual and Random DGB-1– Random DGB-10), indicating that there is a much stronger preference for constructing Actual DGB sites on lower slope than lower elevation. Maintaining a constant elevation of DGB sites is not interpreted as having played a significant role in site selection, as Actual DGB sites are within 1-σ of the mean standard deviation of all DGB sites analyzed in this study. Therefore, principles of intervisibility of DGB sites to each other, the visibility of DGB sites to places on the landscape (and vice versa) as well as the secondary factor of low-slope topography (which may have facilitated and enhanced the first two factors) are interpreted as prime factors in determining the locations of Actual DGB sites in Mandara Mountains of northern Cameroon.

Discussion

Based on the GIS and related statistical analyses of the locations of DGB sites, the null hypotheses (H_0–the locations of DGB sites are randomly located above 700 m.a.s.l.) can be rejected. The statistical evidence demonstrates that there is >99% probability that good intervisibility of DGB sites and/or the surrounding landscape was considered when deciding where to construct sites. The GIS analysis confirms previous research hypotheses [23] that the inferred intervisibility of DGB sites was significant, possibly for maintaining relations of cooperation and/or competition among the communities farming the adjacent hill-slopes. DGB sites retain community ritual and patrimonial functions today [65], and the archaeological evidence suggests that their earliest functions were related at least in part to ritual use [19].

As cited above, intervisibility studies of point features on archaeological landscapes are increasingly common because of their relative ease using GIS computational techniques. We employ relatively novel statistical measures for archaeological research (Cochran's Q, McNemar's, Bayesian logistic regression) to crosscheck the effectiveness of our stratified randomization of potential DGB sites within a circumscribed study area. The advantage of two-way tests over K–S tests is that they are designed to compare binary tables rather than means of distribution. Since the analytical design of the intervisibility study was to compare identical distributions of points, means testing are not suitable to test the hypotheses.

In this study, we reject the null hypothesis that the Actual DGB sites are randomly located, but based on the Cochran's Q analysis, we must also reject the null hypothesis for the Random DGB sites. However, we argue that the virtue of the stratified random sample augmented the inherent intervisibility of the sites in the sample area. For this analysis, we stress that the actual value of the Cochran's Q test provides the significance of the difference between the Actual and Random DGB sites. In this case, given the relative degrees of freedom analyzed, inclusion of the Actual DGB sites yielded Cochran's Q values 5–7× higher than when only Random DGB-1– Random DGB-10 were analyzed. McNemar's contingency tests individually comparing the Actual DGB sites to each of the Random DGB sites confirms that the actual distribution of sites is significantly different than random sites, while the same test performed on Random DGB sites to one another did not show significant differences. These results are further supported with Bayesian logistic regression, which tests whether there is a mathematical preference for site construction within highly visible portions of the landscape.

Weighted means and Spearman's-ρ analysis of the cumulative viewshed shows that DGB sites were either positioned to see or be seen from the surrounding landscape. Stratified random permutations of DGB sites above 700 m.a.s.l. have significantly smaller cumulative viewsheds than the Actual DGB sites. The Ripley's K analysis shows clustering of DGB sites ≤1000 m, and the clusters appear to have been oriented to maximize the cumulative as well as intervisible viewshed.

Possible consideration for a high viewshed is the fact that sites are located on low slope benches of the Mandara massif, which also coincidentally commanded better views of the surrounding areas than any of the Random DGB sites. Site slope has been previously determined to be a high predictor of archaeological settlement, because people do not like to construct settlements in high-slope environments [66–68]. Although this explanation suffices for explaining cumulative viewsheds, it does not explain the intervisibility of the features to one another.

These analyses do not exclude other selective factors, such as geology, soil fertility, or social boundaries, but they do demonstrate that seeing and/or being seen by other DGB sites and from points on the surrounding landscape featured prominently in the choice of locations for these sites. As already noted, there are a number of modern ceremonial cycles still practiced in the region that encompass multiple communities, and where intervisibility could well have played a part in some aspects of the ritual, especially in determining when the beginning of the ceremony was 'passed' from one community to another [28]. While the details of, for example, the modern *maray* ceremonial cycle do not fit with the architectural details of the DGB sites, *maray* at least provides a conceptual framework for understanding interacting, community-based rituals in the area. In addition, it is quite likely that at least some of the sites, and especially the DGB-1/DGB-2 complex, played a sociopolitical role involving conspicuous display – and conspicuousness is, of course, closely related to visibility and intervisibility.

Conclusions

Cumulative viewshed and two-way analyses of archaeological sites from northern Cameroon compared to 10 permutations of stratified random sites above 700 m.a.s.l. within a restricted study area concludes that placement of the sites was non-randomly situated to enhance intervisibility between sites, and visibility between the sites and the surrounding landscape. A Bayesian logical regression model also rejects the null hypothesis that site construction location was random according to principles of enhanced visibility. Although other contingent settlement factors cannot be excluded, the locations of the sites were ideally located for maximizing the viewshed of DGB sites, and this likely relates to the interpreted social functions of the sites.

Ripley's K analysis shows a clustering ≤200 m (above the 95% confidence envelope) and moderate to weak degree of clustering 200–1000 m (within the 95% confidence envelope) reflecting the placement of the sites to maximize the view of intervening areas within the Mandara Mountains. Spearman's-ρ analyses and weighted means of 30-×-30-m grid cells to Actual DGB sites is ~2-σ above the mean for all 11 permutations of cumulative visibility analyzed. Furthermore, intervisibility of the sites was demonstrated using binary statistical techniques directly comparing the Actual DGB sites vs. permutations of Random DGB sites. Both Cochran's Q and McNemar contingency tests indicate that Actual DGB sites have significantly better intervisibility than Random DGB sites in all of the test cases, which is further supported using a logistic regression model.

It is not clear how many of the DGB sites were contiguously utilized, although the available radiocarbon dates indicate some degree of contemporaneity. Intervisibility of some of the features does seem to have affected their placement within the Mandara Mountains along with placement of the sites on lower slope portions of the inselbergs compared to Random DGB sites. This information will be important in continuing analysis of the functioning of the DGB sites. Available archaeological data indicate that that functioning was complex, probably involving both ritual and sociopolitical roles, and the sites probably acted within both local and regional systems of interaction and competition. The placement of the sites in locations that increase visibility and intervisibility certainly reinforces our appreciation of their significance in cultural systems in the Mandara Mountains in the mid-second millennium AD.

Supporting Information

Table S1 Results of Cochran's Q tests. (a) Actual DGB and Random DGB-1–10, (b) Actual DGB and Random-1–2, (c) Random DGB-1–10, (d) Random DGB-1–3. Cochran's Q tests are binary with the number of non-intervisible sites coded as 0, while the number of intervisible sites are coded as 1.

Table S2 McNemar's test comparing (a) Actual DGB sites vs. 10 independent permutations of DGB Random sites and (b) permutations of Random DGB sites against each other.

Acknowledgments

We thank the people of the Mandara Mountains who have provided generous insights into their unique lives to SM over the years. GIS laboratory facilities were provided by the Department of Archaeology and Art History in the College of Humanities at Seoul National University. We thank reviewer Mal Ridges and the editors for consideration of the manuscript.

Author Contributions

Conceived and designed the experiments: DKW JL. Performed the experiments: DKW JL. Analyzed the data: DKW SM JL. Contributed reagents/materials/analysis tools: DKW JL. Contributed to the writing of the manuscript: DKW SM JL. Designed and conceived the geospatial experiments and GIS methods: DKW. Conducted fieldwork identifying sites used in this manuscript: SM. Designed and conceived logistic regression experiment: JL.

References

1. Parsons JR (1972) Archaeological Settlement Patterns. Annual Review of Anthropology 1: 127–150.
2. Binford LR (1980) Willow Smoke and Dogs' Tails: Hunter-Gatherer Settlement Systems and Archaeological Site Formation. American Antiquity 45: 4–20.
3. Hassan FA (1981) Demographic archaeology. New York: Academic Press.
4. Hodder I (1991) Post-modernism, post-structuralism and post-processual archaeology. In: Hodder I, editor. The meanings of things: material culture and symbolic expression. London: Unwin Hyman. pp. 64–78.
5. Earle TK, Preucel RW, Brumfiel EM, Carr C, Limp WF, et al. (1987) Processual Archaeology and the Radical Critique [and Comments and Reply]. Current Anthropology 28: 501–538.
6. O'Brien M, Lyman RL (2004) History and Explanation in Archaeology. Anthropological theory 4: 173–197.
7. Pauketat TR (2001) Practice and history in archaeology. Anthropological theory 1: 73–98.
8. Kirch PV (1990) Monumental architecture and power in Polynesian chiefdoms: A comparison of Tonga and Hawaii. World Archaeology 22: 206–222.
9. Sinopoli CM (1994) The Archaeology of Empires. Annual Review of Anthropology 23: 159–180.
10. Brück J (2001) Monuments, power and personhood in the British Neolithic. Journal of the Royal Anthropological Institute 7: 649–667.
11. Bonanno A, Gouder T, Malone C, Stoddart S (1990) Monuments in an island society: The Maltese context. World Archaeology 22: 190–205.
12. Dillehay TD (2004) Social landscape and ritual pause: uncertainty and integration in Formative Peru. Journal of Social Archaeology 4: 239–268.
13. Letesson Q, Vansteenhuyse K (2006) Towards an Archaeology of Perception: 'Looking' at the Minoan Palaces. Journal of Mediterranean Archaeology 19: 91–119.
14. Garlake P (2002) Early Art and Architecture of Africa. Oxford: Oxford University Press.
15. Shaw T, Sinclair P, Andah BW, Okpoko AI (1993) Archaeology of Africa: foods, metals and towns. London: Routledge.
16. Connah G (1987) African Civilizations Precolonial Cities and States in Tropical Africa: An Archaeological Perspective. Cambridge: Cambridge University Press.
17. Hildebrand EA (2013) Is monumentality in the eye of the beholder? Lessons from constructed spaces in Africa. Azania: Archaeological Research in Africa 48: 155–172.
18. McIntosh SK, editor (1999) Beyond chiefdoms: pathways to complexity in Africa. Cambridge: Cambridge University Press.
19. MacEachern S, David N (2013) Monumental architecture in mountain landscapes: the diy-geδ-bay sites of northern Cameroon. Azania: Archaeological Research in Africa 48: 241–262.
20. MacEachern S (2012) Wandala and the DGB sites: political centralisation and its alternatives north of the Mandara Mountains, Cameroon. Azania: Archaeological Research in Africa.
21. MacEachern S (1993) Selling the iron for their shackles: Wandala-montagnard interactions in northern Cameroon. Journal of African History 34: 247–270.
22. MacEachern S (2002) Beyond the belly of the house: space and power in the Mandara Mountains. Journal of Social Archaeology 2: 179–219.
23. David N (2008) Performance and agency: the DGB sites of Northern Cameroon. Oxford: British Archaeological Reports.
24. Fitton JG, Dunlop HM (1985) The Cameroon line, West Africa, and its bearing on the origin of oceanic and continental alkali basalt. Earth and Planetary Science Letters 72: 23–38.
25. Elangwe HN (1979) Carte Géologique de la République Unie du Cameroun. Yaounde: Minister of Mines and Power.
26. Letouzey R, Fotius G (1985) Carte Phytogéographique du Cameroun. Toulouse: Institut de la Carte Internationale de la Végétation.
27. MacEachern S (2012) The Holocene history of the southern Lake Chad Basin: archaeological, linguistic and genetic evidence. African Archaeological Review: 1–19.
28. von Graffenried C (1984) Das Jahr des Stieres, ein Opferrituel der Zulgo und Gemjek in Nordkamerun. Fribourg: Universitätsverlag.
29. Conolly J, Lake M (2006) Geographic Information Systems in Archaeology. Cambridge: Cambridge University Press.
30. Kvamme KL (1990) One-Sample Tests in Regional Archaeological Analysis: New Possibilities through Computer Technology. American Antiquity 55: 367–381.
31. Wheatley D (1995) Cumulative viewshed analysis: a GIS-based method for investigating intervisibility, and its archaeological application. In: Lock GR, Stančič Z, editors. Archaeology And Geographic Information Systems: A European Perspective. London: Taylor and Francis. pp. 171–186.
32. Sevenant M, Antrop M (2007) Settlement models, land use and visibility in rural landscapes: Two case studies in Greece. Landscape and Urban Planning 80: 362–374.
33. Fisher P, Farrelly C, Maddocks A, Ruggles C (1997) Spatial Analysis of Visible Areas from the Bronze Age Cairns of Mull. Journal of Archaeological Science 24: 581–592.
34. Razali NM, Wah YB (2011) Power comparisons of Shapiro-Wilk, Kolmogorov-Smirnov, Lilliefors and Anderson-Darling tests. Journal of Statistical Modeling and Analytics 2: 21–33.
35. Ogburn DE (2006) Assessing the level of visibility of cultural objects in past landscapes. Journal of Archaeological Science 33: 405–413.
36. Llobera M (2001) Building Past Landscape Perception With GIS: Understanding Topographic Prominence. Journal of Archaeological Science 28: 1005–1014.
37. Lake MW, Woodman PE, Mithen SJ (1998) Tailoring GIS Software for Archaeological Applications: An Example Concerning Viewshed Analysis. Journal of Archaeological Science 25: 27–38.
38. O'Sullivan D, Turner A (2001) Visibility graphs and landscape visibility analysis. International Journal of Geographical Information Science 15: 221–237.
39. Jones EE (2006) Using Viewshed Analysis to Explore Settlement Choice: A Case Study of the Onondaga Iroquois. American Antiquity 71: 523–538.
40. Jones EE (2010) An analysis of factors influencing sixteenth and seventeenth century Haudenosaunee (Iroquois) settlement locations. Journal of Anthropological Archaeology 29: 1–14.
41. Sakaguchi T, Morin J, Dickie R (2010) Defensibility of large prehistoric sites in the Mid-Fraser region on the Canadian Plateau. Journal of Archaeological Science 37: 1171–1185.
42. Kay S, Sly T (2001) An application of Cumulative Viewshed Analysis to a medieval archaeological study: the beacon system of the Isle of Wight, United Kingdom. Archeologia e Calcolatori 12: 167–179.
43. Llobera M, Fábrega-Álvarez P, Parcero-Oubiña C (2011) Order in movement: a GIS approach to accessibility. Journal of Archaeological Science 38: 843–851.
44. Bauer A, Nicoll K, Park L, Matney T (2004) Archaeological site distribution by geomorphic setting in the southern lower Cuyahoga River Valley, northeastern Ohio: Initial observations from a GIS database. Geoarchaeology 19: 711–729.
45. Krist FJ, Brown DG (1994) GIS modeling of Paleo-Indian period caribou migrations and viewsheds in northeastern Lower Michigan. Photogrammetric Engineering and Remote Sensing 60: 1129–1137.
46. Chapman HP (2003) Rudston 'Cursus A'– Engaging with a Neolithic Monument in Its Landscape Setting Using GIS. Oxford Journal of Archaeology 22: 345–356.
47. Bradley R, Harding J, Mathews M (1993) The siting of prehistoric rock art in Galloway, south-west Scotland. Proceedings of the Prehistoric Society 59: 129–145.

48. Howey MCL (2007) Using multi-criteria cost surface analysis to explore past regional landscapes: a case study of ritual activity and social interaction in Michigan, AD 1200–1600. Journal of Archaeological Science 34: 1830–1846.
49. Lambers K, Sauerbier M (2006) GIS-based visibility studies of the Nasca geoglyphs at Palpa, Peru. In: Baltsavias M, Gruen A, van Gool L, Paterak M, editors. Recording, Modeling and Visualization of Cultural Heritage: Proceedings of the International Workshop, Centro Stefano Franscini, Monte Verita, Ascona, Switzerland, May 22–27, 2005. London: Taylor & Francis. pp. 249–261.
50. Bongers J, Arkush E, Harrower M (2012) Landscapes of death: GIS-based analyses of chullpas in the western Lake Titicaca basin. Journal of Archaeological Science 39: 1687–1693.
51. Fraser D (1983) Land and Society in Neolithic Orkney. Oxford: British Archaeological Reports 117.
52. Vanacker V, Govers G, van Peer P, Verbeek C, Desmet J Jr, et al. (2001) Using Monte Carlo Simulation for the Environmental Analysis of Small Archaeologic Datasets, with the Mesolithic in Northeast Belgium as a Case Study. Journal of Archaeological Science 28: 661–669.
53. Crema ER, Bevan A, Lake MW (2010) A probabilistic framework for assessing spatio-temporal point patterns in the archaeological record. Journal of Archaeological Science 37: 1118–1130.
54. Bevan A, Conolly J (2006) Multiscalar Approaches to Settlement Pattern Analysis. In: Lock G, Molyneaux B, editors. Confronting Scale in Archaeology: Springer US. pp. 217–234.
55. Scarsini M (1984) On measures of concordance. Stochastica 8: 201–218.
56. Patil KD (1975) Cochran's Q Test: Exact Distribution. Journal of the American Statistical Association 70: 186–189.
57. McNemar Q (1947) Note on the sampling error of the difference between correlated proportions or percentages. Psychometrika 12: 153–157.
58. Cox DR (1970) Analysis of Binary Data. London: Methuen.
59. Chan IS (2003) Proving non-inferiority or equivalence of two treatments with dichotomous endpoints using exact methods. Statistical Methods in Medical Research 12: 37–58.
60. Guastaferro F, Castrignanò A, Benedetto D, Sollitto D, Troccoli A, et al. (2010) A comparison of different algorithms for the delineation of management zones. Precision Agriculture 11: 600–620.
61. Dixon PM (2006) Ripley's K Function. In: El-Shaarawi AH, Piegorsch WW, editors. Encyclopedia of Environmetrics. New York: John Wiley & Sons. 1796–1803.
62. Goreaud F, Pélissier R (1999) On explicit formulas of edge effect correction for Ripley's K-function. Journal of Vegetation Science 10: 433–438.
63. Tang M, Tang S, Lei X, Zhang H, Hong L, et al. (2003) Edge correction of Ripley's K(d) function on population spatial pattern analysis. Acta Ecologica Sinica 23: 1533–1538.
64. Kass RE, Raftery AE (1995) Bayes Factors. Journal of the American Statistical Association 90: 773–795.
65. Datouang Djoussou JM (2011) Archaeology and cultural heritage in Cameroon: the case of the DGB sites. Antiquity.
66. Warren RE, Asch DL (2000) A predictive model of archoaelogical site location in the eastern Prairie Peninsula. In: Wescott K, Brandon RJ, editors. Practical Applications of GIS for Archaeologists: A Predictive Modelling Toolkit. London: Taylor & Francis. pp. 6–36.
67. Kohler TA, Parker S (1986) Predictive models for archaeological resource location. In: Schiffer M, editor. Advances in Archaeological Method and Theory, Vol 9. New York: Academic Press. pp. 397–452.
68. Kvamme KL (1990) The fundamental principles and practice of predictive archaeological modeling. In: Voorrips A, editor. Mathematicsand Information Science in Archaeology: A Flexible Framework. Bonn: Studies in Modern Archaeology, vol. 3. Holos-Verlag. pp. 257–295.

11

Spatio-Temporal Trends and Risk Factors for *Shigella* from 2001 to 2011 in Jiangsu Province, People's Republic of China

Fenyang Tang[2]⁹, Yuejia Cheng[1]⁹, Changjun Bao[2], Jianli Hu[2], Wendong Liu[2], Qi Liang[2], Ying Wu[2], Jessie Norris[3], Zhihang Peng[1]*, Rongbin Yu[1], Hongbing Shen[1], Feng Chen[1]

1 Department of Epidemiology & Biostatistics, School of Public Health, Nanjing Medical University, Nanjing, Jiangsu, China, **2** Jiangsu Province Center for Disease Control and Prevention, Nanjing, Jiangsu, China, **3** National Center for AIDS/STD Control and Prevention, Chinese Center for Disease Control and Prevention, Beijing, China

Abstract

Objective: This study aimed to describe the spatial and temporal trends of *Shigella* incidence rates in Jiangsu Province, People's Republic of China. It also intended to explore complex risk modes facilitating *Shigella* transmission.

Methods: County-level incidence rates were obtained for analysis using geographic information system (GIS) tools. Trend surface and incidence maps were established to describe geographic distributions. Spatio-temporal cluster analysis and autocorrelation analysis were used for detecting clusters. Based on the number of monthly *Shigella* cases, an autoregressive integrated moving average (ARIMA) model successfully established a time series model. A spatial correlation analysis and a case-control study were conducted to identify risk factors contributing to *Shigella* transmissions.

Results: The far southwestern and northwestern areas of Jiangsu were the most infected. A cluster was detected in southwestern Jiangsu ($LLR = 11674.74$, $P < 0.001$). The time series model was established as ARIMA (1, 12, 0), which predicted well for cases from August to December, 2011. Highways and water sources potentially caused spatial variation in *Shigella* development in Jiangsu. The case-control study confirmed not washing hands before dinner ($OR = 3.64$) and not having access to a safe water source ($OR = 2.04$) as the main causes of *Shigella* in Jiangsu Province.

Conclusion: Improvement of sanitation and hygiene should be strengthened in economically developed counties, while access to a safe water supply in impoverished areas should be increased at the same time.

Editor: Joel Mark Montgomery, Global Disease Detection-Kenya, United States of America

Funding: This work was supported in part by Natural Science Foundation Key Project of Jiangsu Province, China (BK2010079), "National S&T Major Project Foundation of China" (2011ZX10004-902), Jiangsu Province Health Development Project with Science and Education (NO.ZX201109), and the Priority Academic Program Development of Jiangsu Higher Education Institutions (PAPD). The funders had no role in study design, data collection and analysis, decision to publish, or preparation of the manuscript.

Competing Interests: The authors have declared that no competing interests exist.

* E-mail: zhihangpeng@njmu.edu.cn

⁹ These authors contributed equally to this work.

Introduction

Shigella is a significant global public health concern, especially in developing countries like the People's Republic of China. *Shigella* deaths have been steadily decreasing over time, but this has come about through markedly reduced fatality rates rather than lower incidence [1]. Currently, a considerable burden of *Shigella* still exists among the younger generations in China [2]. A recent study showed that in Jiangsu Province, China, children under five years old and adults over 60 years old accounted for the greatest proportion of *Shigella* infections [3]. Furthermore, caring for sick children and the elderly can result in the loss of work-days among family members, potentially leading to socioeconomic problems. Due to vaccine innovations that have not yet proven effective, a successful prophylaxis against *Shigella* is currently unavailable [4]. The alarming speed of antibiotic resistance is also complicating treatment with traditional and inexpensive medications [5]. In considering the health consequences of the prevalence of *Shigella*, the inefficacious vaccine and the inadequate therapy options, public awareness of *Shigella* (based on a convincing estimation of *Shigella* disease burden and risk factors) is crucial in controlling transmission.

Spatial and temporal trends are classic methods of estimating disease burden. Statistical models are irreplaceable in forecasting short-term trends and have previously been performed successfully on *Shigella* cases in China [6]. Compared with traditional models such as linear regression or correlation coefficients, autoregressive integrated moving average (ARIMA) models take changing trends, periodic changes and random disturbances in time series into consideration. Earnest et al found that an ARIMA model was easier to fit in terms of the parameters and ran more quickly in forecasting notifiable infectious diseases through time-series models [7]. ARIMA models can eliminate seasonal patterns and are widely applied in predictions for epidemics like hemorrhagic fever with renal syndrome (HFRS), influenza, malaria, etc [8,9]. At the same time, disease maps produced by geographic information system (GIS) tools can display patchy maps of epidemics [10]. The development of spatial models, which describe spatial autocorrelation as well as geographical trends of epidemics, has also become increasingly popular [11,12]. Spatial and temporal models alone, however, are insufficient for exploring transmission modes. Risk factors can also be used to determine how interventional work should be conducted. Enteric diseases such as *Shigella* appear as complex epidemics related to changes in biology, socioeconomic status, culture/behavior and environment over space and time [13]. As a result, *Shigella* transmission modes that combine climatic, socioeconomic and human behavioral factors remain worthy of study. In recent years, the issue of climatic impact on health has drawn increasing public attention; Zhang et. al. recently announced that under the current trends of climate change, Years Lost due to Disabilities (YLDs) caused by bacillary *Shigella* would double by 2020 in some parts of China [14]. The climate both directly and indirectly determines the features that lead to transmission by influencing the speed of pathogen variation, the accumulation of susceptible hosts, and other environmental indices. A GIS application can illustrate climatic impact on epidemics through spatial correlation. In addition, a case-control study can help serve to explain the socioeconomic statuses or behaviors associated with *Shigella* transmission. Identification of the geographical risk factors and case-control studies are hoped to concomitantly provide information on the complex transmission modes of *Shigella*.

This is the first research study targeted at the spatial and temporal characteristics of *Shigella* in Jiangsu Province over the last decade (2001 to 2011). The authors aimed to illustrate the spatial variation and temporal trends of *Shigella* through use of GIS and the ARIMA model. Furthermore, the use of GIS coupled with a case-control study design facilitated risk factors analysis. The results and conclusions of this study aim to explore appropriate models for *Shigella* surveillance, identify hotspots and risk factors for infection, and provided evidence-based policy advice for interventions.

Materials and Methods

Study Area

Jiangsu Province is located in southeast China between longitudes 116°21′–121°54′E and latitudes 30°46′–35°08′N. Affected by the East Asian monsoon climate, the province has an annual mean temperature between 13.6 and 16.1 degrees Celsius and an annual mean precipitation ranging from 704 to 1250 mm. Jiangsu Province is customarily separated into three geographic regions by the 13 cities of Jiangsu: south (Suzhou, Wuxi, Changzhou and Zhenjiang), middle (Nanjing, Yangzhou, Taizhou and Nantong) and north (Xuzhou, Lianyungang, Suqian, Huai'an and Yancheng). Areas belonging to the same region share similar geographic parameters and economic statuses. These three geographic regions are also collectively composed of 103 counties. In this study, the county is defined as the primary sampling unit.

Data Resources

In China, a reporting system was established based on the Law on Prevention and Control of Infectious Disease that includes all health-care facilities at village, town, county and city levels [15]. In this study, county-level incidence rates and monthly cases of *Shigella* from 2000 to 2011 were obtained from the reporting system by the Jiangsu Province Center for Disease Control and Prevention (CDC). Climatic and environmental data were collected from the Jiangsu Province Meteorological Bureau and consisted of the annual mean temperature, moisture content, distribution of rivers and lakes, distribution of railways and highways, and Normalized Difference Vegetation Index (NDVI) of Jiangsu Province.

Data Analysis

Spatial variation analysis. Spatial analysis was aimed at detecting geographic variation through use of Geographic Information System (GIS, ArcGIS software, version 9.3 ESRI, Redlands, CA, USA) and SaTscan software (version 9.1.1, Boston, MA, USA).

With the help of GIS, a trend surface analysis was applied to detect scattered observations and geographical anomalies [16]. Estimation of incidence rates of unknown nodes was based on the Ordinary Least Square (OLS) method. This ensured minimized squared deviations by all incidence rates (the dependent variable was denoted as Z). Independent variables were longitudes (denoted as X) and latitudes (denoted as Y).

A spatio-temporal cluster analysis (conducted through SaTScan software) and autocorrelation analysis (conducted through GIS) were both applied to detect clusters of *Shigella* within Jiangsu Province. The spatio-temporal statistical method used cylindrical moving windows to scan inside Jiangsu Province in order to detect clustering areas and years. The base diameter and height of the moving window represented the underlying clustering areas and years, respectively. A Log Likelihood Ratio (*LLR*) method inspected the results by comparing real incidence rates with expected ones [17]. The area confirmed as statistically significant in a Monte Carlo test was defined as the most likely cluster.

At the same time, a local autocorrelation analysis was carried out to identify spatial autocorrelations inside the province. Under autocorrelation theory, the closer two locations approach each other, the more likely they are to impact each other's incidence rates [18]. In this study, the Local Moran's Index (LM_i) and the Local Getis-Ord G index (LG_i) were both applied. The Local Moran's Index helped to classify the autocorrelations into positive and negative ones. If incidence rates had similar high values or low values, they were defined as having positive autocorrelation (represented as High-High or Low-Low autocorrelation). If the attributes held opposing high and low values, they were considered to have negative autocorrelation (represent as High-Low or Low-High autocorrelation). The local Getis-Ord G index, moreover, denoted hotspots among positively autocorrelated areas ($P<0.05$). If the G index identified a county as statistically significant, that county might have higher risks of *Shigella* infection than any other county that was positively autocorrelated.

Time-series analysis. Autoregressive integrated moving average (ARIMA) models are known to have better veracity and practicability in describing and forecasting epidemic prevalence [19]. Peng et al previously applied an ARIMA model successfully in confirming *Shigella* transmission with seasonal patterns in China [20]. An ARIMA model is defined as ARIMA (p, d, q), which is decided by components p, the order of autoregression; d, the degree of difference; and q, the order of moving average. Because the monthly data on *Shigella* infections showed seasonal fluctuations, a 12-step finite difference method was initially applied to smooth the temporal sequence. These parameters (p, d, q) were decided by an autocorrelation function (ACF) and a partial autocorrelation function (PACF) [21]. The Ljung-Box test was used to verify the goodness-of-fit of the models and compare them with the Akaike information criterion (*AIC*) to determine the final time series model [22,23]. It measured the ACF of the residuals at the significance level of $P = 0.05$.

Geographic factors detection. The goal for this analysis was to describe *Shigella* development under the influence from each risk factor. In this study, data on the annual mean temperature, moisture content, railway or highway distributions, river or lake distributions and NDVI were obtained from the database of the Jiangsu Province Weather Bureau. Each of the risk factors was overlaid with *Shigella* incidence rates on incidence maps through GIS application. ANOVA analysis was conducted to confirm the correlation with each risk factor. The coefficient of determination (R^2) was used to measure the goodness-of-fit by each risk factor. Furthermore, the risk factors were applied using stepwise regression to establish a multivariate regression model on incidence rates. Risk factors were not included unless $P \leq 0.05$ in the results of the model. After eliminating collinearity from the factors, the best-fitted model was established as the one with the highest R^2 of all models.

Case-control study. The researchers also conducted a matched case-control study to investigate associated health or lifestyle behaviors among those exposed to the infection. The case-control study was performed in 29 clinics representing every city of Jiangsu. Cases were diagnosed according to the diagnostic criteria under the Law on Prevention and Control of Infectious Disease of China (patients diagnosed with a microscopic stool examination). Controls were healthy or non-enteric infected patients who were closest in age to the cases (±three years) and lived in the same household with them. In cases for which no control was available in the same household, a person in the neighborhood who met the criteria of a control (closest in age and healthy or non-enteric infected) was selected instead. The study ensured that cases had the same geographical distribution as controls. Each case and control pairing was compared for sex and age. In total, 1200 cases and 1270 controls were included, including 1296 males (53.5%) and 1174 females (47.5%). Statistics ensured no significant difference between cases and controls in sex, age or education.

Data were collected through pre-coded questionnaires. Questions addressed the socioeconomic status of the family, household environment and health-related behaviors. Socioeconomic status of the family was determined by "number of family members" and "family income". Household environment was determined by "household sanitation", "drinking water hygiene", "presence of a separate kitchen or toilet from the rest of the house", and "frequency of garbage collection". Health-related behaviors were composed of "frequency of dining out" and "habit of washing hands before dinner or after defecation".

The case-control study was approved by the ethical committee and IRB (Institutional Review Board) of Nanjing Medical University. χ^2 analysis was carried out by SAS 9.0 software on the case-control study. The Odds Ratio (*OR*) of each risk factor estimated the power of association between the risk factor and *Shigella* infection. Point estimation provided a 95% confidence interval (*CI*), which was considered statistically significant at $P = 0.05$ if 1 was not included.

Results

Spatial variation analysis

Incidence maps shown in Figure 1 were made through a GIS application of the Inverse Distance Weighted (IDW) method for interpolation (Figure 1). A deeper color in the maps represented higher incidence rates. Compared with the northern part of Jiangsu, the incidence rates of the southern region have remained at a relatively high level since 2001. The southern region's incidence rate continually increased after that point. Areas with high levels of incidence rates increased (the dark areas expanded) in the southern region. In 2004 and 2005, *Shigella* infection in the south became the most serious. After 2007, the counties with high incidence rates in the southern part gradually shrank to comprise only the southwestern region. According to these maps, the western counties were the most infected areas over the most years within the southern region. Despite this, the incidence rates of the far southeastern region were even higher in 2005. The middle and northern areas of Jiangsu each had relatively low incidence rate levels each year. However, the far northeastern and northwestern counties shared similar incidence rate levels with the most infected southwestern counties in every year except 2005. Comparing the accumulated incidence rates over the 11 years of the study (2001 to 2011) in each county, Xuzhou (in the northwest), Nanjing (in the southwest), Wuxi and Suzhou (in the southeast) had the greatest accumulated incidence rates among all counties, with Nanjing having the highest of the four (346.04 infections per 100,000 residents in 11 years).

Trend surface analysis. The geographic incidence trends of *Shigella* around Jiangsu Province fluctuated from 2004 to 2006, consistent with the incidence maps (Table 1). The authors chose two locations expressing trends in four directions (Figure 2): the west to east direction (x axis), the south to north direction (y axis) (Figure 2a), the southwest to northeast direction, and the southeast to northwest direction (Figure 2b).

Incidence rates increased from west to east in years before 2004, but this trend was less visible after 2004. In 2002, 2003, 2007, 2008 and 2010, *Shigella* incidence rates decreased from the south to the north, but formed a U-shaped trend in another four years (2001, 2004, 2006 and 2011); that is, they decreased from the south to the middle and increased from the middle to the north. Incidence rates decreased when the direction moved from the southwest to the northeast before 2007. For trends in the southeast-northwest direction, in five years (2002, 2003, 2008, 2009 and 2010) the incidence rates rose from northwest to southeast, but in another five (2001, 2004, 2006, 2007 and 2011) they manifested a U-shaped trend. In 2005, trends of all directions were more distinctive than in any other year: they formed an inverted U shape, indicating that southern and northern incidence rates were even lower than those in the middle region.

Spatio-temporal cluster analysis and local autocorrelation analysis. Spatio-temporal cluster analysis was applied to observe the incidence rates of *Shigella* from 2001 to 2011 in Jiangsu Province. Incidence rates were aggregated through space and time. A most likely cluster was observed in the southwestern region of the province from 2001 to 2004. This cluster was made up of 30 counties in the southwestern region, including Liyang county, Lishui County, Gaochun County, Jintan County, Yixing County, Jurong County, Danyang County, Changzhou County, Zhenjiang County, Yangzhong County, Wuxi County, Nanjing County, Jiangyin County, Suzhou County, Jingjiang County, Taixing County, Yizheng County, Yanzhou County, Taizhou County, Wujiang County, Jiangdu County, Zhangjiagang County, Changshu County, Jiangyan County, Rugao County, Nantong County, Kunshan County, Gaoyou County, Hai'an County and Taicang County. The cluster window was centered at 31.4205 N, 119.3564 E, in a county named Liyang. The base diameter of the window was 169.96 km, with a Relative Risk (*RR*) of 2.61. The Log Likelihood Ratio for the analysis was 1674.74, $P < 0.001$.

Results of the local autocorrelation analysis are shown in Table 2 (Table 2). In 2001, the only detected hotspot was in Suzhou County, but it disappeared in 2002. In 2003, two negatively

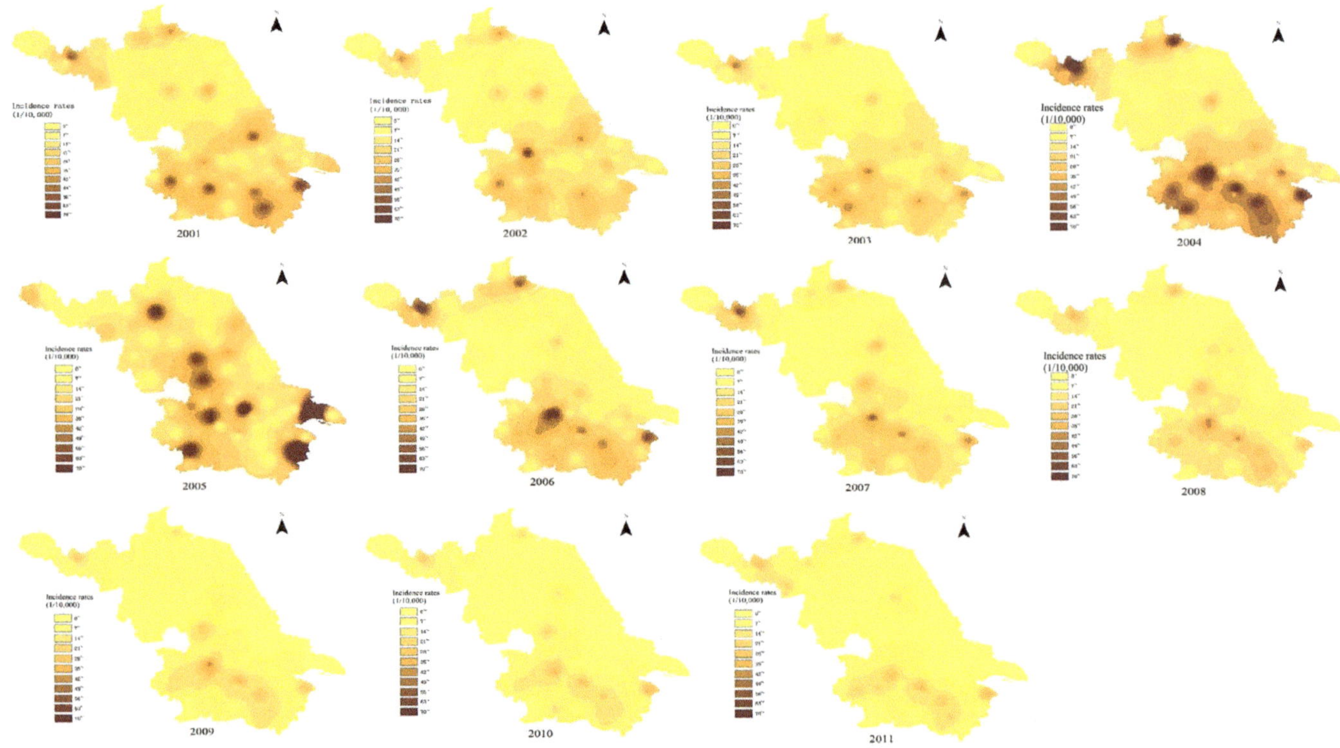

Figure 1. Incidence maps of Shigella in Jiangsu Province from 2001 to 2011.

autocorrelated counties were detected, between which Xuzhou had the higher incidence rate and was surrounded by low incident counties. In 2005, hotspots lying at the southeast region were replaced by hotspots in the southwestern region. In this year, most of the total hotspots were located in the southwestern counties. In 2007, the southeastern hotspots Wuxi and Suzhou were again detected as hotspots, along with other southwestern hotspots. Among all 15 hotspots during the 11 years, Wuxi county and Suzhou County remained hotspots the greatest number of times (six times each). Danyang County, Zhenjiang County and Jurong County were detected as hotspots five, four and two times, respectively. Each of the other 10 counties appeared only once.

Time series analysis

Generally speaking, the number of *Shigella* cases decreased over the years studied. An obvious periodicity by season was observed. The epidemic incidence rates peaked every summer and autumn of the years studied (July to August). Incidence rates decreased from 2001 (25.80 cases per 100,000 residents) to 2003 (19.07 cases per 100,000 residents) and peaked in 2004 (30.60 cases per 100,000 residents). A notable decrease occurred in 2005. In 2006, however, incidence rates rose to a level equal to those in 2001, then kept decreasing after 2006. In 2010, the incidence rate reached its lowest point: 11.46 cases per 100,000 residents. In 2011, the incidence rate was 12.23 cases per 100,000 residents, less than half of what it was in 2001.

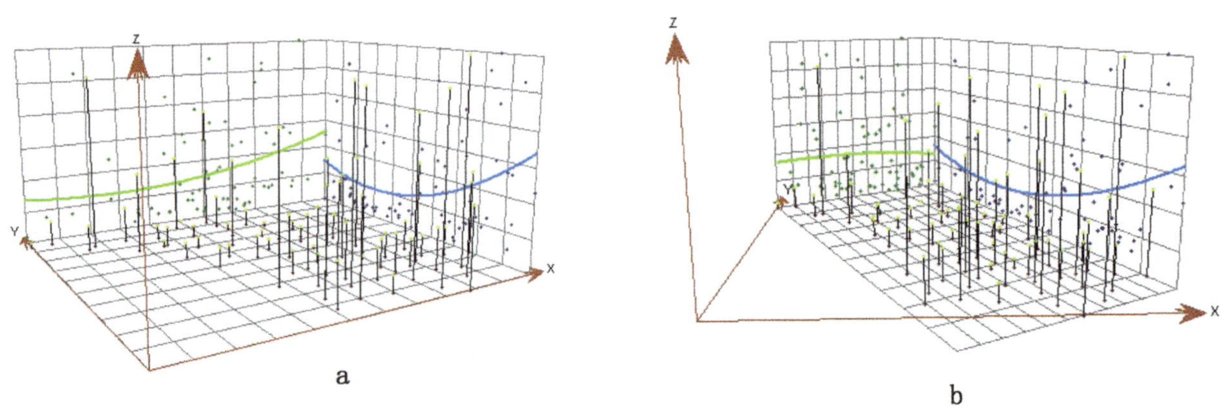

Figure 2. Trend surface graphs of Shigella incidence rates in 2001.

Table 1. Trends for Shigella incidence rates.

Year	West-East	South-North	Southwest-Northeast	Southeast-Northwest
2001	Increase	U shape	Decrease	U shape
2002	Increase	Decrease	Decrease	Decrease
2003	Increase	Decrease	Decrease	Decrease
2004	No trend	U shape	Decrease	U shape
2005	Inverted U shape	Inverted U shape	Inverted U shape	Inverted U shape
2006	No trend	U shape	Decrease	U shape
2007	No trend	Decrease	Inverted U shape	U shape
2008	No trend	Decrease	Inverted U shape	Decrease
2009	No trend	No trend	Inverted U shape	Decrease
2010	No trend	Decrease	Inverted U shape	Decrease
2011	No trend	U shape	Inverted U shape	U shape

Analysis was performed by SAS with significance level $P=0.05$, based on the monthly data of Jiangsu Province from January 2001 to July 2011. After the 12-step difference operation, the trend sequence was flat. Results of the autocorrelation function (ACF) and partial autocorrelation function (PACF) suggest that it was proper to establish an AR(1) model with no constant included (Figure 3). The AIC turned out to be 1787.77. The autocorrelation check for the residuals ensured $P>0.05$ for each lag, which indicated the model fitted well for the data (Table 3). The final model was established as ARIMA (1, 12, 0).

Based on the ARIMA model, the study attempted to predict cases from August to December, 2011 (Table 4). When compared with the observed data, the predicted data agreed very closely with the observed data (Figure 4). Predicted monthly numbers of cases from August to December 2011 were 1172.97, 1039.40, 717.16, 382.24 and 292.06, respectively.

Risk factor detection

Geographic factors detection. Layout charts of the univariate analysis were displayed (Figure 5a–5g). The climatic factors considered were annual mean temperature and moisture content. Distances to railways, highways, lakes, and rivers, as well as NDVI, comprised the environmental factors. A redder color indicates a lower mean temperature (Figure 5a). The moisture content was between 2.10 to 2.99 g/cm^2 (Figure 5b). Incidence rates tended to rise in exponential trends with both annual mean temperatures and moisture content. However, the trends were not obvious with R^2 equal to 0.16, $P=0.001$ (annual mean temperature) and R^2 equal to 0.14, $P=0.002$ (moisture content), respectively.

Buffer zones were established for each railway, highway, river or lake to estimate distances from them (Figure 5c–5f). Incidence rates increased with decreasing distances to railways ($R^2=0.99$, $F=576.61$, $P<0.001$), rivers ($R^2=0.98$, $F=143.68$, $P<0.001$) and lakes ($R^2=0.99$, $F=11411.51$, $P<0.001$) and they peaked at an average of 3 km from the rivers. However, they also peaked at an average 15 km from the highways ($R^2=0.99$, $F=2582.18$, $P<0.001$). No convincing evidence of an association between the NDVI and the incidence rate was observed, with $R^2=0.08$, $P=0.128$ in spring, $R^2=0.04$, $P=0.118$ in summer, $R^2=0.05$, $P=0.367$ in autumn and $R^2=0.17$, $P<0.001$ in winter (Figure 5g).

Based on the above results, risk factors of distances from railways, highways, rivers and lakes were chosen for stepwise regression. However, railway distance was excluded because $P=0.26$ in the first step of regression. A collinearity diagnosis ensured no collinearity existed. The final model was comprised of all three factors, with the highest coefficient of determination R^2 being 0.53, $R=0.73$. It was established as:

$$y = -2.74E-005x_1^3 - 0.021x_1^2 + 0.0763x_1 - 3.52E-006x_2^3 + 6.73E-006x_2^2 + 0.00209x_2 - 0.00133x_3^2 + 0.005312x_3 - 3.3467$$

where y represents the *Shigella* incidence rate, and x_1, x_2 and x_3 stand for distances to the river, the lake and the highway, respectively. ANOVA analysis confirmed $F=140.56$, $P<0.001$.

Case-control study. Results of the univariate analysis in the case-control study are listed (table 5). According to the results, controls tended to have a higher family income ($\chi^2=35.71$, $P<<0.011$) and fewer family members ($\chi^2=33.77$, $P<<0.011$) than cases. Also, controls were more likely to have good household hygiene ($\chi^2=59.12$, $P<<0.011$), use a separate kitchen ($\chi^2=25.47$, $P<<0.011$) and toilet ($\chi^2=65.52$, $P<<0.011$) from the rest of the house, and have access to safe drinking water ($\chi^2=74.54$, $P<<0.011$). Health-related behaviors such as washing hands before dinner ($\chi^2=240.62$, $P<<0.011$) and after defecation ($\chi^2=131.41$, $P<<0.011$), and the frequency of treating household garbage ($\chi^2=14.44$, $P<<0.011$) also differed between cases and controls.

Univariate analysis illustrated that socioeconomic status (number of family members and family income) was negatively associated with *Shigella* infection. In the study, 379 cases and 502 controls reported having fewer than four members in their households (*OR*, 1.75 [*CI*, 1.44–2.13]). A total of 434 cases and 327 controls reported having more than six family members (*OR*, 1.51 [*CI*, 1.24–1.84]). With regard to family income, 454 cases and 371 controls belonged to the lowest group level of 'less than 2000 Yuan/Month' (*OR*, 1.21 [*CI*, 1.01–1.47]). Household environment also turned out to impact *Shigella* infection. This was observed in people not having a separate kitchen or toilet from the rest of the house, with *OR*'s being 1.50 and 1.93, respectively. "Poor household hygiene" or "lack of a safe drinking water supply" were additional risk factors for *Shigella* infection (in Table 5). Washing hands before dinner was strongly associated with *Shigella* infection, with $OR=3.64$, *CI* [3.08–4.03]. Furthermore, the authors also detected that disposing of household garbage every

Table 2. Local autocorrelation analysis for Shigella in Jiangsu Province.

Year	County	LG_i Z Score	LGi P Value	LM_i Index	LM_i Z Score	LM_i P Value	Correlation Type
2001	Suzhou County	2.80	0.01	<0.01	3.91	<0.01	HIGH-HIGH
2003	Tongshan County	1.39	0.16	<0.01	−2.62	0.01	LOW-HIGH
2003	Xuzhou County	0.74	0.46	<0.01	−3.12	<0.01	HIGH-LOW
2004	Suzhou County	2.45	0.01	<0.01	2.67	0.01	HIGH-HIGH
2004	Wuxi County	2.21	0.03	<0.01	2.76	0.01	HIGH-HIGH
2005	Siyang County	2.21	0.03	<0.01	2.19	0.03	HIGH-HIGH
2005	Danyang County	3.24	<0.01	<0.01	3.49	<0.01	HIGH-HIGH
2005	Zhenjiang County	4.86	<0.01	<0.01	4.25	<0.01	HIGH-HIGH
2005	Yangzhou County	4.50	<0.01	<0.01	2.66	0.01	HIGH-HIGH
2005	Taizhou County	4.20	<0.01	<0.01	6.33	<0.01	HIGH-HIGH
2005	Jiangyan County	3.71	<0.01	<0.01	4.68	<0.01	HIGH-HIGH
2005	Taixing County	3.63	<0.01	<0.01	3.44	<0.01	HIGH-HIGH
2005	Yizheng County	2.11	0.04	<0.01	2.18	0.03	HIGH-HIGH
2005	Jiangdu County	4.47	<0.01	<0.01	3.78	<0.01	HIGH-HIGH
2005	Jingjiang County	2.39	0.02	<0.01	2.30	0.02	HIGH-HIGH
2005	Yangzhong County	4.93	<0.01	<0.01	16.03	<0.01	HIGH-HIGH
2006	Danyang County	2.83	<0.01	<0.01	2.73	0.01	HIGH-HIGH
2006	Zhenjiang County	2.64	0.01	<0.01	3.98	<0.01	HIGH-HIGH
2006	Jurong County	3.25	<0.01	<0.01	4.38	<0.01	HIGH-HIGH
2007	Wuxi County	1.98	0.05	<0.01	2.18	0.03	HIGH-HIGH
2007	Danyang County	2.50	0.01	<0.01	3.24	<0.01	HIGH-HIGH
2007	Zhenjiang County	1.83	0.07	<0.01	2.34	0.02	HIGH-HIGH
2008	Suzhou County	2.54	0.01	<0.01	3.17	<0.01	HIGH-HIGH
2008	Wuxi County	2.84	<0.01	<0.01	4.32	<0.01	HIGH-HIGH
2008	Danyang County	2.35	0.02	<0.01	2.59	0.01	HIGH-HIGH
2008	Jurong County	3.05	<0.01	<0.01	2.77	0.01	HIGH-HIGH
2009	Suzhou County	2.29	0.02	<0.01	2.50	0.01	HIGH-HIGH
2009	Wuxi County	2.97	<0.01	<0.01	4.93	<0.01	HIGH-HIGH
2009	Changzhou County	2.38	0.02	<0.01	2.18	0.03	HIGH-HIGH
2009	Danyang County	2.74	0.01	<0.01	3.22	<0.01	HIGH-HIGH
2009	Zhenjiang County	2.28	0.02	<0.01	2.64	0.01	HIGH-HIGH
2010	Suzhou County	2.16	0.03	<0.01	2.27	0.02	HIGH-HIGH
2010	Wuxi County	2.63	0.01	<0.01	3.71	<0.01	HIGH-HIGH
2011	Suzhou County	2.06	0.04	<0.01	1.98	0.05	HIGH-HIGH
2011	Wuxi County	2.81	<0.01	<0.01	4.26	<0.01	HIGH-HIGH
2011	Nanjing County	2.14	0.03	<0.01	2.29	0.02	HIGH-HIGH

day or every other day potentially protected against the chance of *Shigella* infection (*OR*, 0.73 [*CI*, 0.63–0.86]).

Discussion

The primary aim of this research was to illustrate spatial and temporal trends of *Shigella* in Jiangsu Province. Incidence maps indicated the far western areas of the southern and northern regions were most vulnerable to *Shigella* infections. Although infections in the southern region continually accounted for the greatest proportion, the high incident area of the northern region expanded its range in the latter years of the decade (after 2004). The total incidence rate was observed decreasing in Jiangsu Province, which was confirmed by the time-series analysis, but this decrease in total incidence rates may obscure the fact that some parts of the northern region may have had increases in *Shigella* incidence rates. Trend surface analysis confirmed the conclusions of the incidence maps. In contrast with the continuously decreasing trend in the south-to-north direction, the trend curves changed to a U shape in some years (2001, 2004, 2006 and 2011). This indicated that northern incidence rates might have approached parity with those of the southern region.

Suzhou and Wuxi were the two most prominent counties for *Shigella* transmissions in the southeastern region; they were identified as hotspots the greatest number of times among all counties (in 2001, 2007, 2008, 2009, 2010, 2011). However, a cluster was also detected at the southwestern part of Jiangsu from 2001 to 2004. The hotspots detected in the autocorrelation

```
                                   Autocorrelations

Lag   Covariance   Correlation   -1 9 8 7 6 5 4 3 2 1 0 1 2 3 4 5 6 7 8 9 1   Std Error

  0       178847       1.00000   |                    |********************|          0
  1       136307       0.76215   |                  . |***************     |   0.088736
  2    98133.338       0.54870   |                .   |***********         |   0.130466
  3    69475.709       0.38846   |               .    |********            |   0.147522
  4    39605.301       0.22145   |               .    |****  .             |   0.155368
  5    20696.302       0.11572   |               .    |**    .             |   0.157834
  6    10448.762       0.05842   |               .    |*     .             |   0.158501
  7     5032.990       0.02814   |               .    |*     .             |   0.158670
  8     3575.557       0.01999   |               .    |      .             |   0.158709
  9   -11589.711       -.06480   |               .   *|      .             |   0.158729
 10   -13284.145       -.07428   |               .   *|      .             |   0.158937
 11   -23706.573       -.13255   |               . ***|      .             |   0.159210
 12   -31012.362       -.17340   |               . ***|      .             |   0.160077
 13   -21721.168       -.12145   |               .  **|      .             |   0.161549
 14   -22095.896       -.12355   |               .  **|      .             |   0.162267
 15   -20677.234       -.11561   |              .   **|       .            |   0.163006
 16   -17412.237       -.09736   |              .   **|       .            |   0.163650
 17   -19967.017       -.11164   |              .   **|       .            |   0.164105
 18   -16243.105       -.09082   |              .   **|       .            |   0.164702
 19   -17893.842       -.10005   |              .   **|       .            |   0.165096
 20   -17402.751       -.09731   |              .   **|       .            |   0.165573
 21   -10062.616       -.05626   |              .    *|       .            |   0.166023
 22    -3827.419       -.02140   |              .     |       .            |   0.166173
 23      835.932       0.00467   |              .     |       .            |   0.166194
 24    -5588.582       -.03125   |              .    *|       .            |   0.166195

                          "." marks two standard errors
```

```
                         Partial Autocorrelations

Lag   Correlation   -1 9 8 7 6 5 4 3 2 1 0 1 2 3 4 5 6 7 8 9 1

  1       0.76215   |                 .  |***************     |
  2      -0.07674   |                 . **|    .              |
  3      -0.00799   |                 .   |    .              |
  4      -0.12601   |                 .***|    .              |
  5       0.02140   |                 .   |    .              |
  6       0.01474   |                 .   |    .              |
  7       0.01729   |                 .   |    .              |
  8       0.01083   |                 .   |    .              |
  9      -0.20449   |                 ****|    .              |
 10       0.11311   |                 .   |**  .              |
 11      -0.17768   |                 ****|    .              |
 12       0.03457   |                 .   |*   .              |
 13       0.10905   |                 .   |**  .              |
 14      -0.12498   |                 . **|    .              |
 15       0.02549   |                 .   |*   .              |
 16      -0.05899   |                 .  *|    .              |
 17      -0.00866   |                 .   |    .              |
 18       0.01023   |                 .   |    .              |
 19      -0.03380   |                 .  *|    .              |
 20      -0.01258   |                 .   |    .              |
 21       0.01841   |                 .   |    .              |
 22       0.08269   |                 .   |**  .              |
 23      -0.08668   |                 . **|    .              |
 24      -0.10462   |                 . **|    .              |
```

Figure 3. Results of ACF and PACF for time series analysis.

Table 3. Autocorrelation check of residuals in time series analysis.

To lag	χ^2	DF	*p*	Autocorrelations					
6	2.65	5	0.75	0.05	−0.02	0.07	−0.08	−0.07	−0.04
12	15.11	11	0.18	−0.03	0.15	−0.14	0.08	−0.04	−0.19
18	18.23	17	0.37	0.09	−0.03	−0.03	0.06	−0.07	0.05
24	21.71	28	0.54	−0.03	−0.08	0.02	0.02	0.12	−0.03

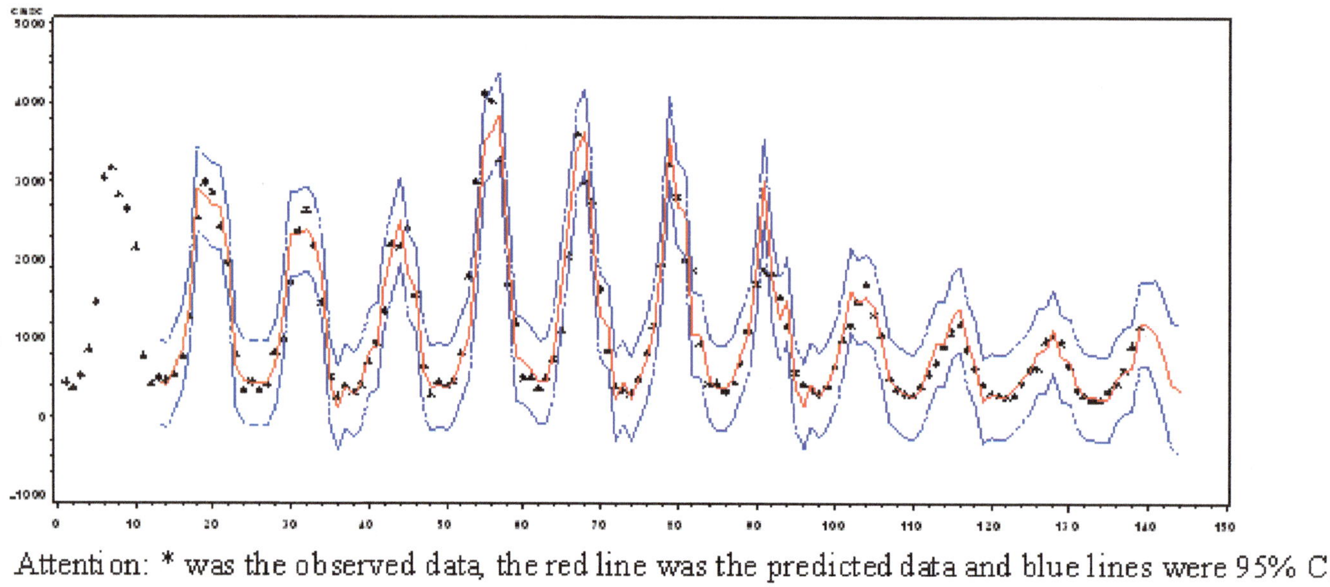

Figure 4. Prediction for Shigella cases from August to December in 2011.

analysis indicated that hotspots were aggregated in the southwestern region after 2005. This indicates that the southwestern region may be becoming a new target for *Shigella*. Notable fluctuations also appeared from the years 2004 to 2006. The ARIMA model confirmed these fluctuations and successfully established a predictive model concerning the time series. Researchers have previously found that the ARIMA model was very effective and reliable in providing decision makers with clear indications of variability among future observations [24]. Establishment of the ARIMA model could help to forecast *Shigella* developments, as well as warn epidemiological monitors of potential abnormalities in the future.

Spatial and temporal trends functioned as a first step to determine pathways for epidemic development and discover potential target areas. Another important aim of this study was to identify potential risk factors. Many researchers have confirmed that geographical factors such as ambient temperatures and relative humidity are driving *Shigella* transmissions [25,26]. However, these conclusions were mostly based on temporal analyses using monthly data. In this article, geographical risk factors were obtained to explore the geographical reasons causing spatial variability among counties. According to the results, temperature and moisture content were not the primary reasons for spatial variations in incidence rates in Jiangsu Province; the distances to railways, highways, rivers and lakes contributed substantially more. Transportation may have led to more frequent interactions among cities or provinces, which concurrently increased the chances of *Shigella* transmission. Since railways and highways are the two main methods of long distance travel in Jiangsu, this indicates that sanitation policies around railways or highways should be well enforced.

The stepwise regression model helped to correlate results with environmental conditions. Study results confirmed that distances to highways, rivers and lakes facilitated *Shigella* transmission, and also indicated the importance of reducing *Shigella* infections transmitted through water systems. Since water sources were the leading reason for spatial variation, improving the imbalanced quality of water supplies in the northern and southern regions should be the next target for *Shigella* prevention. The case-control study confirmed a safe water supply as an important protective factor against *Shigella* transmission. Inadequate water and sanitation not only increased the chances of diarrheal infection, but were also associated with poorer child health status in the long term [27]. In the less developed northwestern areas of Jiangsu Province, the sanitation and hygiene issues drew the authors' greatest concern. The penetration of access to a safe water supply was much lower here than in the southern and middle regions; in 2009, Zhu et. al. announced that in Xuzhou and Lianyungang (two cities of northern Jiangsu), the penetration levels of access to a safe water supply were only 37.3% and 63.2%, respectively (compared to over 92% in other regions) [28]. Furthermore, the rates of treatment of domestic garbage in the northern region were only

Table 4. Forecasts for variable cases of 2011.

Observations	Forecast	Standard Error	95% Confidence Limits	
August	1172.97	275.24	633.51	1712.43
September	1039.40	346.61	360.16	1718.84
October	717.16	382.28	−32.10	1466.41
November	382.24	401.71	−404.99	1169.68
December	292.06	412.66	−516.75	1100.86

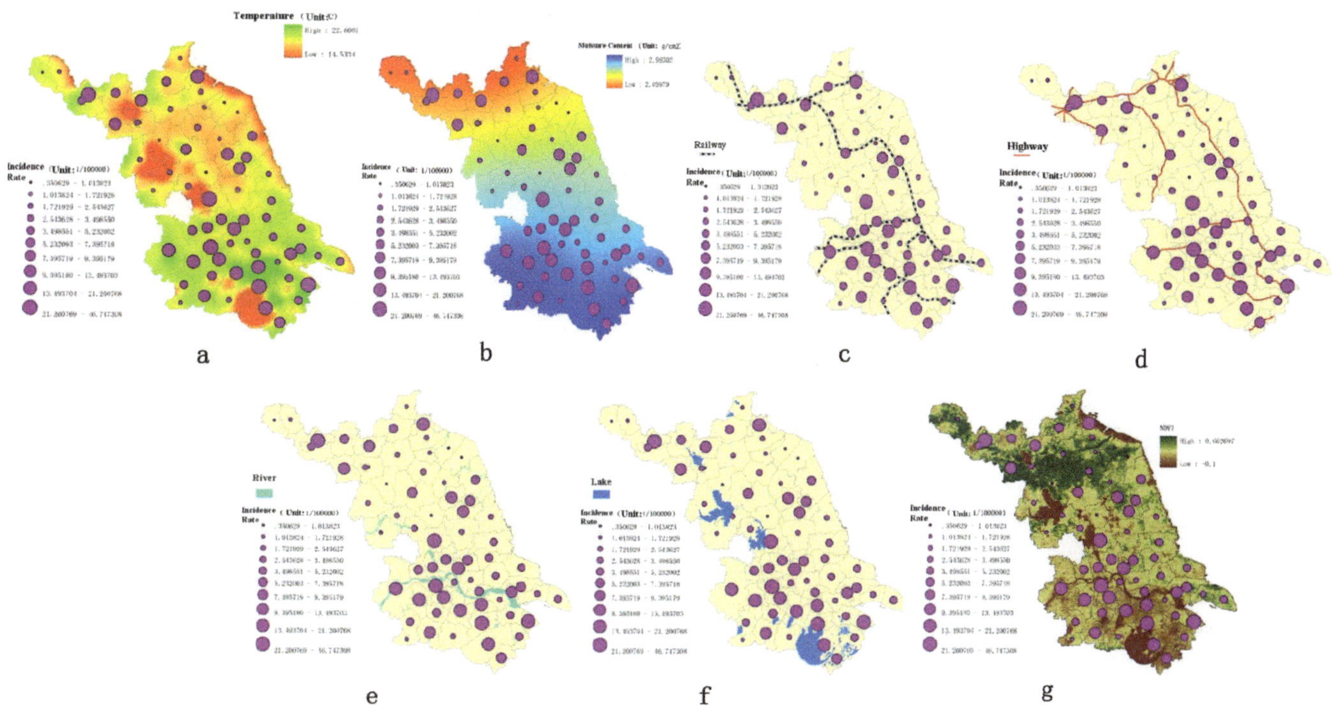

Figure 5. Layout charts concerning risk factors and incidence rates.

about one-sixth those in the southern region. Lack of sanitation protection in the northern counties may have increased the transmission of enteric diseases like *Shigella*. Garbage disposal and hand washing have previously been identified as two protective behavioral factors preventing *Shigella* transmission; Rego et. al. declared that children exposed to garbage in their environment were nearly four times more likely to contract diarrhea [29]. Given the presence of organic waste in family garbage, the surrounding

Table 5. Univariate case-control analysis between Shigella infection and risk factors.

Variable	Groups	Cases N	Controls N	χ^2	*P*	*OR*	95%CI	
Family members	More than six	434	327	33.77	<<0.011	1		
	Between four and six	387	441			1.51	1.24	1.84
	Less than four	379	502			1.75	1.44	2.13
Family income per capita	<2000 Yuan/Month	454	371	35.71	<<0.011	1		
	2000~4000 Yuan/Month	425	421			1.21	1.01	1.47
	>4000Yuan/Month	321	478			1.82	1.49	2.22
Household sanitation	Not good	671	531	59.12	<<0.011	1		
	Good	529	739			1.76	1.51	2.07
Safety of drinking water supply	Not Good	803	632	74.54	<<0.011	1		
	Good	397	638			2.04	1.73	2.40
Separate kitchen	No	669	579	25.47	<<0.011	1		
	Yes	531	691			1.50	1.28	1.76
Separate toilet	No	718	553	65.52	<<0.011	1		
	Yes	482	717			1.93	1.64	2.26
Disposing of household garbage	Daily/every other day	563	693	14.44	<<0.011	1		
	More	637	577			0.73	0.63	0.86
Washing hands after defecation	No	677	991	131.41	<<0.011	1		
	Yes	523	279			0.36	0.30	0.42
Washing hands before dinner	No	799	449	240.62	<<0.011	1		
	Yes	401	821			3.64	3.08	4.30

environment is prone to attracting disease vectors. This increases the chance of young children being exposed to pathogens through playing with garbage, which presents a challenge to maintaining good family hygiene.

In the case-control study, a higher family income and fewer family members both indicated better socioeconomic status. This group also had a lower incidence rate. In 2008 Ferrer et al. confirmed economic factors as one of the factors influencing enteric burden [30]. Low family income may have predisposed families to a lower nutritional status and a lower awareness of hygiene, thus leading to a higher probability of disease. Also, a crowded environment, in which household hygiene is always hard to enforce, would facilitate person-to-person transmission. However, when compared with the conclusions from the case-control study, socioeconomic factors worked differently on *Shigella* transmissions when pointing to regional variations. The highly developed southern region of Jiangsu apparently had more infections than the middle and northern regions. The southern region of Jiangsu, including Suzhou and Wuxi, currently attracts large amounts of migrants due to its prosperous economic development. Living in a transitory environment and having a precarious lifestyle, the migrant population might not have prioritized food safety and sanitation issues, which made enteric infection the largest threat to their health status. At the same time, the rapid economic development brought a large expansion in business activity and increased the frequency of dining outside of the home. It also increased the possibility of exposure to *Shigella* dysenteriae from polluted food and drinking water. These phenomena brought on by economic development may have led to the high incidence level of the southern areas.

As is well-known, macroscopic descriptions or analyses can easily lead to ecological fallacy problems. Discussions about geographic risk factors in this study were carried out province-wide and conclusions were made based on county-level groups of infections. However, individuals might behave differently than the groups did. As mentioned above, economic problems affect individuals and groups in different ways. In order to avoid a possible ecological fallacy, the case-control study, with its more microscopic focus on individuals, provided supplementary explanations for geographical risk factors, especially regarding access to a safe water supply and other socioeconomic factors. Both the macroscopic and microscopic discussions were combined in this research to analyze geographical risk factors and behavior-related risk factors. In this way, the authors hope that conclusions based on the results might be more scientifically robust than a simple one-way analysis.

Based on these results, future interventions should focus on different points addressing the transmission patterns and underlying risk factors of different regions. Improvements in sanitation and hygiene should be strengthened at the county level, while access to safe water supplies in impoverished areas should be increased at the same time. Furthermore, strengthening public education towards prevention of *Shigella* should be incorporated with interventional policies to eliminate or reduce *Shigella* infection, such as advocating washing hands correctly and regularly, or using hygiene facilities in the household. Srtina et al. concluded that hygienic behaviors were associated with a family's individual predisposition rather than the neighborhood's [31]. In this way, emphasizing efforts at the domestic level to the application of hygiene facilities is of great importance.

Although this study was aimed at detecting spatial variations and causal risk factors, as well as establishing an ARIMA model for surveillance, the limitations of this research must also be acknowledged. Among all 11 years, the incidence rate was found to increase from 2001 to 2004 but hotspots continued to exist even after 2005, and some of them appeared only once. There is a possibility that some hotspots might have been detected by chance, in which case more surveillance work should be carried out around the province to monitor these hotspots or potential risk clusters. GIS is powerful in predicting and visualizing models combining risk factors, and is gradually being applied more and more to explore epidemic risk factors in recent years [32,33]. However, in spite of the statistic estimations, visualized incidence maps may easily lead to specious conclusions. In addition, limited data resources make it hard to widely explore environmental risk factors. Therefore the results in this study should be supported by continuous health surveillance programs to see how they function under real-world conditions.

Author Contributions

Conceived and designed the experiments: ZHP FYT CJB HBS JLH. Performed the experiments: WDL YW RBY FC. Analyzed the data: YJC ZHP. Contributed reagents/materials/analysis tools: YJC WDL ZHP QL. Wrote the paper: FYT YJC JN ZHP.

References

1. Bardhan P, Faruque AS, Naheed A, Sack DA (2010) Decrease in shigellosis-related deaths without Shigella spp.-specific interventions, Asia. Emerg Infect Dis 16: 1718–1723.
2. Wang XY, Tao FB, Xiao DL, Lee H, Deen J, et al. (2006) Trend and disease burden of bacillary dysentery in China (1991–2000). Bull World Health Organ 84: 561–568.
3. Hu JL, Qian HM, Zhuang L, Dong C, Tan ZM, et al. (2011) [Analysis on epidemic characteristics of bacillary dysentery in Jiangsu Province from 2004 to 2010] in Chinese. Zhong Hua Yu Fang Yi Xue Za Zhi 12: 419–422.
4. Abu-Elyazeed RR, Wierzba TF, Frenck RW, Putnam SD, Rao MR, et al. (2004) Epidemiology of Shigella-associated diarrhea in rural Egyptian children. Am J Trop Med Hyg 71: 367–372.
5. Gu B, Cao Y, Pan S, Zhuang L, Yu R, et al. (2012) Comparison of the prevalence and changing resistance to nalidixic acid and ciprofloxacin of Shigella between Europe-America and Asia-Africa from 1998 to 2009. Int J Antimicrob Agents 40: 9–17.
6. Yan W, Xu Y, Yang X, Zhou Y (2010) A hybrid model for short-term bacillary dysentery prediction in Yichang City, China. Jpn J Infect Dis 63: 264–270.
7. Earnest A, Tan SB, Wilder-Smith A, Machin D (2012) Comparing statistical models to predict dengue fever notifications. Comput Math Methods Med 2012: 758674.
8. Loha E, Lindtjørn B (2010) Model variations in predicting incidence of Plasmodium falciparum malaria using 1998–2007 morbidity and meteorological data from south Ethiopia. Malar J 9: 166–174.
9. Liu Q, Liu X, Jiang B, Yang W (2011) Forecasting incidence of hemorrhagic fever with renal syndrome in China using ARIMA model. BMC Infect Dis 11: 218–225.
10. Ali M, Goovaerts P, Nazia N, Haq MZ, Yunus M, et al. (2006) Application of Poisson kriging to the mapping of cholera and dysentery incidence in an endemic area of Bangladesh. Int J Health Geogr 5: 45.
11. Li L, Wang J, Wu J (2012) A spatial model to predict the incidence of neural tube defects. BMC Public Health 12: 951–961.
12. Cheng YJ, Norris J, Bao CJ, Liang Q, Hu JL, et al. (2012) Geographical information systems-based spatial analysis and implications for syphilis interventions in Jiangsu province, People's Republic of China. Geospat Health 7: 63–72.
13. Emch M (1999) Diarrheal disease risk in Matlab, Bangladesh. Soc Sci Med 49: 519–530.
14. Zhang Y, Bi P, Sun Y, Hiller JE (2012) Projected Years Lost due to Disabilities (YLDs) for bacillary dysentery related to increased temperature in temperate and subtropical cities of China. J Environ Monit 14: 510–516.
15. Yang GH, Strong DF, Thacker SB (1997) National public health surveillance in China: implications for public health in China and the United States. Biomed Environ Sci 10: 1–13.
16. Zhou XN (2009) Spatial Epidemiology. Beijing: Science Press. 241–242 p.
17. Kulldorff M, Athas WF, Feurer EJ, Miller BA, Key CR (1998) Evaluating cluster alarms: a space-time scan statistic and brain cancer in Los Alamos, New Mexico. Am J Public Health 88: 1377–1380.

18. Liu XN, Huang F, Wang P (2008) Spatial analysis theories and methods on GIS. Beijing: Science Press. 189–194 p.
19. Bruce L, Bowerman, O'Connell (2003). Forecasting and Time Series: An applied approach (Third Edition). Beijing: China Machine Press. 117–156 p.
20. Peng WJ, Zhu Y, He Q, Wang J (2011) [The application of multiple seasonal ARIMA model on prediction of the incidence of bacillary dysentery] in Chinese. Zhong Guo Wei Sheng Tong Ji 28: 645–647.
21. Zhang Y, Bi P, Hiller JE (2008) Weather and the transmission of bacillary dysentery in Jinan, northern China: a time-series analysis. Public Health Rep 123: 61–66.
22. Ljung GM, Box GEP (1978) On a measure of lack of fit in time series model. Biometrika 65: 297–303.
23. Claeskens G, Hjort NL (2008) Model selection and model averaging. Cambridge: Cambridge University Press.
24. Allard R (1998) Use of time-series analysis in infections disease surveillance. Bull World Health Organ 76: 327–333.
25. D'Souza RM, Becker NG, Hall G, Moodie KB (2004) Does ambient temperature affect food borne disease? Epidemiology 15: 86–92.
26. Huang D, Guan P, Guo J, Wang P, Zhou B (2008) Investigating the effects of climate variation on bacillary dysentery incidence in northeast China using Ridge Regression and Hierarchical Cluster Analysis. BMC Infect Dis 8: 30–39.
27. Checkley W, Gilman RH, Black RE, Epstein LD, Cabrera L, et al. (2004) Effect of water and sanitation on childhood health in a poor Peruvian Peri-urban Community. Lancet 363:112–118.
28. Zhou SD, Zhu HG, Ge JH, Wang YX, Wu XW (2009) [Comprehensive evaluation of the rural development of index system in Jiangsu Province] in Chinese. Jiangsu Social Science 2: 212–218.
29. Rego RF, Moraes LR, Dourado I (2005) Diarrhoea and garbage disposal in Salvador, Brazil. Trans R Soc Trop Med Hyg 99: 48–54.
30. Ferrer SR, Strina A, Jesus SR, Ribeiro HC, Cairncross S, et al. (2008) A Hierarchical Model for studying risk factors for childhood diarrhoea: a case–control study in a middle-income country. Int J Epidemiol 37: 805–815.
31. Strina A, Cairncross S, Barreto ML, Larrea C, Prado MS (2003) Childhood diarrhea and observed hygiene behavior in Salvador, Brazil. Am J Epidemiol 157: 1032–1038.
32. Forrer A, Sayasone S, Vounatsou P, Vonghachack Y, Bouakhasith D, et al. (2012) Spatial distribution of, and risk factors for, Opisthorchis viverrini infection in southern Lao PDR. PLoS Negl Trop Dis 6: e1481.
33. Vinceti M, Fiore M, Signorelli C, Odone A, Tesauro M, et al. (2012) Environmental risk factors for amyotrophic lateral sclerosis: methodological issues in epidemiologic studies. Ann Ig 24: 407–415.

Neighbourhood, Route and Workplace-Related Environmental Characteristics Predict Adults' Mode of Travel to Work

Alice M. Dalton[1,3]*, Andrew P. Jones[2,3], Jenna R. Panter[3,4], David Ogilvie[3,4]

1 School of Environmental Sciences, University of East Anglia, Norwich, United Kingdom, **2** Norwich Medical School, University of East Anglia, Norwich, United Kingdom, **3** UKCRC Centre for Diet and Activity Research (CEDAR), Institute of Public Health, Cambridge, United Kingdom, **4** Medical Research Council Epidemiology Unit, Institute of Public Health, Cambridge, United Kingdom

Abstract

Objective: Commuting provides opportunities for regular physical activity which can reduce the risk of chronic disease. Commuters' mode of travel may be shaped by their environment, but understanding of which specific environmental characteristics are most important and might form targets for intervention is limited. This study investigated associations between mode choice and a range of objectively assessed environmental characteristics.

Methods: Participants in the Commuting and Health in Cambridge study reported where they lived and worked, their usual mode of travel to work and a variety of socio-demographic characteristics. Using geographic information system (GIS) software, 30 exposure variables were produced capturing characteristics of areas around participants' homes and workplaces and their shortest modelled routes to work. Associations between usual mode of travel to work and personal and environmental characteristics were investigated using multinomial logistic regression.

Results: Of the 1124 respondents, 50% reported cycling or walking as their usual mode of travel to work. In adjusted analyses, home-work distance was strongly associated with mode choice, particularly for walking. Lower odds of walking or cycling rather than driving were associated with a less frequent bus service (highest versus lowest tertile: walking OR 0.61 [95% CI 0.20–1.85]; cycling OR 0.43 [95% CI 0.23–0.83]), low street connectivity (OR 0.22, [0.07–0.67]; OR 0.48 [0.26–0.90]) and free car parking at work (OR 0.24 [0.10–0.59]; OR 0.55 [0.32–0.95]). Participants were less likely to cycle if they had access to fewer destinations (leisure facilities, shops and schools) close to work (OR 0.36 [0.21–0.62]) and a railway station further from home (OR 0.53 [0.30–0.93]). Covariates strongly predicted travel mode (pseudo r-squared 0.74).

Conclusions: Potentially modifiable environmental characteristics, including workplace car parking, street connectivity and access to public transport, are associated with travel mode choice, and could be addressed as part of transport policy and infrastructural interventions to promote active commuting.

Editor: Harry Zhang, Old Dominion University, United States of America

Funding: The Commuting and Health in Cambridge study was developed by David Ogilvie, Simon Griffin, Andy Jones and Roger Mackett, and initially funded under the auspices of the Centre for Diet and Activity Research (CEDAR), a UKCRC Public Health Research Centre of Excellence. Funding from the British Heart Foundation, Economic and Social Research Council, Medical Research Council, National Institute for Health Research and the Wellcome Trust, under the auspices of the UK Clinical Research Collaboration, is gratefully acknowledged. The study is now funded by the National Institute for Health Research Public Health Research programme (project number 09/3001/06: see http://www.phr.nihr.ac.uk/funded_projects). David Ogilvie is also supported by the Medical Research Council (Unit Programme number MC_U106179474). The views and opinions expressed herein are those of the authors and do not necessarily reflect those of the NIHR PHR programme, NHS or the Department of Health. The funders had no role in study design, data collection and analysis, decision to publish, or preparation of the manuscript.

Competing Interests: The authors have declared that no competing interests exist.

* E-mail: a.dalton@uea.ac.uk

Introduction

People who engage in physical activity are less likely to suffer from a range of preventable chronic diseases [1]. However, most people do not meet government guidelines for physical activity [2], which contributes to the burden of disease in the UK [3] and globally [4]. One way of increasing population activity levels may be to build it into daily routines, for example in the form of active travel (cycling and walking) for part or all of the journey to work [5]. This has the potential to make a significant contribution to physical activity, as commuting to and from work constitutes 15% of the journeys made in the UK and the United States [6,7] and 19% of the distance travelled [7]. There has been a steady decline in cycling and walking to work during the last few decades as the car has become a more popular choice [8] and people have tended to live further from work [9]. Nevertheless, there is evidence that regular active travellers are physically fitter and have higher levels of mental wellbeing and lower sickness absence from work [10,11]. In addition, public transport may provide an alternative to the car for those living further from work and offer health as well as environmental benefits, with US research finding that people walking to transit stops can accrue around 30% of their recommended daily activity levels by doing so [12]. A recent modelling study has suggested that a step-change in the prevalence of walking and cycling in England and Wales could save the National Health Service approximately £17bn in healthcare costs associated with chronic conditions such as type-2 diabetes, cancer, heart disease and depression over the next 20 years [13].

Features of the built and natural environments within which people live and work may influence the mode of travel used [1], such as for journeys to work, by either supporting or acting as a barrier against walking and cycling. In order to develop environments more supportive of a modal shift towards more active travel at the population level, we therefore need to better understand the importance of characteristics of the home, work and journey environments [14,15]. Whilst there is good evidence that modal choice is associated with travel distance, evidence is more equivocal regarding the importance of many other characteristics of the physical environment including street network connectivity, urban design, walkable destinations (including schools, shops, leisure and cultural facilities), land use, infrastructure for walking or cycling, and the availability of or access to public transport [1,16,17]. It is also noteworthy that despite the constraints of distance, some people do use active modes of travel even if they live a relatively long way from work.

In part, our limited understanding of the correlates of active travel may reflect methodological limitations of some previous studies. A recent systematic review of the environmental correlates of cycling noted that few studies had measured the environment objectively and none of the 21 studies identified could be regarded as methodologically strong [16]. Reasons included failing to control for confounders, such as gender and age; failing to carry out analysis at the individual level; and using inappropriate statistical methods. Many studies have examined environmental attributes, public transport (transit) accessibility, or restrictions on car use singly but not in combination, meaning that the relative importance of these factors is not known. For these reasons, a recent review concluded that empirical research either does not include or provides inconclusive evidence on the influence of the transport environment on travel behaviour [1]. In addition, research has tended to treat walking and cycling as a singular behaviour, whereas they are distinct behaviours likely to have different correlates [18,19]. A further limitation is that characteristics of the home neighbourhood have often been the focus for analysis, yet the workplace environment and that of the route between home and work may also be important [20,21,22]. Although some research has considered distances to public transport (transit) stops [23], overall accessibility and the ease by which that transport may be used to reach work has rarely been considered [16,24]. In general, previous studies on active travel suggest that existing models fail to account for a significant proportion of the variance in travel behaviour. Environmental attributes have been shown to explain some of this variance, but prior empirical research has typically tested a limited range of independent variables in rather environmentally homogeneous, often urban, settings.

Using a much wider range of objective environmental measures than have commonly been tested, the aim of this study is to investigate the extent to which features of the environment are associated with modal choice amongst a sample of commuters in Cambridge, UK. We focus on the individual level and assess correlates for travel by all modes. In doing so, we aim to identify modifiable environmental characteristics that might form the targets of future interventions to increase the prevalence of active travel.

Methods

Study design and setting

This research analysed cross-sectional data obtained from a sample of commuters taking part in the Commuting and Health in Cambridge study in Cambridge, UK. The details of this study have been outlined previously [25]. Participants were aged 16 and over, working in Cambridge and living within 30 km of the city. They were sampled using a workplace recruitment strategy that targeted a variety of workplaces and employers in a range of geographical locations across the city centre and urban fringe.

The data used for this analysis were collected between May and November 2009 using postal questionnaires [26]. Participants reported their recent physical activity (at home, at work and for recreation), general health, and travel to and from work and for other purposes. Personal characteristics such as age, gender, weight, height and highest educational attainment were also reported. 1168 respondents returned questionnaires, 1155 of whom provided valid postcodes which could be used to identify their home and work locations.

Ethics statement

Ethical approval for the study was obtained from the Hertfordshire Research Ethics Committee (reference number 08/H0311/208) and written informed consent was provided by each participant. No minors/children participated in this study.

Determining modal choice

Two sections of the survey questionnaire were relevant to this analysis. The first comprised the Recent Physical Activity Questionnaire (RPAQ), a validated instrument that measures physical activity in the previous four weeks [27]. RPAQ was used to classify participants according to their usual mode of travel to work. The survey listed four modes (car/motor vehicle, works or public transport, bicycle, or walking) and participants were asked to specify if they 'always', 'usually', 'occasionally' or 'never or rarely' travelled to work by each mode. Most participants selected the option of 'always' for one mode of transport, and their usual mode was classified accordingly. For the 178 cases in which more than one mode was identified, rules were established to classify participants according to their predominant modal choice based on the most frequently reported travel mode: if participants stated that they always or usually walked and/or cycled but that they also always or usually used the car, they were coded as using the car as it was presumed this would constitute the main component of the journey (n = 92); if they stated they always or usually walked and/or cycled but also always or usually used public transport, they were coded as using public transport (n = 39); if they reported always or usually walking and cycling they were coded as cycling (n = 9); and if they stated that they sometimes used the car and sometimes the bicycle, they were coded as using the car (n = 3).

For 35 participants, no predominant mode could be determined at this stage. In these cases, a question from the second relevant section of the questionnaire, '*About your travel to and from work in the last seven days*', was used to identify the predominant mode or modes used in the previous seven days. Of those always reporting more than one mode per journey, 19 using the car and bus were classified as car users, one reporting walking and using the train was classified as using the train, and one using the train and the car was classified as using the train. For those reporting a mixture of different modes which varied throughout the week, six used the car most frequently and one used the train most frequently. One participant did not answer the RPAQ question but responded as only using the car to the 'last seven days' question. Six participants could not be attributed to one predominant mode because they reported using a mixture of different modes of transport throughout the week with equal frequency, and were therefore not included in subsequent analyses.

Environmental measures, neighbourhood delineation and route identification

For each participant, environmental measures were calculated for the home neighbourhood, the work neighbourhood, and the route between home and work. Table 1 presents a total of 30 different measures together with the data sources that were used to calculate them, grouped into themes of 'roads and routes', 'public transport' and 'land use'.

In order to generate neighbourhoods, participants' home and work postcodes were georeferenced using the Ordnance Survey (OS) Address Layer 2®database [28]. A pedestrian route network dataset was constructed in the ArcGIS 9.3 [29] geographic information system (GIS) software package by combining road data, excluding motorways, from the OS MasterMap® Integrated Transport Network™ (ITN) database with local authority rights-of-way data (public footpaths, bridleways and byways), cycle route information from the charity Sustrans [30], and other informal pathways recorded on OpenStreetMap.com. Using this information, home neighbourhoods were delineated to represent areas that could be accessed within an approximate ten-minute walking time (equating to 800 m along the pedestrian network). This distance has been used in previous accessibility analysis to represent a practically walkable neighbourhood area [19,31,32].

To delineate the route between home and work for each participant, ArcGIS was used to identify sections of the pedestrian network that comprised the shortest route between the two locations. Subsequently, the characteristics of the environment within a 100 m distance of the route were quantified.

The 'roads and routes' environmental variables that were computed included a measure of distance to work (length of shortest route between home and work), route directness (ratio of route network to straight line distance), whether participants were travelling into or out of the city centre on their journey from home to work (a participant was defined as working in the city if they worked within 2 km of Cambridge central bus station, and defined as living in the city if they lived in the Cambridge urban area according to the 2001 census [33]) and the proportion of route length that was along A or B (major) roads. In addition, five measures of the walkability of the streets in the home neighbourhood were calculated, including road density, junction density, road connectivity, existence of A class (major) roads, the proportion of foot/cycle paths, and the general 'walkable area' (defined as the area walkable within 800 m along the route network buffer from participants' home location divided by the area within an 800 m straight line distance).

Public transport variables were derived using the route network dataset in combination with the National Public Transport Access Nodes (NAPTAN) and Data Repository (NPTDR) datasets [34,35]. Firstly, it was established whether or not there was a bus service passing through the neighbourhood that would take the participant to work, either directly or with one or more changes. The other variables calculated for both the home and work locations were the distance to the nearest bus stop, bus service frequency, the number of bus stops, the number of bus routes served and the distance to the nearest railway station. These indicators used the pedestrian route network, the frequency of bus services from the nearest bus stop on a typical weekday, the number of serviced bus stops present, and the number of bus routes available in the neighbourhood based on a count of unique service numbers.

Land use variables included an indicator of land use mix - the Herfindahl-Hirschmann Index (HHI) [36] - and building density at the home location and on the route to work. A count of the number of destinations (schools, eating and drinking establishments, stage and screen venues, sports complexes, and retail units) was taken after they were mapped along the route to work and in the home and work neighbourhoods. A measure of local area deprivation was calculated for the home neighbourhood, based on the proportion of people in lower socioeconomic classes (semi-routine occupations, routine occupations, never-worked and long-term unemployed categories) living within the relevant Census Output Area. In addition, the availability of workplace car parking (none, charged or free) reported by each participant in their questionnaire was used.

Data analysis

Unadjusted associations between usual travel mode (car, public transport, cycling or walking) and home, work, and route characteristics were examined using chi-square tests and analyses of variance. Continuous variables were categorised as tertiles or using other appropriate groupings. Prior to model fitting, multi-collinearity was managed using a pair-wise correlation matrix to identify variables that were highly associated, defined as having a Pearson's correlation coefficient of >0.55 based on previous empirical research [37]. Only one variable from each correlated pair was added into each model, the chosen one being that with the strongest association in the expected direction with the outcome variable. Multinomial regression models were then fitted to examine adjusted associations, using commuting by car as the reference category for the outcome variable. The models were fitted in a number of stages. First, a personal model was created by entering all potential individual and household-level covariates into a multinomial logistic regression. Prior empirical evidence has shown mixed associations between individual factors and active travel [17], therefore all such variables were included at this stage. After all these potential predictors were added, those for which $p>0.1$ were removed in a backwards stepwise manner leaving only those which were statistically significant at $p<0.1$.

Next, in order to establish the potential environmental correlates, three further models were fitted using the same method for the measures grouped into the 'roads and routes', 'public transport' and 'land use' categories. The variables that remained statistically significant at $p<0.1$ in each of the three models were then combined with those from the individual model, and non-significant independent variables were dropped in a stepwise manner using a threshold of $p>0.05$. The resultant model contained those variables that remained statistically significant at $p<0.05$. All analyses were undertaken using PASW Statistics 18 [38].

Results

Sample characteristics

Of the 1168 respondents who completed the questionnaire, 13 were excluded as postcode data for home location, work location or both were missing or invalid. Of the remaining 1155, six could not be attributed to one predominant mode for commuting, while two worked at more than one site and therefore had no 'usual' place of work. Of the remaining sample, 23 participants had missing values for one or more of the covariates adjusted for and

Table 1. Prevalence of travel mode by levels of environmental exposure variables.

Variable	Data source	Classification	% prevalence				
			Car	PT	Bike	Walk	p[1]
Roads and routes							
Distance to work (km)	A	Shortest	4.8	3.5	72.8	18.8	<0.001
		Middle tertile	36.6	8.8	53.6	1.1	
		Longest	76.5	19.7	3.7	0.0	
Route directness to work (ratio)	A	Most direct	52.8	19.0	24.7	3.5	<0.001
		Middle tertile	42.8	8.8	44.1	4.3	
		Least direct	22.7	4.3	61.1	12.0	
Direction of travel (category)	B	Lives and works in city	5.9	2.9	76.6	14.7	<0.001
		Lives in city, works outside	15.1	7.0	66.6	11.4	
		Lives outside, works in city	58.8	26.1	15.1	0.0	
		Lives and works outside city	77.5	8.8	13.7	0.0	
Proportion of A & B roads (%, route)	A	Lowest	28.7	7.8	52.3	11.3	<0.001
		Middle tertile	31.0	7.2	56.2	5.6	
		Highest	58.6	17.1	21.4	2.9	
Road density (ratio, home)	A	Highest	27.1	12.8	50.3	9.8	<0.001
		Middle tertile	35.6	11.8	46.5	6.1	
		Lowest	55.6	7.5	33.2	3.7	
Junction density (ratio, home)	A	Highest	26.6	15.4	48.1	9.8	<0.001
		Middle tertile	34.0	10.2	48.8	7.0	
		Lowest	57.6	6.4	33.2	2.9	
Road connectivity (ratio, home)	A	Most connected	22.9	8.1	55.8	13.2	<0.001
		Middle tertile	42.8	11.2	41.5	4.5	
		Least connected	52.3	12.7	32.9	2.1	
Existence of A roads (category, home)	A	None[2]	53.3	13.1	30.2	3.4	<0.001
		Some	29.3	8.9	52.8	8.9	
Proportion foot/cycle paths (%, home)	A	Highest	52.2	10.8	32.3	4.8	<0.001
		Middle tertile	36.4	14.0	44.9	4.7	
		Lowest	29.8	7.2	52.8	10.2	
Effective walkable area (ratio, home)	A	Most walkable	24.8	12.5	49.9	12.8	<0.001
		Middle tertile	40.8	10.1	45.6	3.5	
		Least walkable	52.7	9.4	34.5	3.5	
Public transport							
Bus service to work (category)	C, D	Direct	30.6	9.5	51.1	8.8	<0.001
		Change	65.0	14.6	20.1	0.4	
		None	46.9	6.3	43.8	3.1	
Distance to nearest bus stop (m, home)	C	Closest	35.1	13.0	45.5	6.4	0.043
		Middle tertile	41.2	12.0	39.3	7.5	
		Furthest	42.0	7.0	45.2	5.9	
Distance to nearest bus stop (m, work)	C	Closest	44.8	11.5	36.3	7.5	0.001
		Middle tertile	31.8	12.5	48.2	7.5	
		Furthest	41.1	7.9	46.3	4.7	
Bus service frequency (count, home)	C, D	Most frequent	11.1	7.4	68.3	12.9	<0.001
		Middle tertile	41.6	11.7	41.9	4.8	
		Least frequent	62.9	12.6	21.8	2.6	
Bus service frequency (count, work)	C, D	Highest	35.3	11.2	47.6	5.9	0.005
		Middle tertile	35.2	14.2	44.4	6.2	
		Lowest	45.4	7.8	39.3	7.4	
Nearest railway station (m, home)	C	Closest	24.3	14.1	50.1	11.5	<0.001

Table 1. Cont.

			% prevalence				
Variable	**Data source**	**Classification**	**Car**	**PT**	**Bike**	**Walk**	**p[1]**
		Middle tertile	23.7	6.7	61.5	8.1	
		Furthest	69.8	11.1	18.8	0.3	
Nearest railway station (m, work)	C	Closest	26.8	13.4	50.1	9.7	<0.001
		Middle tertile	45.0	9.3	39.2	6.4	
		Furthest	46.7	9.3	41.0	3.0	
Number of bus stops (count, home)	C	Highest	20.1	10.2	58.4	11.3	<0.001
		Middle tertile	40.4	13.0	42.5	4.2	
		Lowest	60.7	9.2	26.8	3.3	
Number of bus stops (count, work)	C	Upper half[2]	34.5	12.5	45.7	7.3	0.006
		Lower half	44.2	8.9	41.0	5.9	
Number of bus routes (count, home)	C	Highest	23.4	10.4	53.8	12.5	<0.001
		Middle tertile	41.1	12.7	42.3	3.9	
		Lowest	57.5	8.0	31.4	3.0	
Number of bus routes (count, work)	C	Upper half[2]	31.0	12.2	48.8	8.0	<0.001
		Lower half	46.4	9.4	38.8	5.4	
Land use							
Land use mix (score, route)	E, F, G	Most mixed	45.6	11.5	39.5	3.5	<0.001
		Middle tertile	28.5	9.3	54.9	7.2	
		Least mixed	44.1	11.2	35.6	9.1	
Land use mix (score, route)	E, F, G	Most mixed	39.1	11.8	42.1	7.0	0.197
		Middle tertile	38.0	13.2	42.5	6.3	
		Least mixed	41.1	7.0	45.4	6.5	
Building density (%, route)	G	Highest	8.6	4.6	73.2	13.7	<0.001
		Middle tertile	40.3	9.8	43.8	6.1	
		Lowest	69.3	17.6	13.1	0.0	
Building density (%, home)	G	Highest	16.9	10.7	59.8	12.6	<0.001
		Middle tertile	42.6	12.0	39.6	5.9	
		Lowest	58.7	9.3	30.7	1.3	
Number of destinations (count, route)	H	Highest	20.7	6.7	61.8	10.8	<0.001
		Middle tertile	37.3	14.4	42.0	6.3	
		Lowest	60.4	10.8	26.1	2.7	
Number of destinations (count, home)	H	Highest	17.1	10.5	59.6	12.9	<0.001
		Middle tertile	42.0	9.8	43.4	4.8	
		Lowest	59.9	11.7	26.4	1.9	
Number of destinations (count, work)	H	Upper half[2]	31.3	13.3	47.5	8.0	<0.001
		Lower half	47.2	8.2	39.4	5.2	
Deprivation (%, home)	I	Lowest	26.3	4.8	57.4	11.5	<0.001
		Middle tertile	42.9	11.2	40.5	5.3	
		Highest	48.9	16.0	32.2	2.9	
Car parking availability (category, work)	J	Yes - pay to park	45.4	9.7	40.0	4.9	<0.001
		Yes - free	47.6	8.2	39.9	4.3	
		No	24.2	14.4	50.6	10.8	

Notes: n = 1124. PT: public transport.
[1]Pearson chi-square.
[2]Data divided into two categories using median value (none/some, upper/lower half) when tertiles produced uneven numbers due to multiple participants working in the same location. *Data sources*: A Ordnance Survey (OS) road centre lines [50], Cambridge County Council, Sustrans [30], OpenStreetMap [51] (various dates) and manually digitised; B questionnaire 2009 [26]; C DfT 2010 [35]; D DfT 2009 [34]; E CeH 2000 [52]; F OS 2010 [53]; G Natural England [54], OS [53] and OpenStreetMap [51] (various dates); H PointX Ltd 2010 [55]; I ONS 2001 (proportion of people in semi-routine occupations, routine occupations, never worked and long-term unemployed categories) [56]; J questionnaire 2009 [26].

Table 2. Univariate associations between personal and environmental characteristics and main mode of travel to work (reference category is 'car').

		Odds Ratio (95% CI)		
		Public transport	**Bike**	**Walk**
Personal characteristics *(pseudo $r^2 = 0.400$)*				
Age		1.001 (0.980, 1.023)	0.996 (0.981, 1.011)	1.014 (0.987, 1.042)
Gender	Female (ref)	1.00	1.00	1.00
	Male	1.301 (0.777, 2.179)	2.450 (1.734, 3.462)***	1.591 (0.870, 2.910)
Limiting illness	No (ref)	1.00	1.00	1.00
	Yes	0.465 (0.223, 0.972)**	0.369 (0.218, 0.627)***	0.605 (0.245, 1.492)
Deprivation	Least deprived (ref)	1.00	1.00	1.00
	Middle tertile	1.748 (0.916, 3.338)*	0.499 (0.340, 0.733)***	0.410 (0.214, 0.787)***
	Most deprived	1.751 (0.936, 3.274)*	0.335 (0.226, 0.497)***	0.166 (0.077, 0.356)***
Education	Degree (ref)	1.00	1.00	1.00
	No degree	1.210 (0.761, 1.926)	0.544 (0.381, 0.777)***	0.438 (0.217, 0.884)**
Homeownership	Homeowner (ref)	1.00	1.00	1.00
	Not a homeowner	1.695 (0.975, 2.948)*	1.582 (1.038, 2.412)**	2.559 (1.261, 5.195)***
Children in household	Yes (ref)	1.00	1.00	1.00
	No	0.927 (0.555, 1.550)	0.702 (0.498, 0.989)**	1.339 (0.643, 2.788)
Car ownership	No car (ref)	1.00	1.00	1.00
	1 car	0.058 (0.017, 0.204)***	0.067 (0.020, 0.222)***	0.028 (0.008, 0.101)***
	More than 1 car	0.010 (0.003, 0.038)***	0.012 (0.004, 0.041)***	0.006 (0.001, 0.022)***
Type of work	Sedentary (ref)	1.00	1.00	1.00
	Standing	0.955 (0.521, 1.751)	1.247 (0.822, 1.893)	1.059 (0.499, 2.247)
	Manual	0.760 (0.230, 2.518)	0.988 (0.433, 2.253)	2.809 (0.861, 9.164)*
Roads and routes characteristics *(pseudo $r^2 = 0.630$)*				
ROUTE				
Distance to work (km)		0.990 (0.967, 1.014)	0.725 (0.694, 0.757)***	0.277 (0.210, 0.365)***
HOME NEIGHBOURHOOD				
Junction density (ratio)	Highest (ref)	1.00	1.00	1.00
	Middle tertile	0.524 (0.316, 0.867)**	0.572 (0.346, 0.947)**	0.446 (0.210, 0.947)**
	Lowest density	0.217 (0.124, 0.378)***	0.498 (0.294, 0.843)***	0.373 (0.142, 0.978)**
Effective walkable area (ratio)	Most walkable (ref)	1.00	1.00	1.00
	Middle tertile	0.615 (0.367, 1.029)*	0.908 (0.553, 1.490)	0.550 (0.238, 1.272)
	Least walkable	0.520 (0.307, 0.882)**	0.530 (0.321, 0.874)**	0.496 (0.205, 1.200)
Public transport characteristics *(pseudo $r^2 = 0.390$)*				
HOME NEIGHBOURHOOD				
Distance to nearest railway station (m)	Closest (ref)	1.00	1.00	1.00
	Middle tertile	0.382 (0.212, 0.687)***	0.762 (0.512, 1.135)	0.408 (0.222, 0.749)***
	Furthest distance	0.304 (0.187, 0.493)***	0.163 (0.110, 0.241)***	‡
Distance to nearest bus stop (m)	Closest (ref)	1.00	1.00	1.00
	Middle tertile	0.792 (0.486, 1.291)	0.783 (0.531, 1.153)	1.086 (0.559, 2.108)
	Furthest distance	0.438 (0.253, 0.761)***	0.763 (0.519, 1.121)	0.782 (0.391, 1.564)
Nearest bus stop service frequency (count)	Highest (ref)	1.00	1.00	1.00
	Middle tertile	0.463 (0.245, 0.874)**	0.256 (0.165, 0.399)***	0.166 (0.083, 0.333)***
	Lowest frequency	0.279 (0.148, 0.526)***	0.080 (0.050, 0.127)***	0.052 (0.023. 0.118)***
WORK NEIGHBOURHOOD				
Distance to nearest railway station (m)	Closest (ref)	1.00	1.00	1.00
	Middle tertile	0.363 (0.221, 0.597)***	0.365 (0.249, 0.535)***	0.293 (0.159, 0.539)***
	Furthest distance	0.382 (0.221, 0.659)***	0.434 (0.289, 0.653)***	0.171 (0.075, 0.390)***
Land use characteristics *(pseudo $r^2 = 0.410$)*				

Table 2. Cont.

		Odds Ratio (95% CI)		
		Public transport	**Bike**	**Walk**
ROUTE				
Building density (%)	Highest (ref)	1.00	1.00	1.00
	Middle tertile	0.611 (0.303, 1.232)	0.140 (0.091, 0.217)***	0.113 (0.060, 0.216)***
	Lowest density	0.626 (0.323, 1.214)	0.024 (0.015, 0.039)***	‡
WORK NEIGHBOURHOOD				
Destinations within walking distance (count)	Highest (ref)	1.00	1.00	1.00
	Lowest	0.475 (0.280, 0.804)***	0.819 (0.559, 1.201)	0.888 (0.452, 1.746)
Car parking availability (category)	No (ref)	1.00	1.00	1.00
	Yes - pay to park	0.627 (0.333, 1.178)	0.633 (0.390, 1.025)*	0.357 (0.157, 0.811)**
	Yes - free parking	0.362 (0.215, 0.609)***	0.729 (0.483, 1.098)	0.411 (0.206, 0.820)**

Notes: n = 1124. P values reflect difference from reference category *p<0.1, **p<0.05;
***p<0.01; ‡ coefficient suppressed where n<3.

were also excluded, resulting in a final sample of 1124 for analysis. There were no statistically significant differences between those included and excluded from analysis in terms of age, gender, limiting illness, body mass index (BMI), type of work, education, home ownership or having children in the household.

The mean age of the sample was 42.3 (range 17.6 to 71). 31.7% were male, whilst 60.5% were of normal weight, 27.6% overweight and 9.4% obese. 80.2% were engaged in sedentary work as opposed to standing (16.3%) or manual (3.6%) work, and 59.4% reported excellent or very good health. 10.0% reported a limiting illness, 72.0% were educated to degree level, 72.4% were homeowners, 29.6% had children under the age of 16 living in the household and 85.7% owned at least one car. Based on the classification by Bibby and Shepherd [39], 65.9% of participants lived in urban areas, 19.4% in town and fringe locations, 12.5% in villages and 2.1% in hamlets and isolated dwellings. In terms of predominant mode of travel to work, 39.4% used the car, 10.7% public transport and 43.3% the bicycle, while 6.6% walked. Therefore, nearly half of respondents (n = 561) predominantly used an active mode of travel for commuting.

Unadjusted prevalences of travel modes by levels of environmental exposure variables

Table 1 shows the unadjusted percentage prevalence of travel modes according to the three domains of environmental variables. There were statistically significant differences in travel mode prevalence across the categories of all these exposure variables, with the exception of land use mix in home neighbourhoods. Walking was most prevalent in those individuals whose modelled routes to work were less direct, involved the lowest proportion of A or B roads, had many destinations along them, had a direct bus service, passed through areas of high building density and low land use mix, and started and ended within the city. In terms of home neighbourhoods, walking prevalence was highest in areas which were walkable and highly connected, had frequent bus services and a larger number of bus stops and bus routes, had a railway station nearby, a high building, road and junction density, some A roads and few official foot/cycle paths, had low levels of deprivation, and contained many amenities/destinations. For the work location, a higher prevalence of walking was associated with proximity to a railway station, with access to a higher number of bus stops and amenities/destinations, but fewer bus services, within the work neighbourhood, and with not having access to free workplace car parking. Patterns in the percentage prevalence of cycling were similar to those for walking except in the case of bus service frequency at the work location. The prevalence of cycling was highest in areas of high bus service frequency, opposite to the direction of association for walking.

Multinomial modelling

The multivariable personal model contained nine explanatory variables, with a pseudo r-squared value of 0.40 (Table 2). The three further regression models including the grouped objective environmental criteria had pseudo r-squared values of: roads and routes 0.63; public transport 0.39; land use 0.41.

The full final multinomial model is shown in Table 3. The odds of cycling were higher in men and those without a limiting illness and lower in those with no degree, with children in the household, and owning one or more cars. The odds of walking were higher in men and those undertaking manual work, and decreased with increasing car ownership. The odds of using public transport also decreased with increasing car ownership. As expected there was a strong decline in walking and a less strong but still statistically significant decline in cycling with increasing route length. Using the reciprocal of the distance-to-work odds ratios in Table 3, the estimated odds of walking or cycling (relative to driving) were 3.9 times and 1.3 times lower, respectively, for each additional kilometre between home and work.

Some associations with environmental exposures remained significant after adjustment. Four environmental measures of the home neighbourhood were found to be associated with modal choice. Low junction density (indicating poor street connectivity) and a greater distance to a railway station were associated with lower odds of walking, cycling, and public transport use. A greater distance to the nearest bus stop and a lower bus frequency were associated with lower odds of public transport use, with the odds of cycling also being reduced amongst those living in neighbourhoods with fewer bus services. Only two measures of the work neighbourhood were associated with modal choice. Those working in areas with fewer destinations were less likely to use public transport or cycle, whilst the availability of free parking was associated with lower odds of using public transport, cycling or

Table 3. Adjusted associations (from best-fit multivariable model) between personal and environmental characteristics and main mode of travel to work (reference category is 'car') (pseudo $r^2 = 0.738$).

		Odds Ratio (95% CI)		
		Public transport	**Bike**	**Walk**
Personal characteristics				
Age		0.996 (0.974, 1.019)	1.002 (0.983, 1.022)	1.014 (0.982, 1.046)
Gender	Female (ref)	1.00	1.00	1.00
	Male	1.099 (0.603, 2.001)	3.952 (2.403, 6.499)***	2.618 (1.185, 5.783)**
Limiting illness	No (ref)	1.00	1.00	1.00
	Yes	0.516 (0.227, 1.169)	0.295 (0.141, 0.618)***	0.418 (0.125, 1.396)
Deprivation	Least deprived (ref)	1.00	1.00	1.00
	Middle tertile	1.327 (0.638, 2.758)	0.894 (0.533, 1.500)	0.886 (0.383, 2.047)
	Most deprived	1.553 (0.769, 3.137)	0.685 (0.398, 1.179)	0.399 (0.147, 1.082)*
Education	Degree (ref)	1.00	1.00	1.00
	No degree	1.255 (0.756, 2.083)	0.577 (0.359, 0.929)**	0.634 (0.265, 1.515)
Children in household	Yes (ref)	1.00	1.00	1.00
	No	0.723 (0.410, 1.277)	0.548 (0.341, 0.879)**	1.468 (0.595, 3.622)
Car ownership	No car (ref)	1.00	1.00	1.00
	1 car	0.047 (0.012, 0.176)***	0.140 (0.038, 0.509)***	0.056 (0.013, 0.236)***
	More than 1 car	0.011 (0.003, 0.042)***	0.039 (0.010, 0.144)***	0.029 (0.006, 0.136)***
Type of work	Sedentary (ref)	1.00	1.00	1.00
	Standing	1.124 (0.578, 2.187)	1.225 (0.686, 2.187)	1.063 (0.405, 2.792)
	Manual	0.977 (0.271, 3.528)	0.923 (0.301, 2.830)	9.249 (1.768, 48.38)***
Roads and routes				
Distance to work (km)		0.996 (0.974, 1.019)*	0.753 (0.716, 0.793)***	0.255 (0.185, 0.352)***
Junction density (ratio, home neighbourhood)	Highest (ref)	1.00	1.00	1.00
	Middle tertile	0.744 (0.412, 1.342)	0.593 (0.335, 1.051)*	0.409 (0.173, 0.967)**
	Lowest density	0.450 (0.235, 0.862)**	0.483 (0.259, 0.901)**	0.217 (0.070, 0.666)***
Public transport				
Distance to nearest railway station (m, home)	Closest (ref)	1.00	1.00	1.00
	Middle tertile	0.474 (0.236, 0.949)**	0.504 (0.284, 0.893)**	0.591 (0.259, 1.352)
	Furthest distance	0.382 (0.216, 0.676)***	0.528 (0.298, 0.934)**	‡
Distance to nearest bus stop (m, home)	Closest (ref)	1.00	1.00	1.00
	Middle tertile	0.623 (0.353, 1.101)	0.651 (0.385, 1.101)	0.889 (0.376, 2.102)
	Furthest distance	0.388 (0.209, 0.718)***	0.806 (0.485, 1.342)	1.580 (0.625, 3.992)
Nearest bus stop service frequency (count, home)	Highest (ref)	1.00	1.00	1.00
	Middle tertile	0.450 (0.215, 0.946)**	0.428 (0.241, 0.761)***	0.275 (0.112, 0.674)***
	Lowest frequency	0.322 (0.143, 0.727)***	0.432 (0.226, 0.827)**	0.607 (0.199, 1.847)
Land use				
Destinations within walking distance (count, work)	Highest (ref)	1.00	1.00	1.00
	Lowest	0.405 (0.220, 0.748)***	0.364 (0.214, 0.621)***	0.483 (0.201, 1.160)
Car parking availability (category, work)	No (ref)	1.00	1.00	1.00
	Yes - pay to park	0.789 (0.380, 1.640)	1.334 (0.690, 2.578)	0.919 (0.323, 2.613)
	Yes - free parking	0.351 (0.191, 0.647)***	0.548 (0.317, 0.948)**	0.239 (0.097, 0.592)***

Notes: n = 1124. All associations adjusted for all other variables listed in the table. P values reflect difference from reference category *p<0.1, **p<0.05;
***p<0.01; ‡ findings suppressed where n<3.

walking. The Nagelkerke pseudo r-squared value for the final model was 0.74, suggesting that the outcomes were generally well predicted. This value fell to 0.58 when distance to work was removed from the model, highlighting the importance of that characteristic, and to 0.51 when car ownership was also removed. When distance and all other environmental variables were removed from the final model, retaining only the individual characteristics (including car ownership), the value fell to 0.39.

Discussion

In expanding the range of potential correlates considered, this research sought to address gaps in our understanding of the relative importance of environmental factors that may contribute to explaining the prevalence of active travel, and therefore serve as potential targets for future interventions. Using questionnaire data from the Commuting and Health in Cambridge study, Goodman et al. [40] have previously illustrated how car commuting allows people to accommodate other lifestyle goals, such as home ownership and children, with their daily work commitments. Using the same dataset, Panter et al. [26] have highlighted the importance of distance as a correlate of time spent walking or cycling to work, showing that if people live near enough to work, they may choose to walk or cycle despite having access to a car. As distance is less amenable to intervention than some other environmental characteristics, this study has built on those findings by establishing a range of objective environmental factors other than distance that are associated with modal choice. It has thereby highlighted a number of environmental features that may be amenable to modification and could therefore represent targets for interventions to increase the prevalence of active commuting.

As anticipated, we found travel distance and car ownership to be strong predictors of mode of travel to work. After adjustment we found that public transport provision was associated with its use, whilst neighbourhoods with higher street connectivity (measured by junction density) were associated with higher odds of walking and cycling. The absence of free car parking at work was associated with a markedly higher likelihood of walking, cycling, and public transport use. Many other variables shown to influence modal choice in previous research (e.g. land use mix, the availability, type or directness of roads and paths, or area-level measures of social deprivation) were not found to be significant predictors in the final multivariable model for this study. We do not know the reasons for this, but it may reflect differences in measurement methods employed by different studies, or the fact that we adjusted for a wider range of covariates than is common in the literature. It may also be that the influences on active travel, other than distance, are quite context-specific.

By testing associations with a diverse set of measures based on the objective characterisation of the environment in a well-characterised population cohort, we believe this study has a number of methodological strengths. In particular we examined the potential importance of public transport availability in addition to features of the built environment, thus addressing evidence gaps highlighted by Mackett et al. [22]. We also modelled associations with characteristics of the environment outside the home neighbourhood, which was highlighted as an empirical gap by several papers [20,21,22], and the fact we found differences in associations with the home, work and route environments highlights the importance of doing so. This study also analysed walking and cycling behaviour separately rather than combining the two, as recommended by a number of authors [18,19], and we found some differences in the pattern of associations between them. In addition, the study participants lived in a mixture of urban and rural environments.

There are a number of limitations to this study. Our study cohort sample worked in the city of Cambridge, UK, and lived within a 30 km radius of the centre. It may therefore not be representative of other areas or indeed of the general population of the study area, as illustrated by the fact that 44% of study participants predominantly used the bicycle to travel to work compared with 18.3% of people working in Cambridge according to the 2001 UK Census [41]. The prevalence of cycling in Cambridge is substantially higher than the national average of 3.1%, although a benefit of the high prevalence of cycling in our sample is that it provided us with the statistical power to model this outcome separately from that of walking. Given the high prevalence of cycling in Cambridge, we do not know how readily our results may be generalizable to other locations where cycling culture is less strongly embedded. However, it appears unlikely that the associations we have observed would be completely different in cities that are otherwise similar, and the replicability of our findings could be tested in future studies drawing on samples with greater geographical heterogeneity.

Further limitations include the absence of participants reporting walking as their predominant mode of travel outside the city, which meant that we could not test associations with certain potentially important environmental attributes such as urban-rural status. Almost 40% of our participants worked in one particular campus site, which may have limited the heterogeneity of workplace neighbourhood exposures within the sample, although these participants nonetheless reported varied car parking provision. We were not able to adjust our analyses for income at the individual or household level, for which data were not available, but educational attainment, home ownership and area-level deprivation were all included in the models. The large number of statistical tests we have performed raises the possibility that some of the associations detected could have arisen by chance, and the cross-sectional study design means that it is not possible to determine if the associations are causal. We therefore cannot tell if modifying the characteristics we have found to be associated with our target behaviours would lead to changes in travel behaviour, although the evidence from studies such as this can generate hypotheses to be tested in subsequent intervention studies. A further limitation is that we do not know the motivations for the choice of residential location in our sample, and it may be that certain individuals had migrated to neighbourhoods that were supportive of their preferred travel behaviours. Nevertheless, recent research findings suggest that environmental influences still remain associated with active travel and physical activity even after controlling for this potential self-selection bias [14,42,43]. We defined neighbourhood size as 800 m based on an approximate ten-minute walk, which may not necessarily be the most appropriate scale. However there is evidence to suggest that that the choice of neighbourhood size or scale does not strongly affect associations between urban form and travel behaviour [44] or health outcomes [45], even though measures of the environment do vary with scale. Lastly, we used modelled routes to represent the journey to work and these may not necessarily reflect the routes actually taken.

Conclusions

Because of the constraints of distance, it may be unrealistic to expect a modal shift towards walking and cycling as the predominant modes of travel to work amongst a substantial proportion of the commuting population. However, our findings suggest that the provision of good public transport is associated with higher levels of public transport use, and from a physical activity perspective, this may still be advantageous amongst those living further from work. Our results also suggest that discouraging the provision of free parking in and around the workplace might help to promote walking and cycling to work. In combination with the findings of other recent cross-sectional analyses of environmental factors associated with commuting behaviour [26,40,46,47], our results will help guide the selection of environmental exposure variables for longitudinal analyses of the determinants of behaviour change over time, as well as the design and specification of future intervention studies.

The implications for future transport policy are evident. In response to the growing demand for housing, new urbanised areas – sometimes remote and inadequately linked [48] – are being created where providing frequent public transport connections close to people's homes may not always be economically or politically viable [49]. Therefore, it may be more prudent to encourage people to use existing public transport networks to break their car journeys, for example by encouraging people to 'park-and-ride' at public transport intersections or places where they are able to change to active modes of travel for part of their journey. Our results also suggest that future initiatives seeking to modify environmental features that may promote more active travel behaviours should not only focus on residential neighbourhoods but also encompass characteristics of workplace surroundings as well as transport corridors.

Acknowledgments

Various data providers have made this analysis possible, including Sustrans; the Ordnance Survey; the Centre for Ecology and Hydrology; the County Councils of Bedford, Cambridge, Norfolk and Northamptonshire; and the Department for Transport and Thales. We also thank the study participants for their cooperation and the staff of the MRC Epidemiology Unit Functional Group Team, in particular for study coordination and data collection (led by Cheryl Chapman) and data management.

Author Contributions

Conceived and designed the experiments: AMD APJ. Performed the experiments: AMD. Analyzed the data: AMD. Contributed reagents/materials/analysis tools: AMD APJ JRP DO. Wrote the paper: AMD APJ. Guidance regarding the design of the analysis: APJ JRP DO. Review and significant amendments to manuscript: AMD APJ JRP DO.

References

1. Bauman AE, Reis RS, Sallis JF, Wells JC, Loos RJF, et al. (2012) Correlates of physical activity: why are some people physically active and others not? Lancet 380: 258–271.
2. Department of Health (DH) (2011) Start Active, Stay Active: a report on physical activity from the four home countries' Chief Medical Officers. London: Department of Health.
3. Murray CJL, Richards MA, Newton JN, Fenton KA, Anderson HR (2013) UK health performance: findings of the Global Burden of Disease Study 2010. Lancet 381: 997–1020. Available: http://dx.doi.org/10.1016/S0140-6736(13)60355-4.
4. Lim SS, Vios T, Flaxman AD, Danaei G, Shibuya K, et al. (2012) A comparative risk assessment of burden of disease and injury attributable to 67 risk factors and risk factor clusters in 21 regions, 1990–2010: a systematic analysis for the Global Burden of Disease Study 2010. Lancet 380: 2224–2260.
5. Department of Health (DH) (2009) Be Active, Be Healthy: a plan for getting the nation moving. London: Department of Health.
6. Brownson RC, Boehmer TK, Luke DA (2005) Declining rates of physical activity in the United States: what are the contributors? Annu Rev Public Health 26: 421–443.
7. Department for Transport (DfT) (2011) Personal travel factsheet: commuting and business travel. London: DfT.
8. Pooley CG, Turnbull J (2000) Modal choice and modal change: the journey to work in Britain since 1890. J Transp Geog 8: 11–24.
9. Cavill NKahlmeir SRacioppi F, (2006) Physical activity and health in Europe: evidence for action. Denmark: World Health Organisation.
10. Hendriksen IJM, Simons M, Garre FG, Hildebrandt VH (2010) The association between commuter cycling and sickness absence. Prev Med 51: 132–135.
11. Oja P, Vuori I, Paronen O (1998) Daily walking and cycling to work: their utility as health-enhancing physical activity. Patient Educ Couns 33: S87–S94.
12. Besser LM, Dannenberg AL (2005) Walking to public transit: steps to help meet physical activity recommendations. Am J Prev Med 29: 273–280.
13. Jarrett J, Woodcock J, Griffiths UK, Chalabi Z, Edwards P, et al. (2012) Effect of increasing active travel in urban England and Wales on costs to the National Health Service. 379: 2198–2205.
14. McCormack G, Shiell A (2011) In search of causality: a systematic review of the relationship between the built environment and physical activity among adults. Int J Behav Nutr Phys Act 8: 125. doi: 10.1186/1479-5868-8-125
15. Pucher J, Buehler R (2008) Making cycling irresistible: lessons from The Netherlands, Denmark and Germany. Transport Rev 28: 495–528.
16. Fraser SDS, Lock K (2010) Cycling for transport and public health: a systematic review of the effect of the environment on cycling. Eur J Public Health 21: 738–743.
17. Panter JR, Jones AP (2010) Attitudes and the environment as determinants of active travel in adults: what do and don't we know? J Phys Act Health 7: 551–561.
18. Forsyth A, Krizek K (2010) Promoting walking and bicycling: assessing the evidence to assist planners. Built Environ 36: 429–446.
19. Panter J, Jones A, Van Sluijs E, Griffin S (2010) Neighborhood, route, and school environments and children's active commuting. Am J Prev Med 38: 268–278.
20. Evans G (2009) Accessibility, uban design and the whole journey environment. Built Environ 35: 366–385.
21. Lachapelle U, Frank LD (2009) Transit and health: mode of transport, employer-sponsored public transit pass programs, and physical activity. J Public Health Policy 30: S73–S94.
22. Mackett RL, Achuthan K, Titheridge H (2007) Conflicts between macro level policy and micro level implementation - the case of social exclusion. Paper presented at the World Conference on Transport Research. Berkley, California.
23. Badland HM, Garrett N, Schofield GM (2010) How does car parking availability and public transport accessibility influence work-related travel behaviors? Sustainability 2: 576–590.
24. Achuthan K, Titheridge H, Mackett RL (2010) Mapping accessibility differences for the whole journey and for socially excluded groups of people. Journal of Maps 2010: 220–229. doi: 10.4113/jom.2010.1077
25. Ogilvie D, Griffin S, Jones A, Mackett R, Guell C, et al. (2010) Commuting and health in Cambridge: a study of a 'natural experiment' in the provision of new transport infrastructure. BMC Public Health 10: 703 doi: 10.1186/1471-2458-10-703
26. Panter J, Griffin S, Jones A, Mackett R, Ogilvie D (2011) Correlates of time spent walking and cycling to and from work: baseline results from the commuting and health in Cambridge study. Int J Behav Nutr Phys Act 8: 124. doi: 10.1186/1479-5868-8-124
27. Besson H, Brage S, Jakes R, Ekelund U, Wareham N (2010) Estimating physical activity energy expenditure, sedentary time, and physical activity intensity by self-report in adults. Am J Clin Nutr 91: 106–114.
28. Ordnance Survey (OS) (2010) OS MasterMap Address Layer 2. Southampton: OS. Available: http://www.ordnancesurvey.co.uk/oswebsite/products/os-mastermap/address-layer-2/index.html. Accessed 2011 Nov 25.
29. ESRI (2009) ArcGIS 9.3. California: ESRI.
30. Sustrans (2012) About Sustrans. Bristol: Sustrans. Available: http://www.sustrans.org.uk/about-sustrans. Accessed 2012 Oct 3.
31. Van Dyck D, Deforche B, Cardon G, De Bourdeaudhuij I (2009) Neighbourhood walkability and its particular importance for adults with a preference for passive transport. Health Place 15: 496–504.
32. Nicholls S (2001) Measuring the accessibility and equity of public parks: a case study using GIS. Managing Leisure 6: 201–219.
33. Office for National Statistics (ONS) (2001) English Urban Areas 2001. Newport, Wales: ONS.
34. Department for Transport (DfT) (2009) Thales National Public Transport Data Repository (NPTDR). London: DfT. Available: http://nptdr.dft.gov.uk/. Accessed 2011 Jun 6.
35. Department for Transport (DfT) (2010) Thales National Public Transport Access Nodes (NaPTAN) London: DfT. Available: http://nptdr.dft.gov.uk/. Accessed 2010 Aug 10.
36. Forsyth A (2010) NEAT-GIS (Neighborhood Environment for Active Transport-Geographic Information Systems) protocols. Available: http://208.106.193.160/pdfs/GIS_Protocols/NEAT_GIS_V5_0_26Nov2010FIN.pdf. Accessed 2011 Jan 4.
37. Grewal R, Cote JA, Baumgartner H (2004) Multicollinearity and measurement error in structural equation models: implications for theory testing. Market Sci 23: 519–529.
38. Predictive Analytics SoftWare (PASW) (2009) PASW Statistics 18. PASW.
39. Bibby P, Shepherd J (2004) Developing a new classification of urban and rural areas for policy purposes - the methodology. London: ONS.
40. Goodman A, Guell C, Panter J, Jones NR, Ogilvie D (2012) Healthy travel and the socio-economic structure of car commuting in Cambridge, UK: A mixed-methods analysis. Soc Sci Med 74: 1929–1938.
41. Office for National Statistics (ONS) (2001) Method of Travel to Work - Daytime Population (UV37). Newport, Wales: ONS. Available: http://www.neighbourhood.statistics.gov.uk. Accessed 2011 Jul 29.
42. Frank LD, Saelens BE, Powell KE, Chapman JE (2007) Stepping towards causation: Do built environments or neighborhood and travel preferences explain physical activity, driving, and obesity? Soc Sci Med 65: 1898–1914.
43. Van Dyck D, Cardon G, Deforche B, Owen N, De Bourdeaudhuij I (2011) Relationships between neighborhood walkability and adults' physical activity: How important is residential self-selection? Health Place 17: 1011–1014.

44. Reilly MK, Landis J (2002) The influence of built form and land use on travel behaviour. Berkeley, CA: UC Transport Center.
45. Lovasi GS, Moudon AV, Smith NL, Lumley T, Larson EB, et al. (2008) Evaluating options for measurement of neighborhood socioeconomic context: Evidence from a myocardial infarction case control study. Health Place 14: 453–467.
46. Carse A, Goodman A, Mackett RL, Panter J, Ogilvie D (2013) The factors influencing car use in a cycle-friendly city: the case of Cambridge. J Transp Geogr 28: 67–74.
47. Panter J, Desousa C, Ogilvie D (2013) Incorporating walking or cycling into car journeys to and from work: The role of individual, workplace and environmental characteristics. Prev Med 56: 211–217.
48. Hickman R, Seaborn C, Headicar P, Banister D, Swain C (2010) Spatial planning for sustainable travel? Town & Country Planning February 2010 77–82.
49. Shill J, Mavoa H, Crammond B, Loff B, Peeters A, et al. (2012) Regulation to Create Environments Conducive to Physical Activity: Understanding the Barriers and Facilitators at the Australian State Government Level. PLOS ONE 7: e42831.
50. Ordnance Survey (OS) (2010) OS MasterMap Integrated Transport Network (ITN) Layer. Southampton: OS. Available: www.ordnancesurvey.co.uk/oswebsite/products/os-mastermap/itn-layer/index.html. Accessed 2011 Nov 25.
51. OpenStreetMap (2010) OpenStreetMap: the free wiki world map. Available: www.openstreetmap.org. Accessed 2010 Dec 20.
52. Centre for Ecology and Hydrology (CEH) (2000) Land Cover Map 2000 (LCM2000). Wallingford, UK: CEH. Available: www.ceh.ac.uk/LandCoverMap2000.html. Accessed 2010 Nov 30.
53. Ordnance Survey (OS) (2010) OS MasterMap Topography Layer. Southampton: OS. Available: www.ordnancesurvey.co.uk/oswebsite/products/os-mastermap/topography-layer/index.html. Accessed 2011 Nov 25.
54. Natural England (2011) GIS digital boundary datasets: Open Access Land Available: http://www.gis.naturalengland.org.uk/pubs/gis/GIS_register.asp. Accessed 2011 Apr 5.
55. PointX Ltd (2010) Ordnance Survey Points of Interest. Southampton: OS. Available: www.ordnancesurvey.co.uk/oswebsite/products/points-of-interest/index.html. Accessed 2011 Mar 14.
56. Office for National Statistics (2001) Socio-economic classification (NS-SEC, UV31), 2001 Census of Population in England and Wales. Fareham: ONS. Available: http://www.neighbourhood.statistics.gov.uk. Accessed 2011 Oct 19.

13

The Combined Influence of Distance and Neighbourhood Deprivation on Emergency Department Attendance in a Large English Population: A Retrospective Database Study

Gavin M. Rudge[1]*, Mohammed A. Mohammed[2], Sally C. Fillingham[1], Alan Girling[1], Khesh Sidhu[3], Andrew J. Stevens[1]

1 School of Health and Population Sciences, University of Birmingham, Birmingham, United Kingdom, **2** University of Bradford, Bradford, United Kingdom, **3** Welsh Health Specialised Services Team, Caerphilly, Wales, United Kingdom

Abstract

The frequency of visits to Emergency Departments (ED) varies greatly between populations. This may reflect variation in patient behaviour, need, accessibility, and service configuration as well as the complex interactions between these factors. This study investigates the relationship between distance, socio-economic deprivation, and proximity to an alternative care setting (a Minor Injuries Unit (MIU)), with particular attention to the interaction between distance and deprivation. It is set in a population of approximately 5.4 million living in central England, which is highly heterogeneous in terms of ethnicity, socio-economics, and distance to hospital. The study data set captured 1,413,363 ED visits made by residents of the region to National Health Service (NHS) hospitals during the financial year 2007/8. Our units of analysis were small units of census geography having an average population of 1,545. Separate regression models were made for children and adults. For each additional kilometre of distance from a hospital, predicted child attendances fell by 2.2% (1.7%–2.6% $p<0.001$) and predicted adult attendances fell by 1.5% (1.2% –1.8%, $p<0.001$). Compared to the least deprived quintile, attendances in the most deprived quintile more than doubled for children (incident rate ratio (IRR) = 2.19, (1.90–2.54, $p<0.001$)) and adults (IRR 2.26, (2.01–2.55, $p<0.001$)). Proximity of an MIU was significant and both adult and child attendances were greater in populations who lived further away from them, suggesting that MIUs may reduce ED demand. The interaction between distance and deprivation was significant. Attendance in deprived neighbourhoods reduces with distance to a greater degree than in less deprived ones for both adults and children. In conclusion, ED use is related to both deprivation and distance, but the effect of distance is modified by deprivation.

Editor: Bin Jiang, University of Gävle, Sweden

Funding: The lead author was funded by the National Institute for Health Research (NIHR) through the Collaborations for Leadership in Applied Health Research and Care for Birmingham and Black Country (CLAHRC-BBC)programm (www.clahrc-bbc.nihr.ac.uk). This paper presents independent research funded by the National Institute for Health Research (NIHR). The views expressed are those of the author(s) and not necessarily those of the NHS, the NIHR or the Department of Health. The funders had no role in study design, data collection and analysis, decision to publish, or preparation of the manuscript.

Competing Interests: The authors have declared that no competing interests exist.

* E-mail: g.rudge@bham.ac.uk

Introduction

Background

Presentation at Emergency Departments (EDs) whether by self referral or directed by other services, is an important route into acute hospital care.

The manner in which people interact with healthcare services, particularly EDs, has been shown to be strongly influenced by proximity in England [1], [2], [3], Scotland [4], Northern Ireland [5], Canada [6], the US [7], [8] and Sweden [9]. It has also been shown that higher degrees of socio-economic deprivation are associated with increased attendance in EDs [2], [3], [6].

Importance

The provision of ED facilities is a high cost, high volume service. In England, it is estimated that there were over 19 million visits to EDs and Minor Injury Units (MIU) in 2007/08 [10]. In financial year 2008/09, the National Health Service estimated the cost of attendances at English EDs at over £1.3 billion [11]. Historically, hospitals with EDs were given fixed sums of money each year to provide the service by the local budget holders who financed the health needs of their populations, regardless of the actual demands on the service. Since 2006, ED attendances have come under the England's 'Payment by Results' (PbR) system [12] in which they are individually billed to the budget holder on a fee for service basis. Given the costs involved in providing the service, there has been considerable scrutiny of the extent to which EDs are being inefficiently or "inappropriately" used for less urgent needs instead of making an appointment with the General Practitioner (GP) that the patient is registered with. Also the ability of EDs to process patients quickly is seen as an indicator of performance, particularly in England where specific targets relating to ED waiting times have been centrally set by Government [13]. For these reasons it is important that those who plan, fund or manage EDs understand the factors that affect the demand for this service.

Goals of this investigation

The study has three objectives. 1) To explore the effects of distance and deprivation on ED attendance using a larger and more heterogeneous population than has been studied before. 2) To determine if the relationship between distance and ED attendance varies by deprivation. 3) To see if proximity to Minor Injury Units (MIU) affects the extent to which populations use Emergency Departments in acute hospitals.

Materials and Methods

Study Design

We obtained counts of attenders at EDs for one financial year (1st April 2007 to 31st March 2008) for the West Midlands region of England. These were aggregated into the 3,482 LSOAs that formed small and relatively homogenous units of analysis. We used a negative binomial regression model to investigate the relationship between the number of attenders and distance from the nearest ED, income deprivation, and distance from the nearest minor injuries unit. We also explored interactions between these variables and adjusted for the demographic structure of the neighbourhoods in the study.

Setting

The study area is the Government Office Region of the West Midlands in which approximately 5.4 million people were resident during the period the study was set [14]. This is a highly heterogeneous region covering large cities such as Birmingham, Stoke-on-Trent, Coventry and Wolverhampton as well as relatively sparsely populated rural areas such as Herefordshire and parts of North Staffordshire.

Population units

Proximity and deprivation data were based on Lower Level Super Output Areas (LSOAs). The LSOA is a census geography and is commonly used as a unit to explore small area variation in British populations [15]. They are similar in size and function to census tracts used in the U.S [16]. They are neighbourhoods which at the time of the study, had a mean population of 1560 residents. In the region in question, LSOAs are generally small with a median area of 0.45 km^2 (interquartile range of 0.30 km^2 to 1.00 km^2).

Selection of Participants

Visits captured by the Accident and Emergency Commissioning Data Set (A&CMDS) were included in the study if the attender was resident in the region and attended a "type one" ED. The A&EMDS is a nationally implemented common data set that is captured when someone presents at an ED seeking treatment in the NHS. A type one ED is defined as "a consultant led 24 hour service with full resuscitation facilities and designated accommodation for the reception of accident and emergency patients [17]. This is the typical service model provided by NHS acute hospitals for people needing emergency care, be they walk-ins or ambulance arrivals, and it is attendances at these units that are the subject of this study. MIUs are NHS units which provide a service specifically aimed at people with minor injuries and other less serious conditions. Typically these are provided on the sites of former general hospitals which no longer provide a full range of acute care, but still offer other services such as minor elective surgery or out-patient clinics under the NHS. MIUs are not necessarily led by a consultant in accident and emergency medicine and many do not offer a 24 hour service. Also, they will routinely divert more serious cases (in the event of patients who self-present) to the nearest type one ED.

At the time of our research, there were twenty-two sites providing type one ED services in the Region, including one specialist children's unit (see figure 1). As we were using a National data set we were also able to capture the (relatively few) ED visits made by residents to EDs outside the region and include them in our analysis.

Attenders were attributed to LSOAs by using their postcode. The postcode is a small unit of geography used to facilitate the delivery of mail. Postcode areas typically cover a few tens of dwellings. We calculated the median size of postcode areas in the West Midlands region as 9,678 m^2, approximately 2.4 acres. Each one has a mapped polygon determined nationally and digitized for use in geographical information systems (GIS). A national look-up file is maintained by the Office for National Statistics in co-operation with the Royal Mail to attribute postcodes to higher geographies [18]. This is derived by calculating the longitude and latitude of the centre of each polygon and determining in which higher geography, (such as an LSOA) it falls into. This is then used by the NHS to convert patient postcodes captured at or soon after presentation into higher geographies such as LSOAs or local administrative boundaries, in their national data sets. This attribution is done at source and the study used data with the LSOAs already appended to them.

Methods of Measurement

Attendances. ED attendance was determined for the financial year 2007/2008, the most recent year of stable service provision. We had access to later data but chose this year as there were no major reconfigurations or relocations of type 1 EDs in the region during this period and no changes to MIU provision. This was not the case in subsequent years. As we are looking at spatial relationships, we needed these to be constant during the period of observation. We used records of attendance captured by the A&ECDS. Income deprivation was obtained using the Indices of Deprivation 2007 income domain score [19] for each of the LSOAs. This represented the proportion of people in the LSOA who were estimated to live in an income deprived household. Income deprivation is measured as a household having less than 60% of the median national income and/or being in receipt of a number of specified welfare benefits. Rather than using these scores as a continuous variable, which would assume a linear relationship between them and our response variable, we put them into quintiles based upon the rank of deprivation in the region which we treated as categorical variables.

Proximity to nearest ED and nearest MIU. The proximity of the LSOAs to the nearest Emergency Department was calculated using Arc GIS software (version 9.3), measuring the centroid of an LSOA which was population weighted centroids of an even smaller geography a Census Output Area. These are the smallest level of geography for which we have population characteristics and aggregated census data. For example, a roughly circular LSOA with a concentration of residents in Census Output Areas at the edge of its northern quadrant and fewer people living elsewhere, would have its distance to hospital estimate pulled closer to that part of the LSOA than the geometric centre. The Ordnance Survey, the UK's national mapping agency, produces digitised transport network data for the country [20]. These data, combined with the GIS software used in the project, were used to calculate the shortest road distances between each of the population weighted centroids described above and the

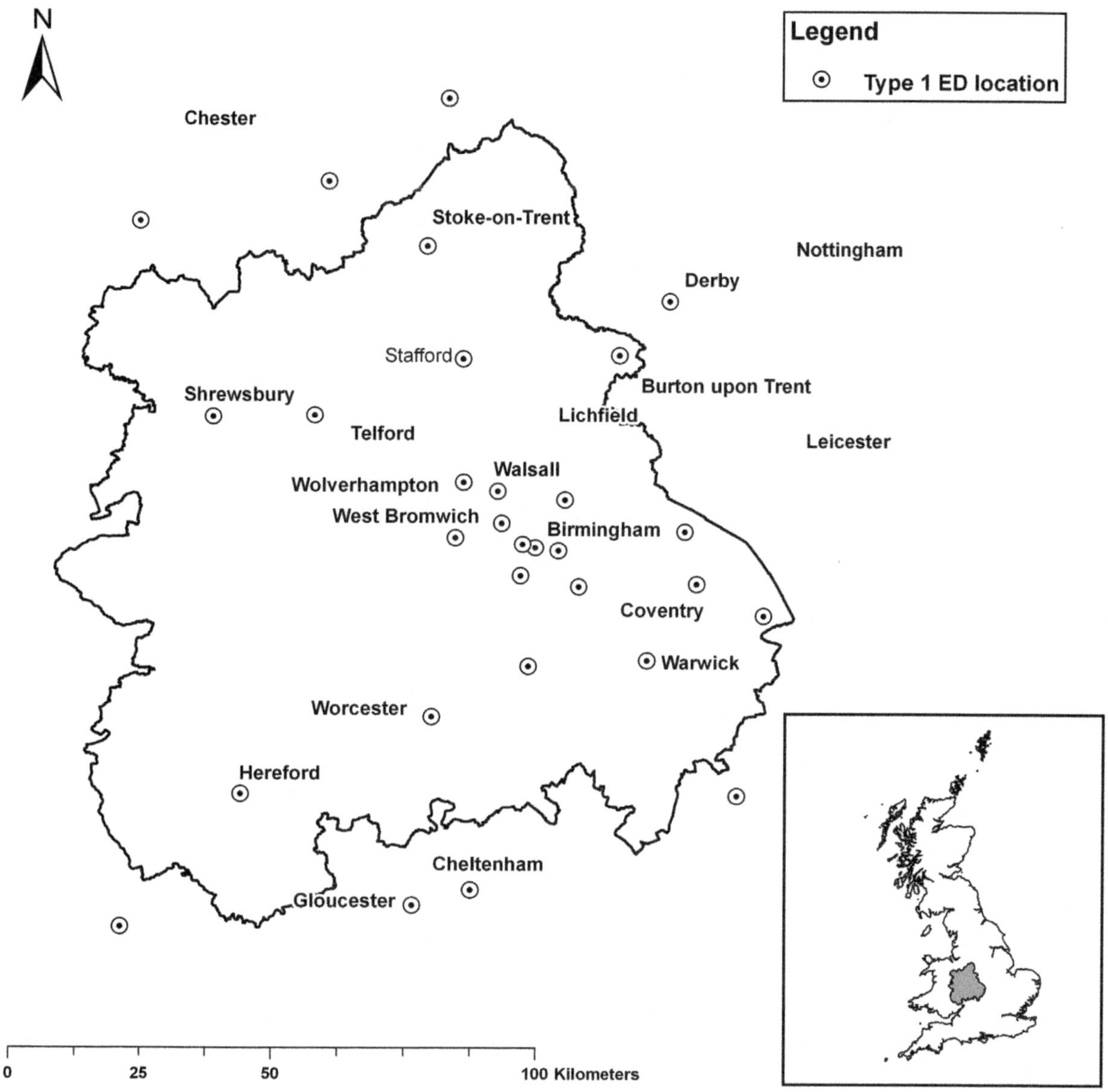

Figure 1. Map of the West Midlands region showing the location of Emergency Departments close to its population in financial year 2007–2008 (inset: the location of the West Midlands region in Great Britain).

nearest ED and nearest MIU. The researchers obtained the longitude and latitude of all of the EDs in the region using data supplied directly from the NHS Information Centre. We used the "NHS Choices" website [21] to identify and locate all MIUs within a 40 mile buffer of our region. This was deliberately chosen to be larger than necessary to guarantee that the closest unit to any of our LSOAs was captured. We then followed the same methods as used to derive proximity to nearest ED.

Statistical Analyses

We used two negative binomial models, one for children under fifteen years and one for people aged fifteen and over. For the purposes of this paper we use the term "adult" to refer to people aged fifteen years and over. We chose fifteen years as the cut-off, as this is one of the quinary age group break points in the small area population estimates (provided by the small area population unit at the Office of National Statistics). Attendance counts have a Poisson distribution, with a degree of over-dispersion, so negative binomial regression was chosen as the preferred method. As well as the variables described above we also adjusted for the age composition of the LSOAs by putting the proportion of residents in five year age bands into the model. We also included the proportion of males in each population as a variable and a number of interaction terms. These included the interaction between deprivation and distance and, in the case of adults, the interaction between proportion male and deprivation. We used the overall population of the LSOA as an offset term. All statistical analyses were undertaken using Stata version 11.

Results

Characteristics of study subjects

There were a total of 1,413,363 attendances to type 1 EDs which were attributable to residents of the West Midlands. Of these, 288,931 (20.4%) were of children under the age of fifteen years and the remaining 1,124,432 (79.6%) were by those aged 15 years and over.

For the 15 years and over group, 73.9% of all visits were to the nearest provider. For the under 15 years group this was slightly lower at 68.6%, but this was expected, as some activity is directed to the specialist children's ED in Birmingham rather than the nearest ED. Only a very small proportion of visits were made outside of the region of residence; just 1.4% for patients aged 15 and over and 1.2% of visits by children under 15. These percentages did not include the small number of people living at the edges of the region for whom an extra-regional ED was the nearest choice.

Main results

The model showed that deprivation and distance from hospital had significant effects on attendance. Tables 1 and 2 show the regression correlation coefficients for the two models, one for children under 15 and one for attenders aged 15 and over. In each case the coefficients are expressed as incidence rate ratios (IRR). In both tables the reference group comprises attenders in the least deprived income deprivation quintile.

For adult attendances, our results show that, having adjusted for the other variables, there is a significant reduction in attendance associated with increasing distance from an ED, of about 1.5% per kilometre in the reference group. This is a large effect size given the variation that exists in distance from EDs across the region, especially in rural areas. We also see a significant effect of deprivation. Using the least deprived quintile as a reference group, each subsequent quintile of deprivation is associated with higher attendances than the last, with attendance in the most deprived quintile being more than twice as large as that in the least deprived adjusting for the other variables (IRR 2.26, 2.01–2.55, $p<0.001$).

Children's attendances decreased with distance from ED more than those of adults, with one kilometre of extra distance being associated with a fall of 2.2% in attendance in the reference group; although the effect of deprivation was broadly similar.

The interaction between deprivation and distance was highly significant in both models at all levels of deprivation ($p = <0.001$). Attendance in deprived neighbourhoods reduces with distance to a much greater degree than attendance in less deprived ones. This is the case for both adults and children. From the IRRs alone this is not straightforward to interpret. To illustrate the effect, two figures have been provided, Figure 2 and Figure 3. These plots show the predicted effect on attendance with distance at the five deprivation quintiles, holding the distance to MIU constant at its median, and the other variables at their mean. This shows the modeled net effect of distance, increasing at higher levels of deprivation. In both adults and children we see the predicted levels of attendance being higher at each level of deprivation, but then we see the decrease in attendance with distance being much greater in more deprived populations. This is more marked in children than in adults, particularly in the attendance of children at the level of the most deprived quintile.

There is a significant positive association between distance to an MIU and attendance at a type one ED in both children and adults. This predicts that for each additional kilometre away from an MIU, type one ED use will increase by 4.2% for adults (3.8%–4.6%, $p<0.001$) and 4.9% for children (4.4%–5.4%, $p<0.001$). Again we see an interaction effect with deprivation. This suggests that the extent to which people in communities with different levels of deprivation use MIUs, differs with respect to distance.

Table 1. Regression co-efficients for a model of child attendance at Emergency Departments.

Variable	IRR	p-value	Lower 95% CI	Upper 95% CI
Deprivation quintile 1	1.000	~	~	~
Deprivation quintile 2	1.344	<0.001	1.199	1.507
Deprivation quintile 3	1.414	<0.001	1.260	1.588
Deprivation quintile 4	1.882	<0.001	1.676	2.115
Deprivation quintile 5	2.198	<0.001	1.904	2.537
Distance to nearest ED (km)	0.978	<0.001	0.974	0.983
Deprivation quintile 2 * distance to ED	0.986	<0.001	0.980	0.992
Deprivation quintile 3 * distance to ED	0.984	<0.001	0.978	0.990
Deprivation quintile 4 * distance to ED	0.972	<0.001	0.965	0.979
Deprivation quintile 5 * distance to ED	0.962	<0.001	0.952	0.972
Distance to nearest MIU (km)	1.049	<0.001	1.044	1.054
Deprivation quintile 2 * distance to MIU	0.996	0.028	0.992	1.000
Deprivation quintile 3 * distance to MIU	0.997	0.103	0.993	1.001
Deprivation quintile 4 * distance to MIU	0.992	<0.001	0.988	0.996
Deprivation quintile 5 * distance to MIU	0.994	0.025	0.989	0.999
Distance to nearest MIU (km)̂2	0.999	<0.001	0.999	0.999
Proportion population aged 5–9	1.288	<0.001	1.150	1.443
Proportion population aged 5–9̂2	0.996	<0.001	0.994	0.998
Proportion population aged 10–14	0.923	<0.001	0.892	0.955
Proportion population aged 10–14̂2	1.001	<0.001	1.000	1.001

Table 2. Regression co-efficients for a model of adult attendance at Emergency Departments.

Variable	IRR	p-value	Lower 95% CI	Upper 95% CI
Deprivation quintile 1	1.000	~	~	~
Deprivation quintile 2	1.346	<0.001	1.228	1.475
Deprivation quintile 3	1.481	<0.001	1.348	1.626
Deprivation quintile 4	1.865	<0.001	1.697	2.049
Deprivation quintile 5	2.259	<0.001	2.006	2.545
Distance to nearest ED (km)	0.985	<0.001	0.981	0.988
Deprivation quintile 2 * distance to ED	0.988	<0.001	0.984	0.993
Deprivation quintile 3 * distance to ED	0.987	<0.001	0.982	0.992
Deprivation quintile 4 * distance to ED	0.980	<0.001	0.975	0.985
Deprivation quintile 5 * distance to ED	0.969	<0.001	0.962	0.977
Distance to nearest MIU (km)	1.042	<0.001	1.038	1.046
Deprivation quintile 2 * distance to MIU	0.996	0.014	0.993	0.999
Deprivation quintile 3 * distance to MIU	0.996	0.010	0.993	0.999
Deprivation quintile 4 * distance to MIU	0.994	0.001	0.991	0.997
Deprivation quintile 5 * distance to MIU	0.996	0.050	0.991	1.000
Distance to nearest MIU (km)̂2	0.999	<0.001	0.999	0.999
Proportion population male	1.007	0.003	1.002	1.011
Proportion population aged 15–19	0.996	0.346	0.987	1.005
Proportion population aged 20–24	0.984	0.039	0.969	0.999
Proportion population aged 25–29	1.053	<0.001	1.025	1.081
Proportion population aged 30–34	0.965	0.001	0.945	0.986
Proportion population aged 35–39	1.001	0.911	0.979	1.024
Proportion population aged 40–44	1.016	0.170	0.993	1.038
Proportion population aged 50–54	0.976	0.053	0.952	1.000
Proportion population aged 55–59	0.950	<0.001	0.928	0.974
Proportion population aged 60–64	1.032	0.002	1.011	1.053
Proportion population aged 65–69	1.014	0.296	0.988	1.039
proportion population aged 70–74	0.962	0.003	0.937	0.987
Proportion population aged 75–79	1.005	0.709	0.979	1.032
Proportion population aged 80–84	1.038	0.009	1.009	1.068
Proportion population aged >=85	0.978	0.002	0.965	0.992

Discussion

We have found that ED use is significantly associated with both deprivation and distance and that attendance in deprived neighbourhoods reduces more with distance from hospital than in less deprived ones. The effect is more marked for children's attendances than those of adults.

Previous studies of distance effects have relied on more limited study areas and are not consistent. Both Hull et al [1] and Walsh [3] found significant associations between attendance and both deprivation and distance. However, both were based on more geographically limited populations than the one in this study, and were limited to adults registered with a GP. The latter study was also limited to patients discharged home on the day of discharge. Other studies investigating the effects of distance and deprivation on attendance have reached different conclusions. McKee et al [5] found that attendance in electoral wards in a rural area of Northern Ireland declined significantly with distance from hospital, but suggested that deprivation did not increase the explanatory power of the regression model they used. Conversely Carlisle et al, [2] in a study set in Nottingham in England, suggested that the apparent effect of distance was nearly wholly explained by deprivation and that distance net of deprivation, had a limited effect. In the Nottingham study subjects were limited to patients registered with a GP who attended out of hours.

The finding regarding the effect of MIU proximity is interesting. It appears that being further away from an MIU is associated with a much greater propensity to attend a type 1 ED however MIUs typically serve more rural areas so those few populations who happen to live in close proximity to them are not typical. There is very little literature on the impact of MIUs in reducing type 1 ED use, evaluations having being more focused on things like patient acceptability and patient satisfaction. However given the proliferation of them, further research on this topic would be timely and useful.

Advantages and limitations of this study

This study is of a large, heterogeneous and geographically contiguous population. We used a data set which is mandated in hospitals nationally and there are no non-state provided EDs in

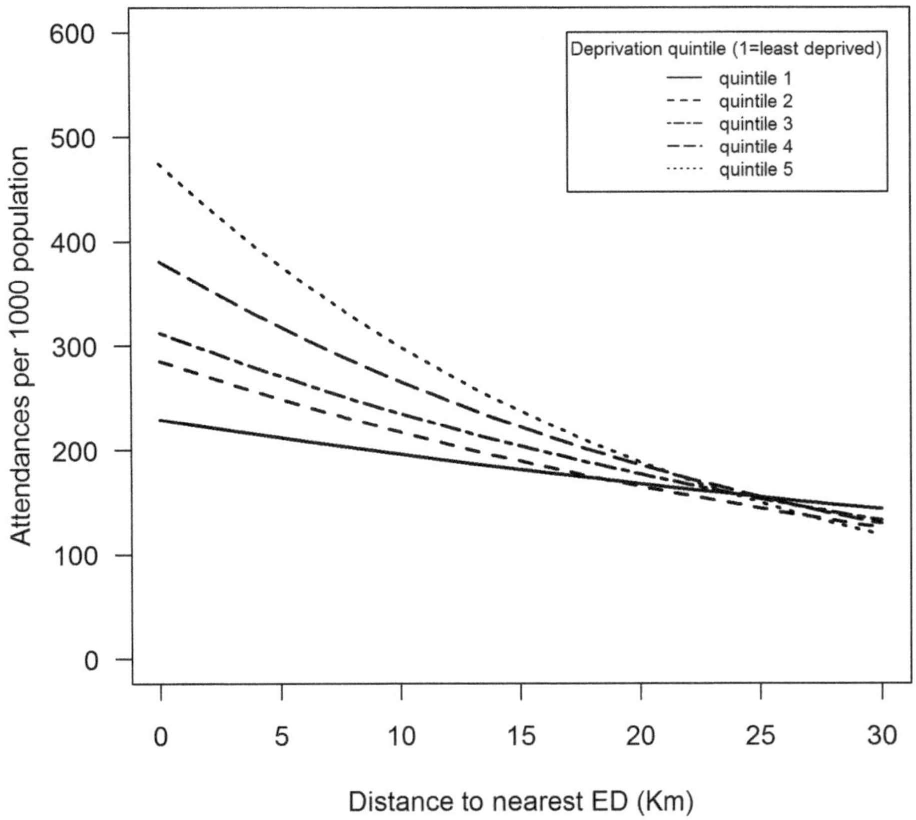

Figure 2. Modelled attendance change with distance at various levels of neighbourhood income deprivation, attenders aged >=15.

the English healthcare system, so our capture of attendance is near complete. The measure of deprivation we used is considered generally robust. As LSOAs are small units designed to facilitate the statistical analysis of the national population, they usefully have a high level of homogeneity. The distance variable was based upon actual road distances rather than Euclidian distances that may have been subject to error, particularly in rural areas. Although the distance variable based upon actual road distances, it used distance from population weighted centroids as a proxy for distance to hospital for all households in a neighbourhood and so has a degree of error, likely to be larger in larger rural LSOAs. We used projected populations, some time from the 2001 census, which need a degree of caution, but represent the best estimate available.

We have not adjusted for case severity. Concerns about the effect of adjusting for case mix, especially using administrative data have been raised in the literature [22], [23], [24]. In addition we had concerns about the quality of capture of specific items we might have used to carry out such an adjustment. The A&ECDS captures up to two presenting complaints using a simple numeric code (although some units use the International Classification of Diseases version 10, to which these codes do not map). The degree to which these data are well captured, is highly variable [25]. To verify this we undertook a sub-analysis of the coding of the reason for attendance and the proportions of visits resulting in an admission. A high degree of variability in these data points would support our decision to avoid case mix adjustment. In this sample the proportion of visits with a missing or invalid diagnosis code in the primary diagnosis on arrival varied between This varied from zero to 65.1%. The proportion of visits with a 'diagnosis unclassifiable' flag varied from zero to 45.7% and the admission rate varied between 14.6% to 36.4% between hospitals.

Interpretation

The reasons why deprived populations and populations close to EDs, use them more often does warrant further study. It is evident that deprived populations have higher incidences of serious illness, resulting in more non-negotiable needs for emergency care. However it is also possible that a higher proportion of attendances in these populations are for less urgent needs, in which case the cost of time and transport may dissuade presentations from further away especially if these problems could normally be addressed through primary care. A Canadian study showed that there was a significant decay with distance for less urgent cases, but not for more urgent ones [6].

In seeking treatment for minor ailments, patients can consult a GP (which is also free of charge in the UK) although getting appointments at short notice or out of normal working hours can be problematic [26]. There is evidence that there are more actual or perceived barriers to accessing good quality primary care in deprived areas. Firstly there are likely to be more people (who are not registered with a GP (for example transients and recent migrants) and who may be unaware that they can access primary care as a temporary resident. For example, in a study of a sample of Eastern European ED attenders (carried out in the same region in which our study was set) it was found that 43% were unregistered, compared to a rate of 7.4% observed in all attenders [27]. There is also evidence that people of lower socio-economic status are less likely to be satisfied with the quality of primary care available to them [28] and that more deprived people have

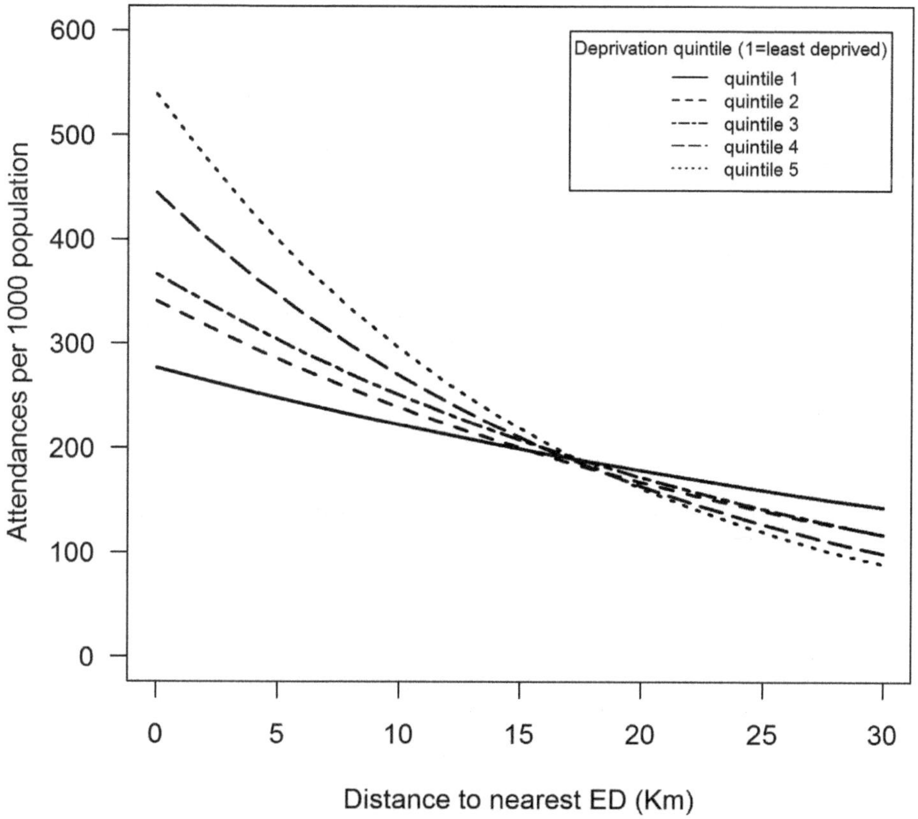

Figure 3. Modelled attendance change with distance at various levels of neighbourhood income deprivation, attenders aged <15.

different expectations of primary care including a lower willingness to travel to access it [29]. Campbell [4] explored this issue and detected a correlation between ED use and heightened dissatisfaction with GP services, but not with other variables such as appointment system functionality. Research using the NHS's Quality and Outcome Framework, which measures aspects of GP practice quality, have found associations between lower quality scores and the deprivation of the population served [30].

More deprived service users or inner-city dwellers who live close to hospitals may access care differently. Studies of ED attenders in Denmark found that migrants were significantly more likely to present at EDs than the indigenous population [31]. However, even if recent migrants do use EDs differently, their relatively small numbers could probably not explain the extreme variation seen in the use of EDs, even at small area level.

The more marked distance effect in children may be due to there being a higher proportion of less serious health needs in this population. The incidence of life threatening medical emergencies is very small in child populations compared with that in adults where events such as strokes and heart attacks, particularly in older people, are more common. A study of injuries in children in Wales showed that there was a significant decrease with distance in injury presentations generally, but in fractures the distance effect was not significant suggesting that more discretionary attendances will be more sensitive to distance [32].

The finding that proximity to MIUs did influence the number of ED visits was expected, given that we would expect minor injury to be a common reason to attend a type one ED. However, we need to be cautious in the interpretation of these results and especially the apparently large effect size. The reference for the intercept of the modelled attendance assumes no distance from an MIU. The populations who live close to MIUs are predominantly rural, less heterogeneous than urban ones, and tend to use EDs less. Whilst we have adjusted for demography and socio-economics, there still may be some unmodelled variation which disproportionately affects the way in which these people use services and which may cause a gradient of increasing use with distance from MIUs. Further research is needed on MIUs however, as they have been set up in large numbers at some cost to the health service, but relatively little is known about how or indeed if, they divert demand from other services.

Conclusions

There are two important policy implications of our findings. Firstly, a number of reconfigurations of hospital service are underway in England. Some of these will result in a step-change in the distance to provider in a densely populated area. These models suggest that this may cause unexpected step-changes in local demand patterns for emergency care, independent of the demographic, socio-economic and even epidemiological conditions extant in that area.

Secondly, if a disproportionate amount of demand for ED services from deprived areas close to hospitals is indeed better dealt with in a primary care setting, it may be possible to modify health seeking behaviour in these areas with a social marketing approach. Campaigns such as "choose well" [33] have been organised by the NHS to encourage more responsible use of EDs. A model such as the one we have developed here could be useful in suggesting locations for more targeted use of such initiatives.

Acknowledgments

The authors would like to acknowledge the kind assistance of Peter Chilton and Marie Crook in the School of Health and Population Sciences at the University of Birmingham, in the preparation of this manuscript.

Author Contributions

Conceived and designed the experiments: GR MAM. Performed the experiments: GR MAM. Analyzed the data: GR SF MAM. Wrote the paper: GR MAM AJS. Undertook data processing cleaning and verification, data linkage and some GIS work: SF. Provided methodological steering and specialist statistical advice and comments on the final draft: AG. Provided raw data and comments on the final draft: KS.

References

1. Hull S, Jones I, Moser K (1990) Factors influencing the attendance rate at accident and emergency departments in East London: the contribution of practice organization population characteristics and distance. J Health Serv Res Policy 2,1: 6–13.
2. Carlisle R, Groom L, Avery A, Boot D, Earwicker S (1998) Relation of out of hours activity by general practice and accident and emergency services with deprivation in Nottingham: longitudinal survey. BMJ 316: 520–523.
3. Walsh M (1990) Geographical factors and A&E attendance. Nurs Stand 5, 8: 28–31.
4. Campbell J (1994) General Practitioner appointment systems, patient satisfaction, and use of Accident and Emergency services – a study in one geographical area. Fam Pract 11,4: 438–445.
5. McKee C, Gleadhill D, Watson J (1990) Accident and emergency attendance rates: variation among patients from different general practices. Br J Gen Pract 40: 150–153.
6. Béland F, Lemay A, Boucher M (1998) Patterns of Visits to hospital-based emergency rooms. Soc Sci Med 47, 2: 1650–179.
7. Henneman P, Garb J, Caparo G, Haiping L, Smithline H, et al. (2009) Geography and travel distance impact emergency department visits. J Emerg Med 40, 3: 333–339.
8. Ingram D, Clarke D, Murdle R (1978) Distance and the decision to visit and emergency department. Soc Sci Med 12, 1: 55–62.
9. Magnusson G (1980) The role of proximity in the use of Hospital Emergency Departments, Sociol Health Illn 2, 2: 202–214.
10. Department of Health (2010) Attendances at accident & emergency departments and minor injury units, NHS organisations in England, 2007–08. Available: http://www.dh.gov.uk/en/Publicationsandstatistics/Statistics/Performancedataandstatistics/AccidentandEmergency/DH_087973.
11. Department of Health (2010). NHS reference costs 2008–2009. Available: http://www.dh.gov.uk/en/Publicationsandstatistics/Publications/PublicationsPolicyAndGuidance/DH_111591.
12. Boyle S (2007) Payment by results in England. Eurohealth 13, 1, 12–16.
13. Department of Health (2000) The NHS Plan.
14. Office for National Statistics (2010) Population estimates for UK, England and Wales, Scotland and Northern Ireland, Mid-year population estimates 2008. Available: http://www.ons.gov.uk/ons/publications/all-releases.html?definition = tcm: 77–22371.
15. Association of Public Health Observatories (2009) Technical Briefing 6, Using small area data in public health intelligence. Available: http://www.apho.org.uk/resource/item.aspx?RID = 74894.
16. United States Census Bureau (2000) Geographic Areas Reference Manual. Available: www.census.gov/geo/www/cen_tract.html. Accessed 2013 Jan 6.
17. Department of Health (2010) NHS Data Model and Dictionary Service. Available: www.datadictionary.nhs.uk/data_dictionary/attributes/a/acc/accident_and_emergency_department_type_de.asp?shownav = 1. Accessed 2013 Jan 6.
18. Office of National Statistics, Postcode Products: Guidance and Methodology. Available: http://www.ons.gov.uk/ons/guide-method/geography/products/postcode-directories/-nspp-/index.html.
19. Noble M, McLennan D, Wilkinson K, Whitworth A, Barnes H, et al. (2008) The English Indices of Deprivation 2007, Department of Communities and Local Government, London.
20. Ordnance Survey (2011) OS MasterMap ® Integrated Transport Network™ Layer Getting started guide. Available: http://www.ordnancesurvey.co.uk/oswebsite/products/os-mastermap/itn-layer/index.html.
21. Department of Health (2009) NHS Choices – Your health, your choices. Available: http://www.nhs.uk/servicedirectories/Pages/ServiceSearch.aspx.
22. Mohammed M, Deeks J, Girling A, Rudge G, Carmalt M, et al. (2009) Evidence of methodological bias in hospital standardised mortality ratios: retrospective database study of English hospitals. BMJ 338, b780.
23. Wright J, Shojania K (2009) Measuring the quality of care (editorial) BMJ 338, b569.
24. Nicholl J (2007) Casemix adjustment in non-randomised observational evaluation: the constant risk fallacy. J Epidemiol Community Health 6: 1010–1013.
25. Downing A, Rudge G (2006) A study of childhood attendance at emergency departments in the West Midlands. Emerg Med J 23,5: 391–393.
26. Wensing M, Vedsted P, Kersnik J, Peersman W, Klingenberg A, et al. (2002) Patient satisfaction with availability of general practice: an international comparison. Int J Qual Health Care 14, 2: 111–118.
27. Leaman A, Rysdale E, Webber R (2006) Use of the emergency department by Polish migrant workers. Emerg Med J 23,12: 918–919.
28. Mercer S, Watt G (2007) The inverse care law: Clinical Primary Care encounters in deprived and affluent areas of Scotland. Ann Fam Med 5, 6: 503–510.
29. Jung H, Baerveldt C, Olesen F Grol R, Wensing M (2003) Patient characteristics as predictors of primary health care preferences: a systematic literature analysis. Health Expect 6: 160–181.
30. Ashworth M, & Armstrong D (2006) The relationship between general practice characteristics and quality of care: a national survey of quality indicators used in the Quality and Outcomes Framework, 2004–5. BMC Fam Pract 7: 68.
31. Norredam M, Krasnik A, Sorenson T, Keiding N, Michaelsen J, et al. (2004) Emergency room utilization in Copenhagen: a comparison of immigrant groups and Danish-born residents. Scand J Public Health 32: 53–59.
32. Lyons R, Lo S, Heaven M, Littlepage B (1995) Injury surveillance in children – usefulness of a centralised database of accident and emergency attendances. Inj Prev 1: 173–176.
33. NHS West Midlands (2010) Press release. Health Bosses say "choose well" this winter. Press release reference number 10121701. Available: http://www.westmidlands.nhs.uk/News/Pressreleases.aspx. Accessed 2013 Jan 6.

14

rAvis: An R-Package for Downloading Information Stored in Proyecto AVIS, a Citizen Science Bird Project

Sara Varela[1,2]*, Javier González-Hernández[3], Eduardo Casabella[4], Rafael Barrientos[5]

1 Department of Ecology, Faculty of Science, Charles University, Prague, Czech Republic, **2** Departamento de Ecologia, Instituto de Ciencias Biologicas, Universidade Federal de Goiás, CxP 131, Goiania, Goiás, Brasil, **3** Toledo, Spain, **4** ACCONTE Group, Actividades de Consultoría y Telecomunicaciones, Tres Cantos, Spain, **5** Departamento de Ciencias Ambientales, Facultad de Medio Ambiente, Universidad de Castilla-La Mancha, Toledo, Spain

Abstract

Citizen science projects store an enormous amount of information about species distribution, diversity and characteristics. Researchers are now beginning to make use of this rich collection of data. However, access to these databases is not always straightforward. Apart from the largest and international projects, citizen science repositories often lack specific Application Programming Interfaces (APIs) to connect them to the scientific environments. Thus, it is necessary to develop simple routines to allow researchers to take advantage of the information collected by smaller citizen science projects, for instance, programming specific packages to connect them to popular scientific environments (like R). Here, we present rAvis, an R-package to connect R-users with Proyecto AVIS (http://proyectoavis.com), a Spanish citizen science project with more than 82,000 bird observation records. We develop several functions to explore the database, to plot the geographic distribution of the species occurrences, and to generate personal queries to the database about species occurrences (number of individuals, distribution, etc.) and birdwatcher observations (number of species recorded by each collaborator, UTMs visited, etc.). This new R-package will allow scientists to access this database and to exploit the information generated by Spanish birdwatchers over the last 40 years.

Editor: Stephanie Ann White, UCLA, United States of America

Funding: Sara Varela and Rafael Barrientos have postdoctoral grants. SV is funded by the European Science Foundation and Czech Republic, from the Education for Competitiveness Operational Programme (ECOP) project "Support of establishment, development and mobility of quality research teams at the Charles University" (CZ.1.07/2.3.00/30.0022)(http://www.cuni.cz/UKEN-88.html). RB received financial assistance from Juan de la Cierva program from the Spanish Ministry of Competitiveness (http://www.idi.mineco.gob.es/portal/site/MICINN/menuitem.dbc68b34d11ccbd5d52ffeb801432ea0/?vgnextoid=fdf669dad3f88310VgnVCM1000001d04140aRCRD). Javier González-Hernández and Eduardo Casabella-Cabana are freelance web-developers that worked on the project during their free time. They were involved in the design, programming and optimization of the R-package ravis and on the web interface of Proyectoavis. The funders had no role in study design, data collection and analysis, decision to publish, or preparation of the manuscript.

Competing Interests: The authors have declared that no competing interests exist.

* E-mail: svarela@paleobiogeography.org

Introduction

During the past several decades, developers have focused their attention on constructing web repositories to store and share biological information. On the one hand, there are online repositories with information generated by scientists, like specimens collected for museums and herbariums, fossil records or genetic data (e.g. GBIF: http://gbif.org, NOW: http://helsinki.fi/science/now/, GeneBank: http://ncbi.nlm.nih.gov/genbank/). On the other hand, there are web sites that store biological information collected by non-scientists, or so-called 'citizen science'.

Citizen science has proven to be an appropriate method to provide researchers with valuable information [1–3], and is increasingly used as an adequate way to sample species occurrences and distributions [4], to collect data to investigate urban ecology [3,5,6], or to collect data on bird biology, ecology and diversity [7–9]. In our case, data stored in Proyecto AVIS, our citizen science project to collect data from amateur Spanish ornithologists, show the same general patterns described by scientists based on their own samples and field experiments. Power law distributions of species/area [10] and species/abundance [11] have been detected (Figure 1), suggesting that the data stored in Proyecto AVIS have similar properties to the data collected by scientists.

One of the main characteristics of the citizen science databases is that they are huge. For instance, birdwatchers' observations stored in the eBird database reached 100,000,000 observations and over 10,000 species (http://ebird.org). As a result, there are terabytes of information about species occurrences (latitude, longitude, altitude, time, habitat, diet, alleles, etc.) stored in online databases that follow different formats and standards of data storage [12], and the challenge now is developing easy strategies to use this information for research [13].

Currently, there are ongoing projects to generate tools to standardize the information stored in those databases (e.g. http://ecodataretriever.org) and to develop R-packages to connect online biological databases to the R-environment (http://ropensci.org/). As a consequence, large international databases are now being made available through R using packages like *rebird* [14], *rfishbase* [15], *rgbif* [16] or *rvertnet* [17] (connecting R with eBird, Fishbase, GBIF and VertNet databases, respectively). All of these new data exponentially increase our capabilities to answer questions about species conservation, global change, macroecology and biogeography.

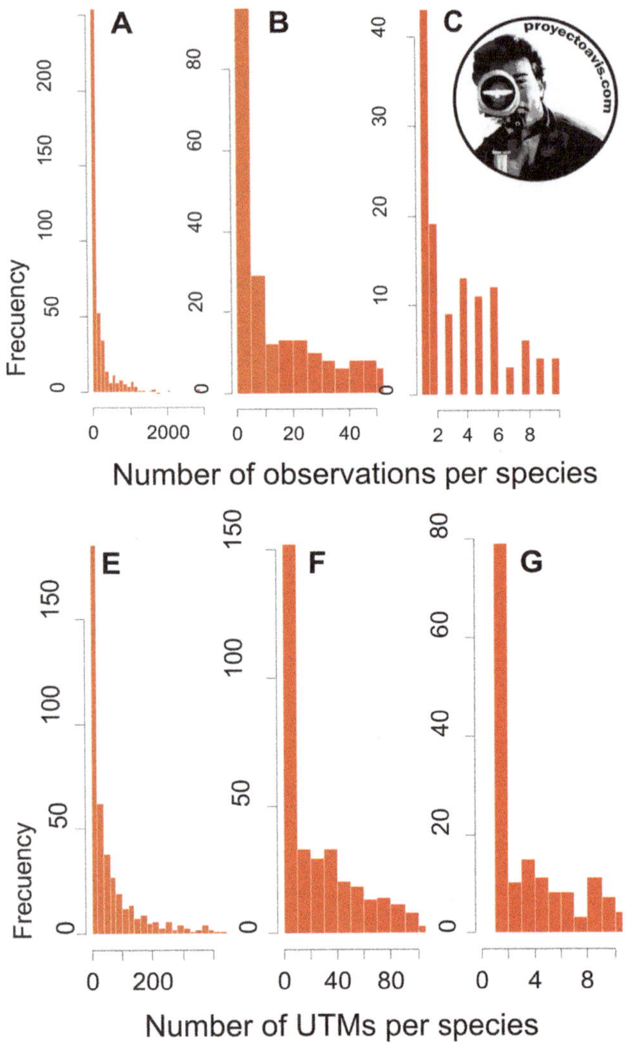

Figure 1. Data collected by amateur birdwatchers and stored in Proyecto AVIS show the same patterns as data collected by scientists, like scale invariant relationship of the frequency distribution of the number of observations per species (A-B-C) and scale invariant relationship of the frequency distribution of the number of UTMs per species (E-F-G).

R is an open source and collaborative framework (http://.r-project.org/), and is one of the most used environments for analyzing biological data and for developing scientific software [18]. Many young scientists are becoming advanced R-users (but see [19]). Thus, R is becoming a standard environment for developing easy-to-use (and re-use) functions and for sharing them with the academic community. For all of these reasons, we decided to build an R package to directly download the information stored in Proyecto AVIS from the R environment, in order to promote the use of the data stored in this database within the growing scientific R-community.

Proyecto AVIS

Each citizen science project stores singular and, consequently, important information [6,9,20–22]. Proyecto AVIS (http://proyectoavis.com) is a citizen science project born in August 2005 with the idea of collecting the data stored in the field notebooks of amateur Spanish ornithologists and sharing them with both other amateur ornithologists and the scientific community. More than one hundred collaborators, including several NGOs, have been actively participating in the project uploading their bird observations. Overall, the database contains records over 40 years (1973–2013), stores 82,503 records, totalling 4,739,171 individuals from 413 species, which represents 90% of the total number of species recorded in Spain. In addition, it contains information from 1,717 different UTMs (squares of 10×10 km), representing 30% of the Spanish territory (query to the database: November 2013).

The Proyecto AVIS database and web page were built using open source software (MySQL, Perl, Apache) and free GIS layers. Proyecto AVIS requires five mandatory fields for each bird observation: 'species', 'number of individuals', 'observation period', 'date' and 'UTM 10×10 km square', plus several optional fields that include variables like 'hour', 'sex', 'age' or 'habitat'. To standardize the taxonomy, the bird species list follows the Bird List of Spain from SEO/BirdLife [23]. Bird occurrences in the Proyecto AVIS database are georeferenced using the projected UTM 10×10 km square system and the MGRS labelling convention (Military Grid Reference System). The UTM/MGRS is the standard system for mapping species occurrences in Spain and is the system used by the Spanish bird atlases [24,25]. To help users identify the UTMs in which they recorded the species, the web application includes an easy-to-use tool to geo-referenced the observations based on a Google Maps ™ routine.

The Proyecto AVIS web page (http://proyectoavis.com) includes several user-friendly tools for exploring the database, like summaries of the bird observations or graphics of the species records throughout the year, and it allows registered users to download detailed information about the species observations to Excel files. However, although the database is already available on the Internet, its use for research has not been properly exploited. Proyecto AVIS lacks a specific package to connect the web repository with the R-environment, and we believe that this fact has prevented scientists from using Proyecto AVIS information.

Description of the package

rAvis exclusively contains R code, which maximizes its portability across platforms, and it works in Unix-like and Windows operating systems. The rAvis functions have been optimized following the standards criteria for software quality [26,27] and they are accessible through GitHub (https://github.com/javigzz/rAvis). Bugs can be reported using GitHub; https://github.com/javigzz/rAvis/issues. rAvis is freely available on the Comprehensive R Archive Network; CRAN (http://cran.r-project.org/) and complete information about rAvis, its functions and their parameters is available in the package help.

rAvis uses functions from other R-packages to get and plot the data stored in Proyecto AVIS. Namely, R-libraries stringr [28], XML [29], tools [30], RCurl [31], scrapeR [32] and gdata [33] are used to download the bird observations; maptools [34], raster [35] and rgdal [36] to plot the GIS files; and, finally, scales [37] is used to plot bird occurrences with a transparency.

Exploring Proyecto AVIS. We developed several functions to explore the database in an easy and visual way and other functions to download the selected information (see Table 1 and run the example). First, *avisSpeciesSummary* allows users to download a table with a summary of the records stored in Proyecto AVIS aggregated by species: number of observations of each species, number of individuals recorded, number of different UTMs (10×10 km) with observations, number of birdwatchers that recorded the species. Second, *avisContributorsSummary* returns a table with a general summary of the records stored in the database aggregated by birdwatcher: number of observations per birdwatcher, number of species observed, number of provinces with data, number of UTMs visited, number of periods of observations. Finally, *avisHasSpecies* checks if a species name exists in Proyecto AVIS and then, *avisMapSpecies* allows users to explore the distribution of the observations of the species by setting the name of the species and selecting the type of map; administrative boundaries ('admin') or physical map ('phys') (Figure 2).

For constructing the plots we used free GIS layers. We downloaded the Spanish administrative map from http://.diva-gis.org/, the Spanish UTM map from the Spanish government online map repository http://bscw.rediris.es/pub/bscw.cgi/524254?client_size = 1366×580, and the Spanish physical map from http://.openstreetmap.org/ using the R- library OpenStreetMap [38].

Advanced queries to Proyecto AVIS. We constructed two main functions to set flexible queries about the species occurrences and the birdwatcher observations: *avisQuerySpecies* and *avisQueryContributor*, respectively. These functions download the information stored in Proyecto AVIS, and are intended to be tuned by the users in relation to their specific objectives. Also, we programmed avisQuery as a flexible function to pass any argument allowed in Proyecto AVIS database. We decided not to predefine queries or to pre-process the data because this would narrow the possibilities for research [12]. Instead, we allow the users to set their own queries to Proyecto AVIS. Arguments include taxonomic levels, like species, family, order; individual characteristics, like age, sex, breeding status; temporal filters, like year and month; or environmental filters, like habitat. Moreover, we added a UTM-latlong conversion to all queries. Thus, the position of the observations is given in two different formats: projected UTMs 10×10 km and geographic coordinates WGS84 (common latitude-longitude coordinates, which are not available in the current web application from Proyecto AVIS). We did not program more specific graphics or statistical analyses because we understand that the purpose of this package is to obtain the biological information stored in Proyecto AVIS and not to re-program statistical algorithms that are already available in other R-packages. We assume that R-users would employ different R-packages for calculating their own statistics and constructing their own plots (see the example).

Example. rAvis could be upgraded in future releases. To download the exact version of rAvis that we used in this example run the function install_github from devtools package as follows: install_github("javigzz/rAvis", ref = "v0.1")

Install rAvis from the CRAN and load the package > install.packages ("rAvis") > library(rAvis) > avisSetup (verbose = FALSE) Check if the target species has records in Proyecto AVIS > avisHasSpecies ("Pica pica") Plot the occurrences of the species to explore the data > avisMapSpecies ("Pica pica", maptype = " phys") Download the occurrrences of the species > Pica_pica<- avisQuerySpecies ("Pica pica") Filter the data using avisQuery. For instance, select only records from forests habitats setting habitat = "bosque" (the database is in Spanish) > Pica_pica_forest<- avisQuery (species = "Pica pica", > habitat = "bosque")

Plot the results using avisMAp > avisMap (Pica_pica_forest, label = "Pica pica; Forest") If interested in several species, explore the database using avisMApSpecies > avisMapSpecies (list("Tyto alba", "Athene noctua", > "Bubo bubo", "Strix aluco"), maptype = "phys") Save the maps individually using the tiff function > directory<- "C:/your_directory" > species<- list(" Tyto alba", "Athene noctua", > "Bubo bubo", "Strix aluco") > for (x in species){ > tiff (file.path (directory, paste ("/", x, ".tiff", sep = ""))) > avisMapSpecies (x) > dev.off() > }

Conclusions

We have programmed rAvis, an R-package designed to help researchers explore and download the information stored in Proyecto AVIS. Thus, biogeographers, macroecologists and ornithologists working in spatial ecology or temporal series, in addition to researchers working on citizen science can easily take advantage of the unique data stored in this database for their own research.

Table 1. Descriptions of the functions of the rAvis R-package.

Functions in rAvis	Description
avisHasSpecies	checks if a species name exists in Proyecto AVIS.
avisSpeciesId	Returns the id of the selected species
avisAllSpecies	Returns a list with the species names and the ids of the species in Proyecto AVIS
avisMap	Renders a map for the observations downloaded using avisQuerySpecies
avisMapSpecies	Renders a map for each of the selected species (could be a list of species)
avisSpeciesSummary	Downloads a table with a summary of the records stored in Proyecto AVIS aggregated by species; number of observations of each species, number of individuals recorded, number of different UTMs (10×10 km) with observations and number of birdwatchers that recorded the species
avisQuery	Downloads information using several filters, like Order, Family, Species, Age, Sex, Habitat, etc.
avisQuerySpecies	Wrapper for avisQuery that allows to perform a query for more than one species at once.
avisContributorsSummary	Returns a table with the observations aggregated by birdwatcher.
avisContributorAggregatedObservations	Downloads the information about the observations of a birdwatcher
avisQueryContributor	Wrapper for avisQuery that allows to perform a query for more than one contributor at once.
avisSetup	Allows the user to turn off the information messages provided by the functions using "verbose = FALSE"

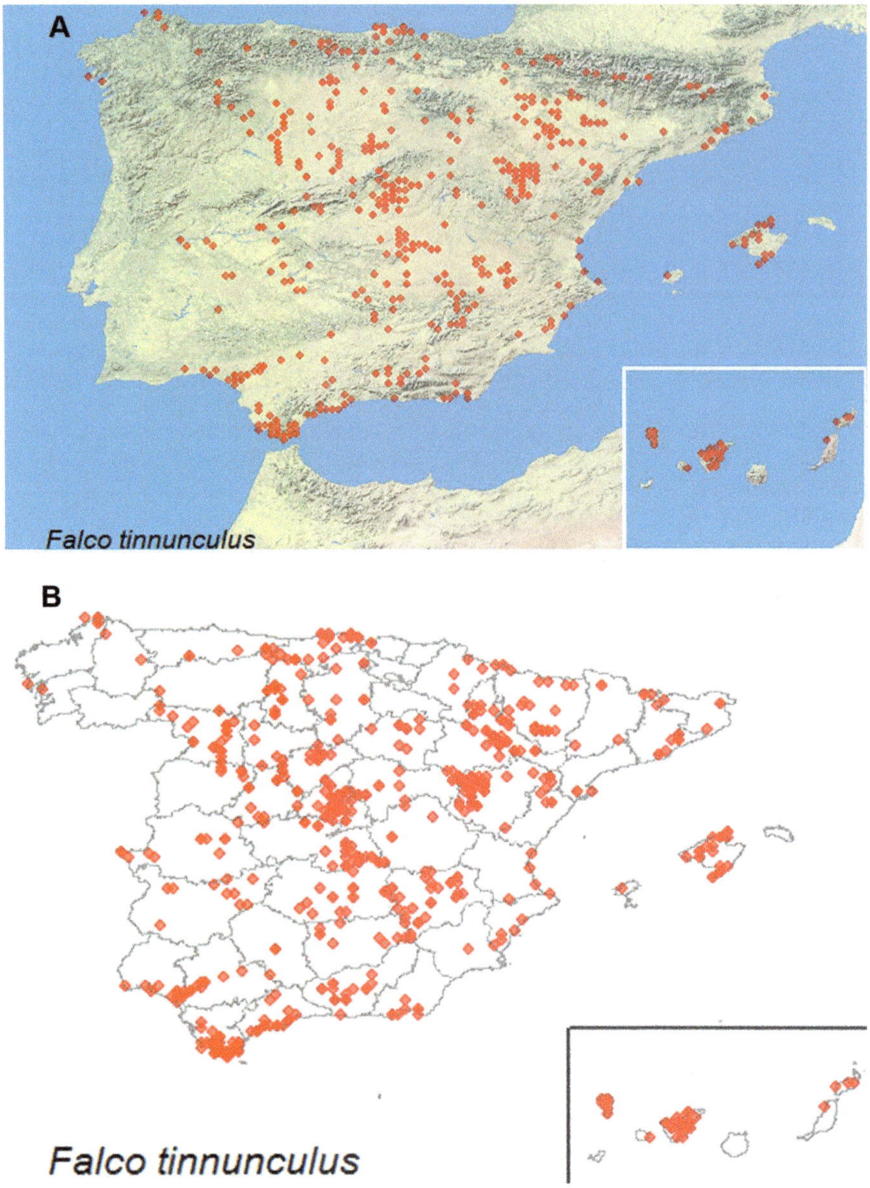

Figure 2. Outputs of the function *avisMapSpecies* setting the parameter map as 'phys' (A), or 'admin' (B) with the *Falco tinnunculus* records as an example.

Acknowledgments

Thanks to Paco Montoya and the ornithologist association Cigüeña Negra from Tarifa (http://cocn.tarifainfo.com) for their important contribution to Proyecto AVIS. Special thanks to Jose Antonio Palomar and Juan Antonio Arce, who decisively contributed to Proyecto AVIS web development at several stages.

Author Contributions

Conceived and designed the experiments: SV JGH. Performed the experiments: JGH SV. Analyzed the data: SV. Contributed reagents/materials/analysis tools: EC RB. Wrote the paper: SV JGH EC RB.

References

1. Devictor V, Whittaker RJ, Beltrame C (2010) Beyond scarcity: citizen science programmes as useful tools for conservation biogeography. Diversity and Distributions 16: 354–362.
2. Oberhauser K, LeBuhn G (2012) Insects and plants: engaging undergraduates in authentic research through citizen science. Frontiers in Ecology and the Environment 10: 318–320.
3. Nagy C, Bardwell K, Rockwell RF, Christie R, Weckel M (2012) Validation of a Citizen Science-Based Model of Site Occupancy for Eastern Screech Owls with Systematic Data in Suburban New York and Connecticut. Northeastern Naturalist 19: 143–158.
4. Tulloch AIT, Possingham HP, Joseph LN, Szabo J, Martin TG (2013) Realising the full potential of citizen science monitoring programs. Biological Conservation 165: 128–138.
5. Weckel ME, Mack D, Nagy C, Christie R, Wincorn A (2010) Using Citizen Science to Map Human-Coyote Interaction in Suburban New York, USA. Journal of Wildlife Management 74: 1163–1171.
6. Mulder RA, Guay P-J, Wilson M, Coulson G (2010) Citizen science: recruiting residents for studies of tagged urban wildlife. Wildlife Research 37: 440–446.
7. Cooper CB, Loyd KAT, Murante T, Savoca M, Dickinson J (2012) Natural History Traits Associated with Detecting Mortality Within Residential Bird

Communities: Can Citizen Science Provide Insights? Environmental Management 50: 11–20.
8. Wiersma YF (2010) Birding 2.0: Citizen Science and Effective Monitoring in the Web 2.0 World. Avian Conservation and Ecology 5: 13.
9. Lepczyk CA (2005) Integrating published data and citizen science to describe bird diversity across a landscape. Journal of Applied Ecology 42: 672–677.
10. Sizling AL, Kunin WE, Sizlingova E, Reif J, Storch D (2011) Between Geometry and Biology: The Problem of Universality of the Species-Area Relationship. American Naturalist 178: 602–611.
11. Tjørve E, Kunin WE, Polce C, Calf Tjørve KM (2008) Species–area relationship: separating the effects of species abundance and spatial distribution. Journal of Ecology 96: 1141–1151.
12. White EP, Baldridge E, Brym ZT, Locey KJ, McGlinn DJ, et al. (2013) Nine simple ways to make it easier to (re)use your data. Ideas in Ecology and Evolution 6: 1–10.
13. Reichman OJ, Jones MB, Schildhauer MP (2011) Challenges and Opportunities of Open Data in Ecology. Science 331: 703–705.
14. Maia R, Chamberlain S (2012) rebird: Interface to eBird. R package version 0.1. Available: http://CRAN.R-project.org/package = rebird.
15. Boettiger C, Lang DT, Wainwright PC (2013) rfishbase: R Interface to FishBASE. R package version 0.2-1. Available: http://CRAN.R-project.org/package = rfishbase.
16. Chamberlain S, Boettiger C, Ram K, Barve V (2013) rgbif: Interface to the Global Biodiversity Information Facility API methods. R package version 0.3.0. Available: http://CRAN.R-project.org/package = rgbif.
17. Chamberlain S, Barve V (2012) rvertnet: Search VertNet database from R. Available: http://CRAN.R-project.org/package = rvertnet.
18. Duck G, Nenadic G, Brass A, Robertson DL, Stevens R (2013) bioNerDS: exploring bioinformatics' database and software use through literature mining. BMC Bioinformatics 14.
19. Joppa LN, McInerny G, Harper R, Salido L, Takeda K, et al. (2013) Troubling Trends in Scientific Software Use. Science 340: 814–815.
20. Moyer-Horner L, Smith MM, Belt J (2012) Citizen science and observer variability during American pika surveys. Journal of Wildlife Management 76: 1472–1479.
21. Worthington JP, Silvertown J, Cook L, Cameron R, Dodd M, et al. (2012) Evolution MegaLab: a case study in citizen science methods. Methods in Ecology and Evolution 3: 303–309.
22. Silvertown J (2009) A new dawn for citizen science. Trends in Ecology & Evolution 24: 467–471.
23. Gutiérrez R, de Juana E, Lorenzo JA (2012) Lista de las aves de España Edición de 2012. Sociedad Española de Ornitología (SEO/BirdLife).
24. Martí R, Moral JCd (2002) Atlas de las aves reproductoras de España. Madrid: Dirección General de Conservación de la Naturaleza - Sociedad Española de Ornitología.
25. SEO/BirdLife (2012) Atlas de las aves en invierno en España 2007–2010. Madrid: Ministerio de Agricultura, Alimentación y Medio Ambiente-SEO/BirdLife.
26. Chambers JM (2008) Software for Data Analysis. Programming in R. Springer.
27. Voulgaropoulou S, Spanos G, Angelis L, Ieee (2012) Analyzing Measurements of the R Statistical Open Source Software. Proceedings of the 2012 Ieee 35th Software Engineering Workshop. pp. 1–10.
28. Wickham H (2012) stringr: Make it easier to work with strings. R package version 0.6.2. http://CRAN.R-project.org/package = stringr.
29. Lang DT (2013) XML: Tools for parsing and generating XML within R and S-Plus.. R package version 3.98-1.1. http://CRAN.R-project.org/package = XML.
30. R-Core-Team (2013) R: A language and environment for statistical computing. R Foundation for Statistical Computing, Vienna, Austria. URL http://.R-project.org/.
31. Lang DT (2013) RCurl: General network (HTTP/FTP/...) client interface for R. R package version 1.95-4.1. http://CRAN.R-project.org/package = RCurl.
32. Acton RM (2010) scrapeR: Tools for Scraping Data from HTML and XML Documents. R package version 0.1.6. http://CRAN.R-project.org/package = scrapeR.
33. Warnes GR, Bolker B, Gorjanc G, Grothendieck G, Korosec A, et al. (2013) gdata: Various R programming tools for data manipulation. R package version 2.13.2. http://CRAN.R-project.org/package = gdata.
34. Bivand R, Lewin-Koh N (2013) maptools: Tools for reading and handling spatial objects. R package version 0.8–27. http://CRAN.R-project.org/package = maptools.
35. Hijmans RJ (2013) raster: Geographic data analysis and modeling. R package version 2.1-62/r2833. http://R-Forge.R-project.org/projects/raster/.
36. Bivand R, Keitt T, Rowlingson B (2013) rgdal: Bindings for the Geospatial Data Abstraction Library. R package version 0.8–11. http://CRAN.R-project.org/package = rgdal.
37. Wickham H (2012) scales: Scale functions for graphics. R package version 0.2.3. http://CRAN.R-project.org/package = scales.
38. Fellows I, Stotz J P (2013) OpenStreetMap: Access to open street map raster images. R package version 0.3.1. http://CRAN.R-project.org/package = OpenStreetMap.

15

Methods for Determining the Uncertainty of Population Estimates Derived from Satellite Imagery and Limited Survey Data: A Case Study of Bo City, Sierra Leone

Roger Hillson[1¤], Joel D. Alejandre[1], Kathryn H. Jacobsen[2], Rashid Ansumana[3,4], Alfred S. Bockarie[3,4], Umaru Bangura[4], Joseph M. Lamin[4], Anthony P. Malanoski[5], David A. Stenger[5]*

1 Information Technology Division, Naval Research Laboratory, Washington, District of Columbia, United States of America, **2** Department of Global and Community Health, George Mason University, Fairfax, Virginia, United States of America, **3** Njala University, Bo, Sierra Leone, **4** Mercy Hospital Research Laboratory, Bo, Sierra Leone, **5** Center for Bio/Molecular Science and Engineering, Naval Research Laboratory, Washington, District of Columbia, United States of America

Abstract

This study demonstrates the use of bootstrap methods to estimate the total population of urban and periurban areas using satellite imagery and limited survey data. We conducted complete household surveys in 20 neighborhoods in the city of Bo, Sierra Leone, which collectively were home to 25,954 persons living in 1,979 residential structures. For five of those twenty sections, we quantized the rooftop areas of structures extracted from satellite images. We used bootstrap statistical methods to estimate the total population of the pooled sections, including the associated uncertainty intervals, as a function of sample size. Evaluations based either on rooftop area per person or on the mean number of occupants per residence both converged on the true population size. We demonstrate with this simulation that demographic surveys of a relatively small proportion of residences can provide a foundation for accurately estimating the total population in conjunction with aerial photographs.

Editor: Abdisalan Mohamed Noor, Kenya Medical Research Institute - Wellcome Trust Research Programme, Kenya

Funding: This work was funded by the Defense Threat Reduction Agency, Joint Science and Technology Office via contract to the United States Naval Research Laboratory. There is no past, present or future Intellectual Property associated with the work described in the paper, and none of the authors have any financial interests or conflicts in the outcome of the study. The funders had no role in study design, data collection and analysis, decision to publish, or preparation of the manuscript.

Competing Interests: The authors have declared that no competing interests exist.

* Email: david.stenger@nrl.navy.mil

¤ Current address: Independent consultant, Fairfax, Virginia, United States of America

Introduction

As Viel and Tran [1] have summarized, "Epidemiology is sometimes considered the science of denominators, because an accurate knowledge of the population at risk is a fundamental requirement for determining rates and deriving meaningful indicators of health status, health services, and health systems." Calculation of the rates of incidence of new disease or the prevalence of existing disease in a population or subpopulation depends on an accurate knowledge of the underlying population denominators, often stratified by age, sex, and even by levels of immunity or exposure [2,3]. Without adequate population denominator data, incidence and prevalence measures must be reported as counts rather than rates, which introduces a bias toward higher apparent risk in areas with higher population density [4]. Although maximum likelihood methods can now be used to estimate both the serial interval and reproductive number of an epidemic from counts of incident cases only [5,6], resource-limited countries often lack data for both the underlying population denominators and the baseline incidence rates. This is the situation addressed in this manuscript.

Our overarching goal with our simulations is to demonstrate approaches for combining readily available satellite imagery with small sample size field surveys as a means of efficiently estimating both population size (denominator data) and health states (numerator data). In this paper, we take a first step towards improving (denominator) measures by estimating total population in a West African urban area using remotely-acquired residential rooftop areas and demographic surveys. Specifically we ask "what number of residential surveys are needed to support a particular level of confidence in the estimated total population size, given a total number of residential structures or a total rooftop area?" We approach this question using a combination of total rooftop areas extracted from satellite imagery; a complete count of residents from 20 neighborhoods in Bo, Sierra Leone, West Africa, which we collected during a household health census; and simulated datasets sampled from our complete demographic dataset to represent small surveys of a subsample of residents.

Although our primary interests are in epidemiological and health care modeling, the simulation methods presented in this paper are not restricted to this domain since "denominator" data are essential for a variety of sectors. In resource-challenged environments, complete information about population size is often unavailable due to infrequent government census initiatives, rapid population growth in cities due to unplanned urbanization and rural-to-urban migration, political and other instability that prevents ongoing data collection and management, and the difficulties associated with securing the time and funding to conduct large-scale data collection activities. Rooftop areas, readily acquired from satellite imagery, and household surveys for a limited number of residences may be combined with the algorithms developed in this paper to produce robust estimates of spatial population distributions.

Background

There are multiple applications for population data, including demographic profiling, the determination of disease incidence and prevalence, the mapping of access to health services, and the estimation of the number of refugees or displaced persons requiring emergency services. When combined with satellite imagery, survey data can facilitate the estimation of population size and population density.

Geospatial Information Systems (GIS) methods are increasingly being used for population estimation purposes. Wu et al. [7] distinguish between aerial interpolation methods that project census data onto a standardized population surface and statistical modeling methods that use socioeconomic variables (such as map layers for urbanicity, land use, built structures, and other physical or human geographic data) to calculate population estimates. Several research groups have used GIS to estimate population sizes. For example, the LandScan project, maintained by the Oak Ridge National Laboratory (ORNL), uses a dasymetric mapping technique that redistributes census-based population counts based on constraints imposed by geographic boundaries, the presence of physical structures, and/or public spaces, bodies of water, and other spatial constraints [8]. The Gridded Population of the World (GPW) and the derivative Global Rural-Urban Mapping Project (GRUMP), both distributed through NASAâ€s Socioeconomic Data and Applications Center (SEDAC), which is hosted by the Center for International Earth Science Information Network (CIESIN) at Columbia University have developed novel methods for estimating the population of urban areas, such as analyzing nighttime satellite images to map city footprints [9] (see [10,11] for applications).

In this paper, we used spatially-directed surveying to gather field data about households in neighborhoods of interest. Using ground truth information about building use and residential occupancy in conjunction with imagery-derived estimates of rooftop areas, we applied a population bootstrap algorithm to estimate the uncertainty of the estimated population as a function of survey sample size. The two estimators used are based on average occupancy per residential structure, or average rooftop area per person. Prior studies have reported findings using comparable methods. Checchi et al. [12] used an occupancy-based measure to estimate the number of displaced persons at 11 different sites. The population was estimated as the product of (1) the *number of structures* and (2) the average *occupancy per structure*, as inferred from published reports. The local population may be overestimated in situations where buildings and tents were not physically distinct from one another, or occupied structures are concealed by trees. Aminipouri [13] used rooftop extraction methods to estimate the total rooftop area for three different slum areas in Tanzania. The total population was estimated as the product of (1) the *total rooftop area* and (2) the *number of persons per rooftop area*, based on an assumption that the majority of dwellings were residential. We compare both estimation results with field data from a household census, and also attempt to quantify the relationship between the uncertainty of the population estimate and the survey size.

Methods

Survey Methodology and Dataset Development

Terminology. In Bo, Sierra Leone's second largest city, a "section" is a formal municipal area with defined boundaries. Inside a section a variety of building structure types may be observed in satellite imagery. We divided structure types into two major categories, residential and non-residential. Residential structures include all buildings where persons sleep at night. Small ancillary structures such as carports, outdoor kitchens, storage sheds, chicken coops, latrines, and small sheds used for micro-enterprise activities which are located on residential properties but not used for sleeping were excluded when we delineated residential structures and rooftop areas. A household was defined as a person or group of persons identifying themselves as a family unit and residing within one structure. Each residence comprises the sleeping space for one or more households. The non-residential category includes governmental, commercial, nonprofit organizational, and religious structures.

Survey Methods and Data Assembly. Our creation of a complete municipal map for Bo, Sierra Leone (central coordinates: 7.959°, −11.740°), using participatory GIS methods has been described elsewhere [14]. Long-term residents and city officials were consulted throughout the mapping and validation process to ensure the accuracy of section boundaries and other map features. Sections in Bo are divided both by natural topographies, such as marshy areas, and by man-made structures such as roads. The central area, which is older, has a more planned layout than the sections toward the edges of the city, where the growth was more informal. Figure 1A shows the satellite image of Bo with the boundaries of sections described in this study marked (red lines). Figure 1B shows the same sections in relation to the overall boundaries of Bo. For surveying, the first two sections (Kulanda Town and Njai Town) were selected for convenience due to proximity to the Mercy Hospital Research Laboratory (MHRL) building, which is located in Kulanda Town. There were 68 sections in Bo at the time of the survey. After surveying Kulanda Town and Njai Town, a random number generator was used to select 18 of the remaining 66 sections for inclusion in the expanded survey. In total, we surveyed 20 of the 68 sections. All field surveyors – MHRL staff and graduate students from Njala University – were residents of the city of Bo. Prior to beginning data collection, all surveyors and interviewers completed several days of training, including instruction on geographic data collection (determining Global Positioning System (GPS) coordinates with a handheld device, a Garmin GPSMAP 62 series), interviewing techniques, and research ethics and regulations. The two sections adjacent to MHRL were surveyed in April 2010. The remaining sections were surveyed between November 2010 and February 2011.

Contact with the households occurred in two stages. First, one adult representative male or female from each residential structure was interviewed about (1) the number of households in the residence, (2) the number of persons in the informant's household, and (3) the age and sex of each person in his or her household. Many residential structures were home to multiple households, even if the residence was a modestly sized structure. Households were usually related, but maintained some independence in function. For example, if two adult siblings were both living in the residence with their minor children, then they were generally reported as being two separate households. When two or more households lived within the same residence, a representative of each household was asked to provide information about the composition of his or her household. Second, the enumerator

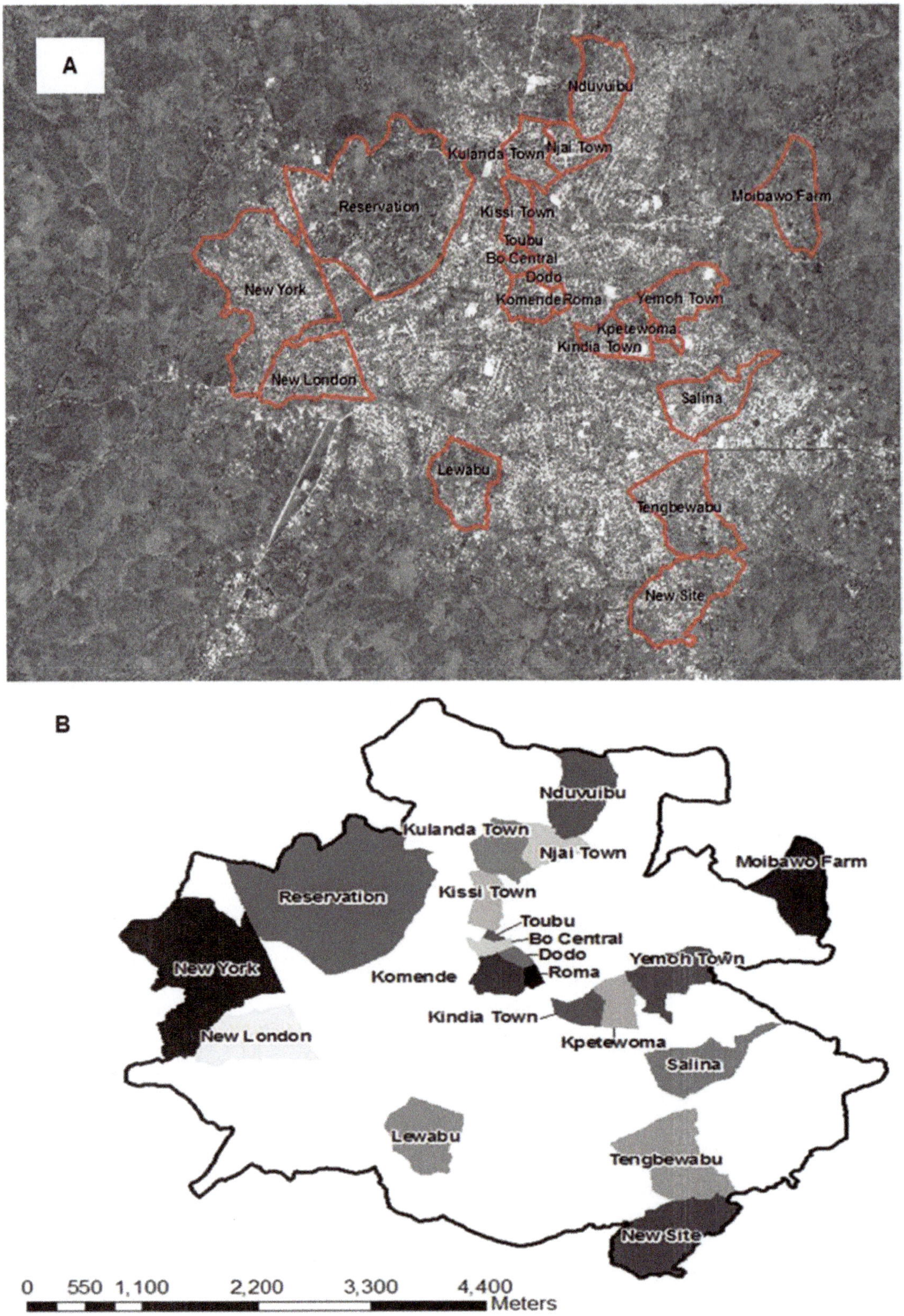

Figure 1. Aerial map of Bo city and municipal sections. (A) Aerial map of Bo, Sierra Leone showing municipal sections (red lines) described in this study and (B) Graphical representation of the same. (Image copyright 2010 DigitalGlobe NextView License. Used with permission.)

Table 1. Summary of survey data for 20 sections of Bo.

Section	Area (km^2)	Structures	Residential Structures*	Households*	Persons*	Population Density†
Bo Central	0.07	103	33	51	273	high
Dodo	0.05	88	26	85	597	high
Kindia Town	0.15	278	102	206	1,160	high
Kissi Town	0.20	287	154	400	2,490	high
Komende	0.20	258	56	175	1,103	high
Kpetewoma	0.20	105	46	94	640	medium
Kulanda Town	0.29	314	197	637	3,882	high
Lewabu	0.48	117	105	170	879	low
Moibawo Farm	0.50	43	17	22	135	low
Nduvuibu	0.49	343	205	439	2,552	medium
New London	0.60	495	208	498	2,873	medium
New Site South (New England)	0.69	194	136	190	1,248	medium
New York	1.51	605	116	176	1,088	low
Njai Town	0.22	269	127	388	2,298	medium
Reservation	2.33	252	66	86	637	low
Roma	0.04	52	4	22	139	high
Salina	0.47	231	59	110	580	low
Tengbewabu	0.68	233	136	185	1,068	low
Toubu	0.02	46	34	88	454	medium
Yemoh Town	0.40	284	152	289	1,858	medium
Total	—	4,597	1,979	4,311	25,954	—

Complete residence and household survey data for 20 municipal sections of Bo, showing the area, the total number of structures and the number of residential structures, households and persons per municipal section.
*From survey data collected from all occupied structures in the section.
†high density: $\geq$11,000 residents per km^2; medium density: 5,500 to 10,999 residents per km^2; low density: $<$5,500 residents per km^2.

sought to interview all mothers of any age who were living in the residence. Only the first of the two stages are relevant for the current analysis.

Handheld GPS units were used to obtain geographical coordinates of all residential structures in the 20 municipal sections. During this process, field staff distinguished between residential and non-residential structures [14]. For two sections where the types and dimensions of all structures (residential and non-residential) were quantized, field surveyors marked the coordinates of all non-ancillary structures and noted whether they were residential or non-residential. Satellite imagery with 0.5-meter resolution (Digital Globe, Satellite: WorldView-1, Image date: 7 February 2010) was used in GIS software, ArcGIS (ESRI, Inc. Redlands, CA) or Quantum GIS (QGIS; www.qgis.org), to digitize rooftops so that individual areas could be determined. The area for each rooftop was calculated only from the two-dimensional coordinates measurable from satellite imagery, irrespective of the total area that would be calculated using full three-dimensional coordinates.

Protection of Human Subjects. All data collection involving human subjects was approved by the institutional review boards of Njala University, George Mason University, and the U.S. Naval Research Laboratory. Written informed consent was obtained from each household representative who participated in the survey, all of whom were adults.

Table 2. The subset of Bo residences with known occupancies and rooftop areas.

Section	Residential Structures with Known Rooftop Areas	Total Residential Structures in Section	Persons in Residential Structures with Known Rooftop Areas
Bo Central	25	33	241
Komende	53	56	1,061
Kulanda Town	193	197	3,809
Njai Town	127	127	2,298
Reservation	66	66	637
Total	464	479	8,046

Subset of structures by neighborhood for which data were available for both the rooftop area and total number of residents in five municipal sections.

Table 3. Summary of the four datasets used for the simulation.

Dataset	***DS*01**	***DS*02**	***DS*03**	***DS*04**
Brief description	All residential structures in the 20 sections	DS01 subset: 5 sections with rooftop areas	Njai Town Composite (Mashup)	Reservation Composite (Mashup)
Total number of residential records (***Tr***)	1,979	464	127	66
Total number of non-residential records (***Tn***)	Not found	Not found	142	186
Total number of structures in dataset ($Ts = Tr + Tn$)	1,979	464	269	252
Total rooftop area (m^2) of residential structures (***Ar***)	Not found	83,605	21,924	16,857
Total rooftop area (m^2) of non-residential structures (***An***)	Not found	Not found	15,434	30,024
Total rooftop area (m^2) of residential and non-residential structures (***As***)	Not found	Not found	37,358	46,881
Number of sections in dataset	20	5	1	1
Total number of individuals	25,954	8,046	2,298	637
Total number of households	4,311	1,313	388	86
Average number of individuals per residential structure	13.1 ($\sigma = 9.1$)	17.3 ($\sigma = 11.7$)	18.1 ($\sigma = 10.4$)	9.7 ($\sigma = 4.7$)
Average rooftop area (m^2) per residential structure	Not found	180.2 ($\sigma = 88.7$)	172.6 ($\sigma = 66.9$)	255.4 ($\sigma = 4.7$)
Average rooftop area (m^2) per non-residential structure	Not found	Not found	108.7 ($\sigma = 70.6$)	161.4 ($\sigma = 91.7$)
Average residential rooftop area (m^2) per individual resident	Not found	16.7 ($\sigma = 18.2$)	13.7 ($\sigma = 13.0$)	32.2 ($\sigma = 22.8$)
Number of records with digitized rooftop areas	464 (5/20) sections	464 (of 479) records with rooftop areas	269	252

Summary of the four datasets used for the simulation.

Bo City Dataset Construction. Complete survey data were collected for 20 of the 68 sections (neighborhoods) in Bo (Table 1). This ground truth Dataset 01 (*DS01*) include survey records for 1,979 residential structures, representing a total population of 25,954 individuals. Five sections – Bo Central, Komende, Kulanda Town, Njai Town, and Reservation, which collectively are home to 479 residential structures and 8,046 individuals (*DS02*) – were selected for manual digitization of the rooftop areas for all of their residential structures (Table 2). For two sections, Njai Town (*DS03*) and Reservation (*DS04*), the rooftop areas of the non-residential structures were also digitized, and composite mashup datasets were constructed that contain records for both residential and non-residential structures (Table 3). Reservation, originally established as a housing area for government officials, was selected as an example of a section with large lots and a relatively high proportion of non-residential buildings. Njai Town was selected as an example of a densely populated residential neighborhood with few commercial structures. The differences between these two sections allowed us to better compare and evaluate population estimation approaches. Cross-matching of the residential rooftop areas to surveyed households was used to identify residential structures; the remaining structures were classified as non-residential. By definition, a non-residential record in *DS03* and *DS04* has a digitized rooftop area and the number of individuals and households is set to 0. Figure 2 summarizes the 4 datasets and all the simulations that were run.

Population Estimation Methods

Two Population Estimators. Using the GIS data summarized above and spatially-linked demographic data we collected in Bo, we are able to simulate the estimation of the population (and the associated uncertainty) as a function of survey size using two different population estimators:

1. An occupancy-based estimator, which is proportional to the product of (1) the average number of persons per residential structure and (2) the estimated total number of residential structures (*Tr*).
2. A rooftop area-based estimator, which is proportional to the product of (1) the average number of persons per m^2 of rooftop area and (2) the estimated total rooftop area of all of the residential structures (*Ar*).

For n surveyed residences of index i and number of individual persons IND_i in the ith residence, and the estimated total population $\hat{P}$ based on the total number of residences is given by:

$$Estimated\_Total\_Population = \hat{P} = \left(\sum_{i=1}^{n} \frac{IND_i}{n}\right) * Total\_Number \quad (1)$$

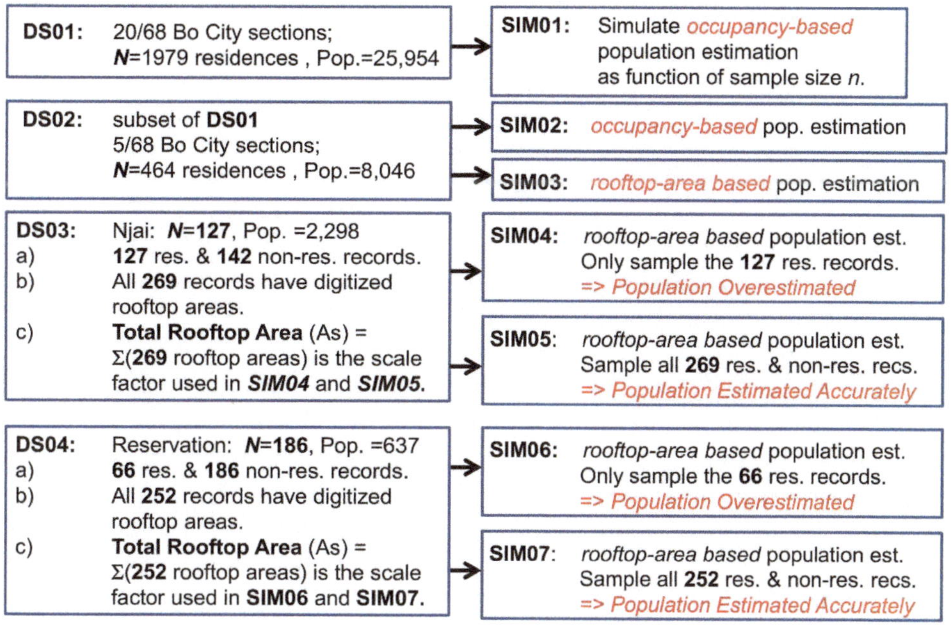

Figure 2. Flowchart illustrating datasets and simulations.

In Equation (1), *Total_Number* is the scale factor for the population estimator, and is either the *Total_Number* of residential structures, or the *Total_Number* of residential and non-residential structures. This scale factor is required because the total number of residence structures $Tr < Ts$, where Ts is the total number of residential plus non-residential structures. If only residential structures are included in the survey, then *Total_Number* = Tr. If the survey includes both residential and non-residential structures, then *Total_Number* should be set equal to Ts, rather than Tr.

Alternatively, the population may be estimated on the total area of the residence rooftops:

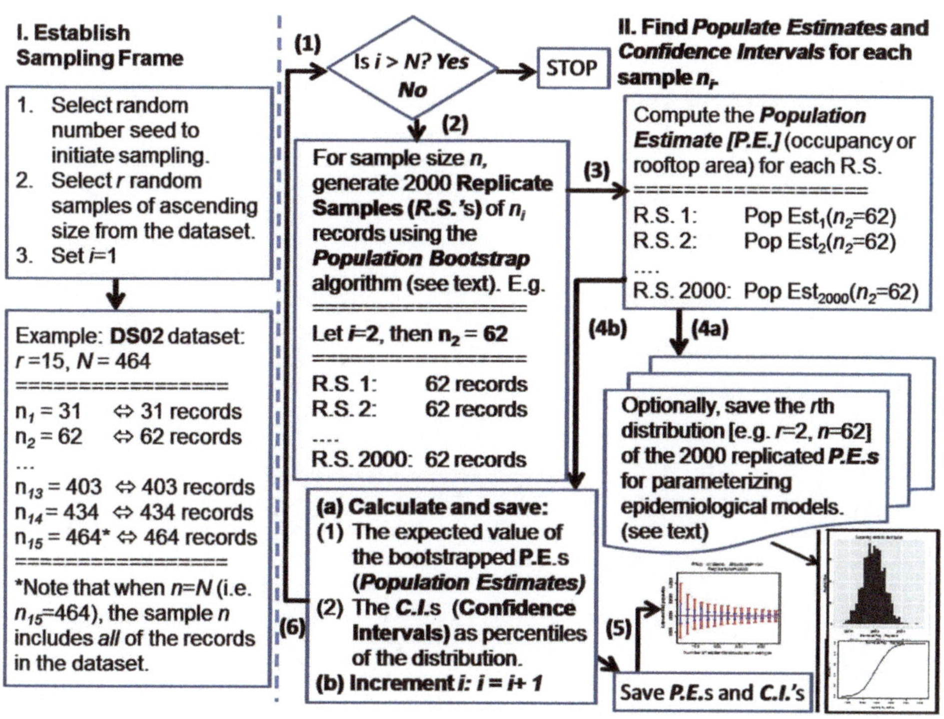

Figure 3. Flowchart for population bootstrap.

Table 4. Summary of the 6 simulations and parameters.

Simulation	*SIM01*	*SIM02*	*SIM03*	*SIM04*	*SIM05*	*SIM06*	*SIM07*
Number of records	1,979	464	464	269	269	252	252
Record type(s)	Residential records only	Residential records only (with rooftop areas)	Residential records only (with rooftop areas)	127 residential records	127 residential & 142 non-residential records	66 residential records	66 residential & 186 non-residential records
Number of Sections	20	5	5	1 (Njai Town)	1 (Njai Town)	1 (Reservation)	1 (Reservation)
Converges to correct population estimate?	Yes	Yes	Yes	No	Yes	No	Yes
Estimator used	Occupancy	Occupancy	Rooftop	Both	Both	Both	Both
Estimator scale factor used	$Tr = 1{,}979$ residential structures	$Tr = 464$ residential structures	$Ar = 83{,}605\ m^2$	$Ar = 21{,}924\ m^2$	$As = 37{,}358\ m^2$	$Ar = 16{,}857\ m^2$	$As = 46{,}881\ m^2$
Number, increment Δ, and *sequence* of the simulated survey samples	15 steps, $\Delta = 132$: [132, 264, ..., 1716, 1848, 1,979]	15 steps, $\Delta = 31$: [31, 62 ... 403, 434, 464]	15 steps, $\Delta = 31$: [31, 62 ... 403, 434, 464]	16 steps, $\Delta = 8$: [8, 16, ..., 112, 120, 127]	15 steps, $\Delta = 18$: [18, 36, ..., 234, 252, 269]	16 steps, $\Delta = 4$: [4, 8, ..., 56, 60, 66]	15 steps, $\Delta = 17$: [17, 34, ..., 221, 238, 252]

Summary of the six simulations run, including all of the scale factors used.

$$Estimated\_Total\_Population = \hat{P} = \left(\frac{\left(\sum_{i=1}^{n} IND_i\right)}{\left(\sum_{i=1}^{n} Res\_Area_i\right)} \right) * Total\_Area \quad (2)$$

In Equation (2), $Total\_Area$ is the scale factor for the population estimator. $Total\_Area$ is either the total rooftop area of the residential structures, or the $Total\_Area$ of the residential and non-residential structures. This scale factor is required because the total rooftop area of residence structures $Ar < As$, where As is the total rooftop area of (residential plus non-residential) structures. If only residential rooftop areas are included in the survey, then $Total\_Area = Ar$. If the survey includes both residential and non-residential structures, then $Total\_Area$ should be set to As rather than Ar.

A population bootstrap algorithm [15] (pages 93–94) is used to estimate the uncertainty of the two estimators as a function of survey size. We will demonstrate that the population of the pooled survey sections can be accurately estimated from small samples of survey participants, given reasonable values for specific parameters such as the total number of residential buildings or the total rooftop area of all residential buildings. The latter parameters can be partially derived from aerial imagery, but residential and non-residential structures cannot be readily distinguished using the imagery alone.

Figure 3 is a flowchart of the bootstrap simulation approach used. Two estimators are tested, the occupancy-based population estimator and the rooftop area-based population estimator. For a given dataset, and for each sample size n, 2,000 replicate bootstrap samples are created (step 2), using the population bootstrap algorithm. The desired population estimate is calculated for each replicate (step 3), along with the associated confidence intervals. The individual 2,000 bootstrap replicate samples themselves can be saved, if desired (step 4a). Alternative estimates of the population and the associated uncertainty can be made from the distributions of replicated samples, as required to parameterize external epidemiological models as a function of sample size. The expected bootstrap estimate of the population, and the associated confidence levels (0.50 and 0.95), are always calculated (step 4b) and saved (step 5). If surveys for all of the sample sizes have not been completed, this process will proceed through the next iteration (step 6).

The estimated total population is the mean value of the 2,000 individual population estimates [15]. The confidence levels (0.50 and 0.95) for the estimate of the total population were calculated as the equal-tailed two-sided bootstrap percentile intervals [16]. The basic bootstrap methodology is summarized in [17], and percentile methods are discussed in [15,18]. The specific samples selected from the pool of 1,979 residences were dictated by the choice of the initial random number seed used for the sampling function, as well as the sequence of contingent bootstrap replicates. Each of the 2,000 replications for a sample of size n requires the random selection of 2,000 samples, each of size n. A random number seed is a number, or vector of numbers, that is used to initialize the pseudorandom number generator used to randomly select the survey records in step 3 of Figure 2. This initial seed was always explicitly specified and saved, which facilitated the uniform replication of any of the simulations presented in this analysis. Changing the random number seed will result in a different sequence of bootstrap replicate samples, and the output from the bootstrap algorithm will vary, as will be discussed.

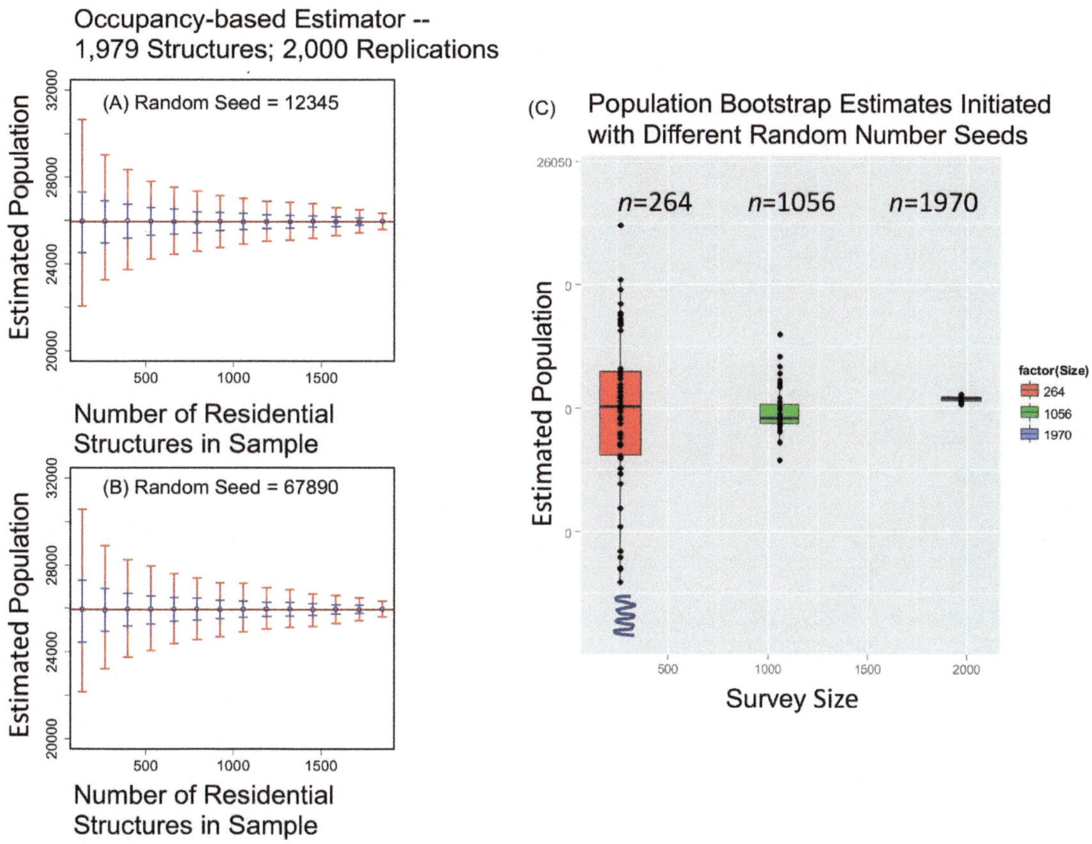

Figure 4. Occupancy-based bootstrap population estimations parameterized by sample size and random number seed. Results for (A) random number seed = 12345 and (B) random number seed = 67890. In both panels, the brown line is the total population count $P = 25{,}954$, the number of individuals counted in the 1,979 residences surveyed for *DS01*. The 0.50 CIs is shown in blue and the 0.95 CI in red. (C) A quartile boxplot shows the variation in the population bootstrap estimates $\hat{P}$ for 25 replicate samples initiated with different random number seeds.

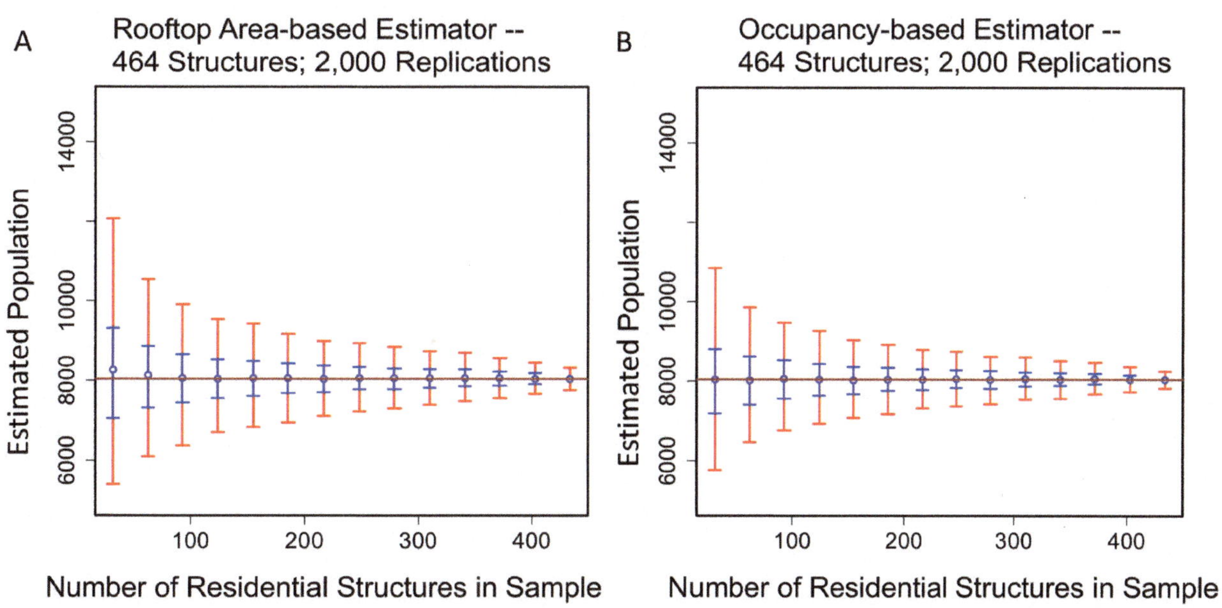

Figure 5. A comparison of occupancy-based and rooftop area-based bootstrap population estimators. (A) Occupancy-based population estimations and (B) rooftop area-based population estimations with 0.50 CIs (blue) and 0.95 CIs (red). There are 464 residences (*DS02*) and 2,000 bootstrap replications per sample. The confidence intervals (CIs) were calculated using bootstrap percentiles.

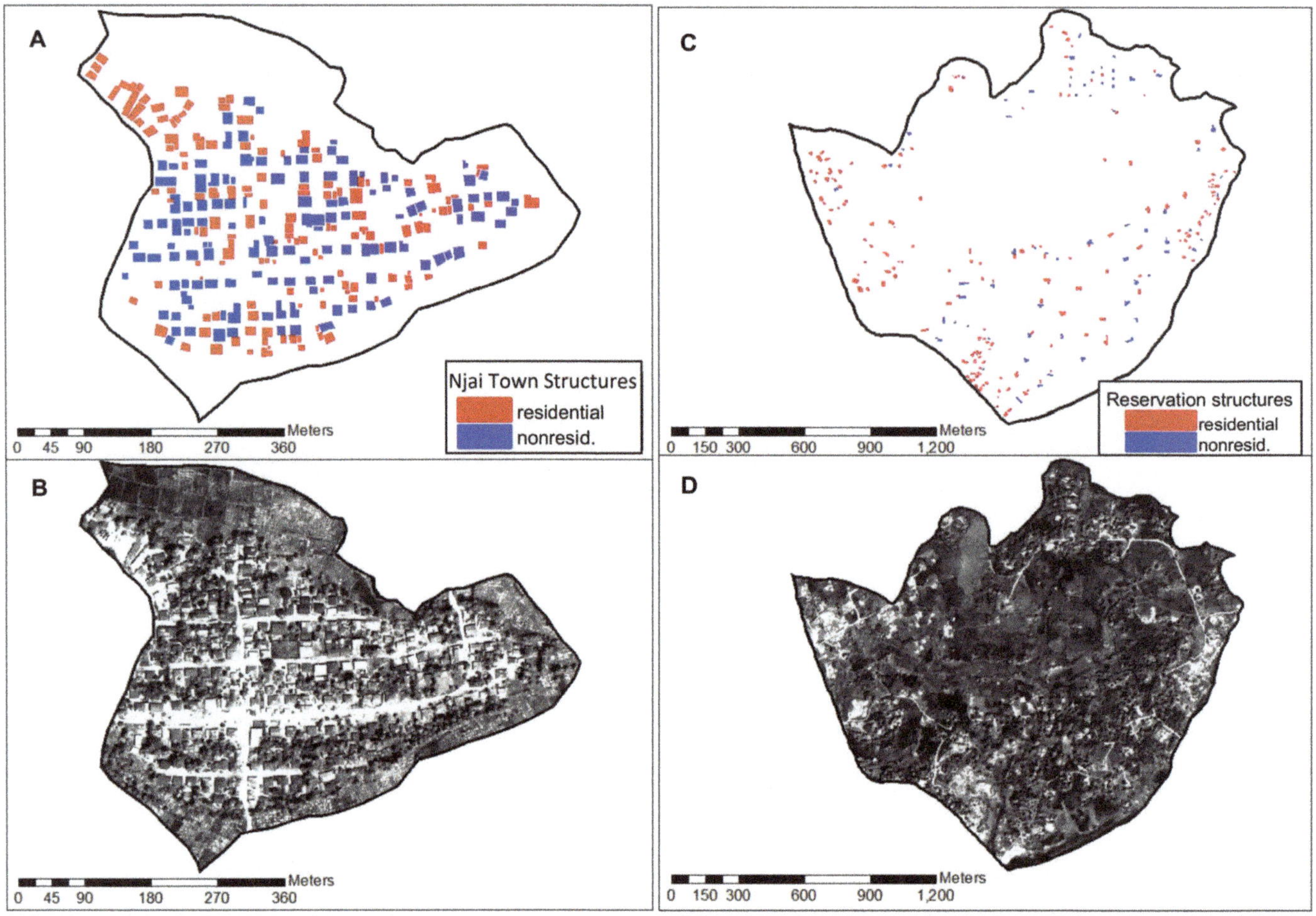

Figure 6. Residential and non-residential structures in Njai Town and Reservation. (A) Njai Town residence structures (blue) and non-residential structures (red), (B) Njai Town satellite imagery, (C) Reservation residential structures (blue) and non-residential structures (red), and (D) Reservation satellite imagery. When comparing either the maps or imagery for Njai Town and Reservation, note the differences in the map scales, which create the illusion that the structures in Reservation are smaller, although the average residential rooftop area in Reservation (255 m^2) is actually greater than that of Njai Town (172 m^2). The total land areas for Njai Town and Reservation are 0.21 km^2 and 2.33 km^2, respectively. (Image copyright 2010 DigitalGlobe NextView License. Used with permission.)

Table 3 shows the values for *Tr*, *Ts*, *Ar*, and *As* as obtained for the 20- and 5-section aggregates and for Njai Town and Reservation individual sections. If any non-residential areas or structures were identified in satellite imagery, they were excluded from consideration. For the 20 and 5 section aggregates, only data for *Tr* were collected. For both Njai Town and Reservation sections, *Tr*, *Ts*, *Ar*, and *As* were completely delineated.

The total rooftop area *As* can be estimated from aerial imagery without distinguishing between residential and non-residential structures. If the ratio of *Ar*/*As* can be estimated, perhaps by collaboratively analyzing the aerial imagery and/or participatory GIS maps with local residents, *Ar* can then be calculated. A similar calculation yields *Tr*:

$$A_r = (A_r/A_s) \times (A_s) \quad (3)$$

$$T_r = (T_r/T_s) \times (T_s) \quad (4)$$

The Population Bootstrap Algorithm. If we had a single sample *n*, drawn from a larger unknown population, then a conventional bootstrap could be used for estimating the uncertainty. In this case, we would repeatedly resample without replacement from *n*, and determine the expected value of the estimator on the distribution of replicated samples – 2,000 replicates, in our examples. In our current study, we are drawing ascending samples of size *n* from a previously measured finite population of size *N*. More specifically, we are resampling from a finite population of either 464 or 1,979 houses (*N*) by drawing increasing samples of size *n*. The conventional bootstrap algorithm would fail to properly account for the decrease in the uncertainty (confidence intervals) that occurs as the sampling fraction (*n*/*N*) becomes large, and the number of possible samples *n* that can be drawn from *N* decrease as C(*N*,*n*). The chapter on finite population sampling in Bootstrap Methods and Their Applications by Davison and Hinkley summarizes this issue succinctly: "The key difficulty with the ordinary bootstrap is that it involves with-replacement samples of size '*n*' and so does not capture the effect of the sampling fraction, which is to shrink the variance of an estimator." [15].

Davison and Hinkley suggest four methods to correct for this effect: (1) modifying the sample size to match the estimator variance (2) matching the original sample size while maintaining

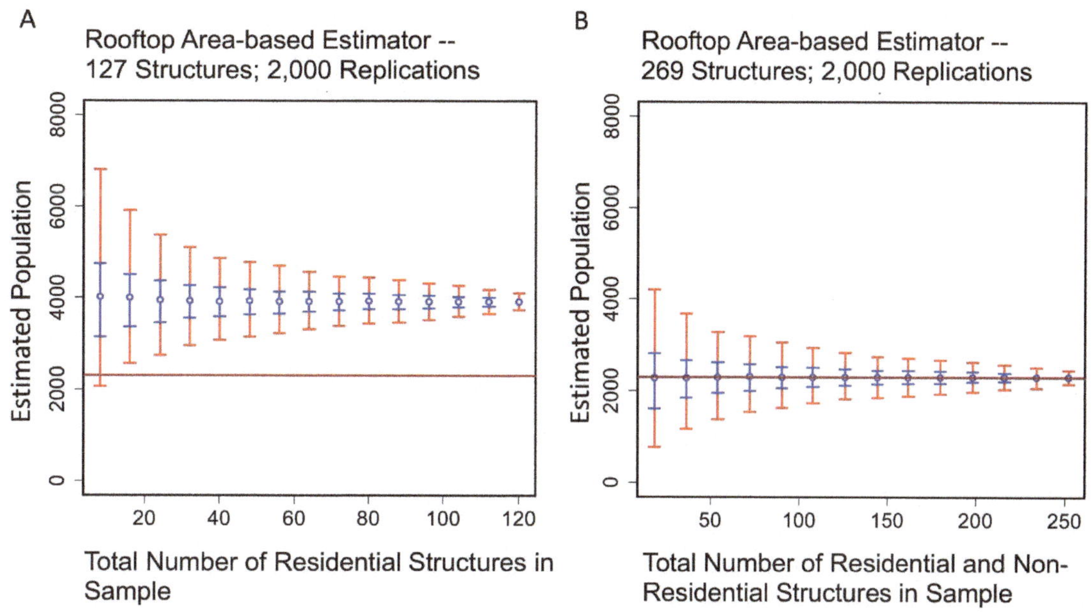

Figure 7. Rooftop area-based population estimations for Njai Town using invalid and valid scale factors. (A) Non-convergent rooftop area-based population estimate with 0.50 (blue) and 0.95 CIs (red). Samples were drawn from the Njai Town dataset of 127 residences only, but the scale factor *Total_Area* was set to the rooftop area for all 269 structures (*As*). (B) Convergent rooftop-based population estimate with 0.50 (red) and 0.95 (blue) CIs. Samples were drawn from the Njai Town dataset of 269 residential and non-residential structures, and *Total_Area* was set correctly set to *As*. Source: residential survey data and digitized rooftop areas, $As = 37{,}358\ m^2$.

the sampling fraction - the mirror-match bootstrap, (3) the superpopulation bootstrap, and (4) the population bootstrap. In a resampling analysis of longitudinal United States city data, the authors conclude (page 97) "that the population and super-population bootstraps are the best of those considered here" [15]. The population bootstrap is easier to implement and executes faster than the superpopulation bootstrap, and is the algorithm used here.

Let $R = 2{,}000$. For each of the R bootstrap replicates, execute the following steps:

1. Randomly select without replacement a sample y_n [$y_1, y_2, \ldots, y_n$] of n records from a dataset of size N. (For example, let $n = 264$ and $N = 1{,}979$.)
2. Create an empty vector Y^* of length N.
3. Copy $f = truncate(N/n)$ consecutive copies of y_n into Y^*, which now contains n*f records.

(a) for example, $f = truncate(1{,}979/264) = truncate(7.49) = 7$
(b) $n*f = 7*264 = 1{,}848$ records

4. Select $m = N\text{-}n*f$ records without replacement from y_n, and append these records to Y^*, which will now contain N records.

(a) $m = 1{,}979–1{,}848 = 131$ records
(b) $n*f+m = 1{,}848+131 = 1{,}979$ records $= N$

5. Select the population bootstrap replicate by randomly selecting n records without replacement from the N records in Y^*.
6. Repeat steps 1 to 5 to create R bootstrap replicates. In our examples, $R = 2{,}000$. Note that each of the 2,000 replicates has the same sampling fraction (n/N). The estimate is exact when $n = N$.

Results

Figure 2 and Table 4 summarize the simulations executed, and the datasets used.

1. *SIM01*. Figure 4AB shows the required convergence [19] of the occupancy-based population estimator as a function of the $N = 1{,}979$ residential structures in *DS01*. The variance (uncertainty) decreases uniformly as the sample size increases. The confidence intervals (CIs) are illustrated by the blue (0.50 CI) and red (0.95 CI) bars. The otherwise identical simulations were initiated with two different random seeds. Changing the random number seed will result in a different sequence of bootstrap replicate samples, and the estimated population $\hat{P}$ as a function of sample size n will vary. Using a different seed has little impact on the variance (CIs) but does result in modest variation in the estimated population $\hat{P}$ as a function of sample size, as shown in the quartile boxplots in Figure 4C. Note that the expanded Y-axis exaggerates the absolute variation relative to the measured value of the population.
2. *SIM02* and *SIM03*. The occupancy-based and rooftop-area based sequences of sample size simulations for dataset *DS02* ($N = 462$) are compared in Figure 5. Both estimators converge properly. The uncertainty of the occupancy-based estimator is about 20% less than for the rooftop-based estimator for both the 0.95 and 0.50 confidence intervals. For a given sample size n and confidence interval, the corresponding confidence interval is defined as the difference between the upper and lower confidence limits. For each estimator, and for each of the two confidence intervals, 14 confidence intervals were calculated. A two-tailed paired t test between the paired differences was statistically significant for both confidence levels ($p < 0.001$).

3. *SIM04* and *SIM05*. Figure 6 shows the imagery and map of the Njai Town and Reservation sections. Residential structures are denoted in red; non-residential structures are denoted in blue. Figure 7 compares the rooftop-area population estimator under two different scenarios, both using *DS03*. In SIM04 only the 127 residential records are sampled – that is, all records drawn have 1 or more individuals as residents – but the total rooftop area *As* is equal to the rooftop area of the combined residential and non-residential structures ($As = 37{,}358\ m^2$). This results in a total population estimate for Njai Town that is much greater than the true value. In *SIM05*, the population estimate draws on both residential and non-residential structures, so in some records the number of individuals is 0. This time the estimator converges correctly.

Thus, we have defined two different rooftop area-based protocols that can be used to estimate the population, with analogous protocols defined for the occupancy-based simulation:

1. Include both residential and nonresidential structures in the n records sampled for each survey size, using the scale factor *As*, knowing that many of the sampled records will have a population of 0.
2. Use the scale factor *Ar*, where $Ar < As$, and sample residential records only, as was done in *SIM02*.

Both methods will converge to the correct population estimate. For brevity, additional results are not graphed, but comparable findings were demonstrated using the occupancy-based population estimator with both the Njai Town and Reservation datasets *DS04* (simulations *SIM06* and *SIM07*).

Discussion

Occupancy-Based Estimation vs. Rooftop Area-Based Estimation

Based on the uncertainty functions, the occupancy-based population estimator is to be preferred. Both the occupancy-based and rooftop-area based estimators converge properly, with decreasing uncertainty as the sample size increases.

Determining which estimator also poses pragmatic considerations, apart from the relative technical merits of the estimators. Consider the following three scenarios:

1. Assume that high-resolution imagery is available, and automated procedures can be applied to extract rooftops and estimate their total area *As*. In this scenario the survey data can be represented spatially and coded to the location of the surveyed structures, if required. (We have not attempted to do so here, as mapping the survey data onto specific residential structures would violate our confidentiality agreement with participants.) If the survey data include both residential and non-nonresidential records, the population can be estimated as shown in Figure 5, using either the occupancy-based or rooftop-area based estimator. If the survey data are only for residential structures, either *Ar* or *Tr* must be estimated for the scale factor (Table 3), perhaps as a percentage of the measured values for *As* or *Ts*.
2. Assume that the individual residential structures can be resolved and counted automatically in the aerial imagery, but the rooftop areas cannot be successfully extracted. In this case, only the occupancy-based population estimator can be used, because the residential rooftop area *Ar* cannot be estimated.
3. Assume that individual structures cannot be resolved in the imagery, but it is possible to estimate the total rooftop area *As* in the region of interest. This may occur when using low-resolution imagery in a densely populated area. In this scenario, the rooftop-area based estimator alone is applicable, provided that either (1) the survey data include proportional numbers of residential and non-residential records, and *As* is the correct scale factor or (2) the survey data include residential records only, but *Ar* can be estimated as a percentage of the measured total rooftop area *As*.

Delineating Structures and Rooftop Areas. In practice, the total rooftop area and/or the total number of individual residences within a large area will likely be extracted from aerial imagery using automated methods, rather than manually, as was done here. Relevant methods for extracting buildings from aerial imagery are described in numerous publications (for example, see [13,20,21]). Object based image analysis (OBIA) provides a particularly powerful set of methods; see [22] for a review of OBIA and a discussion of available commercial tools such as eCognition [23]. Given our current data, it would be instructive to use a commercial program to extract rooftop counts and areas from satellite imagery of Bo, and to compare these estimates with the rooftop areas collected in our limited ground truth datasets. It may also be easier, particularly if only relatively low-resolution imagery is available, to estimate the aggregate rooftop area rather than to extract the rooftop areas of individual structures.

Correcting for Non-Residential Structures. As discussed previously, if data are not available for non-residential structures, it may necessary to estimate either *Tr* or *Ar* from *Ts* or *As*, because only the latter two parameters can be directly estimated from the aerial imagery. In some places it might be possible to estimate the ratio of residential to non-residential structures based on relative locations or other distinguishing features. Very small structures – like ones that are latrine (outhouse) sized – can also be removed from consideration. Census data or published business listings could augment this analysis, if available. As a specific example, Aminipouri [13] identified Tanzanian slum areas, and used rooftop extraction to estimate the total rooftop area, assuming that the majority of dwellings were residential. This approach is likely to be invalid in urban or periurban areas, where buildings may include both family dwellings and non-residential structures, or in rural areas that may have many farm structures per residence.

Figure 6 shows the locations of the residential and non-residential structures in Njai Town and Reservation sections. In both sections, the non-residential structures may include ancillary structures that are on the same property as the residences. Without using auxiliary survey data, it would be difficult to distinguish between the two commingled classes of structures from imagery alone. The mean rooftop area for structures in Njai Town is significantly less than the mean rooftop area for the residential structures alone. The mean rooftop area in Njai Town was 172 m^2 ($\sigma = 67\ m^2$) and the mean non-residential rooftop area was 108 m^2 ($\sigma = 71\ m^2$). In Reservation, these figures were 255 m^2 ($\sigma = 126\ m^2$) and 161m^2 ($\sigma = 92\ m^2$), respectively. These distributions are consistent with our observations that the non-residential structures were often small sheds or outdoor cooking facilities. However, some outbuildings are "residence" sized, so it is not possible to discriminate between residential and non-residential buildings based on rooftop area alone.

The difficulty of estimating the ratio of residential to non-residential buildings or rooftop areas is non-trivial. Based on our analysis of the Njai Town and Reservation datasets, this cannot be accomplished based on simple image analysis, or based on the statistical differences between the rooftop areas of the two groups. One strategy is to exploit expert local knowledge with respect to the distribution of residential and non-residential structures. In this study, having identified all of the residential buildings surveyed in spatially-guided population survey, we were able to estimate both the number (*Ts-Tr*) and total rooftop area (*As-Ar*) of the remaining non-residential buildings in Njai Town and Reservation in the aerial imagery. Participatory GIS (PGIS) is an incredibly valuable tool for the identification of non-residential structures such as markets, businesses, schools, hospitals, nonprofit organizations, and government buildings. Optimally, PGIS can be used in conjunction with analyses of vegetation and other GIS layers predict and validate residential and non-residential zones in other urban and periurban contexts.

Parameter Estimation for Epidemiological Modeling. Given the range and diversity of epidemiological models, it is not possible to determine what minimal sample size is sufficient for all applications. What the present approach enables is the estimated population size as a function of *n*, with confidence intervals of a desired range. For a cross-sectional or cohort study this set of denominator parameters may be sufficient for study design purposes. Each population estimate is the expected value of 2,000 population bootstrap replicates (see Figure 3). (In practice, 200 bootstrap replications are generally considered to be sufficient, but generating 2,000 replicates guarantees the stability of the upper and lower 2.5 percentiles used to estimate the limits of the 0.95 CI [17].) The 2,000 population bootstrap estimates can also be saved for each sample size *n* (Figure 3, step 4a). This provides a number of advantages: Alternative measures of uncertainty can be calculated for each sample size distribution as required, rather than using the percentile confidence intervals. The cumulative distribution function can also be calculated if required from any of the distributions for a given size sample *n* (Figure 3, step 4a). The set of population bootstrap replicate distributions can then be sampled repeatedly to generate the sequence of Monte Carlo inputs required for *Population-Based Microsimulation* (PMS) models (also known as agent-based models), an approach recommended by Sharif et al. [24].

An additional consideration is that the choice of the initial random number seeds will also influence the distribution of the parameters generated in the empirical space [25]. This is demonstrated in Figure 4. Changing the random seed had little impact on the uncertainty, but did cause modest variations in the expected value of the population bootstrap for a given sample size *n*. For PMS models, it is probably best practice to generate and sample multiple distributions of the population bootstrap estimation for a sample of size *n*, each with 2,000 replicates and a different initial random number seed. To achieve even sampling throughout a complex parameter space, Sharif suggests using a Latin hypercube sampling algorithm [24].

Conclusions

Several studies have demonstrated spatial population density estimation using only satellite sensor data [26], acknowledging the limitations of those approaches and recommending combining household data with high resolution remote sensing information [27]. In 2009, no complete municipal maps existed for Bo, Sierra Leone, despite the existence of 68 recognized sections, the boundaries of which were well-known to local populations. In order to establish spatial epidemiology capabilities in Bo, we took an approach from the onset that combines satellite imagery and GIS with local knowledge [14]. The data used in this study came from a larger field survey aimed at obtaining community health data from the 20 municipal sections described in this paper [28–30]. Given the large ground truth dataset afforded by the field survey, we simulated the collection of survey data using samples of varying sizes and a bootstrap population algorithm. Occupancy-based and rooftop-area-per-person-based total population estimators converged correctly on the measured population as the simulated sample size *n* approached the size *N* of the dataset. The uncertainty estimates may be useful designing a resampling study to update the population estimate, while reducing the size of the total number of samples to be collected. These specific findings cannot not be extrapolated to a different set of sections without the making the strong assumption that the new area is demographically comparable with the original survey area, in the sense of having the same underlying distributions of occupants/residential structure. Another advantage of the population bootstrap is that the distributions of the bootstrap population estimates can be saved for each sample of size *n*, and subsequently used to estimate alternative estimates of uncertainty.

The choice of which estimator to use, occupancy based or rooftop area based, will depend upon factors discussed earlier, including (1) the apparent reduction in uncertainty achieved by the occupancy-based estimator to the rooftop area-based estimator; (2) the relative difficulty of estimating the required scale factors (*Ar*, *Tr*, *As*, or *Ts*); and (3) the composition of the survey data.

If only residential buildings are surveyed, as might be dictated by schedule or financing, either *Ar* or *Tr* must be estimated. In this simulation study, both of the residential scale factors, *Ar* and *Tr*, were known for datasets *DS02*, *DS03*, and *DS04*. If both residential and non-residential buildings are surveyed, and represented proportionally in each simulated survey of size *n*, only *As* or *Ts* are required as scale factors. *As* and *Ts* can potentially be inferred directly from aerial imagery of the total area, because it is not necessary to distinguish between residential and non-residential structures.

In this study, we have shown that it is possible to ascertain the appropriate sampling and scale factors by combining local knowledge, satellite imagery, GIS, and surveys of a limited number of households in order to produce robust estimates of total populations. For the type of resource-challenged environment represented in this study, the process consists of the following steps:

1. Visual examination of satellite imagery by persons having local knowledge and removal from consideration of any large non-residential areas (such as commercial or governmental buildings) in the GIS.
2. GIS-directed household surveying of a portion of the residences in a section needed to achieve a desired confidence interval for population estimates.
3. Calculation of the total number of residential structures *Tr*, or the total residential rooftop area *Ar*, to deduce appropriate scale factors (Table 3). Alternatively, if the survey spatially encoded both residential and non-residential structures, the total number of structures *Ts* and/or the total rooftop area *As* provide the scale factors.
4. Use of the scale factors, along with datasets of household surveys, to estimate the population using population bootstrap statistical methods, including the uncertainty in the population estimate as a function of simulated sample size.

We anticipate that refinements of the approach specific to a geographical area will be necessary when extrapolating from an area comprised almost exclusively of single-level residential structures to an area with multiple level ones, as well as those having other unique characteristics. Spatial distributions of population strata, while not estimated here, are a logical extension of total population estimates, and are made possible by defining such strata in the household survey step.

Another potential application of this approach is for augmenting population estimates derived from satellite imagery alone. In [26], Harvey develops regression models for population estimation that aggregate several different remote sensing indicators, but do not use survey data or rooftop extractions directly. The population space is tested against Australian census data. As noted in [26]:

> There was also evidence of estimation bias associated with the somewhat higher population densities in the secondary study area. *This seems to be related to smaller lot sizes and a higher spatial concentration of separate houses, rather than to substantial differences in the distribution of other types of residential structure*.... It is concluded that the potential of this methodology is limited by heterogeneity of both land cover and population density within the individual CDs [collection districts], and that further improvements are in principle unlikely using this approach. *In particular, the sacrifice of detailed spatial information leaves no way to respond to the problem of over-estimation of population in large areas of low density*. (page 2093, italics added)

The first comment emphasizes the importance of collecting local survey data. Although we have not addressed housing density, given the number of resident structures per section and the areas of each section, the density of residential housing can be readily calculated. The second comment is especially applicable to areas like Reservation, where there are a large number of commercial and government buildings and relatively few residential structures. If the residential and non-residential structures can be differentiated in the imagery, or resolved by surveying the area, the scale factors for both population estimators can be estimated. In summary, it appears that some of the biases in population estimates based on satellite imagery alone could be addressed by analyzing supplementary survey data and/or structural-level imagery.

The ability to quickly estimate total population size with reasonable precision based on satellite or aerial imagery, digitization of rooftops, and demographic surveys of a limited sample of households has important applications for epidemiology and public health, the social sciences, and emergency response. For example, in disaster areas the methods described in this paper could be used to identify permanent and temporary shelters from aerial photographs, guide enumerators to a randomly sampled subset of dwellings, and use those household survey results to estimate the total number of affected persons and to identify resource needs. This critical and timely information about affected communities could contribute to improved delivery of goods and services and facilitate better coordination among response agencies, and might ultimately lead to saved lives. Widespread adoption of these techniques in support of relief activities, community development, and social research will not be possible without improved (and less expensive) digitization software, but the field tests we conducted in Sierra Leone suggest that a process of mapping, sampling, surveying, and bootstrapping can be successfully implemented in low-resource settings to generate accurate population estimates.

Acknowledgments

Disclaimer

The views expressed herein are those of the individual authors and do not reflect views of the Department of the Navy or the Department of Defense.

Author Contributions

Conceived and designed the experiments: RH. Performed the experiments: RH. Analyzed the data: RH JDA. Contributed reagents/materials/analysis tools: RA ASB UB JML KHJ APM DAS. Wrote the paper: RH JDA RA KHJ DAS.

References

1. Viel JF, Tran A (2009) Estimating denominators satellite-based population estimates at a fine spatial resolution in a European urban area. Epidemiology 20: 214–222.
2. Gomez-Elipe A, Otero A, van Herp M, Aguirre-Jaime A (2007) Forecasting malaria incidence based on monthly case reports and environmental factors in Karuzi, Burundi, 1997–2003. Malar J 6: 129.
3. Hethcote HW (2000) The mathematics of infectious diseases. SIAM Rev Soc Ind Appl Math 42: 599–653.
4. Katriel G, Yaari R, Huppert A, Roll U, Stone L (2011) Modelling the initial phase of an epidemic using incidence and infection network data: 2009 H1N1 pandemic in Israel as a case study. J R Soc Interface 8: 856–867.
5. White LF, Pagano M (2008) A likelihood-based method for real-time estimation of the serial interval and reproductive number of an epidemic. Stat Med 27: 2999–3016.
6. Archer B, Tempia S, White L, Pagano M, Cohen C (2012) Reproductive number and serial interval of the first wave of influenza A(H1N1)pdm09 virus in South Africa. PLoS One 7: e49482.
7. Wu S, Qiu X, Wan (2005) Population estimation methods in GIS and remote sensing: A review. GIScience Remote Sensing 45: 80–96.
8. Bhaduri B, Bright E, Coleman P, Urban M (2007) Landscan USA: a high-resolution geospatial and temporal modeling approach for population distribution and dynamics. GeoJournal 69: 103–117.
9. Balk D, Deichmann U, Yetman G, Pozzi F, Hay S, et al. (2006) Determining global population distribution: Methods, applications and data. Adv Parasitol 62: 119–156.
10. Hay SI, Guerra C, Gething P, Patil A, Tatem A, et al. (2009) A world malaria map: *Plasmodium falciparum* endemicity in 2007. PLoS Med 6: e1000048.
11. Magalhães RJS, Clements ACA, Patil AP, Gething PW, Brooker S (2011) The applications of model-based geostatistics in helminth epidemiology and control. Adv Parasitol 74: 267–296.
12. Checchi F, Stewart B, Palmer J, Grundy C (2013) Validity and feasibility of a satellite imagery-based method for rapid estimation of displaced populations. Int J Health Geogr 12: 12.
13. Aminipouri M, Sliuzas R, Kuffer M (2009) Object-oriented analysis of very high resolution orthophotos for estimating the population of slum areas, case of Dar-Es-Salaam, Tanzania. ISPRS XXXVII-1-4-7/W5.
14. Ansumana R, Malanoski AP, Bockarie AS, Sundufu AJ, Jimmy DH, et al. (2010) Enabling methods for community health mapping in developing countries. Int J Health Geogr 9: 56.
15. Davison AC, Hinkley DV (1997) Bootstrap Methods and Their Application. Cambridge University Press, 582 pp.
16. Ripley B, Canty A (2013) boot: Bootstrap R (S-Plus) Functions - R package, 1.3-9 edition. Available: http://cran.r-project.org/web/packages/boot/index.html. Accessed 2013 October 11.
17. Efron B, Tibshirani RJ (1993) An Introduction to the Bootstrap. New York: Chapman & Hall, 436 pp.
18. Wildman J, Hollingsworth B (2002) On generating confidence intervals for nonparametric methods.
19. Stine R (1989) An introduction to bootstrap methods: Examples and ideas. Sociol Method Res 18: 243–291.
20. Collins R, Hanson A, Riseman E, Schultz H (1995) Automatic extraction of buildings and terrain from aerial images. Ascona, Switzerland: Proc. Ascona Workshop, Automatic Extraction of Man-Made Objects from Aerial and Space Images, pp. 169–178.
21. Xiao Y, Lim SK, Tan TS, Tay SC (2004) Feature extraction using very high resolution satellite imagery. In: Geoscience and Remote Sensing Symposium, 2004. IGARSS '04. Proceedings. 2004 IEEE International.
22. Blaschke T (2010) Object based image analysis for remote sensing (review article). ISPRS J Photogramm Remote Sens 65: 2–16.
23. Wong TH, Mansor S, Mispan M, Ahmad N, Sulaiman W (2003) Feature extraction based on object-oriented analysis. In: Proceedings of ATC 2003 Conference.
24. Sharif B, Kopec JA, Wong H, Finès P, Sayre EC, et al. (2012) Uncertainty analysis in population-based disease microsimulation models. Epidemiol Res Int 2012: 14.

25. Wolf DA (2001) The role of microsimulation in longitudinal data analysis. Can Stud Popul 28: 313–339.
26. Harvey JT (2002) Estimating census district populations from satellite imagery: Some approaches and limitations. Int J Remote Sens 23: 2071–2095.
27. Wu C, Murray AT (2007) Population estimation using landsat enhanced thematic mapper imagery. Geogr Anal 39: 26–43.
28. Jacobsen KH, Abdirahman HA, Ansumana R, Bockarie AS, Bangura U, et al. (2012) Home birth and hospital birth trends in Bo, Sierra Leone. Acta Obstet Gynecol Scand 91(6): 750–753.
29. Jacobsen KH, Ansumana R, Abdirahman HA, Bockarie AS, Bangura U, et al. (2012) Considerations in the selection of healthcare providers for mothers and children in Bo, Sierra Leone: Reputation, cost, and location. Int Health 4(4): 307–313.
30. Ansumana R, Jacobsen KH, Gbakima AA, Hodges MH, Lamin JM, et al. (2013) Presumptive self-diagnosis of malaria and other febrile illnesses in Sierra Leone. Pan Afr Med J 15: 34.

16

Method Paper – Distance and Travel Time to Casualty Clinics in Norway Based on Crowdsourced Postcode Coordinates: A Comparison with Other Methods

Guttorm Raknes[1]*, Steinar Hunskaar[1,2]

1 National Centre for Emergency Primary Health Care, Uni Health, Uni Research, Bergen, Norway, **2** Department of Global Public Health and Primary Care, University of Bergen, Bergen, Norway

Abstract

We describe a method that uses crowdsourced postcode coordinates and Google maps to estimate average distance and travel time for inhabitants of a municipality to a casualty clinic in Norway. The new method was compared with methods based on population centroids, median distance and town hall location, and we used it to examine how distance affects the utilisation of out-of-hours primary care services. At short distances our method showed good correlation with mean travel time and distance. The utilisation of out-of-hours services correlated with postcode based distances similar to previous research. The results show that our method is a reliable and useful tool for estimating average travel distances and travel times.

Editor: Sergio Gómez, Universitat Rovira i Virgili, Spain

Funding: The authors have no support or funding to report.

Competing Interests: The authors have declared that no competing interests exist.

* E-mail: guttorm.raknes@uni.no

Introduction

Travel distance is a major barrier to the utilisation of all kinds of health services. We have previously shown that increasing distance from the population centroid to a casualty clinic is strongly associated with decreasing contact and consultation rates in a Norwegian out-of-hours district. The association was almost completely independent from important demographic and socio-economic factors registered [1].

Often no data on the exact location of individual patients are available. To examine the effect of geography on the utilisation of health services, several methods have been used to estimate the average travel distance for the entire population of an area. It has previously been shown that estimates of distance to the nearest cancer centre from population centroid is superior to estimates based on geometric centroid and population polygon methods when the exact addresses are unknown [2].

E.g. if a casualty clinic for a population of one municipality is located in another municipality, the distance from the population centroid is fairly representative for the average travel distance. In municipalities hosting their own casualty clinic, however, the population centroid often almost coincides with the location of the clinic. In these cases the distance between the population centroid and the casualty clinic is an underestimation of the travel distance for the average inhabitant.

In order to get more representative measures on distance and travel time at short distance between the casualty clinic and the population centroid of a municipality, we have developed a method that uses postcode coordinates that have been determined by crowdsourcing. The data are freely available for everyone, and since our method uses Google maps, there is no need for complicated and expensive geographic information system (GIS) software. In Norway, the alternative is to get estimates on population centroids or mean distances and travel times from Statistics Norway, the official statistics bureau. This is an expensive and time-consuming process.

Crowdsourcing is an online, distributed problem-solving and production model that involves outsourcing of tasks to a distributed group of people [3]. The contribution from thousands of voluntary people has been important for the development of geographic information systems (GIS) and related online resources [4].

In this paper we describe the method and compare it with other ways of calculating distance and travel time to a casualty clinic when the exact address is unknown. We also apply our method to examine whether distances and travel times are associated with the utilisation of out-of-hours primary care services.

Methods

Ethics Statements

The data used in this study are from three sources. Estimates on mean distance, travel time and population centroids of the municipality are not freely available, and can only be obtained from Statistics Norway for a fee. The postcode database is freely distributed under the terms of the Creative Commons Attribution License 3.0 [5], and is open to all from its website [6]. This database does only contain data on post/ZIP-codes. The Watchtower project database is owned and managed by our institution, the National Centre for Emergency Primary Health Care. The database is not publicly and freely available. These data are anonymised, patient identity is not recorded at any time.

Table 1. Quality of postcode based coordinates by a five point quality scale.

Status (scale value)	N
A. Actual post office location	1 633
B. Coordinate corrected by head of project	1 383
C. Coordinate controlled by a volunteer contributor	1 093
D. Coordinate corrected by a volunteer contributor	685
E. Probably misplaced	0

The Watchtower project is approved by the Regional Committee for Medical and Health Research Ethics and by the Norwegian Social Science Data Services, the Data Protection Official for Research.

Crowdsourcing Process

In 2009 a joint weather service [7] owned by the Norwegian Meteorological Institute and the Norwegian Broadcasting (NRK), wanted to provide weather forecasts for each of the 4 585 Norwegian postcodes. At that time, maps from Norway Post (Posten Norge) displaying the area covered by each postcode were available. Based on these data coordinates for the geometric centroids could be calculated. In a sparsely populated country like Norway, however, these coordinates are in many cases not representative for where people actually live.

To get more representative postcode coordinates, a crowdsourcing project was launched in July 2009. Volunteers were recruited using NRKbeta, the technology website of the Norwegian Broadcasting [8]. The head of the project, Erik Bolstad (EB), created a Google map with approximate locations of each postcode based on Norway Post's downloadable postcode registry. Contributors were able to add more accurate coordinates directly by clicking in the map. Updated coordinates were received by Google forms and transferred to a Google spreadsheet. EB checked all submitted coordinates before updating the project webpage. Before the project started 39% of postcodes were located. After two days, this increased to 74%. Within the first month 4 590 feedbacks from approximately 600 persons were submitted. The capacity of the registration system was then exceeded due to the large amount of information. EB then had to find the location of the remaining 600 postcodes. A five point scale was used to assess the data quality of each location (Table 1). Since the formal termination of the project, the project pages have been open for volunteers to improve the database further. The database also contains data from Statistics Norway on the number of people belonging to each postcode [6].

Casualty Clinics

In Norway, casualty clinics ("legevakt") are emergency primary care centres that handle all kinds of medical out-of-hours inquiries, including life-threatening incidents [9]. Some municipalities have their own casualty clinic (monomunicipal), but casualty clinics serving the population of two or more municipalities (intermunicipal) are also common. Of the 203 out-of-hours districts in Norway in 2012, 84 were monomunicipal and 88 were intermunicipal. The remaining districts were partially monomunicipal [10].

We used the postcode database to estimate average travel distance to the casualty clinic for 18 municipalities participating in the "Watchtower project", a database of a Norwegian sentinel network for monitoring emergency primary health care activity in Norway. The database was established by the National Centre for Emergency Primary Health Care in 2007. The Watchtowers are seven out-of-hours districts that have been selected to be representative for the entire primary health care emergency service in Norway. The development and implementation of this database have been described in detail elsewhere [9].

Calculation of Postcode Based Travel Time and Distance

The travel distance and travel time between each postcode coordinate within the Watchtower municipalities and the casualty clinic was calculated using Google maps [11]. This was then multiplied by the number of inhabitants belonging to the postcode. The average distance to the casualty clinic was then calculated by dividing the sum of person-kilometres or person-minutes by total number of inhabitants in the municipality. Instead of using GIS software to calculate distances and travel times, Google maps was chosen primarily because it is freely available to all. It is also easy for people without knowledge of GIS to calculate distances and travel times since the postcode database website has integrated Google maps functionality.

Validation of the Method

The postcode based distances to casualty clinics for each municipality were compared with mean and median travel distances and travel times and distance from the population centroid of the municipality.

Mean distance and travel time of each municipality to the casualty clinic as calculated by Statistics Norway was considered the main basis of comparison for our method. Statistics Norway used Excel to manually calculate mean and median distance and travel time from all "address points" in the municipality, taking number of persons living at each address point into account. These data were taken from Statistics Norway's Address Points database January 1st 2012. This not freely available database contains the exact location of all addresses in Norway. Population centroids were calculated with the mean centre function in ArcInfo (ArcInfo, ESRI, Redlands, CA). Distances and travel times from population centroids were estimated using OD cost Matrix in ArcInfo. These estimates were commissioned from Statistics Norway for a fee. In addition, we made comparisons with distances and travel times between casualty clinic and the town halls of the municipalities, calculated by use of Google maps.

Comparisons of average absolute deviation from mean distance as calculated by Statistics Norway for each method were performed between mono- and intermunicipal out-of-hours districts, and between municipalities hosting a casualty clinic or not.

Postcode Based Distance and Utilisation of Out-of-hours Services

We used the postcode based distances to examine the association between distance to casualty clinic and the utilisation of out-of-hours services. One municipality (Austevoll) was excluded from these analyses because the out-of-hours service uses more than one casualty clinic. Similar analyses on population centroid based distances have previously been performed by our group in the ten Watchtower municipalities comprising one specific out-of-hours district. These ten municipalities were among the 17 municipalities that were examined in the present study [1].

Utilisation was defined as rate of all first contacts with the communication centre of the out-of-hours district and rates of

Table 2. Baseline data, Watchtower municipalities, sequence by mean distance to casualty clinic.

Municipality	Mean distance to casualty clinic (kilometres)	Hosting a casualty clinic?	Part of intermunicipal out-of-hours district?	Population 2011	Primary care doctors per 10 000 (2011)	Population >80 years (2011, %)
Åsnes	6.1	Yes	Yes	7 597	11.8	8.3
Arendal	6.3	Yes	Yes	42 229	5.6	4.5
Alta	6.6	Yes	No	19 071	11.9	2.7
Nes	7.6	Yes	No	19 049	6.6	4.0
Tromsø	8.8	Yes	No	68 239	10.4	2.7
Kvam	11.0	Yes	No	8 442	12.6	6.9
Froland	12.9	No	Yes	5 127	8.0	3.5
Våler	15.5	No	Yes	3 882	11.7	7.3
Grimstad	20.8	No	Yes	20 823	9.4	3.5
Grue	22.0	No	Yes	5 024	9.9	8.3
Tvedestrand	27.1	No	Yes	5 969	13.3	5.5
Vegårshei	38.5	No	Yes	1 922	8.8	5.9
Risør	47.5	No	Yes	6 871	9.6	6.0
Gjerstad	53.0	No	Yes	2 497	2.4	5.4
Åmli	53.1	No	Yes	1 829	19.2	5.5
Nissedal	96.3	No	Yes	1 405	18.2	6.1
Fyresdal	131.0	No	Yes	1 351	13.5	6.2

face-to-face doctor consultations in each of the Watchtower municipalities. Rates were calculated by dividing aggregated total number of contacts and consultations for each municipality in each year from 2007 to 2011 by population January 1st the same year.

Statistical Methods

IBM SPSS 20 and Excel 2010 software were used to analyse data. The correlations between mean distance and travel time to casualty clinics and postcode based and other methods were examined using Pearson correlation coefficient. The correlation analyses were repeated several times, including municipalities at different maximum distances and travel times. Short distances were defined as less than ten kilometres, and short travel times defined as less than 15 minutes.

The magnitude of difference from mean distance calculated by Statistics Norway was expressed as mean absolute error (MAE) with 95% confidence intervals.

Contact and face-to-face consultations rates are number of events per year, which are assumed to be Poisson distributed. The rates of our main outcomes were so high that a normal distribution will fit the data well. Correlation coefficients, r, and constants and coefficients with 95% confidence intervals of exponential functions (non-linear regression) were used to assess any association between distances from municipality (postcode based) to out-of-hours-centre and contact and consultation rates for the years 2007–2011 for each municipality. Exponential curve fit was chosen because this was found to best describe the association between distance and utilisation of out-of-hours services in a previous study [1].

Table 3. Comparison of different distance measures to casualty clinic.

Distance (kilometres)	N	Postcode		Median		Population centroid		Town hall	
		r	p	r	p	r	p	r	p
<135	18	0.999	<0.001	0.999	<0.001	0.998	<0.001	0.997	<0.001
<50	14	0.999	<0.001	0.993	<0.001	0.992	<0.001	0.994	<0.001
<30	12	0.996	<0.001	0.974	<0.001	0.982	<0.001	0.983	<0.001
<20	9	0.986	<0.001	0.875	<0.001	0.960	<0.001	0.950	<0.001
<10	6	0.868	0.025	0.704	0.12	0.559	0.25	0.287	0.58

Correlation analysis of relationship between mean distance for inhabitants and other distance measures to casualty clinic. The analyses are performed for different maximum distances. r = correlation coefficient.

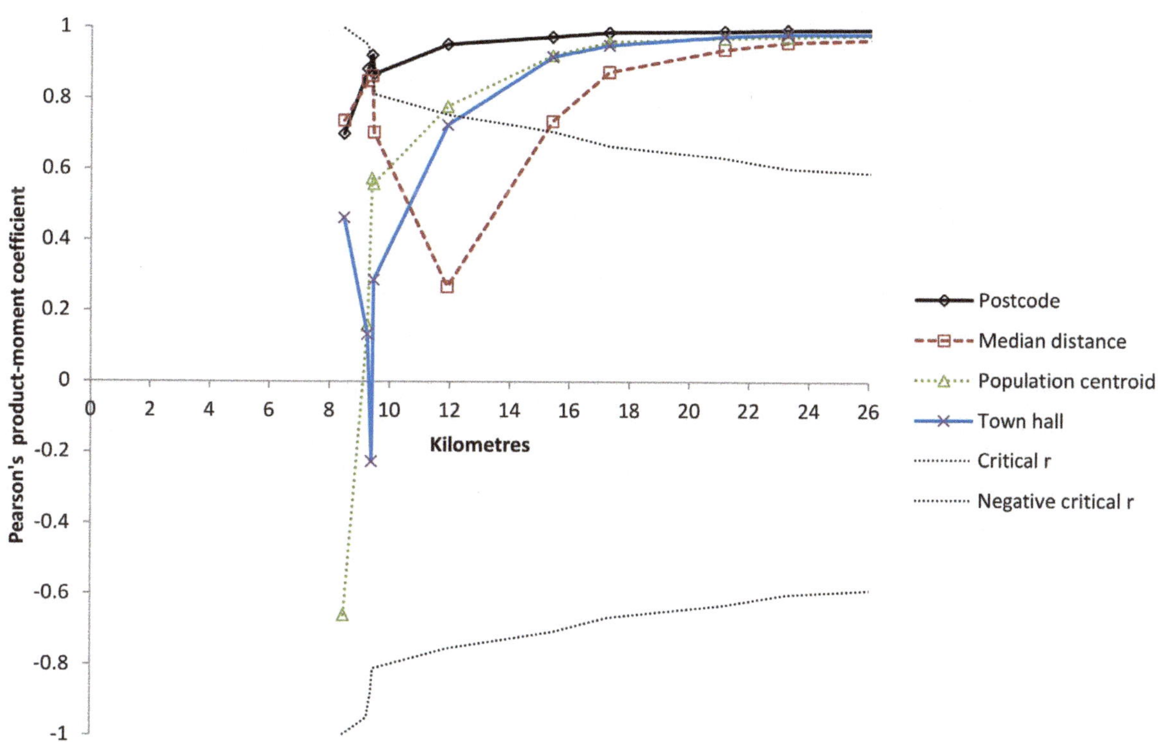

Figure 1. Correlation analysis of different distance measures at short distances by correlation coefficient, r, for the relationship between different average distance measures and mean travel time to casualty clinic. Analyses on municipalities with shorter travel times than indicated on the X-axis (N = 3 to 11). Critical r curves corresponding to p = 0.05. r-values between these curves are statistically non-significant.

Results

449 868 Watchtower contacts were registered from 2007 through 2011. 2 457 contacts were excluded because information about municipality was lacking. 135 770 contacts were excluded because the patient was a non-Norwegian citizen or was living in a municipality outside the Watchtowers out-of-hours districts. 311 641 contacts were eligible for analysis, aggregated to 90 municipality-year observations.

Baseline demographic and socioeconomic data for the municipalities are given in Table 2.

Table 3 and Figure 1 display correlation analyses of the relationship of distance to the casualty clinic from different population centres and mean distance as computed by Statistics Norway. When analysing only municipalities with less than ten kilometres to a casualty clinic, only postcode based distances were significantly correlated with mean distances. Figure 1 shows the superior correlation of the postcode based method at distances less than 20 kilometres.

For travel time to casualty clinic, as shown in Table 4 and Figure 2, there were strong and statistically significant correlations for all time measures at long travel times. At travel times less than 15 minutes, none of the methods were significantly correlated with mean travel time as calculated by Statistics Norway. At longer travel times, the postcode based method showed a correlation to mean travel time similar to the other methods.

Three of the municipalities had very similar mean distances and mean travel times that explain the abrupt change in critical r at 9.5 kilometres (Figure 1) and 11.5 minutes (Figure 2).

For municipalities not hosting a casualty clinic, there were no statistically significant differences in MAE between the distance measures compared with mean travel distance calculated by

Table 4. Comparison of different travel time category measures to casualty clinic.

Time (minutes)	N	Postcode		Median		Population centroid		Town hall	
		r	p	r	p	r	p	r	p
>120	18	0.998	<0.001	0.999	<0.001	0.998	<0.001	0.995	<0.001
<60	16	0.994	<0.001	0.996	<0.001	0.992	<0.001	0.987	<0.001
<45	13	0.980	<0.001	0.986	<0.001	0.980	<0.001	0.960	<0.001
<30	11	0.911	0.004	0.939	0.002	0.937	0.002	0.916	0.004
<15	7	0.626	0.13	0.388	0.39	0.728	0.063	0.676	0.095

Correlation analysis of relationship between mean travel time for inhabitants and other travel time measures to casualty clinic. The analyses are performed for different maximum travel times. r = correlation coefficient.

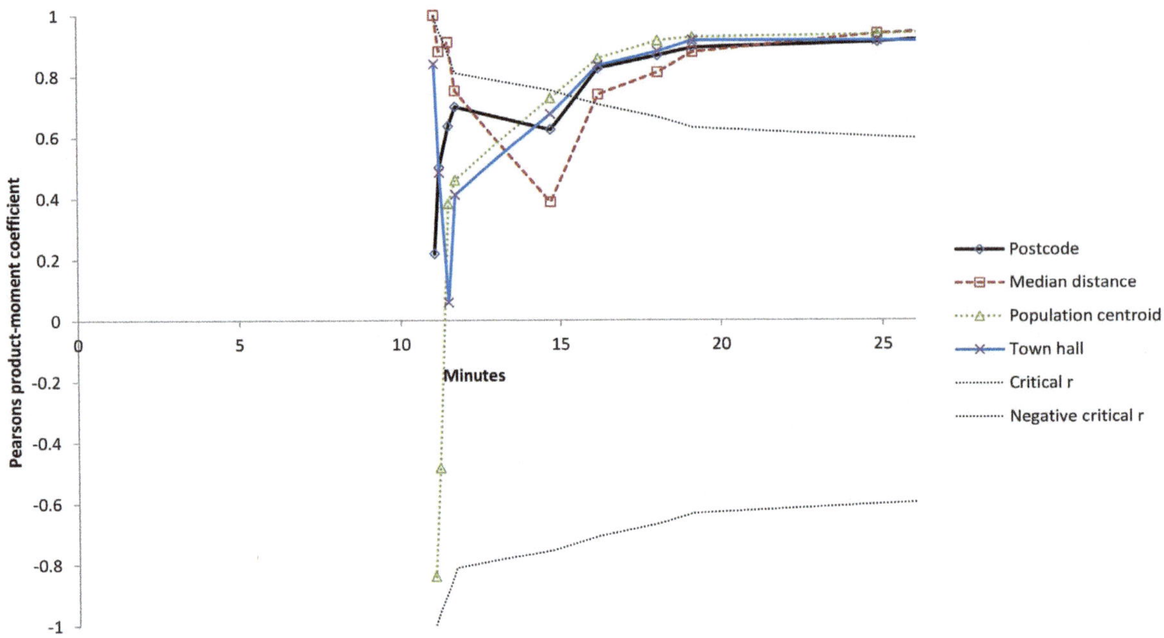

Figure 2. Correlation analysis of different travel time measures at short distances by correlation coefficients for the relationship between different travel time measures and mean travel time as calculated by Statistics Norway to casualty clinic. Analyses on municipalities with shorter distances than indicated on the X-axis (N = 3 to 11). Critical r curves corresponding to p = 0.05. r-values between these curves are statistically non-significant.

Statistics Norway (Figure 3). For distances based on population centroids and town halls, the deviation from mean distance was more than twofold larger in the municipalities where a casualty clinic is located. No statistically significant differences were observed in deviation between hosting and non-hosting municipalities in neither postcode nor median based distances, but the variance of deviation to mean distance was much higher for median based distances (figure 3).

For median and postcode based distances, there was no difference in MAE between mono- and intermunicipal out-of-hours districts. For distances based on population centroid and town hall location, the deviation from mean distance was more

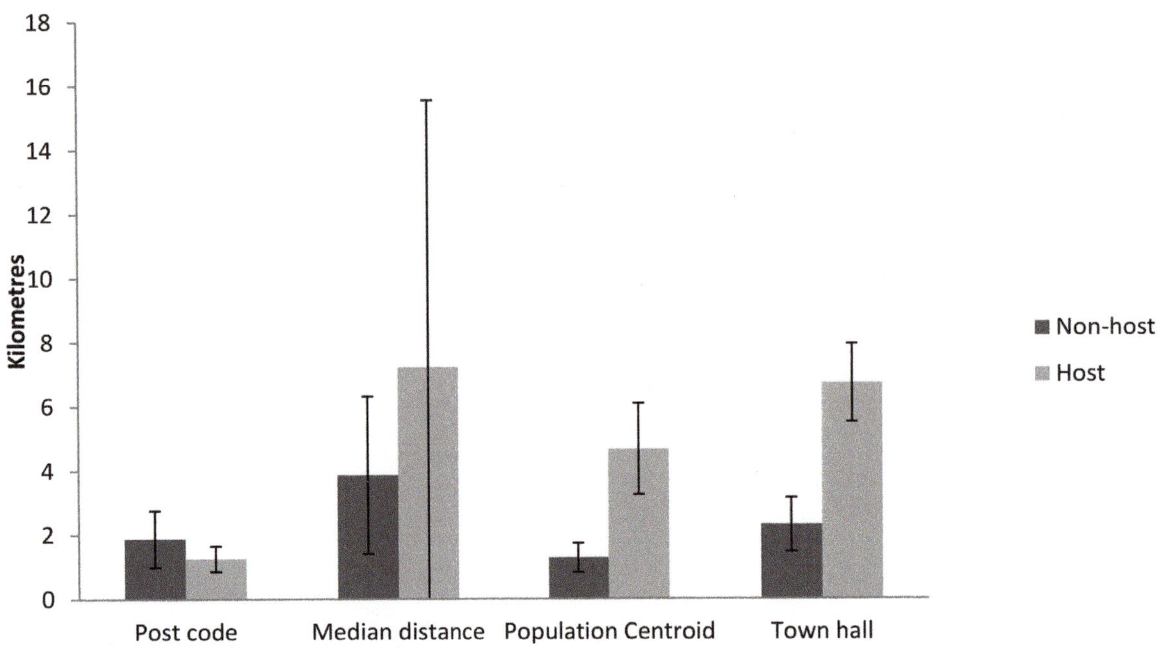

Figure 3. Mean absolute error (MAE) of distance measures in municipalities hosting a casualty clinic or not. Deviations (absolute) from mean distance to casualty clinic (kilometres) as calculated by Statistics Norway for the different distance measures. Error bars indicate 95% confidence intervals.

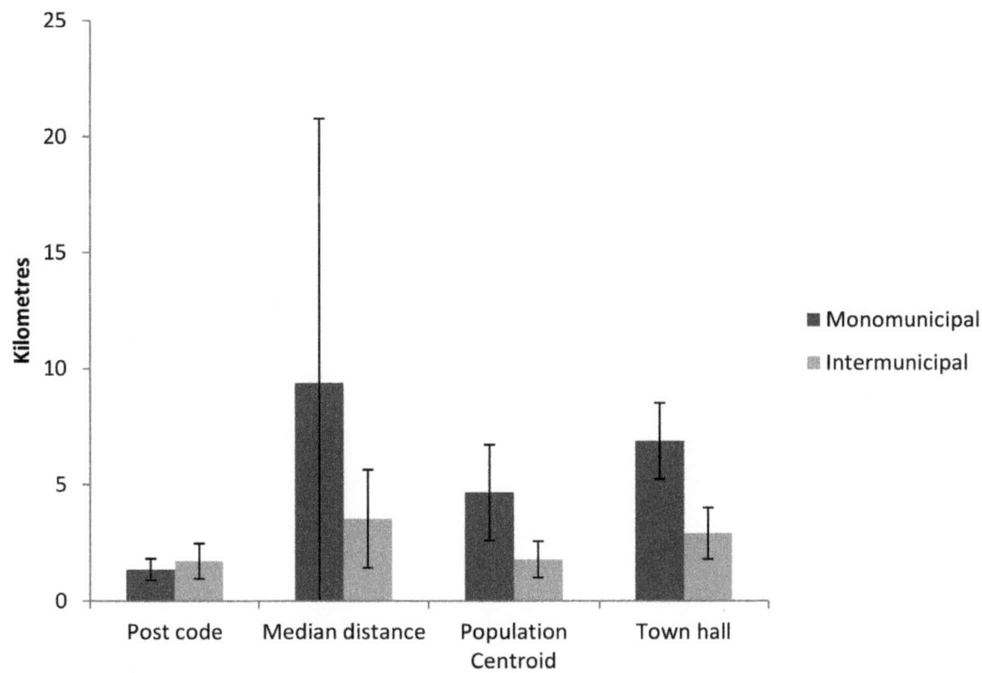

Figure 4. Mean absolute error (MAE) of distance measures in intermunicipal and monomunicipal out-of-hours districts. Deviations (absolute) from mean distance to casualty clinic (kilometres) as calculated by Statistics Norway for the different measures. Error bars indicate 95% confidence intervals.

than twofold higher in municipalities not taking part in an intermunicipal out-of-hours district (Figure 4).

As shown in Tables 5 and 6, there were no significant differences in regression coefficients between different methods when analysing the association between distance and consultation and contact rates, but population centroid and town hall based distances differed more from mean distance, although not statistically significant.

Exponential regression showed that increasing (postcode based) distance was significantly associated with lower contact rate and to a larger extent face-to-face consultation rate (Figures 5 and 6). There was no difference between municipalities participating in intermunicipal or monomunicipal out-of-hours districts.

Discussion

We have described a method for estimating distance and travel time to a casualty clinic based on postcodes determined by crowdsourcing. The method showed good correspondence to the main basis of comparison, which was mean distance and travel time based on address points. The method was at least as good as the other examined methods on both short and long distances. At distances shorter than 20 kilometres, the postcode based measure was superior to distances based on town hall location and population centroid. In contrast to distance and travel time measures based on town hall location and population centroid, the postcode based method showed no difference between mono- and intermunicipal out-of-hours districts, or between municipalities hosting a casualty clinic or not. The association between postcode based distance and contact and consultation rates in the 17 Watchtower municipalities was in accordance with our previous analysis of the 10 municipalities comprising Arendal out-of-hours district [1].

To calculate average travel distance and time from a municipality based on postcodes with the method described in this study, a database containing coordinates and population number for each postcode is necessary. The availability of such data is for most countries more limited than in Norway. There are

Table 5. Distance and contact rates, nonlinear regression.

	Change 0 to 50 km (%)	Constant	Coefficient	95% CI for coefficient	R^2
Postcode	−45.5	464	−0.0121	−0.0133 to −0.0109	0.82
Mean	−45.0	470	−0.0120	−0.0132 to −0.0107	0.81
Median	−45.0	461	−0.0119	−0.0131 to −0.0107	0.81
Population centroid	−44.2	452	−0.0117	−0.0128 to −0.0105	0.82
Town hall	−43.8	444	−0.0115	−0.0126 to −0.0105	0.84

Relative change (%) in contact rates when moving 50 kilometres away from the casualty clinic for different distance measures. Constants and coefficients (with 95% confidence interval (CI)) for exponential functions.

Table 6. Distance and consultation rates, nonlinear regression.

	Change 0 to 50 km (%)	Constant	Coefficient	95% CI for coefficient	R^2
Postcode	−58.8	277	−0.01774	−0.0187 to −0.0168	0.94
Mean	−58.3	283	−0.01751	−0.0185 to −0.0165	0.93
Median	−58.2	275	−0.01745	−0.0184 to −0.0165	0.93
Population centroid	−57.4	267	−0.01706	−0.0179 to −0.0162	0.94
Town hall	−56.4	258	−0.01658	−0.0176 to −0.0156	0.93

Relative change (%) in consultation rates when moving 50 kilometres away from the casualty clinic for different distance measures. Constants and coefficients (with 95% confidence interval (CI)) for exponential functions.

some on-line services in UK that provide coordinates of postcodes [12], [13], but to our knowledge information on the population of each postcode is not readily available. Some commercial providers, like Deutsche Post in Germany provide data based on addresses. Another limitation is that we used Google maps to calculate road distances and travel times. The reliability of Google Maps presumably varies from area to area because map data are acquired from different sources. Crowdsourcing projects with large numbers of contributors result in products that are complete, accurate and up to date, but they are vulnerable to errors and vandalism. Research on Wikipedia has shown that strict coordination of the editing process is crucial for quality [14]. The thorough planning and coordination of the more than 600 contributors of the postcode project suggests that the database is trustworthy.

Conclusion

The results show that distances and travel times for the inhabitants of Norwegian municipalities to a casualty clinic can be easily calculated with standard Internet tools from freely available postcode based coordinates determined by crowdsourcing. The method described here proved valid at both long and short distances, and is more reliable than distance estimates based on population centroid at short distances.

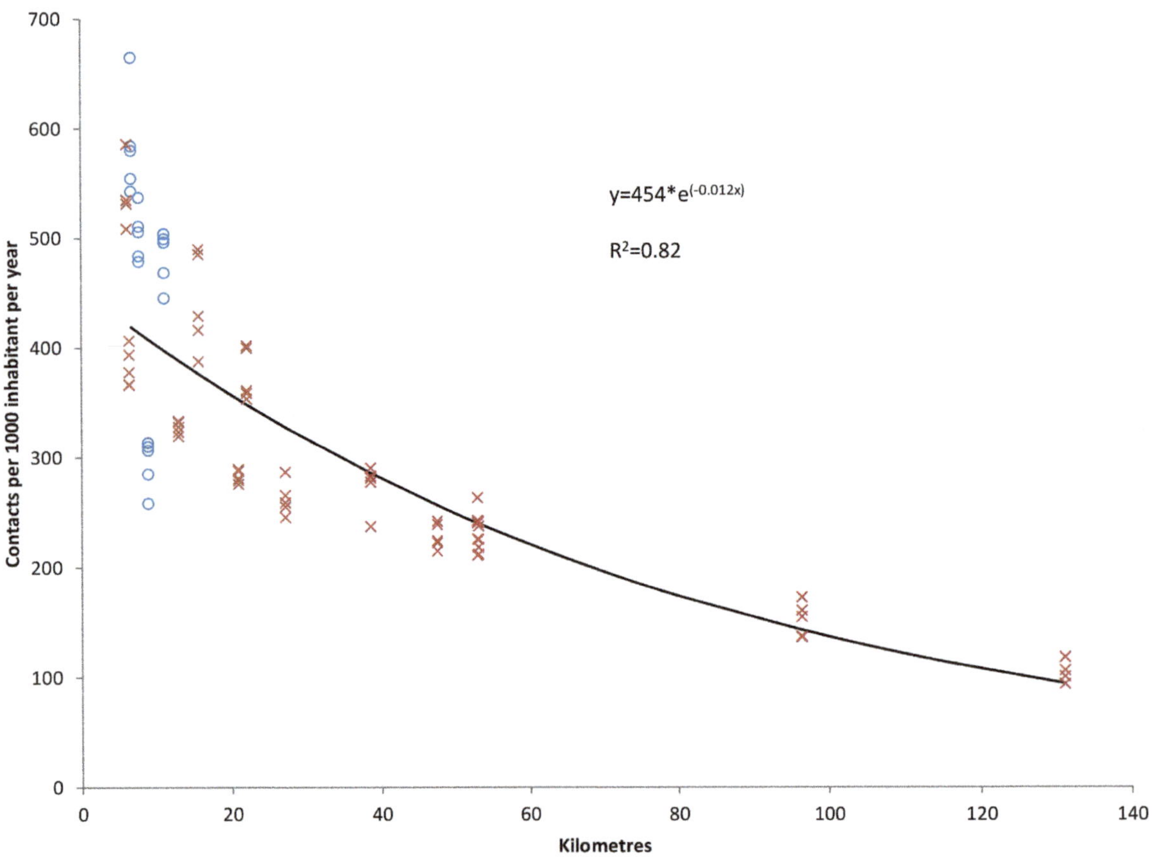

Figure 5. Postcode based distance and consultation rate at the casualty clinic. Relationship between average postcode based distance (kilometres) from Watchtower municipalities to casualty clinic and consultation rate at the casualty clinic (per 1 000 inhabitants per year). Each point represents rate in one municipality for one year (2007–2011). Circles indicate monomunicipal and crosses intermunicipal out-of-hours districts.

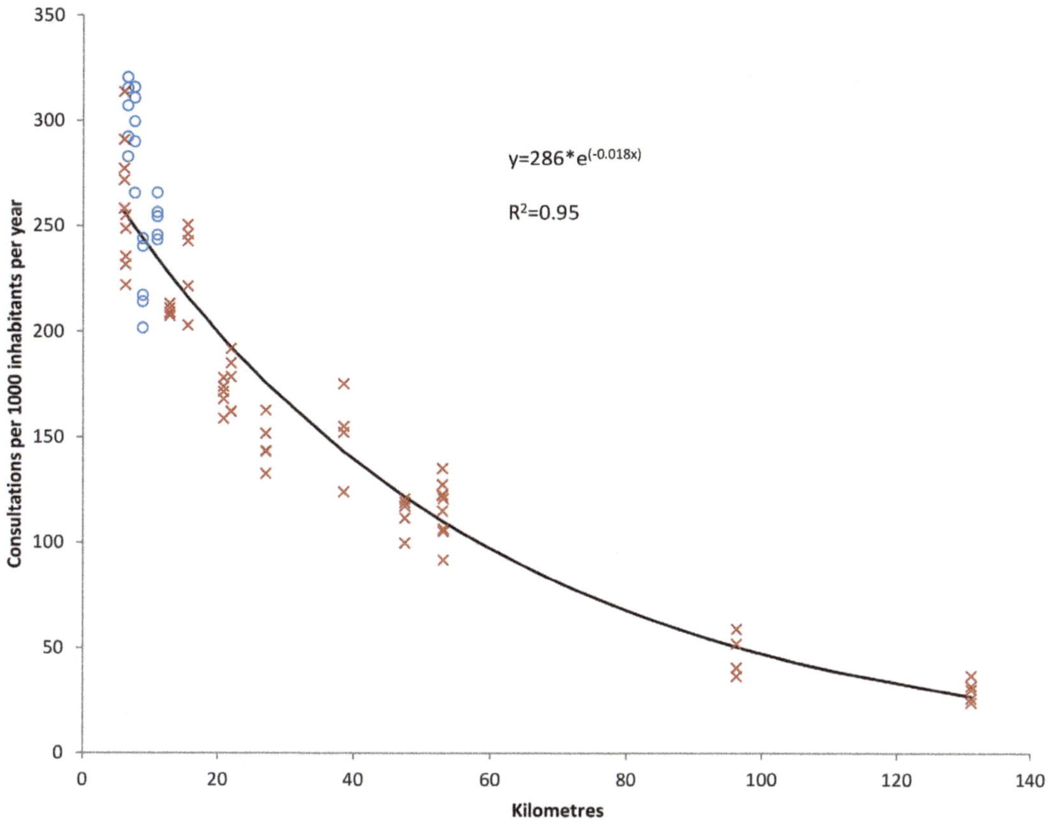

Figure 6. Postcode based travel time and consultation rate at the casualty clinic. Relationship between average postcode based travel time (minutes) from Watchtower municipalities to casualty clinic and consultation rate at the casualty clinic (per 1 000 inhabitants per year). Each point represents rate in one municipality for one year (2007–2011). Circles indicate monomunicipal and crosses intermunicipal out-of-hours districts.

Acknowledgments

We would like to thank Erik Bolstad for all the effort he has put into the postcode project, Ole-Johan Eikeland for preparing the Watchtower data files, Narve Sætre for technical assistance, Åsta Haukås for proofreading the final manuscript, and the staff at the Watchtower casualty clinics for performing the registrations.

Author Contributions

Conceived and designed the experiments: GR SH. Performed the experiments: GR. Analyzed the data: GR. Contributed reagents/materials/analysis tools: GR SH. Wrote the paper: GR SH.

References

1. Raknes G, Hansen EH, Hunskaar S (2013) Distance and utilisation of out-of-hours services in a Norwegian urban/rural district. An ecological study. BMC Health Serv Res 13: 222. doi:10.1186/1472-6963-13-222.
2. Berke EM, Shi X (2009) Computing travel time when the exact address is unknown: a comparison of point and polygon ZIP code approximation methods. Int J Health Geogr 8: 23. doi: 10.1186/1476-072X-8-23.
3. Brabham DC (2008) Crowdsourcing as a Model for Problem Solving: An Introduction and Cases. Convergence: The International Journal of Research into New Media Technologies 14: 75–90. doi: 10.1177/1354856507084420.
4. Boulos MNK, Resch B, Crowley DN, Breslin JG, Sohn G et al. (2011) Crowdsourcing, citizen sensing and sensor web technologies for public and environmental health surveillance and crisis management: trends, OGC standards and application examples. Int J Health Geogr 10: 67. doi: 10.1186/1476-072x-10-67.
5. Bolstad E (2013) Norske postnummer med koordinatar. Available: http://www.erikbolstad.no/geo/noreg/postnummer/. Accessed 5 June 2013.
6. Creative Commons Attribution 3.0 (2013) Available: http://creativecommons.org/licenses/by/3.0/.8.Accessed 5 June 2013.
7. Bolstad E (2013) Yr.no. Available : http://www.yr.no/.Accessed 5 June 2013.
8. Bolstad E (2013) Dugnad på yr.no. Available : http://nrkbeta.no/2009/07/22/postnummer-dugna/.Accessed 5 June 2013.
9. Hansen EH, Hunskaar S (2008) Development, implementation, and pilot study of a sentinel network ("The Watchtowers") for monitoring emergency primary health care activity in Norway. BMC Health Serv Res 8: 62. doi: 10.1186/1472-6963-8-62.
10. Morken T, Zachariassen SM (2012) Legevaktorganisering i Norge. Rapport fra Nasjonalt legevaktregister 2012. Rapport nr. 6–2012. [Organisation of the out-of-hours services in Norway 2012]. Bergen: National Centre for Emergency Primary Health Care, Uni Health, Uni Research. 9.
11. Google (2013) Google Maps. Available : https://maps.google.com/. Accessed 5 June 2013.
12. Foskett M (2008) Convert UK postcodes to longitude & latitude coordinates. Available : http://websemantics.co.uk/online_tools/postcode_to_coordinates_conversion_tool/. Accessed 5 June 2013.
13. Mutton P (2010) UK Postcodes. Finding grid coordinates from area codes. Available : http://www.jibble.org/ukpostcodes/Accessed 5 June 2013.
14. Kittur A, Kraut RE (2008) Harnessing the wisdom of crowds in wikipedia: quality through coordination. In: Proceedings of the 2008 ACM conference on Computer supported cooperative work. New York, ACM. 37–46.

Refining Climate Change Projections for Organisms with Low Dispersal Abilities: A Case Study of the Caspian Whip Snake

Tiberiu C. Sahlean[1,2], Iulian Gherghel[3]*, Monica Papeş[3], Alexandru Strugariu[4], Ştefan R. Zamfirescu[4]

1 Faculty of Biology, University of Bucharest, Bucharest, Romania, **2** Department of Terrestrial Fauna, "Grigore Antipa" National Museum of Natural History, Bucharest, Romania, **3** Department of Zoology, Oklahoma State University, Stillwater, Oklahoma, United States of America, **4** Faculty of Biology, "Alexandru Ioan Cuza" University, Iaşi, Romania

Abstract

Climate warming is one of the most important threats to biodiversity. Ectothermic organisms such as amphibians and reptiles are especially vulnerable as climatic conditions affect them directly. Ecological niche models (ENMs) are increasingly popular in ecological studies, but several drawbacks exist, including the limited ability to account for the dispersal potential of the species. In this study, we use ENMs to explore the impact of global climate change on the Caspian whip snake (*Dolichophis caspius*) as model for organisms with low dispersal abilities and to quantify dispersal to novel areas using GIS techniques. Models generated using Maxent 3.3.3 k and GARP for current distribution were projected on future climatic scenarios. A cost-distance analysis was run in ArcGIS 10 using geomorphological features, ecological conditions, and human footprint as "costs" to dispersal of the species to obtain a Maximum Dispersal Range (MDR) estimate. All models developed were statistically significant ($P<0.05$) and recovered the currently known distribution of *D. caspius*. Models projected on future climatic conditions using Maxent predicted a doubling of suitable climatic area, while GARP predicted a more conservative expansion. Both models agreed on an expansion of suitable area northwards, with minor decreases at the southern distribution limit. The MDR area calculated using the Maxent model represented a third of the total area of the projected model. The MDR based on GARP models recovered only about 20% of the total area of the projected model. Thus, incorporating measures of species' dispersal abilities greatly reduced estimated area of potential future distributions.

Editor: Gil Bohrer, The Ohio State University, United States of America

Funding: The authors have no support or funding to report.

Competing Interests: The authors have declared that no competing interests exist.

* E-mail: iulian.gherghel@okstate.edu

Introduction

Climate is recognized as one of the main factors responsible for shaping large-scale species distributions [1]. Global climate change observed over the past decades has produced shifts in the distribution and abundance of numerous species [2,3] and is responsible for species extinction [4]. Increased levels of global warming are expected to have different effects on species, based on their life-history traits [5,6], dispersal rates [7–9], and habitat requirements [10–12]. Predicted outcomes include range shifts following changing environments, adaptation to novel conditions, isolation to unaffected regions (refugia), and extinction [2,13–16].

Ectothermic animals (such as amphibians and reptiles) are especially prone to be affected by global warming as a result of their particular ecological requirements [17]. Local extinctions are well documented for reptiles [18–21] and amphibians have the highest rate of extinction recorded to date [22–27]. For example, Sinervo et al. [28] predicted that by 2080, 30% of all known lizard species will be locally extinct due of global warming alone. In turtles, Ihlow et al. [29] calculated that changing climates will produce distribution reductions of 86% of all species and 12% of the species will shift completely out of their current range. In snakes, Reading et al. [30] found that, in a period of about 14 years, 11 of 17 populations across Europe, Africa, and Australia have faced abrupt declines. Reading et al. [30] suggested that the reasons for these declines are the same as for the ones observed in other reptile groups but with a special emphasis on global climate change as the root cause.

Ecological niche modeling has become an increasingly popular methodology to study species' potential distributions in recent years and, as a result, several applications have been developed to facilitate generation of such models and distributions [31–34]. Broadly, these tools differ with regards to the type of species' records (presence/absence or presence-only) and predictors utilized (climatic - empirical approach or physiological constrains - mechanistic approach) [35–37]. In their current form, ecological niche models (ENMs) allow us to infer present [38–41], past [42,43], and future potential distribution of species [44–46], speciation scenarios [47–50], to design or refine protected areas for threatened species [51,52], and to predict novel distributions of invasive taxa [53,54] and impact of human-mediated global warming on wildlife [55–57].

As stated, ecological niche modeling is a popular approach to assessing the impact of global climate changes on species' ecological niches [53,56,58–60]. However, resulting predictions of future distributions are sometimes prone to potentially erroneous interpretations if the differences in dispersal ability among taxa are not taken into account. As such, the interpretation of these predictions may be subjective, indicating either species' expansion or contraction of their current ranges, or complete relocation to a new geographic area where the climate remains favorable. Furthermore, if the climatic change occurs rapidly, there is a risk of extinction for certain species, given the concept of niche conservatism, which states that species and clades retain their niches and related ecological traits over time [61,62].

One of the main drawbacks of ecological niche modeling, in its empirical implementation, is its reliance on the correlation between abiotic factors (i.e. temperature, humidity, radiation) and species' presences in what is defined as a "bioclimatic envelope" [46,63]. Other factors are known to play an important role in the distribution of species, such as biotic interactions (i.e. predators, prey, diseases) [64], and dispersal ability or dynamics of vegetation [65]. In the context of climatic shifts, dispersal ability stands out as one of the most important factors not taken into account when generating estimates of species' distributions. It is generally futile to discuss the suitability of future climatic conditions in a certain area that the species may not be able to reach in the first place.

In the present study, we used (i) ENMs to estimate the impact of climate change scenarios on the geographic extent of a narrow range, low dispersal organism (*Dolichophis caspius*) and (ii) cost distance analysis in Geographic Information Systems (GIS) to assess species' dispersal potential into novel, currently unoccupied regions identified as suitable by ENMs. Limitations of this approach are also discussed. Our main goal was to develop an approach to adjust assessments of global warming effects on narrow range, low dispersal organisms such as the Caspian whip snake by incorporating spatial estimates of the species' dispersal capacity on the landscape.

Materials and Methods

Species Account

The Caspian whip snake (*Dolichophis caspius*) is a xerophilous snake species inhabiting primarily steppe open grasslands, Mediterranean scrublands, rocky outcrops, and broad-leaved forest edges at low and medium altitudes (0 to 1600 m ASL) [66,67]. The species is widespread in Eastern Europe, southern Ukraine, the Balkan Peninsula, West Anatolia, Black Sea Coast, east to the Caucasus Mountains, southern Russia, and Kazakhstan [66–68]. In Eastern Europe (Romania, Bulgaria, Greece) the species is possibly one of the most frequent victims of ever-increasing road traffic and its habitat is experiencing ongoing reduction in many parts of the range [69].

Nagy and collaborators [70] recognized two Caspian whip snake haplotype groups separated by the Aegean Sea and the Bosphorus Straight, and estimated to have diverged during the Pleistocene: an eastern group along the Turkish coast and on East Aegean islands, and a western group in the Cyclades islands, Euboea island, and mainland Central and Eastern Europe. The western haplotype may have survived in its current observed range during glacial periods and persisted since, but later rapid recolonization events of Central and southeastern Europe, very likely from the Balkan Peninsula, represents another possible explanation of the current range [70].

Lowland areas such as steppes, forest-steppes, and xeric forests, the preferred habitats for the Caspian whip snake [66–68], are especially fragile and prone to land use changes due to their value as agricultural and grazing fields. In addition, these areas are extremely sensitive to minor variations in humidity and temperature, i.e. to the effects of climate change [71–74]. In the European Union (EU), such ecosystems became a top priority for conservation (EU Habitats Directive 92/43/EEC of 21 May 1992), but in developing countries conservation measures are implemented at a slower pace and may not represent a priority. At the same time, most of the range of *Dolichophis caspius* falls outside of the EU, where such measures are limited or do not exist [75].

Species Occurrence Data

In order to maximize the quantity and quality of the occurrence data used for generating the models, we did an extensive literature review from which we extracted available geographic location information. Imprecise localities such as country, county, coarse resolution UTM grids, as well as locations with uncertainty higher than approximately1 km were excluded from the analysis. Most of the whip snake localities had more than one confirmed record in the past 50 years. The extracted points were manually georeferenced using ArcGIS 10 with populated places and topographic maps as reference layers. For locality descriptions that could not be georeferenced, we contacted the authors to clarify the geographic reference. The georeferenced dataset comprised 338 localities which we further trimmed to retain only spatially unique ones, corresponding to single environmental grid cells using Trim Data function in ENMTools 1.3 [76]. Consequently, only 324 unique records were used to generate ENMs for the Caspian whip snake (Supporting Information S1).

Climate Data

The baseline (current) climatic data used for running the models had a spatial resolution of 30 arc-seconds (approximately 1 km) and was retrieved from the WorldClim database [77,78]. To analyze future climate effects on the potential distribution of the whip snake, we used future climate datasets produced by the Canadian Centre for Climate Modeling and Analysis (CCCma) using the Second Generation Coupled Global Climate Model (CGCM2), for two greenhouse gas emissions scenarios, A2a and B2a. All datasets were downloaded at 2.5 arc-minutes resolution (4.5 kilometers) from the International Centre for Tropical Agriculture website [79]. The chosen emission scenarios follow two opposite views on how the climate will change in the future 70 years. The A2a scenario is considered more liberal and takes into account a high population growth worldwide, increased energy use, land-use changes, and a slow technological advance; the B2a scenario is considered more conservative and simulates a slow human population growth rate, limited land use changes, and reflects a more technologically innovative world [80]. For each scenario we used the projections for 2020, 2050, and 2080 in an effort to forecast time series changes in the climatic niche distribution of *D. caspius*.

Both climatic datasets (baseline and CGCM2) comprise 19 bioclimatic variables (see Hijmans et al. [78] for more details) (Table 1) considered to compute more robust models than monthly temperature and precipitation variables [81] also available in the datasets [78]. However, to obtain robust models, it is necessary to optimize variable use by avoiding highly correlated variables or by selecting those that are increasing model accuracy. Here we opted for the latter to select a subset of the available 19 variables. We ran a first model using all 19 variables (Table 1) in Maxent, selecting for the second and final model runs only the variables that had a contribution above 5% in creating the first model (Table1). The same subset of variables was used for generating the GARP models.

To meet the recommendation that models must be trained only in a region within the known range of the species or within its

Table 1. Variable selection results indicating percent contributions to the initial and last models.

	Contribution	
Variable	**First model**	**Final model**
Mean Temperature of Coldest Quarter	35.2	39.7
Temperature Seasonality	17.1	21.6
Mean Diurnal Range	12.7	16.1
Min Temperature of Coldest Month	6.4	12.3
Precipitation of Driest Month	5	10.3
Temperature Annual Range	4.8*	–
Mean Temperature of Wettest Quarter	3.3*	–
Mean Temperature of Warmest Quarter	3.1*	–
Isothermality	2.8*	–
Precipitation of Driest Quarter	2.6*	–
Annual Mean Temperature	1.7*	–
Precipitation of Coldest Quarter	1.5*	–
Precipitation Seasonality	1.1*	–
Max Temperature of Warmest Month	0.6*	–
Mean Temperature of Driest Quarter	0.6*	–
Annual Precipitation	0.6*	–
Precipitation of Warmest Quarter	0.5*	–
Precipitation of Wettest Month	0.2*	–
Precipitation of Wettest Quarter	0.2*	–

*variables eliminated due to low contribution to model development.

dispersal limits [82,83], we produced the model using as geographic extent only the known distribution range of the *D. caspius*, following the distributional limits given in the literature [66].

Ecological Niche Modeling Procedure

Based on the species records and the bioclimatic variables, we generated models using Maxent version 3.3.3 k [36] and Desktop GARP (Genetic Algorithm for Rule-set Prediction) version 1.1.3 [84]. These two algorithms are among the ones designed for presence-only datasets that produce reliable predictions [85]. Also, previous studies [86–88] have shown that GARP tends to produce wider potential distributions when compared with Maxent, thus we employed both algorithms to assess degree of variation in the potential distribution and the estimated impact of global warming on the species studied.

GARP is a machine-learning algorithm, while Maxent has recently been reclassified as a version of the generalized linear model [89]. Both algorithms are able to produce ecological niche models using presence-only data and environmental predictors [36,90]. Maxent does this by finding the distribution closest to uniform distribution (maximum entropy) constrained by the environmental data input [36]. On the other hand, GARP generates models using rules that are applied to the training data. The changes in the predictive performance between runs are used to evaluate whether a rule is included in a model [32].

To produce the models in Maxent we used the default settings including regularization multiplier = 1 and maximum number of background points = 10,000. Random test percentage was set to 25% of the input species' occurrence records to test the performance of the resulting model. Also, the clamping option was used to downweight areas outside of the range presented by the training data (Supporting Information S2). The clamping function produces a map output that identifies in the projections (i.e., future climate) areas with environmental variable values outside of the minimum and maximum range of values present in the training region (i.e., present-day climate). Model predictions in such areas are deemed uncertain [91]. In GARP, we ran 100 models with a 0.01 convergence limit of model iterations and the maximum itineration number set to 1000. We also activated the internal testing feature and the "best subsets" procedure [92] to select ten best models as the final ones. The selection criteria included omission error (i.e., known occurrences predicted absent) which we set to the lowest 20% of values, and commission error (i.e., areas without known presences predicted present) for which we used the default 50% value. Both Maxent and GARP models were projected onto climate change scenario datasets at the end of the iteration phase. As the final procedure, in ArcGIS 10 (ESRI, Redlands, CA) [93] we summed the best ten GARP models to create a model agreement map and we converted Maxent and GARP models with continuous probability distribution values and model agreement values, respectively, to binary presence-absence potential distribution maps using as threshold 10% omission error of the training presence dataset. These post-modeling procedures were employed for both present-day and future climate potential distributions.

Model Evaluation

The model evaluation was done using three different methods: (1) partial ROC (Receiver Operating Characteristic), (2) omission error calculated using the test occurrence subset and thresholded, presence-absence predictions, and (3) expert's opinion. Generally, the most frequent method to evaluate ENMs is the Area Under the ROC Curve (AUC) [36], but, as it has been shown that this approach may not be the most adequate to evaluate the ecological niche models (see Peterson et al. [92,94] for details), we used a modified version, the partial ROC [80], which calculates the AUC only for a portion of the ROC curve, above an omission error threshold. We set the threshold to 5%. The partial ROC AUC scores were calculated using the partialAUC application developed by N. Barve (University of Kansas). We ran 100 iterations in which the test occurrence data were bootstrapped and we used z tests to assess whether the partial ROC AUC values were above that of a random model.

Post-modeling Analysis

To estimate whether *D. caspius* could disperse into novel areas of suitable climatic conditions predicted by the models, we used the cost distance analysis tool in ArcGIS 10. The cost distance function calculates the "effort" or "resistance" to moving from one point to another on the landscape based on a "cost" raster (GIS grid with cells, or pixels). If no destination point is given, the function automatically calculates "effort" to the edge of the raster. The cost raster is used to explain the difficulty of crossing certain landscape features, topographic or ecological (e.g.: altitude, slope, rivers, ecoregions, human-impacted areas etc.; Supporting Information S3–S5).

To develop the cost raster, we used geomorphological features (altitude and rivers), ecological conditions (ecoregions and presence-absence Maxent and GARP rasters), and the impact of human habitation on the landscape (human footprint). The altitude raster used was included in the standard WorldClim BIOCLIM package and was used to calculate a slope raster in ArcGIS 10. A GIS layer with world ecoregions was downloaded from WWF (http://worldwildlife.org), while the human footprint raster was downloaded from NASA Socioeconomic Data and Applications Center (SEDAC, http://sedac.ciesin.columbia.edu). River data were available in the basic collection of layers offered by ESRI with the license of ArcGIS 10. The above-mentioned rasters were reclassified manually using the "Reclassify" function in ArcGIS 10, while the number of classes varied depending on the specific raster dataset (Supporting Information S5, S6). The values for each interval were assigned based on species' biology and experts' opinion to account for the difficulty of movement (Supporting Information S3–S4). "Cost" rasters were created using the "Mosaic to New Raster" command in ArcGIS 10, which merged all reclassified rasters to a new dataset (Supporting Information S5–S6). To ensure that the cells of the resulting cost raster reflected the highest impediment to movement, priority was given to cells with a higher classification value (i.e., cells associated with higher costs). We generated separate cost rasters for GARP and Maxent model outputs and for each climate projection (two emission scenarios and three time periods; Supporting Information S3–S4). Also, to account for the uncertainty associated with assigning costs based on experts' opinion, we developed three separate scenarios (based on three separate cost rasters; Supporting Information S3–S4): (S1) *a permissive scenario*, in which the populations were assumed to have higher capacity of dispersal, (S2) *a restrictive scenario*, where populations were heavily constrained by conditions outside of the known environmental range, and (S3) *a balanced scenario*, in which the cost values were weighted based on experts' opinion of which environmental conditions are more likely to be suitable for the Caspian whip snake's dispersal and which could impede movement through the landscape.

The cost rasters (Supporting Information S3–S4), along with the occurrence points (Supporting Information S1) gathered from the literature survey, were used as input in the cost distance analysis performed in ArcGIS 10, "Cost Distance" function, which calculates the cumulative cost value for each cell (pixel) on the landscape as an individual would disperse from species' known presence localities. The rasters generated through the cost distance analysis (for both modeling algorithms and climatic scenarios, over all three time periods and dispersal scenarios) were then reduced (thresholded) using the lowest (minimum) cost distance value that connected all species' known presences (Supporting Information S5–S6). The threshold was used since information regarding home range or dispersal capabilities for the Caspian whip snake is not available. We refer to the resulting rasters as Maximum Dispersal Range (MDR). The cost distance analysis was repeated for each cost raster generated for the two ecological niche modeling algorithms and two emission scenarios, for three time periods and three dispersal scenarios. We emphasize that, in addition to ENMs, MDR incorporates variables not directly used in the ecological niche modeling algorithms, such as anthropogenic alterations of the landscape and topography, that would affect dispersal routes for this species.

Results

Model Accuracy Metrics

The partial ROC area under the curve (AUC) ratios had a mean of 1.24 (SD = 0.064) for Maxent and a mean of 1.18 (SD = 0.06) for GARP in 100 replicates, and were statistically significant above the null expectations (z test, $p<0.05$). The rate of false negative records (omission error) was 13% of the total number of presences for Maxent model and 0% for GARP model. From the herpetologist expert's point of view, both algorithms recovered the known distribution and potential distribution of the Caspian whip snake (Fig. 1), with a slight trend of Maxent to underpredict the potential distribution of Caspian whip snake in the northern part of the species' range. The location of presence records predicted absent (omission error) is in agreement with the expert opinion, as these records are at the limits of the species' distribution. The results from all three evaluation methods suggest good predictive power of the models, thus we consider the resulting species' potential distributions under future climate conditions to be reliable estimates of the effects of forecasted climate change.

Present-day Potential Distribution of the Caspian whip Snake

The highest contributing variables (>5%) to both the first and the final models were mean temperature of the coldest quarter, temperature seasonality, mean diurnal range, minimum temperature of the coldest month, and precipitation of the driest month (Table 1). The difference in variable contribution values between the two model runs was not significant ($t=-0.622$, $df=8$, $p=0.551$). Maxent and GARP models predicted most of Balkan Peninsula, Pannonian Plain, Crimean Peninsula, and Western and Southern coast of Caspian Sea, Italy, Syria, Iraq and Iberian Peninsula as climatically suitable for *D. caspius*. Other climatically suitable regions were found in parts of Germany, Turkey, France, Poland, and Baltic and Scandinavian countries.

We observed disagreement in the present-day potential distributions predicted by the two algorithms: large areas predicted suitable only by GARP in Turkey, Romania (excluding the Carpathian Mountains), Moldova, Southern Ukraine, southern European Russia, and the Caspian Sea basin, including large parts of Turkmenistan, Azerbaijan, and Georgia, were only marginally predicted present by the Maxent model. In addition, according to the GARP present-day model, all Western Europe and the Czech Republic are climatically suitable for *D. caspius*, regions only

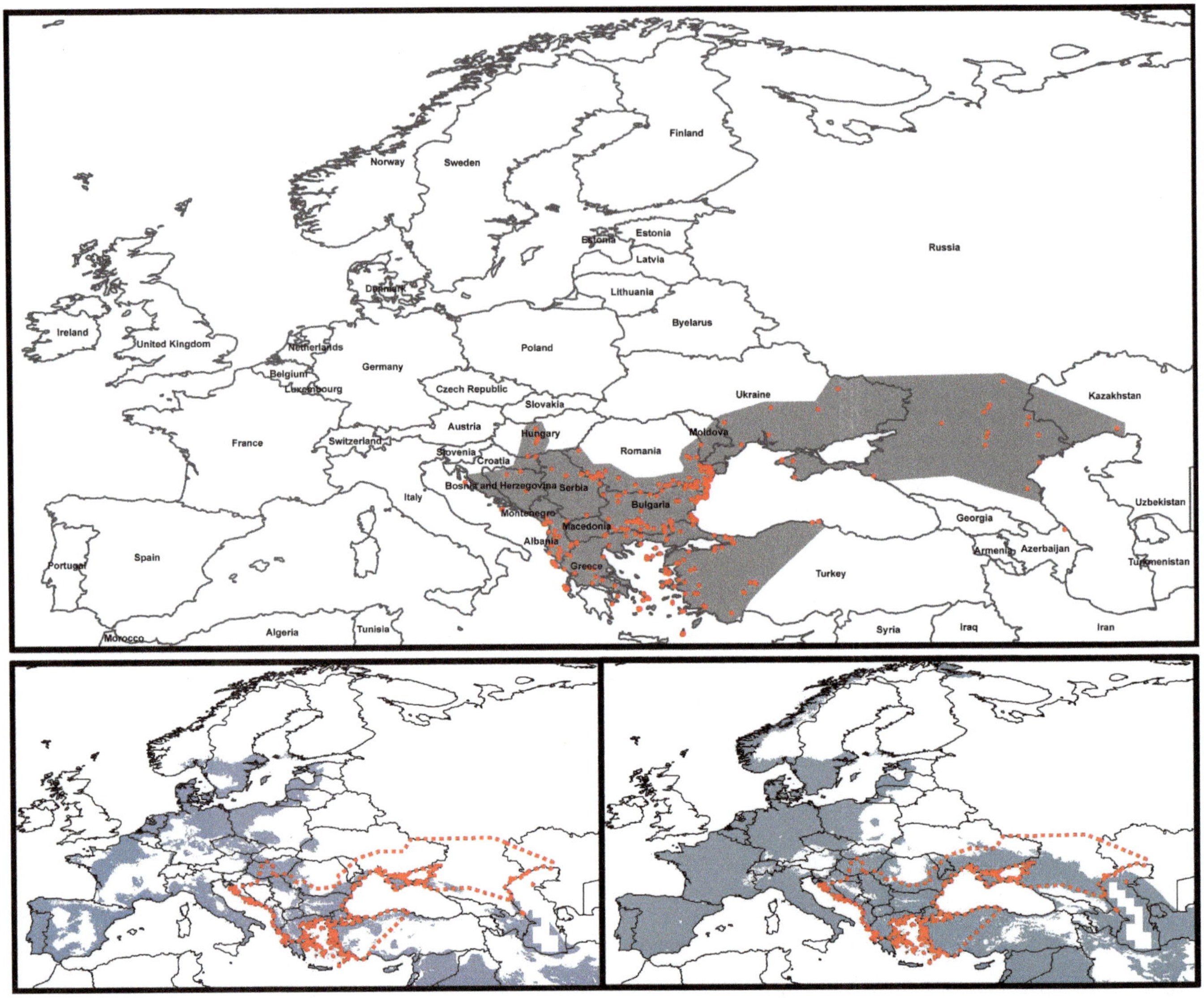

Figure 1. Present-day ecological niche models (green) for *D. caspius* (lower left - model generated by Maxent; lower right - model generated by GARP) in comparison to its known distribution range (upper panel, area in grey) (adapted after [66]) and presence records (red points). The red dotted line in lower panels corresponds to the grey area in upper panel (known distribution range).

partially predicted present by the Maxent model. Generally, discrepancies aside, the models generated by both GARP and Maxent produced pertinent maps of the potential distribution of Caspian whip snake under current climatic conditions.

Moreover, the GARP model recovered almost all of the species' occurrence points used in the analysis (i.e, low omission error) and filled in the gaps in species range in the southern European Russia, Ukraine, and Moldova, where, although anecdotal information about the species' presence exists, to the best of our knowledge, no documented records are available. From the herpetologist' point of view (expert opinion), the prediction generated by GARP is a closer approximation of the current potential distribution of *D. caspius*, with a tendency to overpredict especially at the northern limit of the species' range.

Forecasting the Future Distribution of the Caspian Whip Snake

The clamping results for all future predictions show that the climate conditions in the training region were similar to those in the projected area, across emission scenarios and time frames. The only regions where the climates were different are located in the northeastern Europe, a region in which Caspian whip snake has not been recorded. Thus, we assumed that the models would be reliably transferred to future climates since novel climatic conditions were not identified in the species' geographic range.

The agreement between future projections produced by GARP using the liberal (A2a) and conservative (B2a) emission scenarios was high (>90% congruence between the number of pixels predicted present), for each of the three time periods studied (2020, 2050 and 2080). On the other hand, Maxent projections were more divergent between climate scenarios: over 60% congruence for 2050 and 2080 and 80% for 2020 (Table 2 and Fig. 2). The models produced by Maxent predicted a doubling of

Table 2. Percent of suitable niche overlap between the two emission scenarios (A2a and B2a) for the studied time periods (2020, 2050, 2080).

Algorithm	*2020*	*2050*	*2080*
Maxent	81.85	60.19	66.29
GARP	98.01	93.92	96.07

suitable climate area from the present to future projections under both climate scenarios and suitable regions extended to areas presently unsuitable for *D. caspius*.

The models produced by GARP predicted a more conservative change of the climatically suitable areas, as future projections were more similar to present predictions, adding 10.21% more suitable space by 2080 under scenario A2a and 8.37% more suitable space under scenario B2a (Table 3). The visual analysis of the effects of climate change on *D. caspius* potential distribution using both modeling methods (GARP and Maxent) and both emission scenarios (A2a and B2a) suggests an increase in climatic suitability in the currently known distributional range and also a gradual geographic expansion of the climatic niche of the species farther north. The most significant changes in terms of expansion of suitable climates for *D. caspius* can be observed in the northern part of its range, especially in Poland, Ukraine, the Baltic states, southern and western Russia, and around Caspian Sea (figure 2).

However, while there was consensus between algorithms regarding the northern expansion of the geographic range of favorable conditions, the models generated by GARP under both emission scenarios predicted a small and progressive loss of suitable areas from the species' southern distribution limit, in Turkey, Lebanon, and Syria.

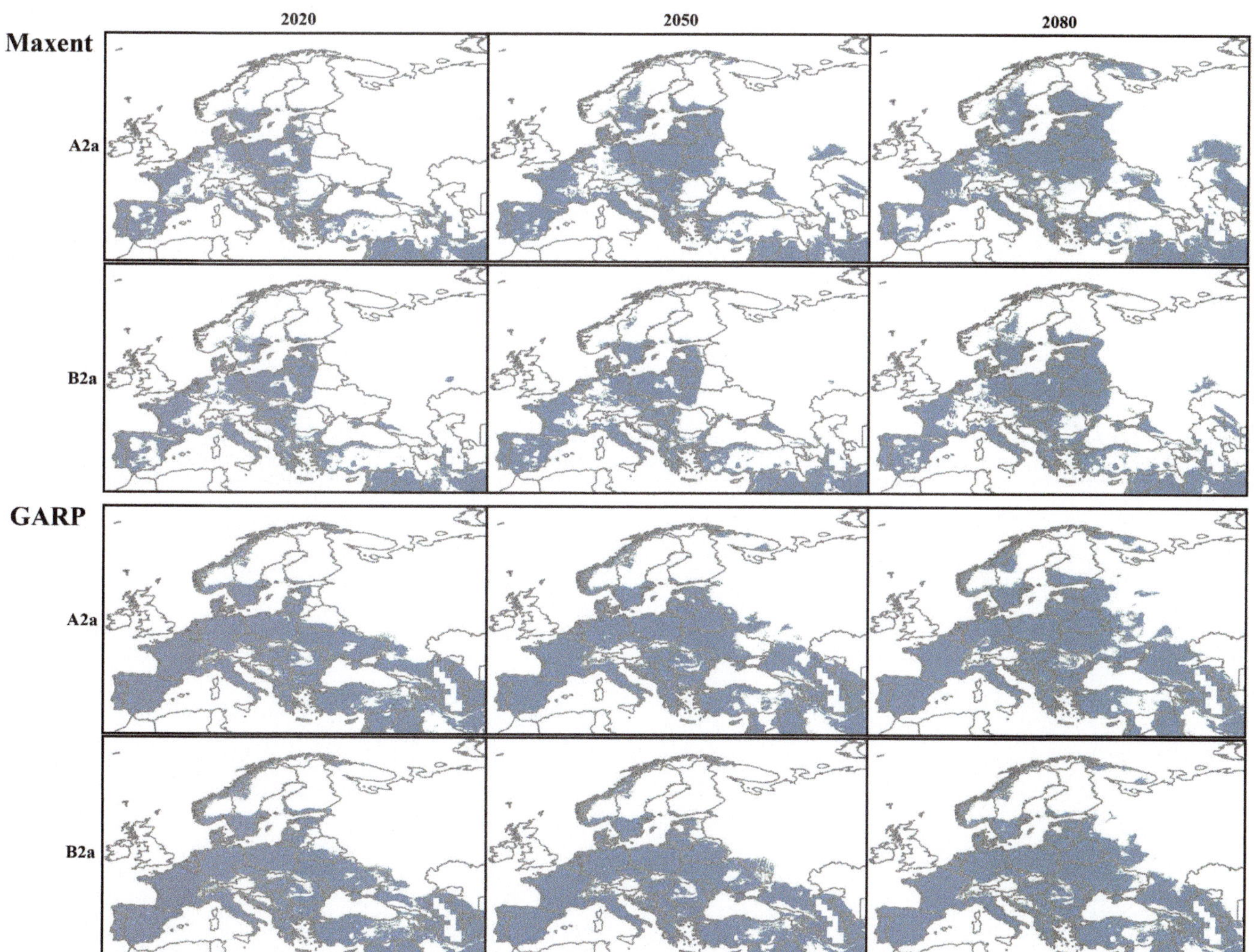

Figure 2. Future climatically suitable ranges for the Caspian whip snake, *Dolichophis caspius*.

Table 3. Percent of climatically suitable areas available for the Caspian whip snake of the total projected space.

		Future climate scenarios					
	Present	A2a			B2a		
Algorithm		2020	2050	2080	2020	2050	2080
Maxent	20.24	24.99	33.54	42.18	25.78	27.65	35.07
GARP	36.85	45.63	46.30	47.06	43.49	43.49	45.22

Maximum Dispersal Range Analysis

The cost distance analysis of the three dispersal scenarios produced different results for the climate projections obtained with the two modeling algorithms (Maxent and GARP) that required different minimum cost distance thresholds to include species' known presences.

In the case of the permissive scenario (S1), the minimum distance threshold that connected all species' known distribution points was 5.26% of the original rasters resulted from the cost distance analysis using Maxent and the liberal A2a emission scenario, and 5.53% using the conservative B2a emission scenario. For GARP, the distance thresholds were 4.65% (A2a) and 4.87% (B2a). For the restrictive scenario (S2), the distance thresholds were 6.84% (A2a) and 7.11% (B2a) using Maxent predictions and 5.24% (A2a) and 5.53% (B2a) using GARP predictions. Finally, in the case of the balanced scenario (S3), distance thresholds were 9% using the liberal emission scenario (A2a) and 8% using the conservative emission scenario (B2a) for Maxent potential distributions, and 4% for both climatic scenarios for GARP potential distributions. This suggests a better ability of the MDRs based on GARP models output to delineate a single area that includes all points used in analysis.

Using the S1 dispersal scenario, the spatial congruence between the MDRs based on GARP was much higher (97%) than in the case of Maxent (88%), whereas using the S2 dispersal scenario the congruence was slightly higher for Maxent (98%) than GARP (96%) predictions. In the balanced S3 scenario, the spatial congruence between the MDR for the two emission scenarios was again higher for the analysis based on GARP outputs (99.16%) than the one based on Maxent outputs (92.67%).

The area defined as MDR using the Maxent projections on the liberal A2a scenario represented 20.4% of the total area of the projected model and 23.27% of the conservative B2a scenario in the case of the permissive scenario (S1, Table 4). The MDRs based on the GARP models for the two scenarios were similar, with 22.35% of the total area of the projected model using the A2a scenario and 23.06% using the B2a scenario (Table 4). In the case of the restrictive scenario (S2), the area defined as MDR using Maxent niche projections were 20.33% using the A2a emissions scenario and 20.81% with the B2a emissions scenario (Table 4). The MDRs based on the GARP projections represented approximately19% (18.72% for A2a and 19.43% for B2a) of the total projected area under both emission scenarios (Table 4). The MDRs from the balanced model (S3) represented 30.34% of the total area of the projected model using the A2a scenario and 28.12% of the conservative B2a scenario projections. The MDRs based on the GARP models for the two emission scenarios featured 21.73% of the total area of the projected model using the A2a scenario and 21.54% using the B2a scenario (Table 4).

Even though the degree of congruence between MDRs within each algorithm (Maxent and GARP) and between climatic scenarios was >90%, slight differences existed across algorithms and scenarios. The MDRs based on GARP predictions showed many more dispersal possibilities for the Caspian whip snake when the analyses were based on the permissive S1 scenario, the MDR covering 98% (A2a) and 97.55% (B2A) of the current distribution range, while the MDRs based on Maxent covered 95.67% (A2a) and 97.20% (B2a) of the current distribution range (Figure 3). By using S2, the MDRs identified similar dispersal options, the MDRs based on GARP covering 98% (A2a) and 97.90% of the current distribution range and the Maxent-based MDRs covered 98.55% (A2a) and 98.89% (B2a) of the Caspian whip snake's current distribution (Figure 3). In S3 the MDRs based on the Maxent predictions indicated many dispersal options, even more so when

Table 4. Percent of the area predicted accessible for *D. caspius* of the total projected space in the context of global warming based on two climate change emission scenarios (A2a and B2a) and the three scenarios of Maximum Dispersal Range (MDR).

Dispersal scenario	*MDR*	*Future climate scenarios*	
		A2a	B2a
S1	Maxent based MDR	20.4	23.27
	GARP based MDR	22.35	23.06
S2	Maxent based MDR	20.33	20.81
	GARP based MDR	18.72	19.43
S3	Maxent based MDR	30.34	28.12
	GARP based MDR	21.73	21.54

taking into account the liberal A2a model, with a 97.61% overlap with the current distribution range of the Caspian whip snake, in contrast to the conservative B2a model with a 94.69% overlap with the current distribution (Figure 3). The MDRs based on GARP predictions overlapped with the current distribution of *D. caspius* in 98.20% of the area for the liberal A2a climatic scenario and 96.09% for the conservative B2a climatic scenario (Figure 3).

In general, using the S1 dispersal scenario, the MDRs based on GARP predict dispersal options for the Caspian whip snake into Austria, Slovakia, the Czech Republic, Poland, Ukraine, Belarus, and Russia towards the north, but also further into Turkey in the southeast and into Georgia, Armenia, Azerbaijan, and Iran in the south. In the case of the MDRs based on the Maxent models, dispersal options are identified as far north as northern Poland and to the west all the way into eastern Germany, but unlike the GARP-based predictions, Ukraine and Belarus are shown only as limited routes for dispersal. As in the case of the GARP-based MDRs, to the south Georgia, Armenia, Azerbaijan, and Iran are presented as likely dispersal routes, but unlike the GARP-based MDRs, the outputs based on Maxent identify a sizable part of Syria and Lebanon as possible dispersal options.

Using the restrictive scenario (S2), the consensus was greater across modeling algorithms and emission scenarios (A2a & B2a), all identifying Slovenia, Austria, the Czech Republic, southern Poland, Ukraine, and Russia as likely dispersal routes towards the north, Georgia, Azerbaijan, and northwestern Iran in the south, and Turkey in the south-east.

In the case of the balanced scenario (S3), as a general agreement between MDRs produced based on Maxent and GARP across climate change emission scenarios, we identified favorable corridors and a relatively low resistance of the landscape for *D. caspius* to disperse to northwestern parts of its range to the Czech Republic, Slovakia, Poland, and Slovenia, assuming no significant changes of the human footprint or ecoregion distribution in the future. Other important areas where *D. caspius* would have the possibility to expand its range are Anatolia and the Caucasus countries, especially Georgia and Azerbaijan. On the other hand, the models predicted reductions in the northern and eastern parts of the species' current range. However, according to our MDRs, large parts of Ukraine, Romania, and Moldova will be climatically suitable and accessible by the Caspian whip snake.

Discussion

Although Maxent model had higher partial ROC AUC values, GARP model performed better in the two other evaluation methods, omission error (no presences predicted absent) and herpetologist expert opinion. These results may be a consequence of the basic differences between GARP and Maxent, as the former tends to produce models with higher commission error than the latter, in other words predict suitable broader areas [88], therefore the herpetologist expert opinion would be in agreement with the models generated by GARP.

Our models identified several environmental variables that had high contribution to generating the potential distribution prediction of *D. caspius* that recovered the current known range as well as identified other, geographically adjacent, climatically suitable areas. As it is the case with all reptiles in general, the Caspian whip snake's large-scale distribution is mainly environmentally dependent, due to its physiological characteristics. Of all 19 variables initially used in the modeling process, the most important ones that we based our final models on, and that best explained the environmental requirements of the species, were four temperature-derived variables and one precipitation-derived variable. Our findings are in agreement with the previous research published on this species that characterized it as xerophilous [66–68]. The species is known to tolerate high temperatures and long dry periods (up to several months) [66], but it cannot tolerate low temperatures during winter, the latter being considered a limiting factor of its distribution, frequently indicated in the herpetological literature [66–68]. This limiting factor was represented in our models by two variables, the mean temperature of coldest quarter and the minimum temperature of the coldest month, which limited the species' distribution northwards of 50° latitude in the Eastern European Plains. The critical 50° latitude is also indicated in literature as the northernmost limit of the species' distribution but here we identify two climate factors that may explain the observed northern distributional limit. On the other hand, *D. caspius* is not able to live in desert and semi-desert environments and this also was captured by our models which indicated as limiting factor the mean diurnal range variable in Near East, Middle East, and Central and Southern Iran due to the very high temperatures during the warm period, which are frequent in the area. In contrast, in most of the Mediterranean basin and Central and Western Europe, *D. caspius* models had no limitations, thus probably other, non-climate variables such as dispersal ability of the species and landscape features play an important role in shaping its distribution.

The projection of models onto future climatic conditions, for both algorithms used and both emissions scenarios (the liberal A2a and the conservative B2a), predicted similar trends in distributional shifts of the *D. caspius*. According to these models, as a result of global warming, suitable climatic conditions for *D. caspius* will be present in geographic areas north of its current range, especially in the Central and Western Europe, but also in the Eastern European

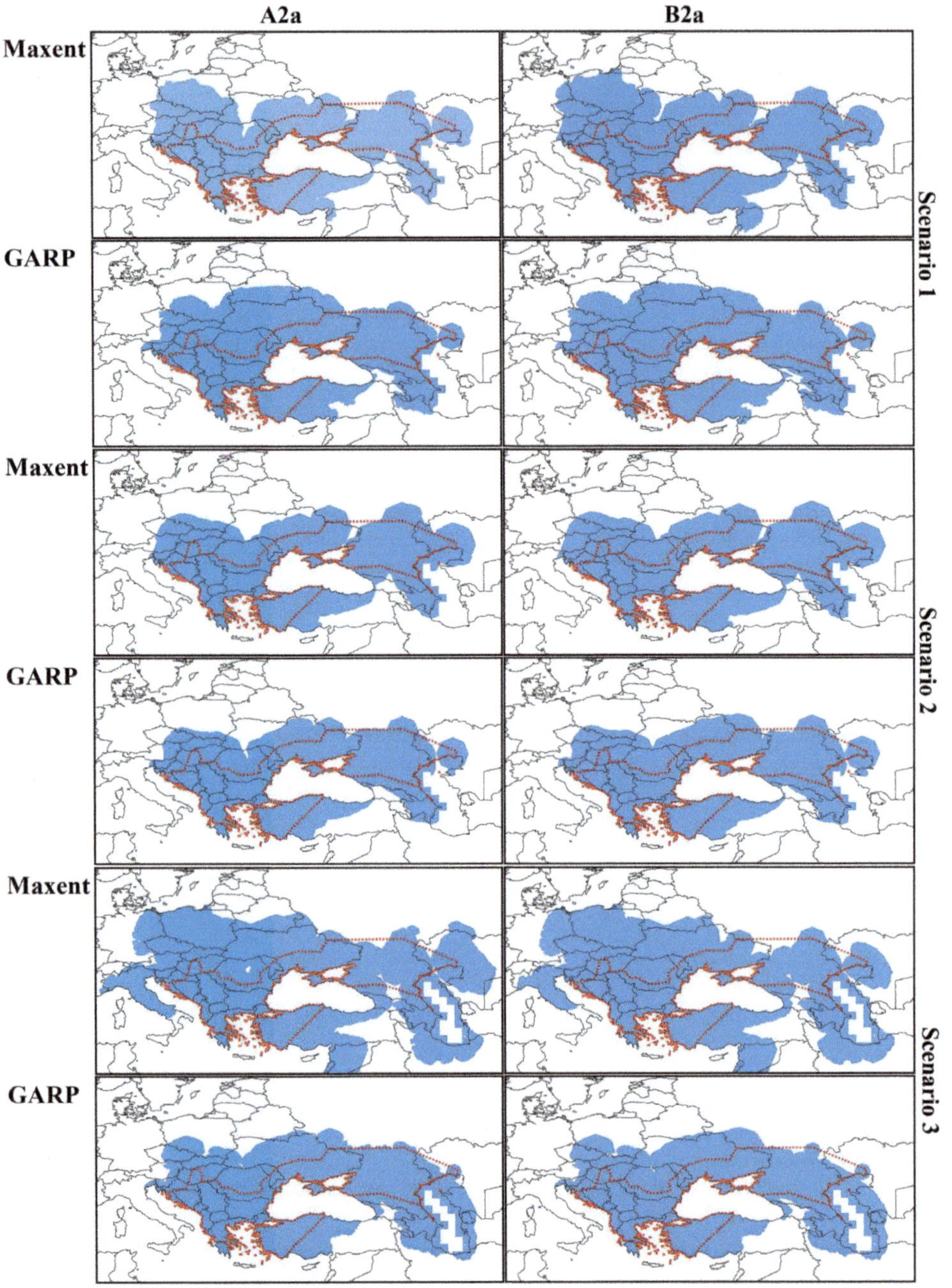

Figure 3. Maximum dispersal range estimates (blue) using Maxent and GARP future potential distributions (A2a – liberal scenario and B2a – conservative scenario) and three dispersal scenarios (S1 - permissive scenario, S2 - restrictive scenario, S3 - balanced scenario). The known species' range is shown in red dotted line.

Plains. The latter region, under current conditions, is considered highly unsuitable in the literature and by our models built on current climatic data. The projected climate changes may present the opportunity for *D. caspius* to migrate to new regions northwards of the climate-driven 50° latitude barrier.

The global warming process was estimated at a rate of 0.2°C per decade for the next two decades in most global climate models [80]. At this accelerated rate of warming, the predicted effects in most cases are range contractions or, in extreme situations, complete relocation to a new geographic area where the climate would become favorable. Studies on the effects of global warming on reptile species have generally predicted negative consequences [28–30,95,96]. In contrast, our models generated using Maxent and GARP show that the climatic space available for *D. caspius* will expand geographically, especially beyond the northern distribution limit.

While expansion is indicated by both algorithms across both climate change scenarios, this geographic expansion of favorable climatic conditions for *D. caspius* does not automatically infer actual distribution expansion. As numerous studies have shown, while the distribution of a species seems to be influenced mostly by climatic conditions at large scales, at finer scales additional factors become essential [37,82,97,98], such as landscape features, ecological communities, predator-prey interactions, and anthropogenic pressures [99,100]. Moreover, as these factors are responsible for the existence of a species at local scales, the large-scale distribution of the species will depend on the existence of favorable climatic conditions plus the dispersal ability. The MDR analysis performed here reveals precisely the importance of dispersal: while novel areas with suitable conditions will arise for *D. caspius* outside its current range, its low dispersal ability will impede the colonization of the distant areas with newly suitable climates. Thus, MDRs are in agreement with current views regarding the effects of global climate change on amphibians and reptiles.

Limitations of the Maximum Dispersal Range Analysis

The MDRs developed in this study estimate the dispersal range by associating a difficulty score to environmental or physical features that individuals may encounter in their migration paths. While these simulations too rely on certain assumptions, such as unchanging human land use patterns and ecological communities, our opinion is that this method provides additional, essential information to refine views on the impact of global warming on species' distribution.

Using the MDR analysis, we expand our ability to assess the impacts of global warming on species' distributions. However, to use this technique, researchers are expected to have advanced knowledge regarding the ecology, habitat requirements, and dispersal potential of the studied species, information that is crucial to creating the cost raster for the MDR analysis. Nevertheless, this method cannot be standardized to all species due to differences in dispersal ability or lack of sufficient knowledge of a certain species' ecology. Thus, we recommend careful selection and prioritization of the parameters for the cost analysis, on a species by species basis.

Another limitation to the method is the omission of adaptation capability of individual species. While certain species (e.g., *Elaphe sauromates*, *Zamenis longissimus*) are known for their sensitivity to environmental changes, especially the human-induced ones [101,102], other species (e.g., *Bufo viridis*, *Lacerta viridis*, *Natrix natrix*) exhibit ecological plasticity [103,104]. Thus generating the MDRs and inferring the results need to be based on knowledge of the species' biology.

Supporting Information

Supporting Information S1 Georeferenced occurrence points for *Dolichophis caspius*.

Supporting Information S2 Maps indicating areas in which used climatic variables were most dissimilar between present and future climatic conditions. Each color corresponds to a variable that in the future will be very dissimilar compared with current conditions. Red corresponds to minimum temperature of the coldest month, Light blue corresponds to mean temperature of the coldest quarter, Yellow corresponds to temperature isothermality, Dark blue corresponds to mean diurnal range, and Purple corresponds to precipitation of driest month.

Supporting Information S3 Maps corresponding to each cost raster used to predict future dispersal (Maximum Dispersal Range) in context of the climate change predictions. Colors represent an gradient of cost, from blue (lowest cost) to red (highest cost).

Supporting Information S4 Environmental variables used in the MDR Analysis, reclassification ranges, "cost" values assigned and sources for the datasets.

Supporting Information S5 Additional information regarding environmental data acquisition.

Supporting Information S6 Additional information regarding the creation of the "cost" layer for the MDR Analysis.

Acknowledgments

T. C. S. is indebted to D. Murariu and G. Chisamera for numerous productive discussions regarding the manuscript. The authors would like to thank D. Ş. Mǎntoiu (Institute of Speleology of the Romanian Academy) for his advices in using ArcGIS.

Author Contributions

Conceived and designed the experiments: IG TCS MP AS SRZ. Performed the experiments: IG MP. Analyzed the data: IG TCS MP. Contributed reagents/materials/analysis tools: MP SRZ. Wrote the paper: IG TCS MP AS SRZ. Proofread and edit the manuscript: TCS MP IG AS SRZ. Literature review: TCS IG. Georeferenced the locality data: TCS.

References

1. Peterson AT, Ortega-Huerta MA, Bartley J, Sanchez-Cordero V, Soberon J, et al. (2002) Future projections for Mexican faunas under global climate change scenarios. Nature 416: 626–629.
2. Parmesan C, Yohe G (2003) A globally coherent fingerprint of climate change impacts across natural systems. Nature 421: 37–42.
3. Root TL, Price JT, Hall KR, Schneider SH, Rosenzweig C, et al. (2003) Fingerprints of global warming on wild animals and plants. Nature 421: 57–60.
4. Pounds JA, Fogden MPL, Campbell JH (1999) Biological response to climate change on a tropical mountain. Nature 398: 611–615.
5. Both C, Visser ME (2005) The effect of climate change on the correlation between avian life-history traits. Global Change Biology 11: 1606–1613.
6. Winkler DW, Dunn PO, McCulloch CE (2002) Predicting the effects of climate change on avian life-history traits. Proceedings of the National Academy of Sciences of the United States of America 99: 13595–13599.
7. Pitelka LF, Gardner RH, Ash J, Berry S, Gitay H, et al. (1997) Plant migration and climate change. American Scientist 85: 464–473.
8. Pearson RG, Dawson TP (2005) Long-distance plant dispersal and habitat fragmentation: identifying conservation targets for spatial landscape planning under climate change. Biological Conservation 123: 389–401.
9. Malcolm JR, Markham A, Neilson RP, Garaci M (2002) Estimated migration rates under scenarios of global climate change. Journal of Biogeography 29: 835–849.
10. Keith DA, Akcakaya HR, Thuiller W, Midgley GF, Pearson RG, et al. (2008) Predicting extinction risks under climate change: coupling stochastic population models with dynamic bioclimatic habitat models. Biology Letters 4: 560–563.
11. Jetz W, Wilcove DS, Dobson AP (2007) Projected impacts of climate and land-use change on the global diversity of birds. PLoS Biology 5: 1211–1219.
12. Travis JMJ (2003) Climate change and habitat destruction: a deadly anthropogenic cocktail. Proceedings of the Royal Society B-Biological Sciences 270: 467–473.
13. Colwell RK, Brehm G, Cardelus CL, Gilman AC, Longino JT (2008) Global warming, elevational range shifts, and lowland biotic attrition in the wet tropics. Science 322: 258–261.
14. Keith K, Berry KH, Weigand JF (2008) When desert tortoises are rare: Testing a new protocol for assessing status. California Fish and Game 94: 75–97.
15. McLaughlin JF, Hellmann JJ, Boggs CL, Ehrlich PR (2002) Climate change hastens population extinctions. Proceedings of the National Academy of Sciences of the United States of America 99: 6070–6074.
16. Parmesan C (2006) Ecological and evolutionary responses to recent climate change. Annual Review of Ecology Evolution and Systematics 37: 637–669.
17. Gibbons JW, Scott DE, Ryan TJ, Buhlmann KA, Tuberville TD, et al. (2000) The global decline of reptiles, Deja Vu amphibians. Bioscience 50: 653–666.
18. Driscoll DA (2004) Extinction and outbreaks accompany fragmentation of a reptile community. Ecological Applications 14: 220–240.

19. Foufopoulos J, Kilpatrick AM, Ives AR (2011) Climate Change and Elevated Extinction Rates of Reptiles from Mediterranean Islands. American Naturalist 177: 119–129.
20. Jaggi C, Baur B (1999) Overgrowing forest as a possible cause for the local extinction of Vipera aspis in the northern Swiss Jura mountains. Amphibia-Reptilia 20: 25–34.
21. Richman AD, Case TJ, Schwaner TD (1988) Natural and Unnatural Extinction Rates of Reptiles on Islands. American Naturalist 131: 611–630.
22. Fisher MC, Henk DA, Briggs CJ, Brownstein JS, Madoff LC, et al. (2012) Emerging fungal threats to animal, plant and ecosystem health. Nature 484: 186–194.
23. Pounds JA, Bustamante MR, Coloma LA, Consuegra JA, Fogden MPL, et al. (2006) Widespread amphibian extinctions from epidemic disease driven by global warming. Nature 439: 161–167.
24. Kiesecker JM, Blaustein AR, Belden LK (2001) Complex causes of amphibian population declines. Nature 410: 681–684.
25. Stuart SN, Chanson JS, Cox NA, Young BE, Rodrigues ASL, et al. (2004) Status and trends of amphibian declines and extinctions worldwide. Science 306: 1783–1786.
26. Harvell CD, Mitchell CE, Ward JR, Altizer S, Dobson AP, et al. (2002) Ecology - Climate warming and disease risks for terrestrial and marine biota. Science 296: 2158–2162.
27. Hof C, Araujo MB, Jetz W, Rahbek C (2011) Additive threats from pathogens, climate and land-use change for global amphibian diversity. Nature 480: 516–U137.
28. Sinervo B, Mendez-de-la-Cruz F, Miles DB, Heulin B, Bastiaans E, et al. (2010) Erosion of Lizard Diversity by Climate Change and Altered Thermal Niches. Science 328: 894–899.
29. Ihlow F, Dambach J, Engler JO, Flecks M, Hartmann T, et al. (2012) On the brink of extinction? How climate change may affect global chelonian species richness and distribution. Global Change Biology 18: 1520–1530.
30. Reading CJ, Luiselli LM, Akani GC, Bonnet X, Amori G, et al. (2010) Are snake populations in widespread decline? Biology Letters 6: 777–780.
31. Nix HA (1986) A biogeographic analysis of Australian elapid snakes. In: Longmore R, editor. Atlas of elapid snakes of Australia. Canberra: Australian Government Publishing Service. 4–15.
32. Stockwell D, Peters D (1999) The GARP modelling system: problems and solutions to automated spatial prediction. International Journal of Geographical Information Science 13: 143–158.
33. Hirzel AH, Hausser J, Chessel D, Perrin N (2002) Ecological-niche factor analysis: how to compute habitat-suitability maps without absence data? Ecology 83: 2027–2036.
34. Phillips S, Anderson R, Schapire R (2006) Maximum entropy modeling of species geographic distributions. Ecological Modelling 190: 231–259.
35. Pearson RG, Dawson TP (2003) Predicting the impacts of climate change on the distribution of species: are bioclimate envelope models useful? Global Ecology and Biogeography 12: 361–371.
36. Phillips SJ, Anderson RP, Schapire RE (2006) Maximum entropy modeling of species geographic distributions. Ecological Modelling 190: 231–259.
37. Elith J, Leathwick JR (2009) Species Distribution Models: Ecological Explanation and Prediction Across Space and Time. Annual Review of Ecology Evolution and Systematics 40: 677–697.
38. Bombi P, Salvi D, Vignoli L, Bologna MA (2009) Modelling Bedriaga's rock lizard distribution in Sardinia: An ensemble approach. Amphibia-Reptilia 30: 413–424.
39. DeMatteo KE, Loiselle BA (2008) New data on the status and distribution of the bush dog (Speothos venaticus): Evaluating its quality of protection and directing research efforts. Biological Conservation 141: 2494–2505.
40. Papeş M, Gaubert P (2007) Modelling ecological niches from low numbers of occurrences: assessment of the conservation status of poorly known viverrids (Mammalia, Carnivora) across two continents. Diversity and Distributions 13: 890–902.
41. Owens HL, Bentley AC, Peterson AT (2012) Predicting suitable environments and potential occurrences for coelacanths (Latimeria spp.). Biodiversity and Conservation 21: 577–587.
42. Puschendorf R, Carnaval AC, VanDerWal J, Zumbado-Ulate H, Chaves G, et al. (2009) Distribution models for the amphibian chytrid Batrachochytrium dendrobatidis in Costa Rica: proposing climatic refuges as a conservation tool. Diversity and Distributions 15: 401–408.
43. Moussalli A, Moritz C, Williams SE, Carnaval AC (2009) Variable responses of skinks to a common history of rainforest fluctuation: concordance between phylogeography and palaeo-distribution models. Molecular Ecology 18: 483–499.
44. Iverson LR, Prasad A, Schwartz MW (1999) Modeling potential future individual tree-species distributions in the eastern United States under a climate change scenario: a case study with Pinus virginiana. Ecological Modelling 115: 77–93.
45. Lee DN, Papes M, Van Den Bussche RA (2012) Present and Potential Future Distribution of Common Vampire Bats in the Americas and the Associated Risk to Cattle. Plos One 7.
46. Dormann CF, McPherson JM, Araujo MB, Bivand R, Bolliger J, et al. (2007) Methods to account for spatial autocorrelation in the analysis of species distributional data: a review. Ecography 30: 609–628.
47. Cowling SA, Maslin MA, Sykes MT (2001) Paleovegetation simulations of lowland Amazonia and implications for neotropical allopatry and speciation. Quaternary Research 55: 140–149.
48. Raxworthy CJ, Ingram CM, Rabibisoa N, Pearson RG (2007) Applications of ecological niche modeling for species delimitation: A review and empirical evaluation using day geckos (Phelsuma) from Madagascar. Systematic Biology 56: 907–923.
49. Wiens JJ (2011) The Causes of Species Richness Patterns across Space, Time, and Clades and the Role of "Ecological Limits". Quarterly Review of Biology 86: 75–96.
50. Wiens JJ (2004) What is speciation and how should we study it? American Naturalist 163: 914–923.
51. Doko T, Fukui H, Kooiman A, Toxopeus AG, Ichinose T, et al. (2011) Identifying habitat patches and potential ecological corridors for remnant Asiatic black bear (Ursus thibetanus japonicus) populations in Japan. Ecological Modelling 222: 748–761.
52. Wilson JJ, Lantz VA, MacLean DA (2010) A benefit-cost analysis of establishing protected natural areas in New Brunswick, Canada. Forest Policy and Economics 12: 94–103.
53. Peterson AT, Papes M, Kluza DA (2003) Predicting the potential invasive distributions of four alien plant species in North America. Weed Science 51: 863–868.
54. Peterson AT (2001) Predicting species' geographic distributions based on ecological niche modeling. Condor 103: 599–605.
55. Araujo MB, Pearson RG, Thuiller W, Erhard M (2005) Validation of species-climate impact models under climate change. Global Change Biology 11: 1504–1513.
56. Peterson AT, Sanchez-Cordero V, Soberon J, Bartley J, Buddemeier RW, et al. (2001) Effects of global climate change on geographic distributions of Mexican Cracidae. Ecological Modelling 144: 21–30.
57. Peterson AT, Shaw J (2003) Lutzomyia vectors for cutaneous leishmaniasis in Southern Brazil: ecological niche models, predicted geographic distributions, and climate change effects. International Journal for Parasitology 33: 919–931.
58. Thuiller W, Lavorel S, Araujo MB (2005) Niche properties and geographical extent as predictors of species sensitivity to climate change. Global Ecology and Biogeography 14: 347–357.
59. Thuiller W, Lavorel S, Araujo MB, Sykes MT, Prentice IC (2005) Climate change threats to plant diversity in Europe. Proceedings of the National Academy of Sciences of the United States of America 102: 8245–8250.
60. Hijmans RJ, Graham CH (2006) The ability of climate envelope models to predict the effect of climate change on species distributions. Global Change Biology 12: 2272–2281.
61. Peterson AT, Soberon J, Sanchez-Cordero V (1999) Conservatism of ecological niches in evolutionary time. Science 285: 1265–1267.
62. Warren DL, Glor RE, Turelli M (2008) Environmental Niche Equivalency Versus Conservatism: Quantitative Approaches to Niche Evolution. Evolution 62: 2868–2883.
63. Pearson RG, Dawson TP, Berry PM, Harrison PA (2002) SPECIES: A Spatial Evaluation of Climate Impact on the Envelope of Species. Ecological Modelling 154: 289–300.
64. Davis AJ, Jenkinson LS, Lawton JH, Shorrocks B, Wood S (1998) Making mistakes when predicting shifts in species range in response to global warming. Nature 391: 783–786.
65. Woodward FI, Beerling DJ (1997) The dynamics of vegetation change: health warnings for equilibrium 'dodo' models. Global Ecology and Biogeography Letters 6: 413–418.
66. Schcherbak NN, Böhme W (1993) *Coluber caspius* Gmelin, 1789– Kaspische Pfeilnatter oder Springnatter. In: Böhme W, editor. Handbuch der Reptilien und Amphibien Europas, Band 3/I, Schlangen (Serpentes) I. Wiesbaden: AULA-Verlag. 83–96.
67. Fuhn IE, Vancea Ş (1961) Reptilia. The fauna of the People's Republic of Romania. Bucharest, Romania: Editura Academiei R.P.R.
68. Zinner H (1972) Systematics and evolution of the species group *Coluber jugularis* Linnaeus, 1758 - *Coluber caspius* Gmelin, 1789 (Reptilia, Serpentes). Jerusalem: Hebrew University. 62 p.
69. Iftime A (2005) Reptile. In: Botnariuc N, Tatole V, editors. Cartea Rosie a vertebratelor României. Bucharest, Romania: Curtea Veche. 173–196.
70. Nagy ZT, Bellaagh M, Wink M, Paunovic A, Korsos Z (2010) Phylogeography of the Caspian whipsnake in Europe with emphasis on the westernmost populations. Amphibia-Reptilia 31: 455–461.
71. Cremene C, Groza G, Rakosy L, Schileyko AA, Baur A, et al. (2005) Alterations of steppe-like grasslands in Eastern Europe: a threat to regional biodiversity hotspots. Conservation Biology 19: 1606–1618.
72. Kovacs-Lang E, Kroel-Dulay G, Kertesz M, Fekete G, Bartha S, et al. (2000) Changes in the composition of sand grasslands along a climatic gradient in Hungary and implications for climate change. Phytocoenologia 30: 385–407.
73. Moreira F, Russo D (2007) Modelling the impact of agricultural abandonment and wildfires on vertebrate diversity in Mediterranean Europe. Landscape Ecology 22: 1461–1476.
74. Mosier AR, Delgado JA, Cochran VL, Valentine DW, Parton WJ (1997) Impact of agriculture on soil consumption of atmospheric CH_4 and a comparison of CH_4 and N_2O flux in subarctic, temperate and tropical grasslands. Nutrient Cycling in Agroecosystems 49: 71–83.
75. FAO (2005) Grasslands of the World; J.M. S, S.G. R, C. B, editors. Rome.

76. Warren DL, Glor RE, Turelli M (2010) ENMTools: a toolbox for comparative studies of environmental niche models. Ecography 33: 607–611.
77. Hijmans RJ, Cameron S, Parra J (2005) WorldClim, Version 1.4. Available: http://www.worldclim.org. Berkeley: University of California.
78. Hijmans RJ, Cameron SE, Parra JL, Jones PG, Jarvis A (2005) Very high resolution interpolated climate surfaces for global land areas. International Journal of Climatology 25: 1965–1978.
79. Ramirez J, Jarvis A (2008) High Resolution Statistically Downscaled Future Climate Surfaces. Cali, Colombia: International Center for Tropical Agriculture (CIAT).
80. Solomon S, Qin D, Manning M, Chen Z, Marquis M, et al. (2007) Contribution of Working Group I to the Fourth Assessment Report of the Intergovernmental Panel on Climate Change, 2007. Cambridge, United Kingdom and New York, NY, USA.
81. Beaumont LJ, Hughes L, Poulsen M (2005) Predicting species distributions: use of climatic parameters in BIOCLIM and its impact on predictions of species' current and future distributions. Ecological Modelling 186: 250–269.
82. Soberón J, Peterson AT (2005) Interpretation of models of fundamental ecological niches and species' distributional areas. Biodiversity Informatics 2: 1–10.
83. Barve N, Barve V, Jimenez-Valverde A, Lira-Noriega A, Maher SP, et al. (2011) The crucial role of the accessible area in ecological niche modeling and species distribution modeling. Ecological Modelling 222: 1810–1819.
84. Scachetti-Pereira R (2001) Desktop GARP. Available: http://wwwlifemapperorg/desktopgarp/. Accessed 2012 September 14.
85. Elith J, Graham CH, Anderson RP, Dudik M, Ferrier S, et al. (2006) Novel methods improve prediction of species' distributions from occurrence data. Ecography 29: 129–151.
86. Hernandez PA, Graham CH, Master LL, Albert DL (2006) The effect of sample size and species characteristics on performance of different species distribution modeling methods. Ecography 29: 773–785.
87. Pearson RG, Thuiller W, Araujo MB, Martinez-Meyer E, Brotons L, et al. (2006) Model-based uncertainty in species range prediction. Journal of Biogeography 33: 1704–1711.
88. Peterson AT, Papes M, Eaton M (2007) Transferability and model evaluation in ecological niche modeling: a comparison of GARP and Maxent. Ecography 30: 550–560.
89. Renner IW, Warton DI (2013) Equivalence of MAXENT and Poisson Point Process Models for Species Distribution Modeling in Ecology. Biometrics 69: 274–281.
90. Elith J, Kearney M, Phillips S (2010) The art of modelling range-shifting species. Methods in Ecology and Evolution 1: 330–342.
91. Elith J, Phillips SJ, Hastie T, Dudik M, Chee YE, et al. (2011) A statistical explanation of MaxEnt for ecologists. Diversity and Distributions 17: 43–57.
92. Anderson RP, Lew D, Peterson AT (2003) Evaluating predictive models of species' distributions: Criteria for selecting optimal models. Ecological Modelling 162: 211–232.
93. ESRI (2011) ArcGIS Desktop: Release 10. 10 ed. Redlands: Environmental Systems Research Institute, Inc.
94. Peterson AT, Papes M, Soberon J (2008) Rethinking receiver operating characteristic analysis applications in ecological niche modeling. Ecological Modelling 213: 63–72.
95. Araujo MB, Thuiller W, Pearson RG (2006) Climate warming and the decline of amphibians and reptiles in Europe. Journal of Biogeography 33: 1712–1728.
96. Hays GC, Broderick AC, Glen F, Godley BJ (2003) Climate change and sea turtles: a 150-year reconstruction of incubation temperatures at a major marine turtle rookery. Global Change Biology 9: 642–646.
97. Thuiller W, Araujo MB, Lavorel S (2004) Do we need land-cover data to model species distributions in Europe? Journal of Biogeography 31: 353–361.
98. Wiens JA (1989) Spatial Scaling in Ecology. Functional Ecology 3: 385–397.
99. Austin MP, Van Niel KP (2011) Improving species distribution models for climate change studies: variable selection and scale. Journal of Biogeography 38: 1–8.
100. Sillero N (2011) What does ecological modelling model? A proposed classification of ecological niche models based on their underlying methods. Ecological Modelling 222: 1343–1346.
101. Tibu PL, Strugariu A (2007) New record for the blotched snake Elaphe sauromates (Reptilia : Colubridae) in Romania. North-Western Journal of Zoology 3: 62–65.
102. Bonnet X, Guy N, Shine R (1999) The dangers of leaving home: dispersal and mortality in snakes. Biological Conservation 89: 39–50.
103. Sinsch U, Leskovar C, Drobig A, Konig A, Grosse WR (2007) Life-history traits in green toad (Bufo viridis) populations: indicators of habitat quality. Canadian Journal of Zoology-Revue Canadienne De Zoologie 85: 665–673.
104. Strugariu A, Hutuleac-Volosciuc MV, Puscasu MC, Sahlean TC, Gherghel I (2007) Preliminary aspects concerning the herpetofauna from urban and peri-urban environments from North-Eastern Romania: A case study in the city of Suceava. Herpetologica Romanica 1: 31–43.

18

Modeling Fence Location and Density at a Regional Scale for Use in Wildlife Management

Erin E. Poor[1,2]*, Andrew Jakes[3], Colby Loucks[2], Mike Suitor[3,4]

1 Department of Fish and Wildlife Conservation, Virginia Tech, Blacksburg, Virginia, United States of America, **2** Conservation Science Program, World Wildlife Fund, Washington D.C., United States of America, **3** Faculty of Environmental Design, University of Calgary, Calgary, Canada, **4** Environment Yukon, Dawson City, Yukon, Canada

Abstract

Barbed and woven wire fences, common structures across western North America, act as impediments to wildlife movements. In particular, fencing influences pronghorn (*Antilocapra americana*) daily and seasonal movements, as well as modifying habitat selection. Because of fencing's impacts to pronghorn and other wildlife, it is a potentially important factor in both wildlife movement and habitat selection models. At this time, no geospatial fencing data is available at regional scales. Consequently, we constructed a regional fence model using a series of land tenure assumptions for the Hi-Line region of northern Montana – an area consisting of 13 counties over 103,400 km^2. Randomized 3.2 km long transects (n = 738) on both paved and unpaved roads were driven to collect information on habitat, fence densities and fence type. Using GIS, we constructed a fence location and a density model incorporating ownership, size, neighboring parcels, township boundaries and roads. Local knowledge of land ownership and land use assisted in improving the final models. We predict there is greater than 263,300 km of fencing in the Hi-Line region, with a maximum density of 6.8 km of fencing per km^2 and mean density of 2.4 km of fencing per km^2. Using field data to assess model accuracy, Cohen's Kappa was measured at 0.40. On-the-ground fence modification or removal could be prioritized by identifying high fence densities in critical wildlife areas such as pronghorn migratory pathways or sage grouse lekking habitat. Such novel fence data can assist wildlife and land managers to assess effects of anthropogenic features to wildlife at various scales; which in turn may help conserve declining grassland species and overall ecological functionality.

Editor: Justin David Brown, University of Georgia, United States of America

Funding: The work was supported by: Kathryn Fuller Science for Nature Fund, (no grant number) (http://worldwildlife.org/initiatives/fuller-science-for-nature-fund); Alberta Conservation Association, 030-00-40-106-4000 (http://www.ab-conservation.com/); U.S. Bureau of Land Management, 090-50778 (http://www.blm.gov/wo/st/en.html); Montana Fish, Wildlife & Parks, Cooperative Agreement 080018 (http://fwp.mt.gov/); Saskatchewan Ministry of Environment, Permit 08FW111 (http://www.environment.gov.sk.ca/); University of Calgary, (no grant number) (http://www.ucalgary.com/). The funders had no role in study design, data collection and analysis, decision to publish, or preparation of the manuscript.

Competing Interests: The authors have declared that no competing interests exist.

* E-mail: erinpoor@vt.edu

Introduction

Worldwide, most long-distance terrestrial migrations have been lost [1] or greatly reduced [2] largely due to anthropogenic factors creating barriers to wildlife movements [1]. Restricted access to food and water is a primary threat to migrations worldwide [1]. In particular, fencing has played a large role in ungulate population declines where fencing closes off parks, delineates national boundaries and separates rangelands, thus physically cutting off access to necessary resources [3–8,1] and in grassland bird decline, where birds may not be able to see fencing and die due to collision or entanglement [9–11]. In Africa, fences were implemented in the 1950's to prohibit transfer of wildlife disease between livestock and wildlife populations [1]. In Kruger National Park, fencing restricted migrations of wildebeest and the population, cut off from seasonal water sources, declined by nearly 88% [3]. In North America, 75% of seasonal migrations, mostly those of bison (*Bison bison*) and pronghorn antelope (*Antilocapra americana*) have been lost largely due to overhunting and disruption of migration routes [2] and it is hypothesized that fence collisions have contributed to the decline of sage grouse populations [12].

Wire fencing was introduced in the North American west by the homesteaders of the 19th century to avoid importing timber or stone to create barriers around their land [13]. By the late 19th century, barbed wire had been invented and was being produced commercially at the high demand of homesteaders wishing to claim their land [14]. By 1880–1884, barbed wire production reached a peak with an estimated 643,000–965,000 km being produced annually [13]. Eventually ranchers and farmers in the west began using fencing pervasively. Subsequently, access to roads and water holes were inadvertently restricted for cattle [14].

Today much of the fencing erected in the 1800's still exists and is profuse across northern Montana. It is used by land owners to delineate property, section off agricultural fields, demarcate parcel boundaries of the same ownership, corral cattle, and along roads for safety. Fencing type in northern Montana varies with land ownership and operation type. Three and four strand barbed wire are most common, however, mesh woven fences are also common in the Hi-Line region of northern Montana (A. Jakes, University of Calgary, *unpublished data*) and these wire fences can pose significant impediments to the movements of wildlife, thus modifying habitat availability and migration opportunities of various native species [1,3,7]. As such, fences can influence a particular species movement patterns at multiple spatial and temporal scales [15]. Species perceive the surrounding landscape differently from one another to fulfill life-history requirements [16,17]. Appreciating that wildlife may have multi-scaled responses to anthropogenic factors such as fences requires that we too address anthropogenic factors at various scales.

Loss of migrations could result in extirpation of wildlife populations, resulting in overall contraction of species range due to decreased access to forage and safe calving grounds. Thus, to conserve migrations and allow continued movement to optimal seasonal ranges, conservation and maintenance of habitat outside of protected areas is required. Because wildlife generally track high quality forage and water sources through their migrations, erecting fencing along or through migration routes or breeding habitat can restrict access to necessary habitat, resulting in declines in population numbers [1,18,11]. For example, in a study in Colorado and Utah, USA, most mule deer (*Odocoileus hemionus*), elk (*Cervus canadensis*) and pronghorn (*Antilocapra americana*) fence-related mortalities were due to becoming entangled in fence wires [19]. Furthermore pronghorn density and fence-related mortality were positively correlated; indicating fencing may negatively impact pronghorn populations [19]. In addition, increased indirect mortalities of fawns being separated from does was observed from woven wire fences as opposed to plain barbed wire fences and additionally, mortalities increased with increasing fence height [19]. In a region where fencing is a common landscape feature such as northern Montana, migratory opportunities could be diminished and as a result, population numbers of wildlife could suffer declines unless landscape permeability is maintained [20].

Although North American migrations are among the best studied, and despite the importance of fencing information on wildlife movements at multiple scales, regional datasets on fence locations and fence densities do not exist. In this study we create a regional fence location and density model for the Hi-Line area of northern Montana. To aid in future studies identifying wildlife movement pathways or identifying factors affecting seasonal habitat selections we determined it was important to have a readily available predicted fence location and density information for this region. We provide methods for modeling fence locations and density at a large scale, using publicly available information in an effort to encourage the creation of fence spatial datasets for wildlife research and management in other regions.

Methods

Study Area

The study area is comprised of 13 counties in the Montana Hi-Line region of northern Montana (Figure 1). These counties are bounded by the Canadian border on the north and the Marias and Missouri Rivers to the south (Figure 1). The study area was chosen initially due to ongoing research on pronghorn migrations and movements at the northern terminus of pronghorn range (completed under wildlife capture and handling permit #11-2007 from the Montana Fish, Wildlife & Parks Institutional Animal Care and Use Committee). The total area for which fencing was modeled was 103,426 km^2 (Table 1). Privately owned land is the dominate tenure type, more than double public land ownership (Table 1, Figure 1).

Ground Truth Surveys

Transect Identification. A seamless land cover dataset developed by bordering state and provincial wildlife agencies was clipped to the bounds of the study area. From this land cover data, three generalized habitat type regions were delineated by extracting polygons around land cover of similar types. Habitat types included "grassland", "agriculture" and "shrubland". To create a fourth generalized habitat type, we used removed areas not previously defined and classified as "mixed" habitat to include the remaining areas. Next, we identified roads, paved or unpaved, from the roads dataset created by the Northern Sagebrush Steppe Initiative and intersected this layer with the generalized habitat type region within the Montana portion of the Northern Sagebrush Steppe Initiative land cover data. New areas were identified, based on both generalized habitat type and road type, leaving a total of eight different classifications: Grass/Unpaved; Grass/Paved; Agriculture/Unpaved; Agriculture/Paved; Shrub/Unpaved; Shrub/Paved; Mix/Unpaved and; Mix/Paved. Within each of these generalized areas, random points along roads were generated, each which served as the middle point of each 3.2 km road transect for field surveying. This distance was used to capture a majority of land tenure changes within the "checkerboard pattern" of land ownership found within this area. Random points were generated at a minimum distance of 3.5 km apart, to avoid transect overlap. Each random point had a unique transect associated with it and unique numbers were generated and manually entered for each random point. Transect identification took place in ArcGIS 9.3 (ESRI 2009) and random points and transect regions coordinates were defined in WGS 1984.

Field Surveys. During June-August 2009, 2,362 km of fence along roadside transects were sampled in Blain, Phillips, Valley and Daniels Counties (Figure 2). In ArcGIS 9.3 (ESRI 2009), we created random midpoints for 3.2 km long survey transects. At each point along the 3.2 km transect that a roadside fencing ended, appeared, or changed in fence structure, a GPS location was recorded. Locations were also recorded at every interior fence and road intersection. Information on GPS locations, transect number and heading, fence structure, ground cover and road type was recorded. Locations and structure of internal fencing and twinning and tripled roadside fences were also recorded. Fencing had to be within 200 m of the road to be considered along the roadside and therefore within the transect. Additionally, changes in fence structure had to be longer than 100 m to be recorded. GPS waypoints were downloaded after each field day to create an overall layer of fence changes along roadsides or where interior fences converged with roadsides across the area. Complete sampling protocols (Text S1) were created to standardize methodologies for unique fencing schematics found across the landscape and are applicable to all survey areas across the study area.

Fencing Location Modeling

Fence location and density were modeled using private land ownership data provided by the Montana Public Land Ownership dataset in ArcGIS 10.0 (ESRI 2012). We predicted fence presence based on parcel ownership, size and ownership adjacency. We used publicly available free datasets including land tenure data from the Montana Cadastral Database (Montana Department of Administration/Information Technology Services, 2011), pasture data provided by the Bureau of Land Management (Bureau of Land Management, 2011), the 2010 TIGER roads dataset (US Census Bureau, 2010) and land cover from the National Gap Analysis Program Land Cover Dataset (United States Geologic Service 2000) and the 2000 Land Cover for Agricultural Regions of Canada (Canada Agri-Geomatics Service 2000) national land cover dataset.

Historically, most ranches were based on 'sections' (2.59 km^2) in this region of Montana, and we use this unit of measurement as our base unit of area. We began the modeling process with the land ownership polygon dataset. We edited this dataset based on a series of assumptions about where fencing exists around parcels,

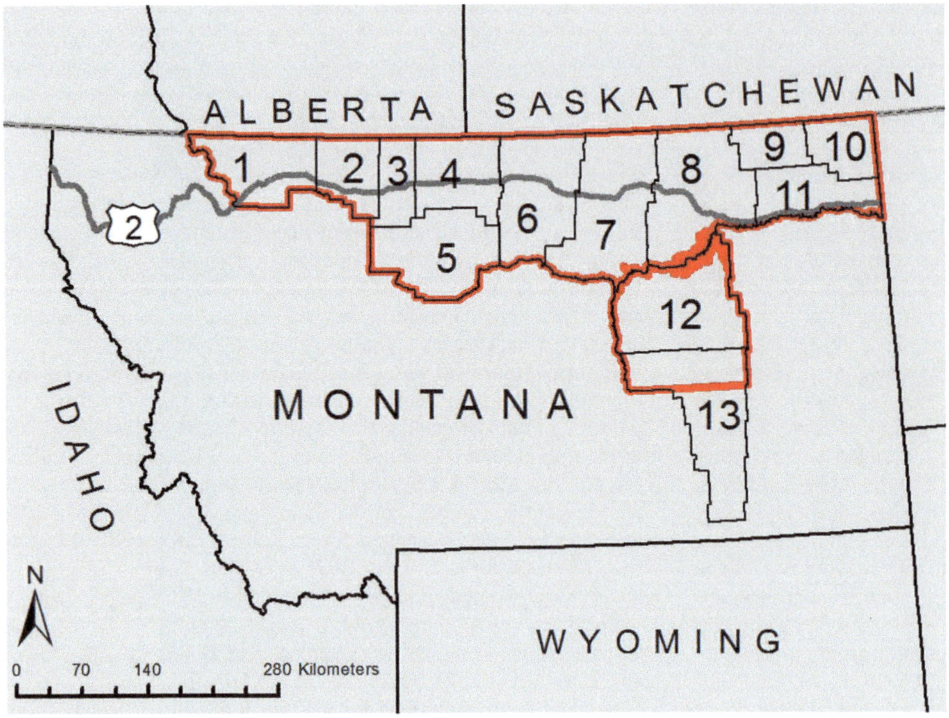

Figure 1. The location of the study area, the Hi-Line region of Montana (red), 2011. Counties included in the study are: 1 Glacier; 2 Toole; 3 Liberty; 4 Hill; 5 Chouteau; 6 Blaine; 7 Phillips; 8 Valley; 9 Daniels; 10 Sheridan; 11 Roosevelt; 12 Garfield; 13 Rosebud (northern half).

along roads and where it defines crop fields (Text S2). The remaining outlines of polygons after applying these assumptions and their associated GIS functions represented potential fencing. For example, we assumed private lands with the same mailing address <½ section in size are not fenced if adjacent to each other, so parcels <½ section that are privately owned are merged together in the GIS polygon dataset and the boundary of the two merged parcels is a fence. To create the assumptions, we consulted a variety of local experts, including BLM, Montana Fish, Wildlife and Parks, World Wildlife Fund, Montana Department of Natural Resources & Conservation, Rocky Boys Indian Reservation, Fort Belknap Indian Reservation and, Fort Peck Indian Reservation local personnel. Fence data for the Charles M. Russell National Wildlife Refuge (CMR) was provided, so this area was not included in the modeling efforts, but was added in for density estimates. We first modeled fencing by dissolving and merging the land tenure parcels based on the land tenure-based assumptions and then combined this dataset with the fencing modeled using the land cover-based assumptions. Within areas of large cropland, all land tenure-based fencing was removed, assuming there would be no parcel fencing within these areas, and fencing in these areas follows different rules (See below). Finally, roads-based fencing was overlaid on the two previously combined datasets.

Table 1. Land ownership within the Hi-Line region.

Ownership	Area (km^2)	% Study Area
Private	67879	65.63
US Bureau of Land Management	13736	13.28
Tribal	9401	9.09
State Government	7105	6.87
US Fish and Wildlife Service	2619	2.53
US National Park Service	1469	1.42
Water	416	0.40
USDA Forest Service	242	0.23
Local Government	213	0.21
US Bureau of Reclamation	205	0.20
Right of Way	85	0.08
US Department of Defense	26	0.03
US Department of Interior	19	0.02
Undetermined	7	0.01
US Government, Other	4	0.004
Total	103,426	

Land Tenure Fence Modeling. To model fencing based on land ownership, we removed all BLM land from the land tenure layer and replaced it with pastures data we received from the BLM. We assumed the pasture polygon outline represented fences and that no additional fencing would exist on the BLM. Next, we eliminated the boundaries of neighboring state lands, assuming that if more than one state-owned parcel was adjacent, they would be fenced together. State-owned lands that were surrounded by BLM lands were dissolved into the BLM lands, assuming that the state lands would be leased to the BLM. Bureau of Reclamation lands were then treated the same as state lands, in that if two or more parcels were adjacent, they would be fenced together. This process was repeated to remove boundaries separating more than one Fish and Wildlife Service parcel. For private and tribal lands, we first merged parcels based on ownership and adjacency (parcels owned by the same party were combined as one if they were adjacent) and then selected resulting parcels that were greater than

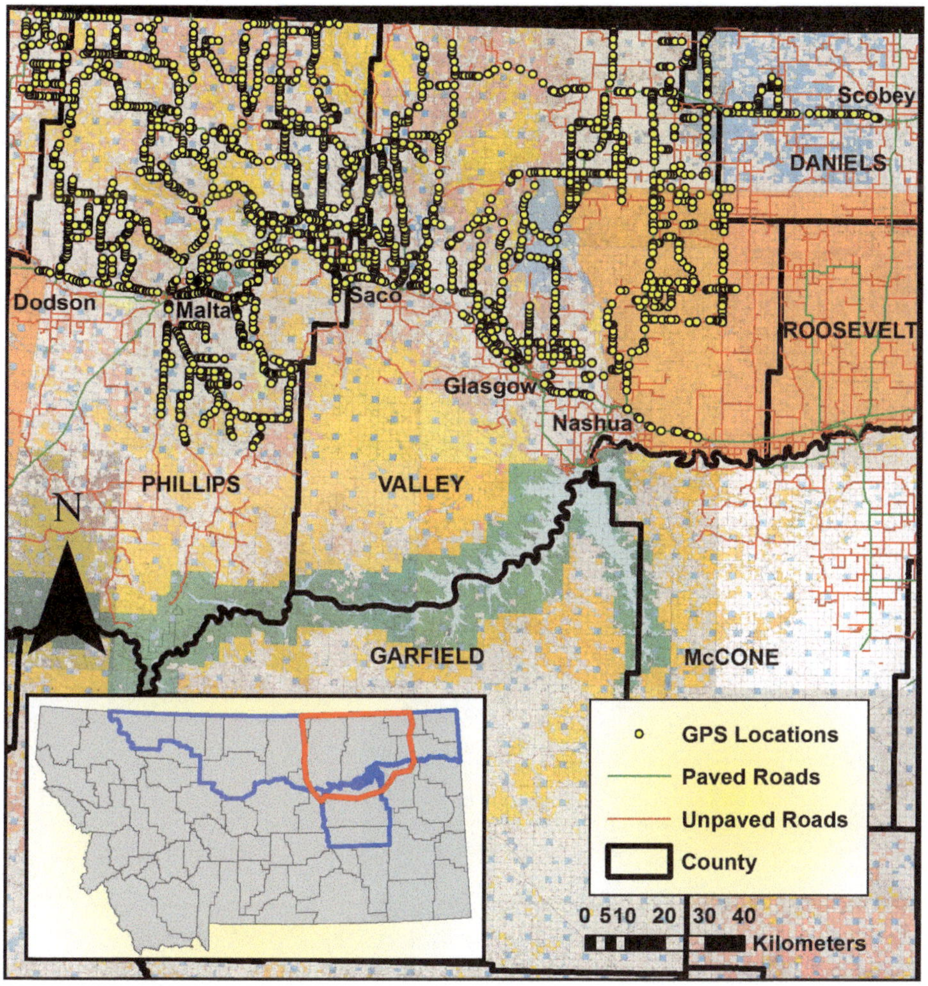

Figure 2. GPS locations of sampled fence transects during Summer 2009. Each GPS location represents a change in fence structure type or addition/deletion of fence along the sampled transect within four counties in northern Montana. Red outline – location of GPS sampling, Blue outline – the study area.

2 sections (5.2 km^2). These parcels were combined with the overall fence layer. The smaller parcels were dissolved into a neighboring parcel of similar size (5.2 km^2) and then combined with the fence dataset. Next, we removed fencing on National Park Service and Forest Service land. This resulted in the final layer representing land tenure fencing.

Road Fence Modeling. We identified primary and secondary roads and buffered them using the estimates of road width as buffer width. These buffer outlines, 19 m wide on primary roads, and 11 m wide on secondary roads then approximated fence lines along roads. These square-ended buffers were dissolved, merged and converted to lines. So fences would not bisect roads, we removed the buffer ends by removing line segments that were exactly 38 m long, in the case of primary roads, and 22 m long in the case of secondary roads. This process left fence lines on either side of these roads. To model fencing along local roads, we first removed roads from the CMR. Next, length was calculated and long local roads (≥1,200 m) were merged with primary and secondary roads. Because local roads were assumed to have fencing only on one side, we used these local road polylines to represent fencing along them. To include local fenced roads in the model without including the misclassified local roads which were two-tracks or driveways we then selected short roads (<1200 m) that intersected long local roads. This began an iterative process where increasingly smaller roads which intersected larger roads were selected and added to the fenced roads classification. This process ensured small line fragments (two tracks, trails, and possible data errors) from the roads dataset were not included in the fence dataset. We merged the fenced short local roads identified through this process with longer local roads. We buffered this combined dataset by 11 m on each side, and converted the buffer to a line. One side of this double line was erased, and the remaining line represented fencing. Again, the remaining buffer ends were removed, and the resulting dataset of one-sided fencing along local roads was combined with the double-sided fencing along primary and secondary roads, defined by the US Census TIGER roads classification.

Land Cover Fence Modeling. The land cover layer was first converted to a polygon shapefile. To remove inconsistencies and small sections of non-cropland within large areas of crop (usually wheat and corn in this area) we identified crops, dissolve these polygons together and then calculated their area. We did the same for non-crop land cover classes. Next, we identified non-crop polygons ≤½ section (1.3 km^2) that intersected large crop (≥3 sections or 7.8 km^2). These non-crop sections were dissolved into the large areas of crop.

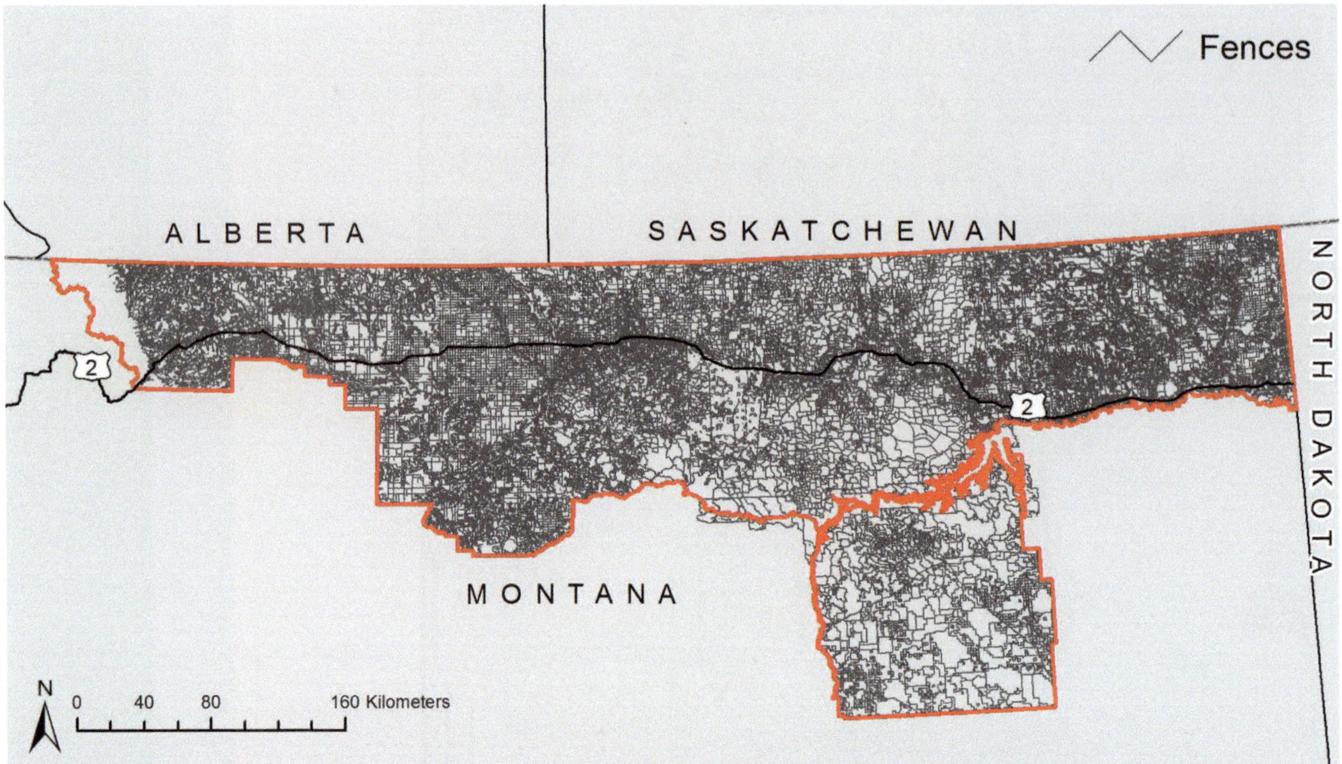

Figure 3. Modeled fences in the Montana Hi-Line region using land tenure, land cover and roads data.

Based on advice we received from experts, we assumed there would always be a fence line between large areas of cropland ($\geq$3 sections or 7.8 km^2) and larger areas of native prairie (half of a section or $>$1.3 km^2). These areas of prairie were identified and erased from the large sections of crop, which created a polygon boundary. This layer of large croplands was then converted to a raster, reclassified, and expanded from within to remove any remaining holes. This was then again converted to a polygon and represented fences around large croplands, and between large croplands and native prairie. We erased all land tenure fencing from the areas of large croplands, assuming the only other fencing on large croplands would be those along roads.

Fence Dataset Synthesis. To combine the three polygon fence layers, roads, land tenure and land cover, we first removed additional fence lines from the land cover layer from the BLM layer, again assuming the polygon outlines of the BLM pastures dataset would represent all of the fencing on this land type. We erased boundaries of large lakes and rivers polygons and lines that intersected study area boundary lines. The polygon layers of parcel fencing and land cover fencing were then combined and this layer was converted to a line dataset. Assuming that there would not be parcel boundary fence parallel to nearby roads fencing, we next removed roads fencing that were completely within a 20 m buffer of other types of fencing. Finally, the remaining roads fencing was merged with the land cover and land tenure fencing.

Model Accuracy Assessment

Because fencing along roads was modeled separately from internal parcel fencing, we completed individual accuracy assessments for these portions of the fence model, as well as a combined accuracy assessment. For the roads accuracy assessment, all fenced transect lines (a fence on either or both sides) were merged and clipped to the accuracy assessment study area to identify the true positives and all non-fenced transects were merged, clipped, and used to identify the true negatives. These line transects were then converted to points and buffered by 30 m to account for spatial error. Only modeled fencing from the roads fencing model were used in this assessment. We identified areas where buffered GPS points representing true positives and true negatives intersected modeled fences along roads, to result in the true positives and false positives, respectively. The number of true positives was subtracted from the total number of fence transect points to result in the number of false negatives. The number of false positives was subtracted from the total number of non-fenced transect points to identify the number of true negatives. Total number of samples was 1,832 for true positives and 469 for true negatives.

For the internal fencing accuracy assessment, all GPS points were merged and clipped to the accuracy assessment study area to identify the true positives. We selected GPS locations that were along fenced transects and these points were then buffered by 30 m. We isolated the modeled internal fencing by removing roads fencing and identifying intersections in the modeled internal fencing. The points of intersection were then buffered by 30 m. To identify true positives, we identified where these buffered modeled intersection points intersected the buffered GPS points. The true positives were then subtracted from the total number of GPS points used to result in the false positives. To identify the true negatives, the buffered areas around the GPS sample points were subtracted from the fenced transects to ensure we used actual areas with fences, but with no internal fences (indicated by GPS locations) for this assessment. On the remaining transect lines, where fencing was not observed, we created random points and buffered them by 30 m. Identifying locations where these buffered

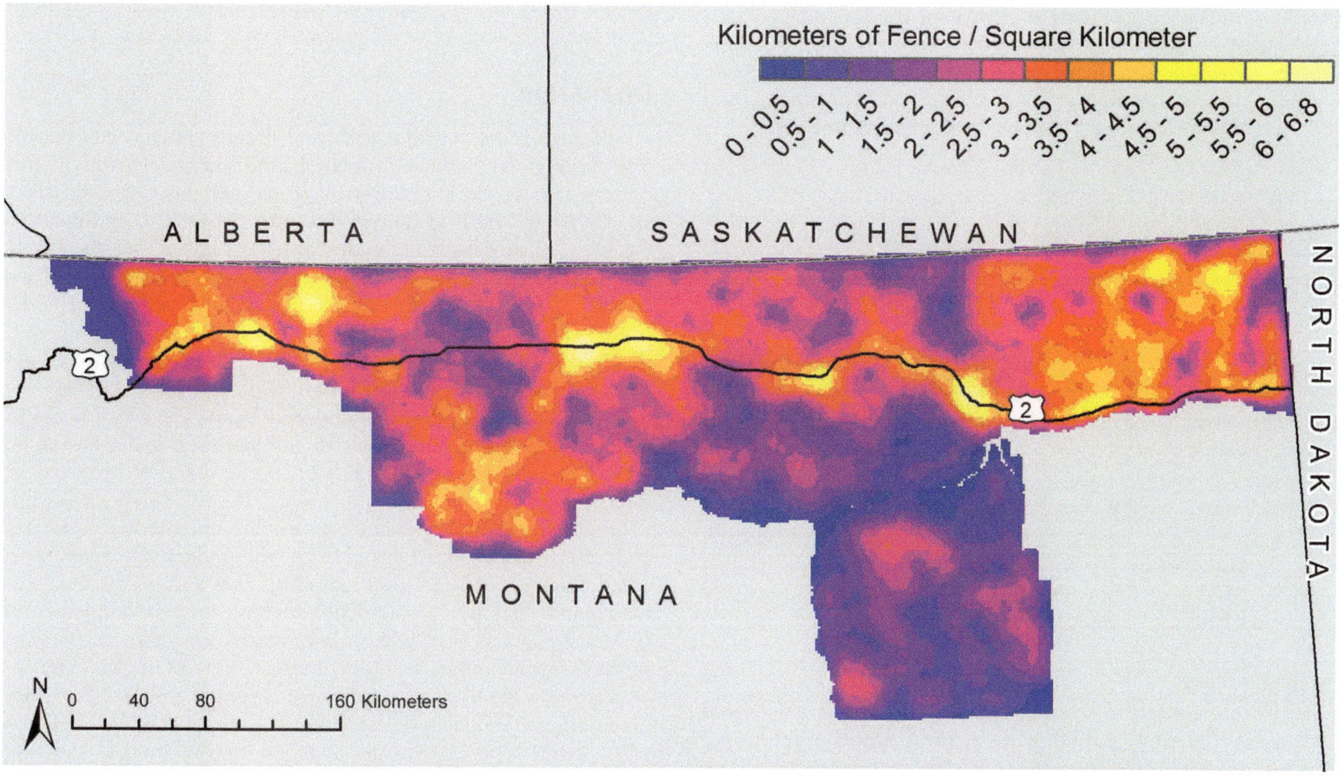

Figure 4. Fence density from modeled fence lines in the Montana Hi-Line region using publically available data.

points intersected modeled internal fences resulted in the false positives. The false positives were then subtracted from the total number of random points to result in the true negatives. Total number of samples was 1,333 for true positives and 1,290 for true negatives.

To calculate the total accuracy of the fence dataset, all GPS locations and the modeled fencing were buffered by 30 m to account for spatial error. Random points were created along the non-fenced transects and similarly buffered by 30 m. GPS locations intersecting modeled fencing resulted in the true positive rate and random points along non-fenced transects intersecting modeled fencing resulted in the false positive rate. The number of true positives were subtracted from the total number of GPS locations to result in the false negative rate and the number of false positives were subtracted from the number of non-fence transect random points to result in the true negatives. There were 1,655 samples for the true positive and the number of samples for true negative as 1,113.

We then calculated Cohen's Kappa, an accuracy measure commonly used in remote sensing applications [21]. The Kappa statistic is the chance agreement subtracted from the observed accuracy divided by chance agreement subtracted from 1. The Kappa statistic can range from −1 to 1, where 1 represents 100% accuracy, and 0 represents accuracy no better than that due to chance. Negative values are rare and generally indicate accuracy worse than random; a mismatch between ground truth locations and modeled data. Cohen's Kappa and confidence intervals were

Table 2. Average fence density by county within the Hi-Line region.

County	Km Fence/Km^2
Roosevelt	3.4
Sheridan	3.16
Toole	3.08
Daniels	2.97
Chouteau	2.81
Hill	2.74
Glacier	2.59
Blaine	2.55
Liberty	2.52
Valley	2.35
Phillips	2.06
Garfield	1.33
Rosebud	1.22

Table 3. Accuracy assessment results for the complete fence layer.

	Total Accuracy	Kappa	Confidence Intervals*
Roads	0.63	0.12	0.07–0.17
Internal	0.64	0.29	0.26–0.33
Total	0.73	0.40	0.36–0.44

*95% confidence intervals.

Table 4. Accuracy assessment results for the fence layer without BLM land.

	Total Accuracy	Kappa	Confidence Intervals*
Roads	0.65	−0.07	−0.14–−0.01
Internal	0.63	0.28	.24–.31
Total	0.60	0.18	0.13–0.22

*95% confidence intervals.

calculated in R statistical software (R Core Development Team 2012), using the FMSB package [22], which tests the null-hypothesis that the agreement between the model is the same as random, with Kappa =0.

Because the advice we received from our experts (local personnel from various organizations) about fencing on large croplands and BLM lands was variable we completed additional accuracy assessments. To identify causes of inaccuracies, we first removed BLM land from the model and repeated the accuracy assessment, and then removed land cover fencing.

Fence Density

After we created the fence model, the density of fences was calculated using ArcGIS 10.0. The density function search radius for the entire study area was 10,000 m and cell size was 1500 m.

Results

Fence Location and Density Modeling

A total of 263,308 km of fencing was predicted (Figure 3). Maximum fence density was 6.79 kilometers of fencing per km^2 (Figure 4). Mean fence density was 2.37 km of fencing per km^2. Fence density was highest along the U.S. Highway 2 corridor. Roosevelt County had the highest average density at 3.40 km/km^2 and Rosebud County had the lowest density at 1.22 km/km^2 (Table 2).

Model Accuracy Assessment

To compare how well our different decision rules reflected true fence lines, we created three fence datasets: all fencing on the landscape; all fencing except fencing associated with BLM lands; and all fencing except fencing associated with large crop lands. From these three layers, we then calculated accuracy for all fence types together; fencing only along roads and fencing only around land parcels. We chose to examine the rules associated with BLM lands and large croplands because our experts suggested these rules may vary greatly across the landscape.

Accuracy for all fencing and all types of fencing within the dataset was more accurate than random (Kappa =0), with a Kappa of 0.40 (Table 3). Within this dataset, the accuracy of the roads was lowest with a total accuracy (proportion of ground truth points that matched modeled fencing) of 0.63 and Kappa of 0.12.

In the dataset excluding fences associated with BLM land, we found a consistent decrease in accuracy (Table 4). When assessing the accuracy of the roads fencing only, Kappa was −0.07. The internal (parcel) fencing was the most accurate within this layer, with a Kappa of 0.28.

Accuracy was highest when large areas of cropland were excluded, in the third fence dataset. In this assessment, Kappa was 0.56. In analyzing accuracy for roads fencing in this dataset, Kappa was 0.27 and Kappa for land tenure fencing was 0.41 in the absence of fencing associated with large croplands (Table5).

Discussion

Although these fence location and density datasets can benefit from improvements in data analysis and sampling methods, this exercise did produce noteworthy results. Higher fence densities appear along Highway 2, where residential areas contributed to the increase in density. Less developed areas, such as the CMR and Glacier National Park contribute to the areas of low fence density (Figure 4). From our accuracy assessment, our model displayed actual fence locations along predicted fence lines at a Kappa of 0.40–0.56, considered moderate agreement between the modeled and ground truth data [23]. Therefore, using this modeling approach offers moderately accurate fence locations and density over a large spatial scale. In addition, regional rules can be created to hone the methodology to specific states or provinces of interest.

Our accuracy varied slightly depending on whether croplands and BLM lands were included in the model. Because these decision rules were based on expert opinion and experience, and our experts were not as confident with large areas of cropland and BLM lands, we believe that adding experts from these areas may further improve model accuracy. In some areas of the West, fences are used on BLM lands. Since we assumed the BLM pasture boundaries were fences on BLM land, accuracy may be increased when other fence types are included in the model. We also assumed that fencing on areas of croplands would be different than non-cropland areas, however, the increase in accuracy gained when removing the croplands from the model may suggest otherwise. Accuracy was lowest when fencing associated with large croplands were included in the fence dataset, but we believe that our model overall is an accurate reflection of fencing on the landscape.

Improvements in our analysis methods may improve our overall accuracy of the fence dataset. Different fence structure types have different effects on wildlife movement and habitat selection [19] and the fence layer created here did not include fence structure data. Fence structure type is difficult to model over large regions and is more a result of private landowner's preference and type of livestock production per ranch. Wire mesh fencing has been used in this study area to corral sheep and may be particularly hard for wildlife to cross. Additional improvements in accuracy of the fence layer may be made by amending the fence sampling protocol. Fence surveyors at times recorded fence locations at as much as 100 m from the actual fence due to railroad right-of-ways and property rights and interior pasture fences off roadways were not ground-truthed. Future sampling forays are planned to sample more of the study area, which will be imperative for future landscape fence permeability analyses for wildlife. Finally, because this data was created with the continued reliance on assumptions

Table 5. Accuracy assessment results for the fence layer without large croplands.

	Total Accuracy	Kappa	Confidence Intervals*
Roads	0.72	0.27	0.22–0.33
Internal	0.72	0.41	0.37–0.45
Total	0.78	0.56	0.53–0.60

*95% confidence intervals.

about neighboring parcels of land, there may be a decrease in accuracy around the edges of the study area. We recognize that individual fences may not be accurately modeled. Although improvements may be made, this novel effort uses extractable methodologies we believe will assist in modeling a key variable towards unraveling wildlife movement and habitat selection.

Fences can exhibit both indirect and direct effects on native wildlife populations worldwide [24–26,11,10]. In North America, indirect effects of fencing such as animal displacement, reduced habitat availability, and habitat fragmentation may have a higher impact on pronghorn and other wildlife populations than direct effects, by altering behavior and movement rates resulting in eventual population decline [,1,19,25,27,28]. Concerning pronghorn, and in particular during sever winter conditions, snow and ice can accumulate over the bottom-most fence wire during winter, thus preventing pronghorn from crawling underneath fences. Because of this, fencing may prevent migrating pronghorn from reaching higher quality habitat, during which time they have expended energy without finding better conditions. Fencing therefore can alter behavior at multiple scales and place North American wildlife in perilous situations as it has in other areas of the world [3,1,10,11,18].

A regional fence layer allows both wildlife and land managers to assess effects to wildlife at various scales, including at home range and within home range level of habitat selection [15], as well as identifying important population-level movement pathways between seasonal ranges during migration. Certainly, it can aid researchers through inclusion into predictive modeling efforts to assess habitat suitability and connectivity at regional level scales. As a priority, on-the-ground management practices could identify high fence densities (here along the developed Highway 2 (Figure 4)), along migratory pathways and within breeding grounds; ecological necessities to sustain North America's dwindling grassland wildlife [29]. Federal, state and provincial agencies, along with non-profit organizations and community organizations all can play an important role by undertaking cooperative projects to modify fences in strategic locations. These could include key-linkage areas which are geographic or anthropogenic areas and/or critical stopover locations along the migration pathway and fawning and/or lek areas. Without planning and the proper data, the cumulative anthropogenic changes to landscapes will continue to erode wildlife habitats and seasonal migration opportunities, reducing effective habitat patch size, potentially leading to sustained population declines and contraction of overall species range.

Acknowledgments

We thank Lee Smith and Carrie Breneman in Alberta, and Samantha Howlett in Montana for their many hours in the field and the numerous people we consulted to develop our fencing assumptions: Kelvin Johnson, Scott Thompson, Mark Sullivan, Adam Messer, MT FWP; John Carlson, Steve Zellmer, Abby Hall BLM; Dennis Jorgensen, WWF; Randy Matchett, USFWS; Paul Jones ACA; Matt Lopez, US BIA; John Doney, Ft. Peck Indian Reservation; Hoyt Richards, MT DNRC. We also thank our two anonymous reviewers at PLoS ONE.

Author Contributions

Conceived and designed the experiments: EEP AJ CL MS. Performed the experiments: EEP AJ. Analyzed the data: EEP AJ. Contributed reagents/materials/analysis tools: EEP AJ CL MS. Wrote the paper: EEP AJ CL MS.

References

1. Harris G, Thirgood S, Hopcraft JGC, Cromsigt JPGM, Berger J (2009) Global decline in aggregated migrations of large terrestrial mammals. Endang Sp Res 7: 55–76.
2. Berger J (2004) The last mile: how to sustain long-distance migration in mammals. Conserv Biol 18: 320–331.
3. Whyte IJ, Joubert SCJ (1988) Blue wildebeest population trends in the Kruger National Park and the effects of fencing. S Afr Wildlife Res 18: 78–87.
4. Williamson D, Williamson J (1985) Botswana's fences and the depletion of Kalahari wildlife. Parks 10: 5–7.
5. Berry HH (1997) Aspects of wildebeest Connochaetes taurinus ecology in the Etosha National Park – a synthesis for future management. Madoqua 20: 137–148.
6. Spinage CA (1992) The decline of the Kalahari wildebeest. Oryx 26: 147–150.
7. Ben-Shahar R (1993) Does fencing reduce the carrying capacity for populations of large herbivores? J Trop Ecol 9: 249–253.
8. Estes R (1967) Territorial behavior of the wildebeest (Connochaetes taurinus Burchell, 1823). Zool Tierpshych 26: 284–370.
9. Stevens BS, Connelly JW, Reese KP (2012) Multi-scale assessment of greater sage-grouse fence collision as a function of site and broad scale factors. J Wild Manage 76: 1370–1380.
10. Baines D, Summers RW (1997) Assessment of bird collisions with deer fences in Scottish forests. J Appl Ecol 34: 941–948.
11. Moss R (2001) Second extinction of capercaillie (Tetrao urogallus) in Scotland? Biol Cons 101: 255–257.
12. Knick ST, Hanser SE, Miller RF, Pyke DA, Wisdom MJ, et al. (2011) Ecological influence and pathways of land use in sagebrush. Studies in Avian Biology 38: 203–251.
13. Hayter EW (1939) Barbed wire fencing: A prairie invention: Its rise and influence in the Western States. Agr Hist 13:189–207.
14. Tufford W (1960) The wire that tamed the west. Montana 10: 62–67.
15. Johnson DH (1980) The Comparison of Usage and Availability Measurements for Evaluating Resource Preference. Ecology 6: 65–71.
16. Kie JG, Bowyer T, Nicholson MC, Boroski BB, Loft ER (2002) Landscape heterogeneity at differing scales: Effects on spatial distribution of mule deer. Ecology 83: 530–544.
17. Schaefer JA, Bergman CM, Luttich SN (2000) Site fidelity of female caribou at multiple spatial scales. Landscape Ecol 15: 731–739.
18. Boone RB, Hobbs NT (2004) Lines around fragments: effects of fencing on large herbivores. Afr J of Range For Sci 21: 147–158.
19. Harrington JL, Conover MR (2006) Characteristics of ungulate behavior and mortality associated with wire fences. Wildlife Soc B 34: 1295–1305.
20. Bolger DT, Newmark WD, Morrison TA, Doak DF (2007) The need for integrative approaches to understand and conserve migratory ungulates. Ecol Lett 10: 1–15.
21. Congalton RG (1981) The use of discrete multivariate analysis for the assessment of Landsat classification accuracy. MS Thesis, Virginia Polytechnic Institute and State University, Blacksburg, VA. 111 pp.
22. Nakazawa M (2013) Functions for medical statistics book with some demographic data. CRAN. Available: http://minato.sip21c.org/msb. Accessed 2013 May 15.
23. Landis J, Koch G (1977) The measurement of observer agreement for categorical data. Biometrics 33: 159–174.
24. Berger J, Young JK, Berger KM (2008) Protecting Migration Corridors: Challenges and Optimism for Mongolian Saiga. PLOS Biol 6:1365–1367.
25. Sheldon DP (2005) Pronghorn Movement and Distribution Patterns in Relation to Roads and Fences in Southwestern Wyoming. Department of Zoology and Physiology. Laramie, University of Wyoming. Masters of Science: 126 pp.
26. Thirgood S, Mosser A, Tham S, Hopcraft G, Mwangomo E, et al. (2004) Can parks protect migratory ungulates? The case of the Serengeti wildebeest. Anim Conserv 7: 113–120.
27. Ockenfels RA, Alexander A, Ticer CLD, Carrel WK (1994) Home ranges, movement patterns, and habitat selection of pronghorn in Central Arizona. Arizona Game and Fish Department Technical Report 13, Phoenix, vi +80 pp.
28. Sawyer H, Kauffman MJ, Middleton AD, Morrison TA, Nielson RA, et al. (2013) A framework for understanding semi-permeable barrier effects on migratory ungulates. J Appl Ecol 50: 68–78.
29. Laliberte AS, Ripple WJ (2004) Range contractions of North American carnivores and ungulates. Bioscience 54: 123–137.

19

Quantifying Spatial Variability of Selected Soil Trace Elements and Their Scaling Relationships Using Multifractal Techniques

Fasheng Zhang[1,2]*, Guanghua Yin[1], Zhenying Wang[1], Neil McLaughlin[2], Xiaoyuan Geng[2], Zuoxin Liu[1]*

1 Institute of Applied Ecology, Chinese Academy of Sciences, Shenyang, China, **2** Eastern Cereal and Oilseed Research Centre, Agriculture and Agri-Food Canada, Ottawa, Canada

Abstract

Multifractal techniques were utilized to quantify the spatial variability of selected soil trace elements and their scaling relationships in a 10.24-ha agricultural field in northeast China. 1024 soil samples were collected from the field and available Fe, Mn, Cu and Zn were measured in each sample. Descriptive results showed that Mn deficiencies were widespread throughout the field while Fe and Zn deficiencies tended to occur in patches. By estimating single multifractal spectra, we found that available Fe, Cu and Zn in the study soils exhibited high spatial variability and the existence of anomalies ($[\alpha(q)_{max}-\alpha(q)_{min}]\geq 0.54$), whereas available Mn had a relatively uniform distribution ($[\alpha(q)_{max}-\alpha(q)_{min}]\approx 0.10$). The joint multifractal spectra revealed that the strong positive relationships ($r\geq 0.86$, $P<0.001$) among available Fe, Cu and Zn were all valid across a wider range of scales and over the full range of data values, whereas available Mn was weakly related to available Fe and Zn ($r\geq 0.18$, $P<0.01$) but not related to available Cu ($r=-0.03$, $P=0.40$). These results show that the variability and singularities of selected soil trace elements as well as their scaling relationships can be characterized by single and joint multifractal parameters. The findings presented in this study could be extended to predict selected soil trace elements at larger regional scales with the aid of geographic information systems.

Editor: Helmut Ahammer, Medical University of Graz, Austria

Funding: This work was funded by the National Natural Science Foundation of China (41201212), the National Food Production Project of China (2011BAD16B12) and the National Major Technology Program of China (2012BAD09B02). The funders had no role in study design, data collection and analysis, decision to publish, or preparation of the manuscript.

Competing Interests: The authors have declared that no competing interests exist.

* E-mail: fasheng.zhang@yahoo.com (FSZ); liuzuoxin@iae.ac.cn (ZXL)

Introduction

Although trace elements are needed by plants in only very small quantities [1,2], their deficiencies in soils may result in severe plant malnutrition as well as animal and human health problems [3,4]. Quantifying the spatial variability of soil trace elements enables prediction of trace elements at intermediate locations where measurements have not been made [5]. The thematic map produced by the prediction is critical to understanding, assessing and managing trace element-deficient soils [6].

Plant availability of trace elements in soils have been shown to have a high degree of spatial variability [7,8]. This arises mainly from the existence of anomalies in their distributions, which is a consequence of local extreme soil and weather conditions [3] such as alkalinity, soil infertility, erosion, drought, etc. From a practical standpoint, anomalies may carry valuable information for identifying specific trace element deficiencies. Including anomalies is essential when developing models that characterize the spatial variability of soil trace elements.

Under an assumption of intrinsic stationarity, the semivariogram has been the most common method for quantifying the spatial variability of soil trace elements [7–11]. In modern geostatistics, anomalies in a trace element dataset are usually transformed or eliminated in advance to avoid erratic behavior of a semivariogram [12]. As a second-order moment statistic, semivariogram cannot differentiate subtle variations in a trace element dataset. These facts reveal that current semivariogram models may not be able to fully characterize the spatial variability of soil trace elements, especially the singularities of anomalies.

Over a given area, plant availability of soil trace elements is commonly associated with similar soil and environmental conditions. An understanding of the relationships among soil trace elements can help in the analysis of the underlying processes controlling their spatial variability [7,8]. Thus, there is a need to examine the spatial variability of soil trace elements in a way that takes account of the relationships among them.

Factors and processes affecting soil properties including plant availability of trace elements operate at different spatial scales and influence these properties in different ways [9]. Relationships among soil trace elements may be heterogeneous across nested spatial scales. Although correlation analysis can quantify the relationship among soil trace elements at the measured scale, this relationship may not be valid at other spatial scales. As such, relationships among soil trace elements across a range of spatial scales, i.e., scaling relationships, are needed.

Over the past three decades, single multifractal spectrum has been applied in soil science to quantify the spatial variability of soil properties [13–18] including pH, organic matter content, cation exchange capacity, macro- and moderate-nutrient content, texture, structure, hydraulic properties, etc. The relationships among these properties and their associations with other environmental variables across spatial scales have also been investigated by researchers using a joint multifractal spectrum [18−20]. To date, however, studies examining spatial variability and singularity of soil trace elements and their scaling relationships using multifractal techniques are still rare [21]. Usually, the spatial distributions of pH, organic matter content, macro-nutrient content, texture, and other basic soil properties are relatively uniform. Geostatistical methods are able to characterize their spatial variability well and predict them with high accuracy. However, the distribution of soil trace elements is quite different from that of other soil properties, and is often characterized by high spatial variability and anomalies that are accompanied by singularities. As previously mentioned, semivariogram analysis and correlation analysis may not be able to completely characterize the singularity of soil trace elements and their scaling relationships. Panahi and Cheng pointed out that an exploratory dataset with anomalies may be more likely to follow a multifractal rather than a normal or lognormal distribution [22]. Meanwhile, the multifractal techniques have long been known to be able to characterize singularities and scaling relationships. Thus, the application of multifractal techniques in the analysis of soil spatial variability and their scaling relationships may be more appropriate for spatial analysis of soil trace elements rather than for other soil properties.

The objective of this study was therefore to examine whether the spatial distribution of selected soil trace elements in an agricultural field in northeast China exhibits a multifractal property, and if so, to quantify their spatial variability and scaling relationships using multifractal techniques. We also discuss the practical significance of the findings for prediction of selected soil trace elements.

Theory of Multifractal Analysis

Single Multifractal Spectrum

For a spatial dataset, there are several methods for estimating a single multifractal spectrum (e.g., the moment method, Rényi method). Here we used the moment method discussed by Agterberg [23], which is much easier to apply yet still provides accurate results.

Using this method, a normalized variable, $P_i(\varepsilon)$, is defined first according to:

$$P_i(\varepsilon) = \frac{\varphi_i(\varepsilon)}{\sum_{i=1}^{n} \varphi_i(\varepsilon)}$$

where $\varphi_i(\varepsilon)$ is the measured value of variable φ in the ith segment of scale ε in a plot, and n is the number of segments of scale ε over the whole plot. A partition function, $\chi_q(\varepsilon)$, which varies with the moment order q and scale ε, is then calculated as:

$$\chi_q(\varepsilon) = \sum_{i=1}^{n} [P_i(\varepsilon)]^q$$

For a spatial dataset, the power-law relationship between the partition function and scale is used to determine whether or not it exhibits a multifractal property. This power-law relationship can be confirmed if the logarithm of the partition function varies linearly against a logarithm of scale with a slope of the mass exponent, $\tau(q)$:

$$\log \chi_q(\varepsilon) \propto \tau(q) \cdot \log(\varepsilon)$$

By plotting $\log\chi_q(\varepsilon)$ against $\log\varepsilon$ at many moment orders q, we compute the generalized fractal dimension, $D(q)$:

$$D(q) = \frac{\tau(q)}{q-1} = \frac{1}{q-1} \times \lim_{\varepsilon \to 0} \frac{\log \chi_q(\varepsilon)}{\log \varepsilon}, \quad q \neq 1$$

$$D(q) \approx \lim_{\varepsilon \to 0} \frac{\sum_{i=1}^{n} [P_i(\varepsilon) \cdot \log P_i(\varepsilon)]}{\log \varepsilon}, \quad q = 1$$

Then the Hölder exponent (i.e., singularity index), $\alpha(q)$, with respect to variable φ, is computed by Legendre transformation:

$$\alpha(q) = \frac{d\tau(q)}{dq}$$

The fractal dimension, $f(q)$, with respect to variable φ, is given as:

$$f(q) = q\alpha(q) - \tau(q)$$

Plotting $f(q)$ against $\alpha(q)$ produces the so-called single multifractal spectrum. $\alpha(q)$ can be interpreted as an index of crowding of variable φ at the moment order q when ε approaches zero. $f(q)$ is the fractal dimension of the subsets of variable φ having specific $\alpha(q)$. Thus, the abundance of $\alpha(q)$ reflects the extent of spatial variability in variable φ. More $\alpha(q)$ (i.e., wider spectrum) represent greater spatial variability in distribution of variable φ and vice versa. In addition, with the amplification of high positive or negative q values, the spatial variability controlled by the higher or lower values of variable φ is separated from each other and represented by the $\alpha(q)$ in the left and right parts of the single multifractal spectrum, respectively. In particular, the $\alpha(q)_{max}$ and $\alpha(q)_{min}$ correspond to the singularity strength of the lowest and highest values of variable φ, respectively. The single multifractal spectrum therefore can detect a spatial pattern and separate anomalous variation from the background variation [13,23].

Chhabra and Jensen [24] proposed a method to avoid the difficult task of estimating $\alpha(q)$ by Legendre transformation. In the Chhabra and Jensen method, we define another normalized variable, $\delta_i(q,\varepsilon)$:

$$\delta_i(q,\varepsilon) = \frac{[P_i(\varepsilon)]^q}{\sum_{i=1}^{n} [P_i(\varepsilon)]^q}$$

Then the singularity index and fractal dimension are computed respectively as:

$$\alpha(q) = \lim_{\varepsilon \to 0} \frac{\sum_{i=1}^{n} [\delta_i(q,\varepsilon) \cdot \log P_i(\varepsilon)]}{\log \varepsilon}$$

$$f(q) = \lim_{\varepsilon \to 0} \frac{\sum_{i=1}^{n} [\delta_i(q,\varepsilon) \cdot \log \delta_i(q,\varepsilon)]}{\log \varepsilon}$$

Joint Multifractal Spectrum

For coexisting variables, we need to extend the single multifractal spectrum of each variable to a joint multifractal spectrum for two variables [25]. Two normalized variables, $P_i(\varepsilon)$ and $Q_i(\varepsilon)$, are first defined respectively for variables φ and μ. Then the joint normalized variable for variables φ and μ, $\delta_i(q,t,\varepsilon)$, is calculated as:

$$\delta_i(q,t,\varepsilon) = \frac{[P_i(\varepsilon)]^q [Q_i(\varepsilon)]^t}{\sum_{i=1}^{n} [P_i(\varepsilon)]^q [Q_i(\varepsilon)]^t}$$

where q and t are both moment orders and ε is the scale. The singularity indices, $\alpha(q,t)$ and $\beta(q,t)$, with respect to variables φ and μ, are computed respectively as:

$$\alpha(q,t) = \lim_{\varepsilon \to 0} \frac{\sum_{i=1}^{n} [\delta_i(q,t,\varepsilon) \cdot \log P_i(\varepsilon)]}{\log \varepsilon}$$

$$\beta(q,t) = \lim_{\varepsilon \to 0} \frac{\sum_{i=1}^{n} [\delta_i(q,t,\varepsilon) \cdot \log Q_i(\varepsilon)]}{\log \varepsilon}$$

Then the joint fractal dimension, $f(q,t)$, with respect to variables φ and μ, is given as:

$$f(q,t) = \lim_{\varepsilon \to 0} \frac{\sum_{i=1}^{n} [\delta_i(q,t,\varepsilon) \cdot \log \delta_i(q,t,\varepsilon)]}{\log \varepsilon}$$

Plotting $f(q,t)$ against $\alpha(q,t)$ and $\beta(q,t)$ produces the so-called joint multifractal spectrum, contours of which describe the scaling relationships between variables φ and μ. For positive scaling relationships, the joint distribution of higher $\alpha(q,t)$ and $\beta(q,t)$ values at the upper right part of the joint multifractal spectrum indicates the positive correlation between lower values of variables φ and μ, while the joint distribution of lower $\alpha(q,t)$ and $\beta(q,t)$ values at the lower left part indicates the positive correlation between higher values of variables φ and μ. For negative scaling relationships, higher $\alpha(q,t)$ values correspond to lower $\beta(q,t)$ values and vice versa at the diagonal parts of the joint multifractal spectrum, which indicates that lower and higher values of variables φ and μ are distributed together. If variables φ and μ are not correlated, there will be no correlation between $\alpha(q,t)$ and $\beta(q,t)$ [25,26].

Materials and Methods

Site and Soil

The study site is located in Fuxin County in northeast China (42°08′14″ N, 121°44′21″ E). The climate is typical of a warm temperate zone with continental monsoons, hot summers and cold winters. The mean annual temperature is 7.2 degrees Celsius with an average of 2868 hrs of annual sunshine. This region is a typical dryland farming area with an average annual precipitation of 480 mm, over 60% of which occurs between June and August (SWCP I soil moisture sensor & WITU-WS weather station, WITU Technologies, Inc., China).

The main agricultural soil in the region is cinnamon soil which develops through a combination of calcium carbonate leaching, illuviation and humification. It is characterized by a thin humus layer and medium or thick solum. Soil pH (H_2O) is 7.5−8.5 and soil organic matter content is 10.2 g·kg^{-1}. The cation exchange capacity is 19.2 meq/100 g, and mean concentrations of available Fe, Mn, Cu, and Zn are 9.9 mg·kg^{-1}, 36 mg·kg^{-1}, 1.32 mg·kg^{-1}, and 1 mg·kg^{-1}, respectively (The Second National Soil Survey of China, 1985).

Data Collection and Analysis

A total of 1024 soil samples were collected from a flat agricultural field of 10.24-ha (320 m by 320 m) in Matiyingzi Village, 10 km from the center of Fuxin County, in April 2008 (before annual planting). The soils were sampled over the 20-cm depth within a 10-m grid. The samples were taken with a stainless steel sampler and brought to the laboratory. After removing visible plant debris and air drying, soil samples were ground to pass through a 2-mm plastic sieve. Available Fe, Mn, Cu and Zn in each sample were measured by a DTPA (diethylene triamine pentaacetic acid) soil test [27]. The extractant consisted of 0.005 M DTPA, 0.1 M TEA (triethanolamine), and 0.01 M $CaCl_2$, with a pH of 7.3. Hooda and Alloway reported that DTPA method could provide accurate information about plant availability of trace elements in soils [28]. All trace element concentrations were determined by inductively coupled plasma mass spectrometry (Perkin Elmer Optima 3000). All necessary permits were obtained for the described field studies. The land user and owner, Mingfu Zhang, approved the field-work activities at each sampling location. The field examined in this study is not protected in any way, and the study did not involve any endangered or protected species.

A series of five scales were superimposed on the sample plot: 10-m pixel, 20-m pixel, 40-m pixel, 80-m pixel, and 160-m pixel with 1,024, 256, 64, 16 and 4 samples, respectively. The dataset for the minimum scale was the measured trace element data in the initial 10-m pixel grid, while the dataset for the other scales was the average of the measured trace element data of all initial 10-m pixel grids included within the larger pixel in the respective grid.

The single and joint multifractal spectra parameters for selected soil trace element data were calculated using the equations in the part of theory of multifractal analysis by an Office VBA project and were presented in graphs by SigmaPlot 10.0. The correlation analysis between $\alpha(q,t)$ and $\beta(q,t)$ was conducted using SPSS 16.0.

Results

Descriptive Statistics

A summary of descriptive statistics for selected trace elements in the study soils is presented in Table 1. Available Fe, Cu and Zn varied to a high degree with coefficient of variation (CV) of 82.7%, 41.7% and 89.5%, respectively. The CV is very sensitive to the anomalies in a dataset and high CV's for available Fe, Cu and Zn data suggest that there must be some significant anomalous values in their distributions. Available Mn had a relatively uniform distribution with a CV of only 9.5%. Based on the thresholds recommended by Liu [29], we found that Mn deficiency was widely present throughout the field. Although Fe and Zn deficiencies were not widespread, 25.5% of soil samples were deficient in available Fe, and 13.3% were deficient in available Zn. This indicated that Fe and Zn deficiencies tended to occur in

Table 1. Descriptive statistics for selected soil trace elements in the study agricultural field of northeast China.

Trace elements	Min† (mg•kg^{-1})	Max (mg•kg^{-1})	Mean (mg•kg^{-1})	CV	Deficiency‡
Fe	0.30	73.73	18.93	82.7%	25.5%
Mn	25.53	91.11	69.27	9.5%	100%
Cu	0.18	5.27	1.34	41.7%	1.4%
Zn	0.02	14.81	2.48	89.5%	13.3%

†Min, Max and CV are abbreviations for minimum, maximum and coefficient of variation.
‡Critical levels for available Fe, Mn, Cu and Zn recommended by Liu [29] are 4.5 mg•kg^{-1}, 100 mg•kg^{-1}, 0.3 mg•kg^{-1} and 1.0 mg•kg^{-1} (DTPA extracted). The concentrations of available Fe, Mn, Cu and Zn under the critical levels are recognized as deficiencies.

patches rather than throughout the field. Only a very limited percentage of soil samples were deficient in available Cu.

Empirical Tests of Multifractality

Scatter plots and their fitted relationships between the logarithm of partition function and the logarithm of superimposed scale for selected soil trace elements are shown in Figure 1. The coefficients of determination R^2 for all linear regression lines were larger than 0.99 (F-test, $P<0.001$) which confirms the assumption of the existence of the power-law relationship between partition function and scale. This finding revealed that the spatial distribution of available Fe, Mn, Cu and Zn in the study soils exhibited a multifractal property. Hence, multifractal techniques can be used to characterize the spatial variability of selected soil trace elements and their scaling relationships.

Variables showing a multifractal property usually have a $D(q)$ that demonstrates a monotonic and nonlinear trend with q values. This pattern can be observed when examining the trend of $D(q)$ for available Fe, Cu and Zn in Figure 2. The $D(q)$ values for available Fe, Cu and Zn decreased markedly as q values increased. This is a typical characteristic of a multifractal system, and also indicated that the multifractal property observed here was strong. In contrast, the $D(q)$ values for available Mn decreased very slowly against q values and showed small deviations from the capacity dimension of the sample plot (i.e., 2). This revealed that the distribution of available Mn can be represented by a monofractal or a weak multifractal. Note that a monofractal is a special case of a multifractal.

Variability and Singularity of Soil Trace Elements

To examine the spatial variability of selected soil trace elements, their single multifractal spectra were constructed and shown in Figure 3. The single multifractal spectra for available Fe, Cu and Zn all had large widths ($[\alpha(q)_{max}-\alpha(q)_{min}]\geq 0.54$). This indicated that available Fe, Cu and Zn had quite high spatial variability in studied soils. To characterize such high spatial variability of available Fe, Cu and Zn, a large number of singularity indices ranging from 1.69 to 2.83, 1.81 to 2.35, and 1.60 to 2.38

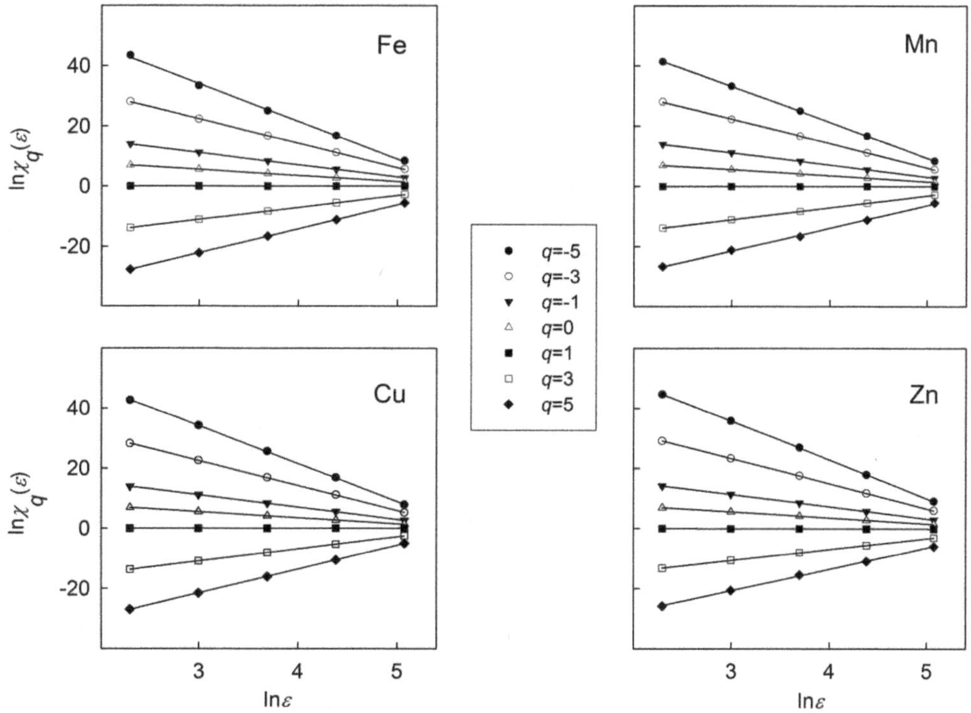

Figure 1. Logarithms of partition functions of selected soil trace elements plotted against scales. The R^2 values of the linear fit for all q considered were always greater than 0.99.

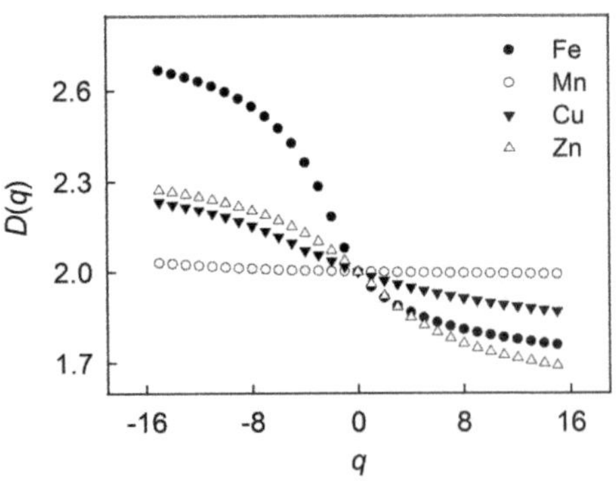

Figure 2. General fractal dimension $D(q)$ for selected soil trace elements at −15 to 15 range of q evaluated at 1 increment.

respectively were needed. In contrast, the single multifractal spectrum of available Mn had a very narrow width ($[\alpha(q)_{max} - \alpha(q)_{min}] \approx 0.10$). This indicated that the spatial variability of available Mn was relatively low. Such low spatial variability can be characterized by a very small set of singularity indices, ranging from 1.99 to 2.09.

The single multifractal spectra of available Fe, Mn, Cu and Zn all had asymmetric tails on either the right or left part (Figure 3), which is another typical characteristic of a multifractal property. The left and right parts of a single multifractal spectrum respectively represent the higher and lower values of a dataset, emphasized by high positive and negative q values. The long left and right tails of the single multifractal spectra for available Fe, Cu and Zn were therefore the result of both extremely high and low values in their distributions. The singularities of these anomalies were characterized by the corresponding singularity indices in the single multifractal spectra of available Fe, Cu and Zn. The single multifractal spectrum of available Mn had a relatively long tail only on the right part. This indicated that there were no anomalies but just some relatively low values in the distribution of available Mn in the study soils.

Scaling Relationships among Soil Trace Elements

The relationships among selected soil trace elements across the spatial scales superimposed on the sample plot were analyzed using the joint multifractal spectra (Figure 4). In general, there were strong positive scaling relationships among available Fe, Cu and Zn, which were evident from the compact joint distribution of their singularity indices at the upper right and lower left parts of

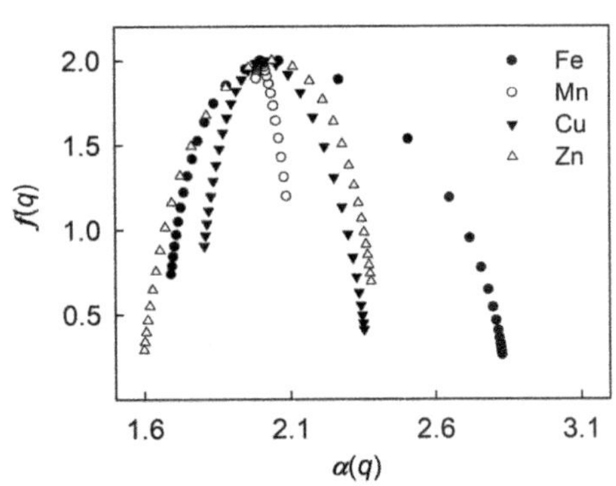

Figure 3. Single multifractal spectrum for selected soil trace elements at −15 to 15 range of q evaluated at 1 increment.

their joint multifractal spectra (Figures 4A, 4B and 4C). The contours in Figure 4B were tightly concentrated in one bundle. This indicated that there were strong positive relationships between available Fe and Zn for both higher and lower values across the superimposed spatial scales. With a small diagonal stretch at the upper right, Figure 4A and Figure 4C had a similar pattern of joint distribution of singularity indices as Figure 4B. This indicated that strong positive relationships also existed between available Cu and available Fe and Zn over a wider spatial scale and for the entire range of data values. In Figures 4D, 4E and 4F, the wide stretch of diagonal contours showed that the singularity indices of available Mn was not correlated tightly with that of available Fe, Cu and Zn. This suggested that available Mn may not be associated, or may be weakly associated, with available Fe, Cu and Zn over the range of measured concentrations and superimposed spatial scales.

To further confirm the scaling relationships among selected soil trace elements, the Pearson correlation coefficients (r) between their singularity indices were analyzed (Table 2). The singularity indices between available Fe and Zn had the highest positive correlation coefficient ($r = 0.98$, $P < 0.001$). The positive correlation coefficients between the singularity indices of available Fe versus available Cu and available Zn versus available Cu were relatively high as well ($r = 0.87$ and 0.86, $P < 0.001$). These results indicated that the positive correlation relationships among available Fe, Cu and Zn were all very strong across superimposed spatial scales. Although the associations between singularity indices of available Fe versus available Mn and available Zn versus available Mn were also significant, their correlation coefficients ($r = 0.19$ and 0.18, $P < 0.01$) were relatively low compared to those among available Fe, Cu and Zn. This suggested the positive correlations between available Mn and available Fe and Zn were weak. The correlation coefficient between singularity indices of available Cu and Mn was not significant ($r = -0.03$, $P = 0.40$), which implied that available Cu and Mn were not correlated across superimposed spatial scales. The strong scaling relationships among available Fe, Cu and Zn were likely due to a common controlling process, which was quite different from that for available Mn.

Discussion

The influences of factors in the CLORPT function or the SCORPAN model on the variations of selected soil trace elements should be relatively consistent over a flat 10.24-ha level field [30,31]. In most soils, available Fe, Mn, Cu and Zn are not directly related to their total contents, but are more related to other soil characteristics such as pH, redox potential, organic matter content, cation exchange capacity, and texture [32]. In agricultural soils, long-term intensive agricultural practices also add another layer of complexity on the distribution of available Fe, Mn, Cu and Zn. Kravchenko et al. [13], Perrier and Bird [14], Caniego et al. [15], Guan et al. [16] and Perrier et al. [33] have reported that pH, redox potential, organic matter content, cation exchange capacity, texture and structure interact with each other and are prone to lead to a nonlinear soil system in which multifractal properties are expressed. The multifractal nature of available Fe, Mn, Cu and Zn observed in this study can be attributed to the complexity of this system.

The spatial variability of selected soil trace elements characterized by single multifractal spectrum was partly consistent with the result of CV analysis. However, the results based on the single multifractal spectrum were obtained over a wider range of spatial scales and moment orders and better reflected the true spatial variability of selected soil trace elements. The single multifractal spectrum also captured the detailed singularity strength of anomalies in available Fe, Mn, Cu and Zn data while the joint multifractal spectrum characterized the scaling relationships among them. These two points demonstrate that multifractal techniques present unique advantages over CV and semivariogram analyses.

Knowledge of soil spatial variability is a prerequisite needed to predict soil properties at unsampled locations. In this study, the spatial variability and singularity of selected soil trace elements quantified by the parameters of single multifractal spectra would have practical significance for prediction of these trace elements. Appropriate prediction procedures for variables with high spatial variability and a large amount of singularity indices are different than that for those with low spatial variability and limited singularity indices. This implication has been summarized in the study by Cheng [34] and justified by Cheng [35] and Xie et al. [36]. In our study, available Mn had a narrow multifractal spectrum with a limited number of singularity indices. Its spatial variability was very low and there were no anomalies within its spatial distribution. For a variable with those distribution characteritics and spatial autocorrelation, a simple prediction procedure such as an ordinary Kriging interpolation may be appropriate. In contrast, available Fe, Cu and Zn in this study's soils had wide multifractal spectra with many singularity indices both at $\alpha(q)>2$ and $\alpha(q)<2$. Their spatial variability was high with both extremely high and extremely low values in their spatial distributions. Under these circumstances, it is possible, but difficult, to apply a prediction procedure. For example, $\alpha(q)>2$ implied that there were some specific locations with extremely low concentrations of available Fe, Cu and Zn. An ordinary interpolation will smooth out the extremely low values of available Fe, Cu and Zn, resulting in the loss of valuable information for identifying Fe, Cu and Zn deficiencies. In this case, incorporating singularity indices $\alpha(q)>2$ into the prediction procedure can help preserve the extremely low values of available Fe, Cu and Zn.

Table 2. Pearson correlation coefficients (*r*) between singularity indices of joint multifractal spectra for selected soil trace elements.

	Fe	Mn	Cu
Mn	0.19 ($P<0.01$)		
Cu	0.87 ($P<0.001$)	−0.03 ($P=0.40$)	
Zn	0.98 ($P<0.001$)	0.18 ($P<0.01$)	0.86 ($P<0.001$)

Similarly, $\alpha(q)<2$ is also needed when predicting available Fe, Cu and Zn with extremely high values at specific locations.

Recent studies have delineated the spatial distribution of selected trace elements in soils using their relationships with environmental data, but those relationships were observed at only a single scale and may not be valid at other scales. Across a regional soil landscape, factors and processes impacting the plant availability of soil trace elements are expected to be much more complicated and heterogeneous than that of the 10.24-ha agricultural field observed in this study. The major controlling processes for different soil trace elements may differ at specific spatial scales, which would further impact the spatial variability of soil trace elements. The results of joint multifractal analysis in this study had demonstrated the ability of joint multifractal spectrum in characterizing the scaling relationships among selected soil trace elements at the field scale. It is well-known that fractals and multifractals are ubiquitous in complex ecosystems [37]. Further study is needed to evaluate the technique in modeling the scaling relationships of soil trace elements at larger regional scales where their variability in relation to the local environments would be expected. Using scale-specific relationships future researchers should be able to improve models when predicting soil trace elements with auxiliary environmental data.

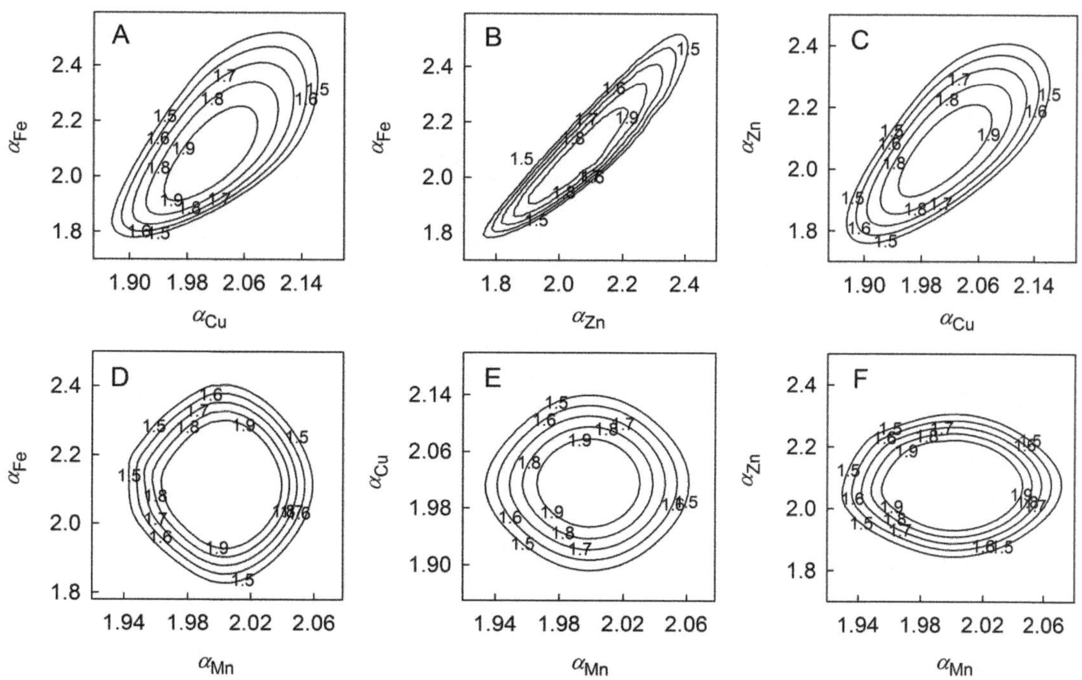

Figure 4. Joint multifractal spectrum for selected soil trace elements at −15 to 15 range of *q* evaluated at 1 increment.

In recent years, geographic information systems (GIS) have facilitated the process of spatial analysis and computation. If further research demonstrates that the multifractal phenomenon exists over larger regional scales, the findings in this study could be incorporated into GIS to improve the prediction of selected trace elements over a range of scales for optimizing soil use and management.

Conclusions

Spatial variability of available Fe, Mn, Cu and Zn and their scaling relationships were examined using multifractal techniques for a 10.24-ha agricultural field in northeast China. The distribution of selected soil trace elements exhibited a multifractal property. The variability and singularities of available Fe, Cu and Zn in the study soils needed to be quantified by multifractal spectra, whereas that of available Mn can be characterized by a monofractal or a weak multifractal spectrum. The joint multifractal spectra and the Pearson correlation analysis of their singularity indices captured the relationships among selected soil trace elements across a wider range of spatial scale and over the full range of data values.

The use of single and joint multifractal parameters has potential to enhance the prediction of selected soil trace elements in future research. We are currently examining the multifractality of soil trace elements at the county scale and are building a model that incorporates the scaling relationships between selected soil trace elements and environments into a prediction procedure. Such studies are essential to validate the applicability and extension of the findings presented here.

Acknowledgments

We thank Guixian Li for providing assistance in carrying out the field experiment.

Author Contributions

Conceived and designed the experiments: FSZ ZXL. Performed the experiments: FSZ GHY ZYW. Analyzed the data: FSZ ZYW. Contributed reagents/materials/analysis tools: FSZ. Wrote the paper: FSZ NM XYG ZXL.

References

1. Havlin JL, Beaton JD, Tisdale SL, Nelson WL (2005) Soil fertility and fertilizers: An introduction to nutrient management. 7th Edition. Upper Saddle River, NJ:Prentice Hall. 515 p.
2. Hood PS (2010) Trace elements in soils. Chichester, UK:Wiley-Blackwell. 616 p.
3. Alloway BJ (2008) Micronutrient deficiencies in global crop production. Dordrecht, The Netherlands:Springer. 353 p.
4. Gupta UC, Wu KN, Liang SY (2008) Micronutrients in soils, crops, and livestock. Earth Science Frontiers 15: 110–125.
5. Heuvelink GBM, Webster R (2001) Modelling soil variation: Past, present and future. Geoderma 100: 269–301.
6. White JG, Zasoski RJ (1999) Mapping soil micronutrients. Field Crops Research 60: 11–26.
7. Wu CF, Luo YM, Zhang LM (2010) Variability of copper availability in paddy fields in relation to selected soil properties in southeast China. Geoderma 156: 200–206.
8. Behera SK, Singh MV, Singh KN, Todwal S (2011) Distribution variability of total and extractable zinc in cultivated acid soils of India and their relationship with some selected soil properties. Geoderma 162: 242–250.
9. Goovaerts P (1999) Geostatistics in soil science: State-of-the-art and perspectives. Geoderma 89: 1–45.
10. Shi J, Xu J, Huang P (2008) Spatial variability and evaluation of status of micronutrients in selected soils around Taihu Lake, China. Journal of Soils and Sediments 8: 415–423.
11. Liu XM, Xu JM, Zhang MH, Si BC, Zhao KL (2008) Spatial variability of soil available Zn and Cu in paddy rice fields of China. Environmental Geology 55: 1569–1576.
12. Gringarten E, Deutsch CV (2001) Teacher's aide: Variogram interpretation and modeling. Mathematical Geology 33: 507–534.
13. Kravchenko AN, Boast CW, Bullock DG (1999) Multifractal analysis of soil spatial variability. Agronomy Journal 91: 1033–1041.
14. Perrier EMA, Bird NRA (2002) Modelling soil fragmentation: the pore solid fractal approach. Soil and Tillage Research 64: 91–99.
15. Caniego FJ, Espejo R, Martín MA, José FS (2005) Multifractal scaling of soil spatial variability. Ecological Modelling 182: 291–303.
16. Guan XY, Yang PL, Ren SM, Li YK (2007) Multifractal analysis of soil structure under long-term wastewater irrigation based on digital image technology. New Zealand Journal of Agricultural Research 50: 789–796.
17. Wang ZY, Shu QS, Liu ZX, Si BC (2009) Scaling analysis of soil water retention parameters and physical properties of a Chinese agricultural soil. Australian Journal of Soil Research 47: 821–827.
18. Biswas A, Cresswell HP, Si BC (2012) Application of multifractal and joint multifractal analysis in examining soil spatial variation: A Review. In: Ouadfeul SA, editor. Fractal analysis and chaos in geosciences. InTech. pp. 109–138.
19. Zeleke TB, Si BC (2005) Scaling relationships between saturated hydraulic conductivity and soil physical properties. Soil Science Society of America Journal 69: 1691–1702.
20. Wang ZY, Shu QS, Xie LY, Liu ZX, Si BC (2011) Joint multifractal analysis of scaling relationships between soil water-retention parameters and soil texture. Pedosphere 21: 373–379.
21. Jesus AP, Mateus A, Gonçalves MA, Munhá J (2013) Multi-fractal modelling and spatial Cu-soil anomaly analysis along the southern border of the Iberian Terrane in Portugal. Journal of Geochemical Exploration 126–127: 23–44.
22. Panahi A, Cheng Q (2004) Multifractality as a measure of spatial distribution of geochemical patterns. Mathematical Geology 36: 827–846.
23. Agterberg FP (1995) Multifractal modeling of the sizes and grades of giant and supergiant deposits. International Geology Review 37: 1–8.
24. Chhabra A, Jensen RV (1989) Direct determination of the $f(\alpha)$ singularity spectrum. Physical Review Letters 62: 1327–1330.
25. Meneveau C, Sreenivasan KR, Kailasnath P, Fan MS (1990) Joint multifractal measures: Theory and applications to turbulence. Physical Review A 41: 894–913.
26. Kravchenko AN, Bullock DG, Boast CW (2000) Joint multifractal analysis of crop yield and terrain slope. Agronomy Journal 92: 1279–1290.
27. Lindsay WL, Norvell WA (1978) Development of a DTPA soil test for zinc, iron, manganese, and copper. Soil Science Society of America Journal 42: 421–428.
28. Hooda PS, Alloway BJ (1994) The plant availability and DTPA extractability of trace metals in sludge-amended soils. Science of the Total Environment 149: 39–51.
29. Liu Z (1996) Microelements in soils of China. Nanjing;Jiangsu Sci. and Technol. Publishing House. 396 p. (in Chinese).
30. Jenny H (1941) Factors of soil formation: A system of quantitative pedology. New York: McGraw-Hill. 281 p.
31. McBratney AB, Mendonça-Santos ML, Minasny B (2003) On digital soil mapping. Geoderma 117: 3–52.
32. Fageria NK, Baligar C, Clark RB (2002) Micronutrients in crop production. Advances in Agronomy 77: 185–268.
33. Perrier E, Bird N, Rieutord T (2010) Percolation properties of 3-D multiscale pore networks: how connectivity controls soil filtration processes. Biogeosciences Discussion 7: 2997–3018.
34. Cheng Q (2005) A new model for incorporating spatial association and singularity in interpolation of exploratory data. In: Leuangthong O, Deutsch CV, editors. Geostatistics Banff, 2004, Quantitative geology and geostatistics. Dordrecht, The Netherlands:Springer. pp. 1017–1025.
35. Cheng Q (2008) Modeling local scaling properties for multiscale mapping. Vadose Zone Journal 7: 525–532.
36. Xie SY, Cheng Q, Xing XT, Bao ZY, Chen ZJ (2010) Geochemical multifractal distribution patterns in sediments from ordered streams. Geoderma 160: 36–46.
37. Solé RV, Bascompte J (2006) Self-organization in complex ecosystems. Princeton:Princeton University Press. 392 p.

20

Spatial and Temporal Patterns of Locally-Acquired Dengue Transmission in Northern Queensland, Australia, 1993–2012

Suchithra Naish[1]*, Pat Dale[2], John S. Mackenzie[3], John McBride[4], Kerrie Mengersen[5], Shilu Tong[1]

1 School of Public Health and Social Work & Institute of Health and Biomedical Innovation, Queensland University of Technology, Kelvin Grove campus, Brisbane, Queensland, Australia, **2** Environmental Futures Centre, Australian Rivers Institute, Griffith School of Environment Griffith University, Brisbane, Queensland, Australia, **3** Australian Biosecurity CRC and Faculty of Health Sciences, Curtin University, Perth, Western Australia, Australia, **4** School of Medicine and Dentistry, James Cook University, Cairns, Queensland, Australia, **5** Mathematical Sciences, Queensland University of Technology, Gardens Point campus, Brisbane, Queensland, Australia

Abstract

Background: Dengue has been a major public health concern in Australia since it re-emerged in Queensland in 1992–1993. We explored spatio-temporal characteristics of locally-acquired dengue cases in northern tropical Queensland, Australia during the period 1993–2012.

Methods: Locally-acquired notified cases of dengue were collected for northern tropical Queensland from 1993 to 2012. Descriptive spatial and temporal analyses were conducted using geographic information system tools and geostatistical techniques.

Results: 2,398 locally-acquired dengue cases were recorded in northern tropical Queensland during the study period. The areas affected by the dengue cases exhibited spatial and temporal variation over the study period. Notified cases of dengue occurred more frequently in autumn. Mapping of dengue by statistical local areas (census units) reveals the presence of substantial spatio-temporal variation over time and place. Statistically significant differences in dengue incidence rates among males and females (with more cases in females) ($\chi^2 = 15.17$, d.f. = 1, $p<0.01$). Differences were observed among age groups, but these were not statistically significant. There was a significant positive spatial autocorrelation of dengue incidence for the four sub-periods, with the Moran's *I* statistic ranging from 0.011 to 0.463 ($p<0.01$). Semi-variogram analysis and smoothed maps created from interpolation techniques indicate that the pattern of spatial autocorrelation was not homogeneous across the northern Queensland.

Conclusions: Tropical areas are potential high-risk areas for mosquito-borne diseases such as dengue. This study demonstrated that the locally-acquired dengue cases have exhibited a spatial and temporal variation over the past twenty years in northern tropical Queensland, Australia. Therefore, this study provides an impetus for further investigation of clusters and risk factors in these high-risk areas.

Editor: Yury E. Khudyakov, Centers for Disease Control and Prevention, United States of America

Funding: Authors would like to acknowledge ARC grant (#DP 110 100 651). The funders had no role in study design, data collection and analysis, decision to publish, or preparation of the manuscript.

Competing Interests: The authors have declared that no competing interests exist.

* E-mail: s.naish@qut.edu.au

Introduction

Dengue is emerging and resurging as a worldwide public health problem since the 1950s, affecting more than 110 countries today, and is a leading cause of hospitalisation and morbidity among children in the tropics and subtropics [1]. It is a mosquito-borne viral disease caused by one of four dengue viruses transmitted by *Aedes* mosquitoes. Currently mosquito control is the only available mitigation strategy but *Wolbachia* and the genetic manipulation of mosquitoes to lead to male sterility, may change this in the intermediate future. *Aedes aegypti* is the most common dengue-transmitting mosquito in the state of Queensland in Australia. *Aedes albopictus* (Asian tiger mosquito) another mosquito able to transmit dengue is currently threatening the Australian mainland having been detected on a number of Torres Strait islands [2]. *Ae. aegypti* typically breeds in human-made container habitats such as water storage jars in and around human settlements including those in dense urban areas [3]. *Ae. albopictus* breeds in the same containers as *Ae. aegypti* but also breeds in natural containers in the bush such as tree holes, cut bamboo, banana trees and coconut shells. However, both these mosquitoes breed in fresh waters, but not in swamps or creeks. The recent arrival of the exotic species *Ae. albopictus* is of great concern because, if *Ae. albopictus* colonises the mainland, it could extend to the southern states due to its tolerance of more temperate conditions [2].

In Australia, dengue re-emerged in north Queensland during 1992–1993, after disappearing for about 10 years, i.e., 1981–1991 [4]. Since then, outbreaks and epidemics of dengue, with locally-acquired cases, were reported in Cairns and Port Douglas [4]. These are located in the north-east (urban tropical) coastal regions in Queensland. From January 1993 to June 2012, a total of 2,398 locally-acquired cases were documented in Queensland by Queensland Health. Outbreaks of dengue occur primarily in areas where *Ae. aegypti* mosquitoes are found. Dengue viruses may be introduced into areas by travellers who become infected while visiting other areas of the tropics where dengue commonly exists [5].

Most mosquito-borne diseases exhibit spatial and temporal variations in their distribution [6]. Spatial analyses using geostatistics such as spatial autocorrelations, variograms, interpolations, and temporal analyses using chi-squared statistics and time series models are commonly applied to highlight patterns of disease incidence [7,8].

Geographic information systems (GIS) play an important role in disease surveillance and control of the mosquito-borne diseases as they assist in the analysis of potential risk factors associated with the disease through the geo-coding processes [8] and facilitate maps that are useful for the identification of spatially and temporally localised areas of potential high-risk populations [7–11]. The visualised information represented in different types of maps based on GIS enables simultaneous observation of both the attribute and geographical relationships [12–14]. Maps also help policy decision-makers and public health officials to communicate with the public and policy decision-makers about complex information in an easily interpretable format [15,16].

GIS can provide not only an opportunity to improve our understanding of the distribution patterns of dengue, but can also provide an environmentally and socially informed platform to develop the elements of an early warning system towards control and prevention of dengue [8]. The advancement of geographical information systems (GIS) and spatial statistics has greatly improved the understanding of dengue dynamics, including its dependence on ecological factors. Hence, in this study, we examined the spatial and temporal patterns of locally-acquired dengue transmission in northern Queensland, Australia using GIS tools and geostatistical techniques.

Methods

Study area

Queensland is the third largest state by population size in Australia (after New South Wales and Victoria), occupying a total area of 1,723,936 km^2 with a total population of 4.56 million people (20% of Australia's total population) and is the fastest growing state with 23.9% of population growth in Australia to June 2012 [17]. Northern Queensland, a tropical region which is 100 km north of the Tropic of Capricorn (Figure 1) is selected as the study area as it had the largest total number of recorded notifications (n = 2,273) in Queensland, and the largest compared any other Australian state and territory, during the period 2005–2012 [18]. Northern Queensland has a tropical climate, with average temperatures in summer ranging from 24–33 degrees Celsius, and in winter 14–26.

Data collection

Dengue data. Dengue is a notifiable disease and all positive cases are required to be reported by laboratories to the state government (Queensland Health), by the Public Health Act 2005. These records are archived by the Data Custodians, Communicable Disease Branch (CDB) unit in Queensland Health under the National Notifiable Disease Surveillance System (NNDSS). The NNDSS was established in 1990 under the auspices of the Communicable Diseases Network Australia. Vector-borne diseases notified to the NNDSS include mosquito-borne diseases caused by alphaviruses, such as Ross River virus (RRV), Barmah Forest virus (BFV) and the flavivirus, Dengue.

In Australia, dengue outbreaks are a combination of locally-acquired and overseas- or imported-cases. According to government of Queensland Health, a dengue outbreak is declared when there is one or more locally-acquired dengue cases are confirmed. An overseas- or imported-case is defined as someone who is infected with dengue overseas (i.e., viraemic traveller) and arrives Australia with the virus in their blood [19]. A locally-acquired case is defined as when a local dengue mosquito bites this overseas dengue infected person and it passes the virus on to other people by biting them [19].

We obtained computerised and anonymous notification data (data that does not contain any identifiers such as name, street and house number or Medicare number or other medical insurance number) on locally-acquired dengue cases from January 1993 to June 2012 (approximately 20 years) for the study area from the CDB, Queensland Health. Dengue data included date of notification, age group (e.g., <1, 1–4, 5–9, 10–14 etc.), gender, post code of residence and statistical local area (SLA) (census unit) name. SLA is an Australian Standard Geographical Classification (ASGC) defined area which consists of one or more collection districts (the smallest geographical unit) in Australian census. Therefore, we have analysed the data based on age group.

Population data. Population data for the SLAs for the national census years 2001, 2006 and 2011 were obtained from the Australian Bureau of Statistics. For the remaining years during 1993 to 2012, the annual population data were estimated based on linear interpolation. We have adjusted SLA boundaries to match earlier censuses.

The study was approved by the Data Custodians, Human Research Ethics Committee under Chapter 6, Part 4, Section 280 of the Public Health Act 2005, CDB of Queensland Health and following the ethical considerations of the Research Ethics Unit, Queensland University of Technology (Number: 1100001110).

Statistical analyses

The study period was divided into four time periods, with five years in each time period for the ease of the analysis and to visualise the spatial and temporal patterns more clearly and precisely: Period 1: 1993–1997, Period 2: 1998–2002, Period 3: 2003–2007 and Period 4: 2008–2012. Population data for each period were attached to each SLA in the maps and these were used as the denominator in the computation of incidence rates. Period-wise distribution maps were produced on dengue cases and incidence rates by SLA. MapInfo Professional (version 11) incorporated with Vertical Mapper (version 3.7) was used to produce the final outputs as tables and maps.

Spatial and temporal analyses

To investigate the spatial and temporal patterns of dengue disease and to determine the risk of dengue disease, monthly incidence rates were calculated at both SLA and state levels. Incidence rates for each age group and gender were also calculated from the total number of dengue cases notified in each age group for each SLA in different time periods, divided by the respective total person-years and then multiplied by 100,000. These incidence rates were expressed as: total number of dengue cases/total population*100,000. Differences between age- and gender-specific incidence rates were tested using chi-square analyses (SPSS version 21).

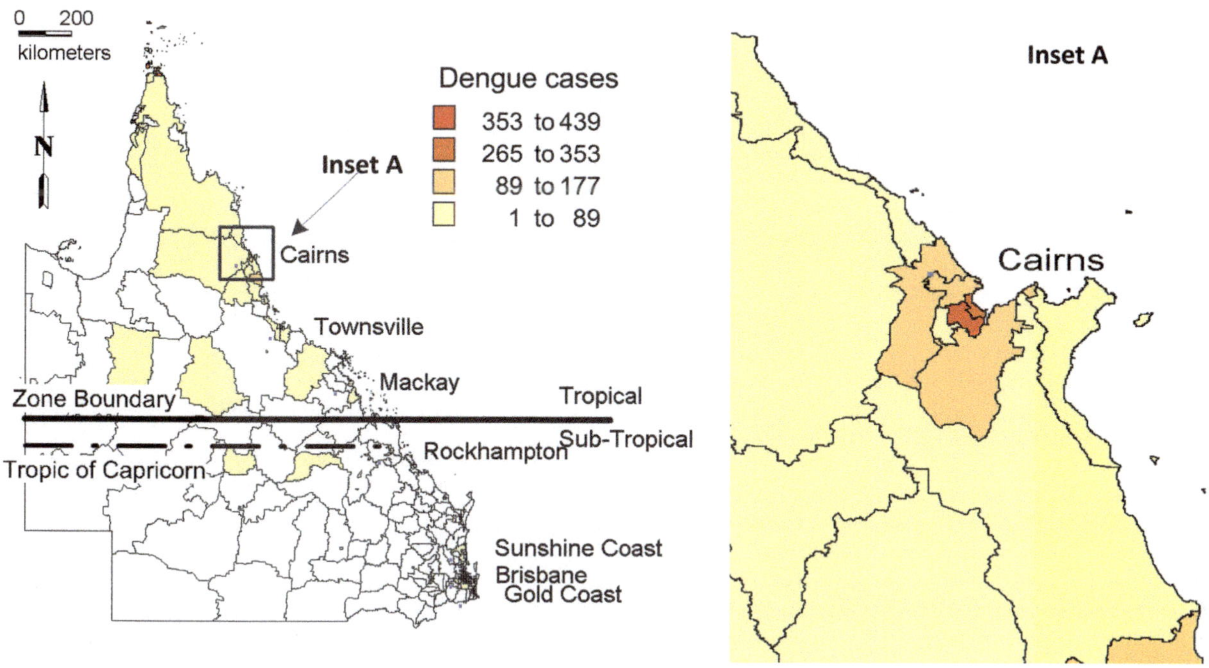

Figure 1. Study area showing the spatial distribution of dengue cases in northern Queensland, Australia, 1993–2012.

Age- and gender-standardised incidence rates (SIRs) were calculated for each SLA, using the direct method (based on Queensland population as a "reference"), adjusted for differences in the age and gender distribution. For example, dengue was high among the 25–29 year old age group so a SLA with a higher proportion of this age group would have a higher overall incidence rate of dengue. The equation for calculating SIR is: $SIR = \frac{\sum e_i}{\sum d_i^{(comparison)}}$ (expressed per 100,000 people)

Where $\sum e_i$ is the total number of expected cases generated using the reference population rates for each SLA; $\sum d_i^{(comparison)}$ is the total population in the comparison group.

Further, the SIR estimates were mapped to identify the spatial differences. A significant difference between the observed and expected number of cases is indicated if the confidence interval (CI) does not contain zero. Differences between observed and expected number of cases for age and gender were tested using chi-square analyses.

Spatial analysis. We conducted spatial analysis comprising three components: 1) spatial autocorrelations, 2) semi-variogram models and 3) interpolations of SIR values (kriging).

Spatial autocorrelations. The global Moran's I test statistic was used to assess the presence of significant spatial autocorrelation of dengue overall incidence rates in each of the four time periods of 1993–1997, 1998–2002, 2003–2007 and 2008–2012. Moran's I ranges from -1 to 1: a value close to 0 indicates spatial randomness while a positive value indicates positive spatial autocorrelation, and a negative one indicates a negative autocorrelation, that is clustering of observations, and similarly a negative value indicates negative spatial autocorrelation, that is, repulsive behaviour between observations. Statistical significance was tested using randomisation based on 999 permutations [20]. The neighbourhood matrix used for the computation of spatial autocorrelation statistics, was based on Queen contiguity and Euclidean distance [20].

Semi-variogram analysis. We used semi-variogram modelling analysis to explore the spatial structure and spatial autocorrelation of SIRs of dengue and age. The semi-variogram is a graphical representation of the variation among observations as the distance between the observations increases. If the variation, measured in terms of semi-variance, is distinctly small for low values of lag distance, it is considered as an indication of positive spatial autocorrelation, i.e., values at short distance from each other are more alike than those at larger distances. The best-fit semi-variogram model was identified by using Vertical Mapper within MapInfo Professional to identify the smallest differences in the model [21].

Kriging interpolation. Since the semi-variogram illustrates the spatial dependency between the observed measurements as a function of the distance between them, it allows us to estimate the SIR value of dengue at any point from the observed data. This interpolation was based on the best-fit semi-variogram using a kriging method [22–24] with inverse distance weighting [25]. Inverse distance weighting interpolation has been employed in other analyses of mosquito-borne diseases [26–28]. The kriged SIR values were obtained using the interpolation method in Vertical Mapper within MapInfo Professional and mapped to better visualise the distribution of spatially related patches of dengue.

Temporal analysis. To examine temporal patterns, epidemic curves were produced by calculating the annual incidence rate of dengue (annual dengue cases for each year divided by total population for each year * 100,000 people) and monthly cases of dengue during the period 1993–2012. Monthly differences between incidence rates for the study period were tested using Chi-square test. The statistical significance was set at <0.05.

Results

Descriptive statistics

Table 1 shows summary descriptive statistics for the dengue cases for the four time periods across SLAs in northern Queensland, Australia. Overall, the average number of dengue cases was 6.82 cases per year in northern Queensland. The SLAs notified with dengue cases varied over the four time periods. There were 13 SLAs with dengue cases (n = 34) in 1993–1997, 11 SLAs in 1998–2002 (n = 130 cases), 58 SLAs in 2003–2007 (n = 1042 cases) and 58 SLAs in 2008–2012 (n = 1192 cases). In all these periods, seasonal differences were observed between the dengue cases (Table 1). During 1998–2002 and 2008 to 2012, higher number of dengue cases (n = 86 and 773, respectively) were documented in summer whereas during 1993–1997 and 2003 to 2007, higher number of dengue cases (n = 29 and 625, respectively) were recorded in autumn.

Figure 2 shows the age- and gender-specific distribution of dengue incidence rates (per 100,000 people) during 1993–2012 for northern Queensland. The age and gender distributions are comparable with the last national population census data in 2011 [29]. The median age of the dengue cases was 44 years (range <1–80 above years). The incidence rate increased steadily with increasing age (children aged 5–9 years to those aged 50–54 years). Females in the younger (20–25 year) age groups had slightly higher incidence rates than males in the same age groups whereas in all other age groups males had slightly higher incidence rates compared with females. This difference was statistically significant ($\chi^2 = 15.17$, d.f. = 1, $p<0.01$). Differences were observed among age groups, but these were not statistically significant.

Temporal analysis

Figure 3 displays the temporal epidemic patterns of dengue cases and incidence rates for four time periods in northern Queensland, including four major outbreaks in 2003, 2004, 2005 and 2009. The annual incidence rates fluctuated from 5.65 per 100,000 people (in 2000) to 82.84 per 100,000 people (in 2005).

The distribution of dengue cases and incidence rates was highly seasonally sensitive over the whole study period and in each of the four time periods. The incidence rates of dengue for the years 1993 to 2012 in northern Queensland indicates a strong seasonal pattern ($\chi^2 = 8.357$, d.f. = 1, $p<0.01$), with a peak in Autumn (i.e., March) and reduction in Winter (i.e., August), with striking differences (Fig 4) among the four time periods. This supports the observed differences in cases in Table 1.

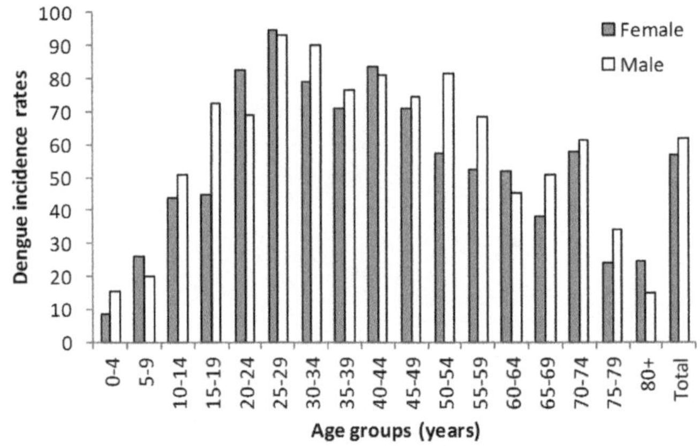

Figure 2. Dengue incidence rates by age and gender in northern Queensland, Australia, 1993–2012.

Spatial analysis

Figure 5 depicts the spatial patterns of dengue incidence rates in the four time periods during 1993–2012. Overall and in each period, the far north Queensland regions had the highest incidence rates: SLAs with highest incidence rates were Torres followed by Cairns City (central suburbs) for the years 2008 to 2012. Other relatively large incidence rates were reported in south Townsville and Cairns central suburbs. In all the time periods, incidence rates were high in Torres, Cairns City (central suburbs), Cairns - Mt Whitfield, South Townsville, Currajong, North Ward - Castle Hill, South Flinders and Rowes Bay - Belgian Gardens. Overall, dengue appears to have spread dramatically from Cairns regions to north Queensland regions during 1993–2012. In general, dengue occurred for 15 consecutive years from 1998 primarily in the same SLAs along the coastal northern Queensland regions and gradually expanded into newer SLAs, with the highest number of cases (n = 956) reported in 2009 during January to April.

Spatial autocorrelation. Table 2 shows the results of the calculation of global autocorrelation statistics for dengue cases for the four time periods in northern Queensland. The results of the global Moran's I tests for dengue incidence rates for all four periods are statistically significant and indicate spatial heterogeneity. There was an increase in spatial autocorrelation over the period 2003–2007, reaching the highest value during 2008–2012.

Standardised incidence rates. Standardised incidence rates (SIRs) of dengue for each SLA in northern Queensland over the the study period were calculated and mapped (Fig 6). Geograph-

Table 1. Descriptive statistics of dengue cases among SLAs in northern Queensland, Australia, 1993–2012.

Indicators	1993-1997	1998-2002	2003-2007	2008-2012	Total
No. of SLAs positive	13	11	58	58	140
Total cases in all SLAs	34	130	1042	1192	2398
In summer	3	86	233	773	1095
In autumn	29	43	625	350	1047
In winter	1	0	50	17	68
In spring	1	1	134	52	188

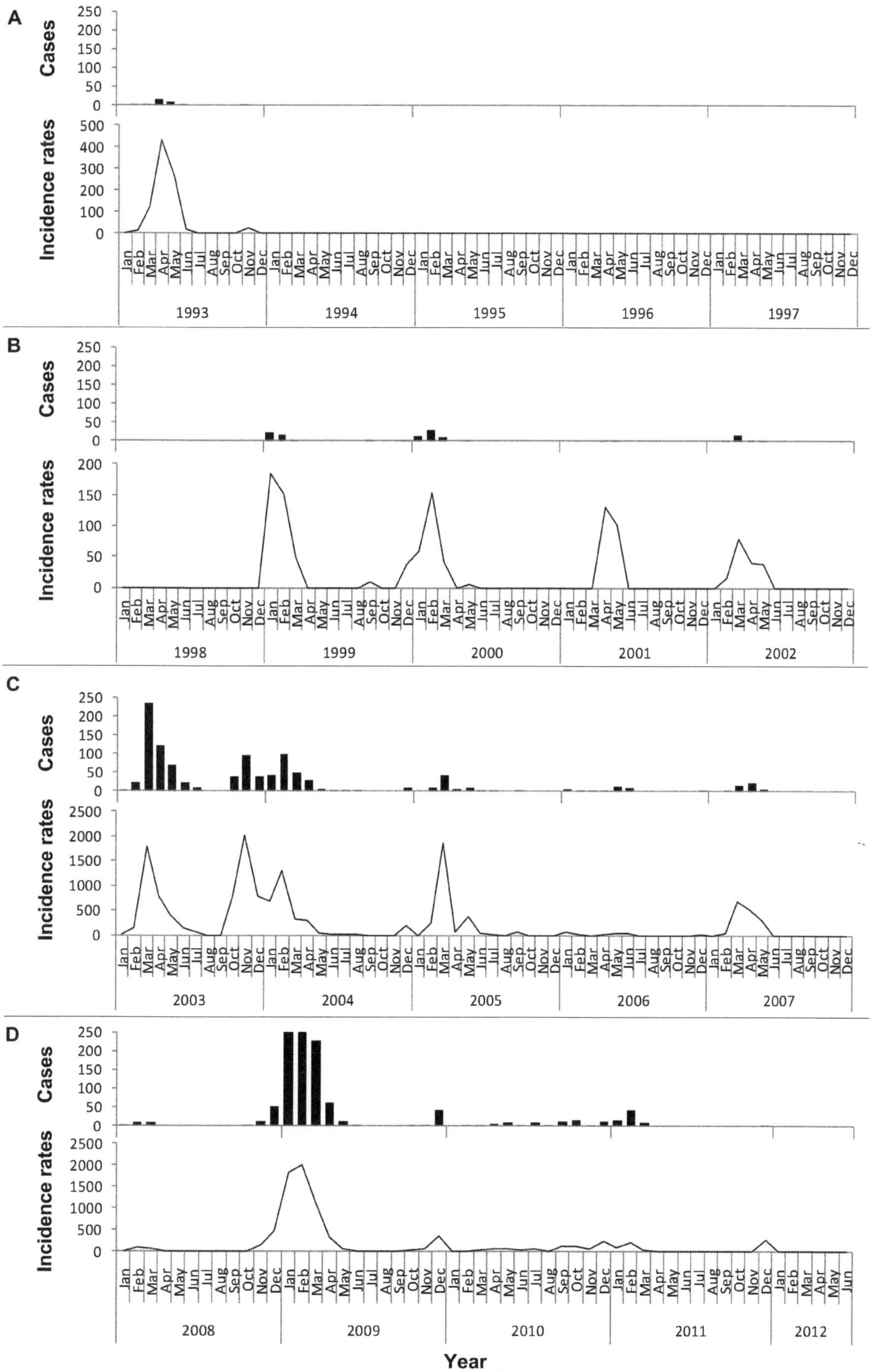
A
Cases
Incidence rates
1993
1994
1995
1996
1997
B
Cases
Incidence rates
1998
1999
2000
2001
2002
C
Cases
Incidence rates
2003
2004
2005
2006
2007
D
Cases
Incidence rates
2008
2009
2010
2011
2012
Year

Figure 3. Temporal distribution of dengue cases (depicting bar) and incidence rates (depicting line) for different time periods (A: 1993–1997, B: 1998–2002, C: 2003–2007 and D: 2008–2012).

ically, the largest SIRs were observed in the coastal areas; with the peak SIR of 180/100,000 for Cairns City (central suburbs) while the average SIR in northern Queensland was 2.06/100,000.

Semi-variogram analysis and kriging. Spatial dependence of dengue SIRs was evaluated using semi-variograms. A quadratic model was fitted to the semi-variogram using a sill and nugget of 400 and 0, respectively and a range of 5 degrees (Fig 7). This best-fit semi-variogram model was then used in the kriging procedure to map the SIRs. The map of the kriged SIR is shown in Figure 7. This provides visual confirmation that the pattern of SIRs of dengue disease is not homogeneous across the northern Queensland.

Discussion

This study reveals the spatial and temporal characteristics of dengue in northern Queensland using GIS tools and geostatistical analytical techniques. These methods have been applied to other vector-borne infectious diseases to study the distribution patterns of the disease, to identify the high-risk areas or hot spots, and to determine the risk factors for the transmission of the disease [26,30–35]. Locally-acquired dengue cases only occurred in northern Queensland where the vector was common and where the virus was commonly introduced by viraemic travellers [36]. Our results indicate that dengue incidence rates had an uneven spatial distribution in northern Queensland and thus the disease was spatially heterogeneous. In addition, the GIS maps clearly revealed spatial expansion of dengue transmission in northern Queensland over recent years (i.e., 2008–2012). For example,

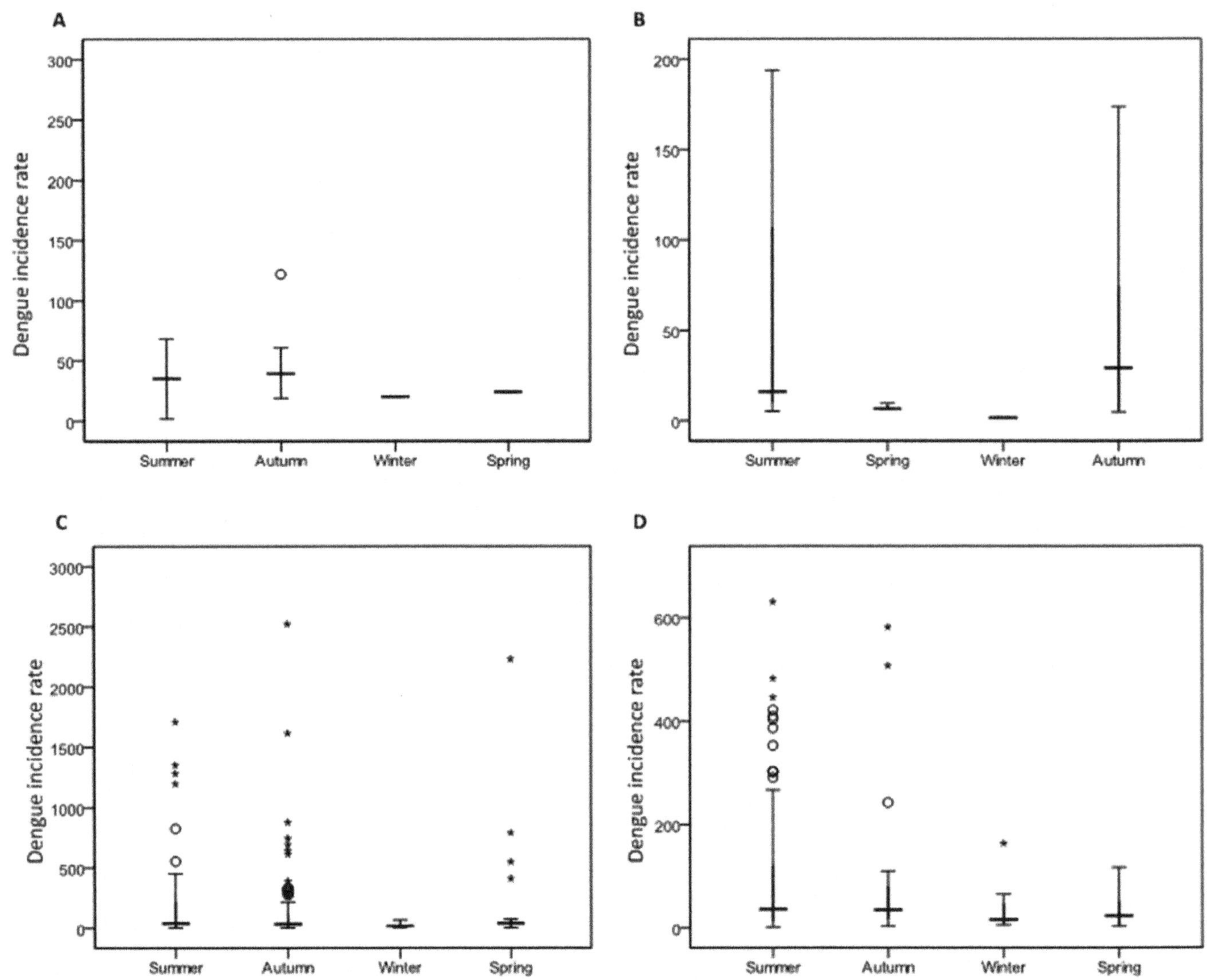

Figure 4. Box-plots showing the seasonal distribution of dengue cases for the different periods: A: 1993–1997, B: 1998–2002, C: 2003–2007 and D: 2008–2012.

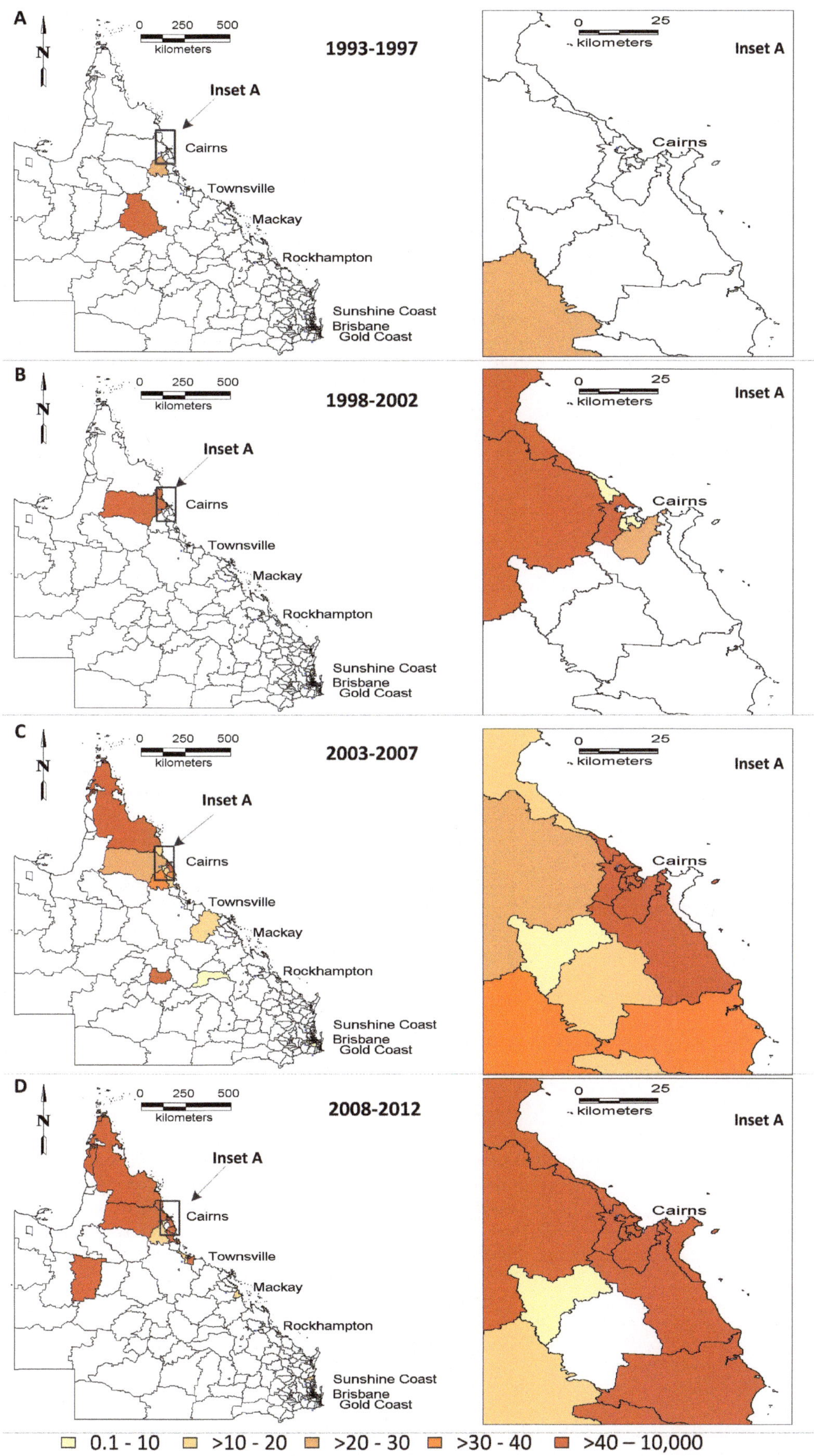
A
1993-1997
B
1998-2002
C
2003-2007
D
2008-2012
N
0 250 500
kilometers
Inset A
Cairns
Townsville
Mackay
Rockhampton
Sunshine Coast
Brisbane
Gold Coast
0 25
kilometers
0.1 - 10
>10 - 20
>20 - 30
>30 - 40
>40 – 10,000

Figure 5. Maps showing the dengue incidence rates by SLA over different periods (A: 1993–1997, B: 1998–2002, C: 2003–2007 and D: 2008–2012).

dengue transmission has increasingly spread from Cairns regions to a wider area.

Temporally, significant differences were noticeable across different time periods in the study area. Our findings demonstrate that the annual incidence rates fluctuated considerably, with the two largest peaks in 2003 and 2009 (see Fig 2). It is important to understand the influence of annual seasons on transmission cycles. Our results also show that there has been an increasing trend in the incidence rates of dengue during the study period. The overall temporal pattern of dengue incidence in northern Queensland indicates two general peaks per annum: the first short peak in summer (i.e. extending for two months in November and December), and a second long peak in autumn (i.e. four months from January to April). These two peaks may be preceded by peaks of mosquito densities in December and May, following the normal typical wet season. Our findings strongly support previous studies that have suggested an increase in mosquito population in summer, with a lagged impact on the seasonal variation of dengue [37,38]. However, whether this corresponds to the peaks of tourists who might introduce the virus is still unknown.

The study results indicate that dengue outbreaks usually occur during the rainy season (i.e., November to April) in northern tropical Queensland, which generally has a lower temperature by a few degrees and higher humidity than during the hot dry season. These small environmental fluctuations may possibly increase the female mosquito longevity and survival, thus increase the likelihood of dengue transmission. Our findings strongly support previous studies that have reported a strong seasonal pattern of dengue [2,37,39].

The spatial and temporal differences may due partly to occurrence of epidemics due to introduced cases, mosquito distribution, local changes in climate and people's culture and behaviour towards personal protection, human adaptation to forecasted drought (e.g., water storage facilities) and housing conditions and life-style activities [2,40] and partly to local vector control programs such as dengue mosquito management plan and dengue alert response team activities [4]. Thus, further research is needed to investigate the social mobilisation, entomological surveillance, vector biology, dengue viruses, local people's behaviour and the local climatic factors on the transmission of dengue in the northern Queensland region.

In this study, locally-transmitted dengue was reported in a small number of SLAs across the study area. The study results reveal that quite a few numbers of SLAs had the highest incidence rates and SIRs (see Figs 4 & 7). This may be due to the several conditions that favour vector, density, distribution, survival and longevity [2,39,41]. A combination of flooding and heavy rainfall has resulted in dengue epidemics across Australia [2,4,40]. In addition, local factors such as the climate and socio-ecological conditions may be more suitable for the local mosquitoes and thus the transmission of dengue. Therefore, it is evident that further studies on the environmental and socio-ecological factors on dengue mosquitoes and dengue virus would assist in identifying the reasons behind this phenomenal variation.

Table 2. Moran's *I* values for the dengue incidence rates in northern Queensland, Australia, 1993–2012.

Period	Moran's *I* value	p-value
1993–1997	0.0116	0.03
1998–2002	0.0019	0.001
2003–2007	0.1362	0.001
2008–2012	0.4639	0.001

In this study, dengue incidence occurred in all age groups, but was highest among males and females of age group 25–29 (see Fig 3). The reasons for the differences in incidence rates among age and gender groups are unknown, but may include different exposure rates or other behavioural risk factors such as increased internal (domestic) travel, immigration, work and leisure related activities [29], however, this kind of information is unavailable in this study. The possible reasons could be increased levels of time spent outdoors for recreation (e.g., fishing and walking) and leisure (e.g., sport and exercise) activities and internal (within the state) travel due to work and family commitments. Clearly, the relationships between the incidence rate of dengue, age and gender needs to be understood in order to better manage and reduce incidence spikes in certain age groups.

Epidemiologists normally use the ratio of case numbers at a particular time to past case occurrences using the mean or median [42]. However, since dengue cases vary from one place to another, the spatial and temporal component must also be taken into consideration. In disease surveillance and public health surveys, spatial and temporal patterns are one of the most important components influencing the distribution of diseases. Although spatial analytical techniques rarely provide reasons for the occurrences of spatial patterns, they do identify the geographical locations of the occurrence of spatial pattern. Within this realm, it provides a useful means to hypothesise about factors that may influence health outcomes or to identify spatial issues that need to be further investigated [43]. The evaluation of spatial distributions as a measure of disease risk may provide etiological insights [44]. Regular time series models are difficult to fit to our data, given the epidemic and irregular seasonal patterns of dengue incidence. Thus, we haven't pursued these models in this study.

In this study, the global Moran's *I* statistic is used to measure the degree of spatial autocorrelation and map the geographic patterns of the areal units. To appropriately use dengue notification data aggregated according to SLAs, it is important to choose the spatial autocorrelation technique for the specification of local neighbourhoods. This is defined by the spatial weights matrix. In general, the spatial autocorrelation may be the strongest between the nearest neighbours. As the neighbourhoods increase in number, this autocorrelation weakens. However, a recognised guide for choosing a proper spatial weight matrix has not yet been developed. In this study, an appropriate spatial weight matrix was chosen after a comparison of the connectivity distributions of neighbours obtained with the distance-based contiguity and the first order Queen's contiguity methods [20].

Spatial autocorrelation and semi-variogram analysis are valuable tools to study spatial patterns over time. The semi-variogram estimators used in this paper directly account for population size, attenuating the influence of less reliable rates recorded in sparsely populated areas. Maps created from kriging interpolation revealed that dengue was spatially and temporally distributed. Further studies of local environmental and socio-environmental factors that operate at fine spatial scales are crucial for improving the

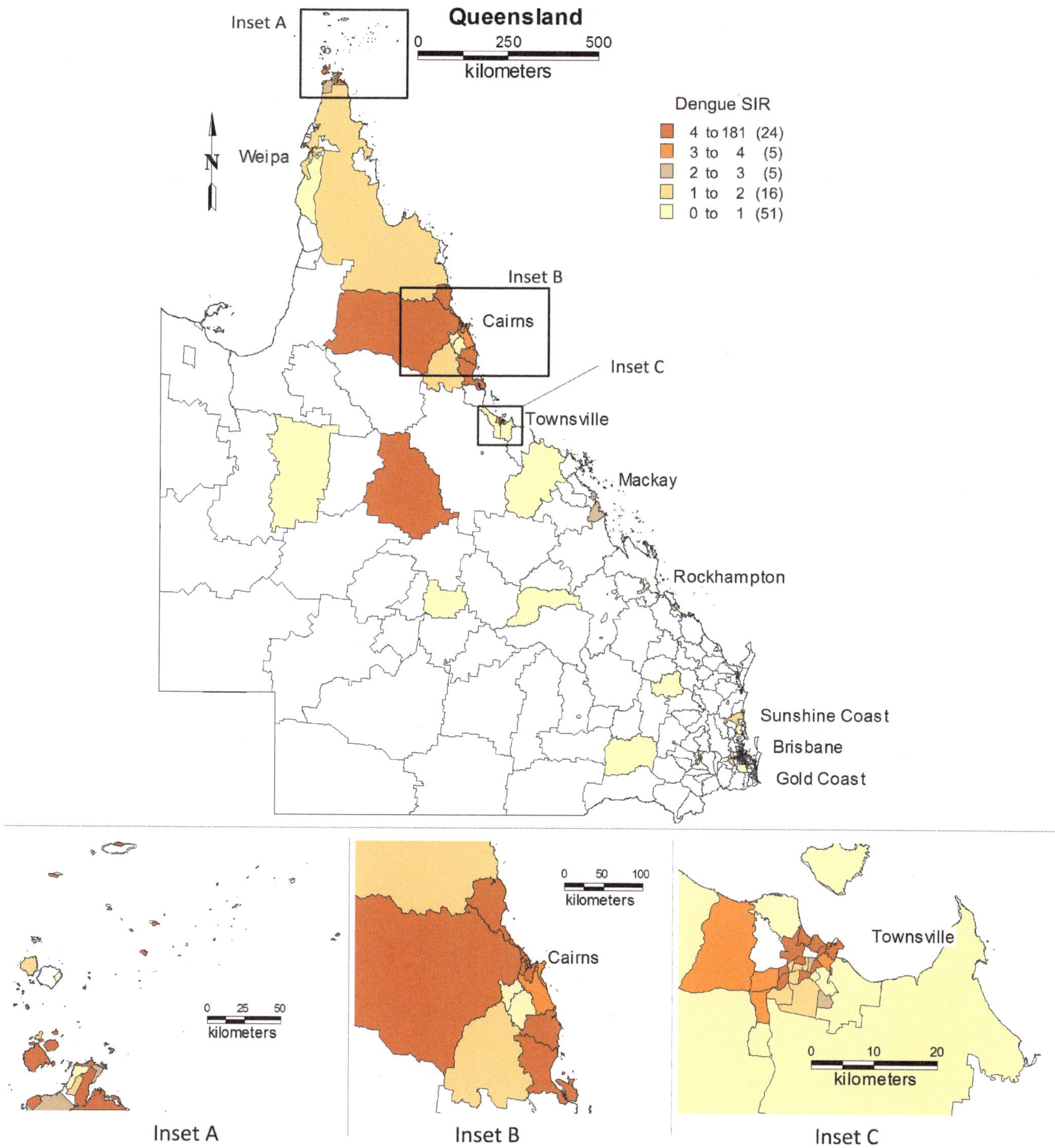

Figure 6. Map showing the standardised incidence rates of dengue by SLA in northern Queensland, Australia, 1993–2012.

understanding of the spatial and temporal patterns of dengue. Moreover, further investigation is warranted to understand the effect of climatic and topographic factors on dengue local transmission in the study area.

Vector-borne infectious diseases such as dengue present complex and dynamic transmission patterns, which include vector related factors such as type, density, distribution, breeding places and human related factors such as population density, behaviour and immunity, and virus related factors such as circulating multiple virus serotypes DENV 1 to DENV 4, and environmental factors such as temperature and rainfall [1,11,45]. Several dengue outbreaks by different serotypes may occur in the same population, and there is a large range of factors in intra-urban areas which may favour the maintenance of potential breeding sites of mosquitoes.

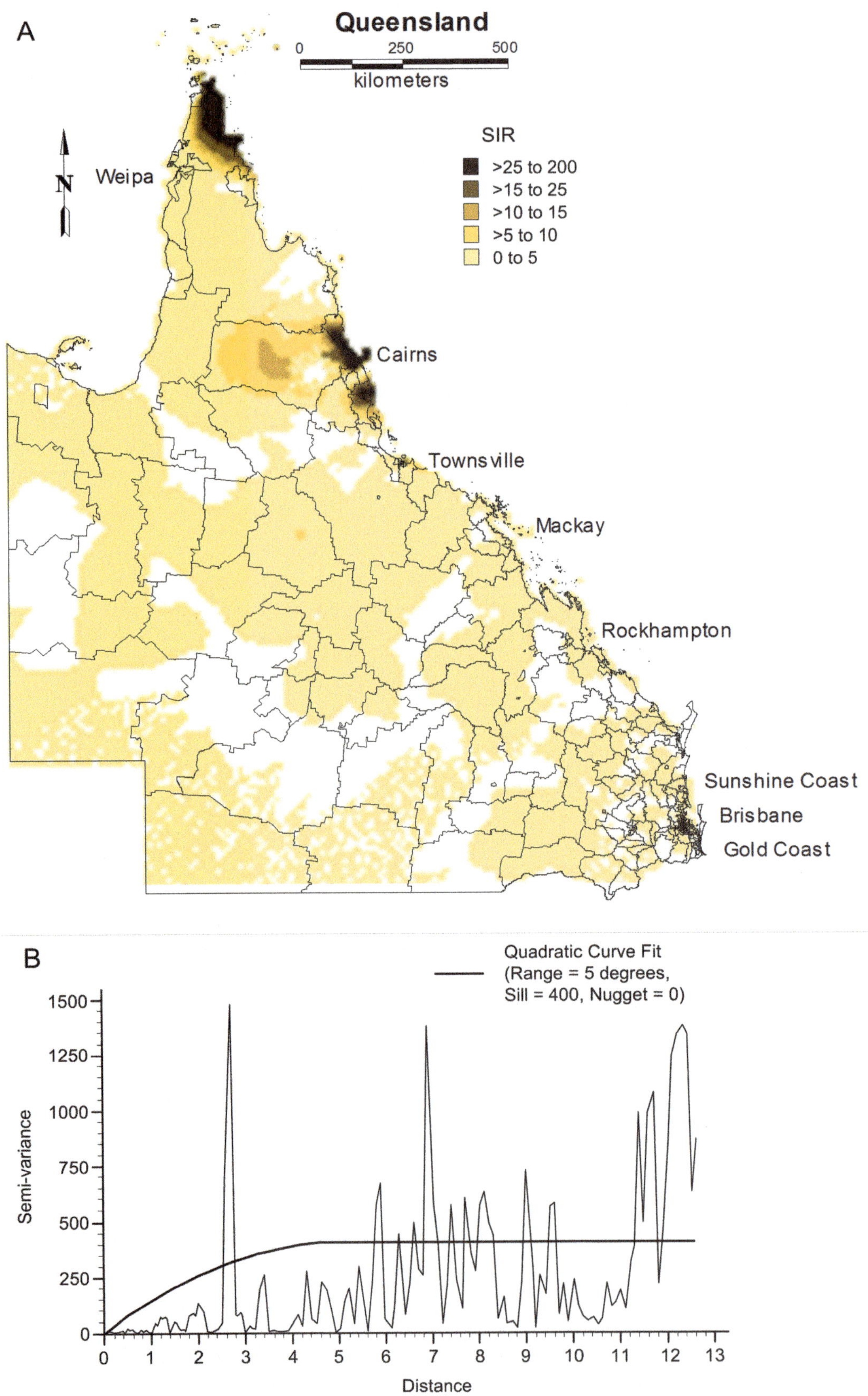

Figure 7. The smoothed map of standardised incidence rates of dengue using kriging (Panel A) and a semi-variogram model (Panel B).

We acknowledge that there could be issues in monitoring and reporting dengue locally-acquired notification data. For this study, the clinically proven cases on dengue were provided by Queensland Department of Health. Dengue is one of the notifiable diseases in Australia, and is required to be reported to the local health, by law. Data reliability issues for mosquito-borne diseases have previously been discussed by Russell [46]. Underreporting is also likely to occur when people infected with dengue do not seek medical attention. For example, it is likely that there was considerable underreporting in 1992 and 1993 but we have evidence to suggest that in more recent epidemics the publicly notified cases are not a gross underestimate of the total number of cases. Nevertheless, these issues cannot entirely account for the geographic distribution of dengue across northern Queensland.

This study has three major strengths. Firstly, this is the first study to examine the geographic variation of dengue across geo-political borders (i.e., SLAs) in northern Queensland using GIS tools and geostatistical techniques. This study is an impetus for future investigations on the spatial and temporal risk factor analysis of dengue including multiple factors on environmental, entomological, ecological and socio-demographic variables. Secondly, the results of this study demonstrate that GIS mapping techniques may be used as a tool to quickly display information and generate maps to highlight dengue risk-prone areas for developing strategies towards dengue management. The maps could be used by policy-decision makers to suggest particular geographical localities or communities where further investigation should be focused, to identify whether increased disease surveillance measures or possible control activities are warranted. Moreover, the corresponding statistical analyses can limit over-reaction or unwarranted action based on purely visual assessment, for example unsubstantiated identification of so-called 'clusters' of cases and putative source identification. Finally, dengue data used in this study are quite comprehensive, covering northern Queensland for approximately 20 years.

The study has also two key limitations. First, the quality of the dengue surveillance system may vary with place and time as the awareness of dengue among medical professionals and public may have increased over recent years. However, spatiotemporal variation of dengue infers that the dengue distribution is unlikely to be entirely accounted for by a detection/surveillance. Second, the exact place/location (i.e., residential address) where dengue cases were notified may vary from those where they were infected/acquired (i.e., exact place/location), particularly during holiday periods. However, all the cases involved in this study were locally-acquired and were not imported.

The maps produced in this study would provide useful information to health authorities and could assist in focusing and implementing control and preventive activities to monitor and control the incidence of dengue precisely and effectively, especially in the event when there is no report on dengue cases. The study suggests that local surveillance teams should be vigilant at all times, particularly after the wet season, not only in the epidemic period, as dengue mosquitoes live with the human population. Furthermore, this study provides a new dimension to the health authorities in northern Queensland, specifically in the potential of using GIS applications to develop locally appropriate and eco-environmentally friendly strategies for the implementation of preventive and control activities, not only for dengue, but also for other vector-borne diseases in Queensland.

In conclusion, this study has revealed that the spatio-temporal patterns of dengue differ significantly in northern Queensland and the study has highlighted that there are different transmission patterns in SLAs between regions. This study has also concluded that the spatial distribution of dengue appears to have expanded over recent years. This is based on the results (see Tables 1 & 2) and on the observation that dengue has spread from Cairns regions to a wider area during 1993–2012. Autocorrelation function can be beneficial for public health officials or policy-developers to visualise and understand the distribution and trends of diffusion patterns of dengue and to prepare warnings over high-risk areas only rather than for a whole state or a whole region. This may save time and cost and make public health authorities efforts more efficient.

Future research should focus on the spectrum of dengue risk factors and the prediction of future dengue transmission which are necessary to improve the effectiveness and efficiency of dengue local prevention and control programs.

Author Contributions

Conceived and designed the experiments: SN. Performed the experiments: SN. Analyzed the data: SN. Contributed reagents/materials/analysis tools: SN. Wrote the paper: SN PD JM JMB KM ST. Wrote the manuscript, reviewed critically and accepted for publication: SN PD JM JMB KM ST.

References

1. WHO (2013) Dengue - Neglected tropical diseases research Available: http://www.who.int/tdr/research/ntd/dengue/en/index.html. Accessed 6 February, 2013.
2. Russell RCCB, Lindsay MD, Mackenzie JS, Ritchie SA, Whelan PI (2009) Dengue and climate change in Australia: predictions for the future should incorporate knowledge from the past. Med J Aust 190: 265–268.
3. Tsuzuki AVT, Higa Y, Nguyen TY, Takagi M (2009) Effect of peridomestic environments on repeated infestation by preadult Aedes aegypti in urban premises in Nha Trang City, Vietnam. Am J Trop Med Hyg 81: 645–650.
4. Queensland Health. (2013) Dengue fever. Available: http://www.health.qld.gov.au/dengue/outbreaks/current.asp. Accessed 20 January, 2013.
5. Schwartz E, Weld LH, Wilder-Smith A, von Sonnenburg F, Keystone JS, et al. (2008) Seasonality, annual trends, and characteristics of dengue among ill returned travelers, 1997–2006. Emerg Infect Dis 14: 1081–1088.
6. Lloyd CD, Atkinson PM (2009) Cartographic veracity in medieal mapping:a-nalyzing geographic variation in the Gough map of Great Britain. Ann Asso Am Geo 99: 27–48.
7. Fotheringham AS, Rogerson PA, editors (2009) The Sage Handbook of Spatial Analysis. London: Sage Publications. 183–206 p.
8. Lloyd C (2010) Spatial data analysis - An introduction for GIS users. Oxford: Oxford University Press.
9. Collinge SK (2010) Spatial Ecology and Conservation. Nature 1: 69.
10. Haining R (2003) Spatial Data Analysis: Theory and Practice. Cambridge: Cambridge University Press.
11. Githeko AK (2012) Advances in developing a climate based dengue outbreak models in Dhaka, Bangladesh: Challenges & opportunities. Indian J Med Res 136: 7–9.
12. Cleveland WS (1993) Visualising data: Summit, N.J. Hobart Press.
13. Indaratna K, Hutubessy R, Chupraphawan S, Sukapurana C, Tao J, et al. (1998) Application of geographical information systems to co-analysis of disease and economic resources: dengue and malaria in Thailand. Southeast Asian J Trop Med Public Health 29: 669–684.
14. Keim DA (2002) Information visualization and visual data mining. IEEE Transaction on Visualization and Computer Graphics 7: 100–107.
15. Kitron U (2000) Risk maps: Transmission and burden of vector-borne diseases. Parasitoly Today 16: 324–325.
16. Rezaeian M (2008) How to Visualize Public Health Data? Part Two: Direct and Indirect Standardization Methods. Middle East Journal of Family Medicine 7: 42–44.
17. Queensland Treasury and Trade (2013) Population growth highlights and trends, Queensland. Available: http://www.oesr.qld.gov.au/products/publications/pop-growth-highlights-trends-qld/pop-growth-highlights-trends-qld-2013.pdf. Accessed 2013, 10 August.
18. National Notificable Disease surveillance system N (2013) Number of notifications of dengue virus infection received from State and Territory health authorities in the period of 1991 to 2012 and year-to-date notifications for 2013. Department of Health and Ageing.

19. Queensland Health (2013) Mosquito-borne diseases. Available: www.health.qld.gov.au. Accessed 2013, 20 November.
20. Anselin L, Syabri I, Kho Y (2006) GeoDa: An introduction to spatial data analysis. Geographical Analysis 38: 5–22.
21. MapInfo Professional Software version 10.5.2. (2011) MapInfo Vertical Mapper, version 3.7, MapInfo Corporation
22. Fotheringham AS, Rogerson PA, editors (2009) The Sage Handbook of Spatial Analysis. London: Sage Publications. 183–206 p.
23. Lloyd CD, Shuttleworth IG (2005) Analysing commuting using local regression techniques: scale, sensitivity and geographical pattterning. Environment and Planning A: 81–103.
24. Longley PA, Goodchild MF, Maguire D, Rhind DW (1999) Geographical Information Systems New York: John Wiley & Sons.
25. Isaaks EH, Srivastava RM, editors (1989) An introduction to applied geostatistics. Oxford, United Kingdom: Oxford University Press. 249–277 p.
26. Naish S, Hu W, Mengersen K, Tong S (2011) Spatio-temporal patterns of Barmah Forest virus disease in Queensland, Australia. PLoS ONE 6: e25688.
27. Quinn HE, Gatton ML, Hall G, Young M, Ryan PA (2005) Analysis of Barmah forest virus disease activity in Queensland, Australia, 1993–2003: Identification of a large, isolated outbreak of disease. J Med Entomol 42: 882–890.
28. Woodruff RE, Guest CS, Garner MG, Becker N (2006) Early warning of Ross River virus epidemics: combining surveillance data on climate and mosquitoes. Epidemiology 17: 569–575.
29. ABS (2012) Census of Population and Housing - Basic community profiles, Commonwealth of Australia, Canberra, Australian Bureau of Statistics, ABS. Available: www.abs.com.au. Accessed 10 December, 2012.
30. Naish S, Hu W, Mengersen K, Tong S (2011) Spatial and temporal clusters of Barmah Forest virus disease in Queensland, Australia. Trop Med Int Health 16: 884–893.
31. Russell RC (1998) Vectors vs. humans in Australia-who is on top down under? An update on vector-borne disease and research on vectors in Australia. J Vector Ecol 23: 1–46.
32. Earnest A, Tan SB, Wilder-Smith A (2012) Meteorological factors and El Niño Southern Oscillation are independently associated with dengue infections. Epidemiology and Infection 140: 1244–1251.
33. Huang FZS, Zhang S, Zhang H, Li W (2011) Meteorological Factors-Based Spatio-Temporal Mapping and Predicting Malaria in Central China. Am J Trop Med Hyg 85: 560–567.
34. Impoinvil DE, Solomon T, Schluter WW, Rayamajhi A, Bichha RP, et al. (2011) The Spatial Heterogeneity between Japanese Encephalitis Incidence Distribution and Environmental Variables in Nepal. PLoS ONE 6: e22192.
35. Rochlin ITD, Gomez F, Ninivaggi DV, Campbell SR (2011) Predicitve mapping of human risk for West Nile Virus (WNV) based on environmentala dn socioeconomic factros. PLoS ONE 6: e23280.
36. Knope KGC (2013) Increasing notifications of dengue in Australia related to overseas travel, 1991 to 2012. Commun Dis Intell 37: E5–E7.
37. Hu W, Clements A, Williams G, Tong S (2010) Dengue fever and El Nino/Southern Oscillation in Queensland, Australia: a time series predictive model. Occup Environ Med 67: 307–311.
38. Hu W, Clements A, Williams G, Tong S, Mengersen K (2012) Spatial patterns and socioecological drivers of dengue fever transmission in Queensland, Australia. Environ Health Perspect 120: 260–266.
39. Beebe NW, Cooper RD, Mottram P, Sweeney AW (2009) Australia's dengue risk driven by human adaptation to climate change. PLoS Negl Trop Dis 3: e429.
40. Williams CR, Bader CA, Kearney MR, Ritchie SA, Russell RC (2010) The Extinction of Dengue through Natural Vulnerability of Its Vector. PLoS Negl Trop Dis 4: e922.
41. Jansen CC, Beebe NW (2010) The dengue vector Aedes aegypti: what comes next. Microbes Infect 12: 272–279.
42. Nakhapakorn KJS (2006) Temporal and spatial autocorrelation statistics of dengue fever. Dengue Bull 30: 177–183.
43. Er ACRM, Asmahani A, Mohamad Naim MR, Harsuzilawati M (2010) Spatial mapping of dengue incidence: A case study in Hulu Langat District, Selangor, Malaysia. International Journal of Human and Social Sciences 5: 410–414.
44. Moore DA, Carpenter TE (1999) Spatial analytical methods and geographic information systems: use in health research and epidemiology. Epidemiologic Reviews 21 143–161.
45. WHO (2012) Global strategy for dengue prevention and control, 2012–2020. Available: Accessed 5 February, 2013.
46. Russell RC (1994) Ross River Virus: disease trends and vector ecology in Australia. Bulletin of Social and Vector Ecology 19: 73–81.

Malaria Prevalence, Spatial Clustering and Risk Factors in a Low Endemic Area of Eastern Rwanda

Stephen Rulisa[1,3]⁹, Fredrick Kateera[2,6*⁹], Jean Pierre Bizimana[5], Steven Agaba[3], Javier Dukuzumuremyi[3], Lisette Baas[3], Jean de Dieu Harelimana[3], Petra F. Mens[2,3,4], Kimberly R Boer[3,4], Peter J de Vries[2,3,7]

1 University Teaching Hospital of Kigali, National University of Rwanda, Kigali, Rwanda, **2** Academic Medical Center, Division of Infectious Diseases, Tropical Medicine and AIDS, Amsterdam, The Netherlands, **3** Amsterdam Institute for Global Health and Development, INTERACT Project, Kigali, Rwanda, **4** Royal Tropical Institute/Koninklijk Instituutvoor de Tropen (KIT), KIT Biomedical Research, Amsterdam, The Netherlands, **5** Geography Department, Faculty of Science, National University of Rwanda, Butare, Rwanda, **6** Medical Research Centre, Rwanda Biomedical Centre, Kigali, Rwanda, **7** Department of Internal Medicine, Tergooiziekenhuizen, Hilversum, The Netherlands

Abstract

Background: Rwanda reported significant reductions in malaria burden following scale up of control intervention from 2005 to 2010. This study sought to; measure malaria prevalence, describe spatial malaria clustering and investigate for malaria risk factors among health-centre-presumed malaria cases and their household members in Eastern Rwanda.

Methods: A two-stage health centre and household-based survey was conducted in Ruhuha sector, Eastern Rwanda from April to October 2011. At the health centre, data, including malaria diagnosis and individual level malaria risk factors, was collected. At households of these Index cases, a follow-up survey, including malaria screening for all household members and collecting household level malaria risk factor data, was conducted.

Results: Malaria prevalence among health centre attendees was 22.8%. At the household level, 90 households (out of 520) had at least one malaria-infected member and the overall malaria prevalence for the 2634 household members screened was 5.1%. Among health centre attendees, the age group 5–15 years was significantly associated with an increased malaria risk and a reported ownership of ≥4 bednets was significantly associated with a reduced malaria risk. At the household level, age groups 5–15 and >15 years and being associated with a malaria positive index case were associated with an increased malaria risk, while an observed ownership of ≥4 bednets was associated with a malaria risk-protective effect. Significant spatial malaria clustering among household cases with clusters located close to water- based agro-ecosystems was observed.

Conclusions: Malaria prevalence was significantly higher among health centre attendees and their household members in an area with significant household spatial malaria clustering. Circle surveillance involving passive case finding at health centres and proactive case detection in households can be a powerful tool for identifying household level malaria burden, risk factors and clustering.

Editor: Abdisalan Mohamed Noor, Kenya Medical Research Institute - Wellcome Trust Research Programme, Kenya

Funding: This study was funded by Netherlands-African partnership for capacity development and clinical interventions against poverty-related diseases (NACCAP) through the Infectious diseases Network for Treatment and Research in Africa (INTERACT) - programme (Rwanda/Uganda) URL: http://www.nwo.nl/nwohome.nsf/pages/NWOA_6LRD4R_Eng. The funders had no role in study design, data collection and analysis, decision to publish, or preparation of the manuscript.

Competing Interests: All authors have declared: no support from any organization for this submitted work; no financial relationships with any organizations with possible interest in work submitted work; no other relationships or activities that could possibly influence this submitted work.

* E-mail: fkkateera@yahoo.com

⁹ These authors contributed equally to this work.

Introduction

From 2005 to 2010, Rwanda achieved the 2005 global community commitment of reducing the malaria burden by at least 50% [1]. During this period, a rapid malaria assessment conducted at 30 out of 40 Hospitals in Rwanda showed reductions of; 74% among confirmed outpatients cases of all ages, 26% in slide positivity rates, 65% among inpatients of all ages, and 55% in malaria deaths [2]. These gains followed rapid scale-up of insecticide-treated mosquito nets (ITNs), indoor residual spraying (IRS), use of artemisinin combination therapies (ACTs) and laboratory confirmation of presumed malaria cases with microscopy (at health facilities) and rapid diagnostic tests (RDTs) (by community health workers) as recommended by WHO's Roll Back Malaria program [1]. Despite these gains, malaria still causes significant morbidity; 7.8% of all febrile patients presenting at the

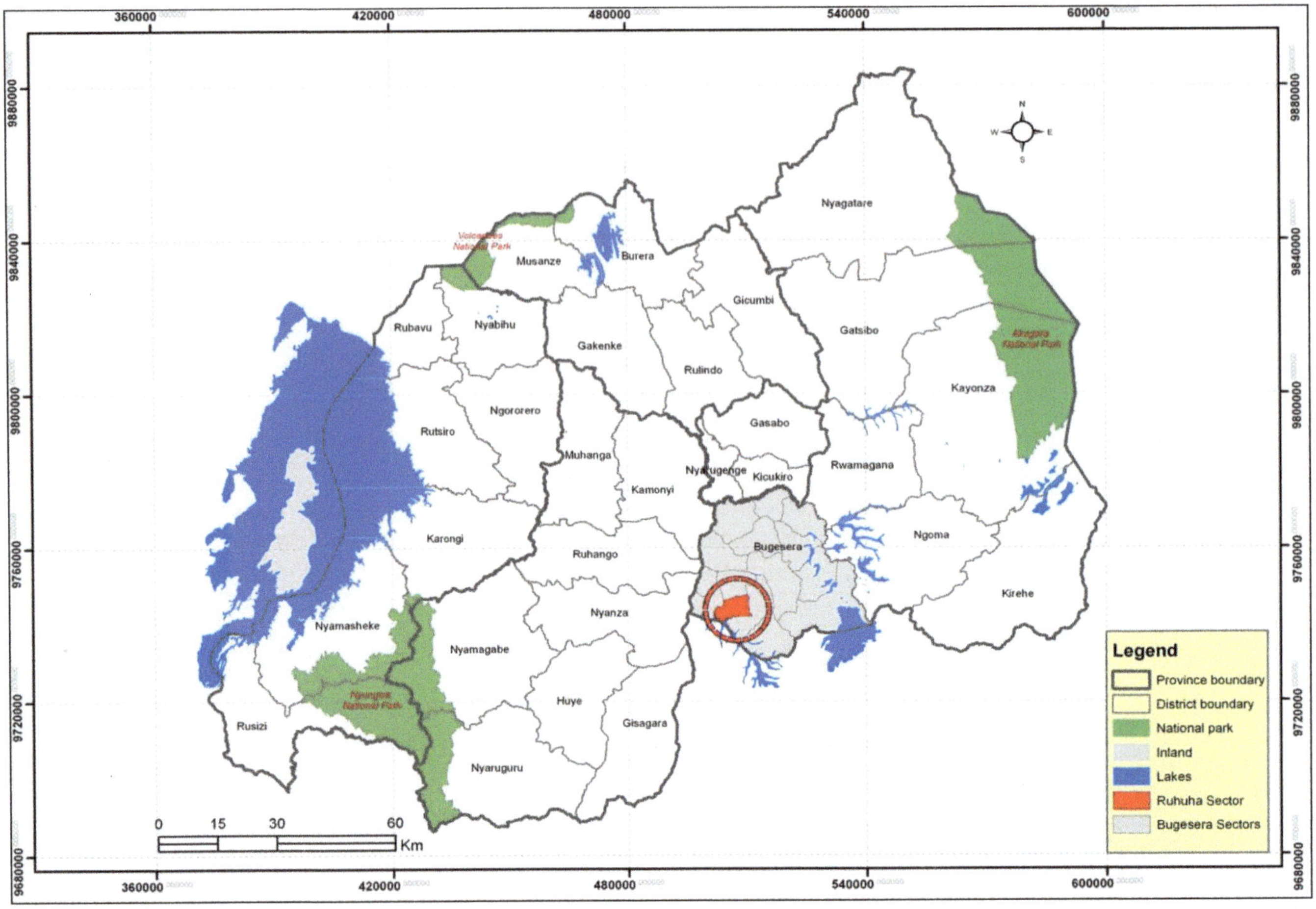

Figure 1. Location of Ruhuha Sector (Red), Bugesera District (Grey) in Rwanda. *Source: MINITRACO/CGIS-NUR, 2001 and NISR 2006.*

health centre (HC) had malaria and 12.9% of all age mortality were malaria associated in 2010, with a malaria resurgence recorded in 2009 [2,3,4]. These observations highlight the fragility of gains in malaria reduction achieved, especially in areas with a high baseline malaria transmission potential.

Current anecdotal Rwandan national routine data suggests a heterogeneous spatial malaria distribution with the entire population remaining at risk with the exception of the very high altitude zones [3,5]. Malaria heterogeneity has been reported across the different malaria endemic settings and has been attributed to risk factors including altitude, climate, occupation and socio-economic status [6,7,8,9,10]. However, at all malaria endemicity levels, and particularly in low incidence areas, malaria tends to cluster in 'hotspots' and 'hot' populations that become sources of continued infection. We defined a 'hotspot' of malaria transmission as 'a geographical part of a focus of malaria transmission where transmission intensity exceeds the average level' [11]. In a community, asymptomatic and minimally symptomatic malaria cases, whose symptoms may not be severe enough to seek care, can serve as significant parasite reservoirs for maintaining transmission [7,8,12]. Active and timely identification of these hotspots and associated risk factors is essential for targeting interventions to optimize malaria control [13].

Risk factors associated with malaria clustering for which we also investigated include distance of households (HHs) from potential mosquito-breeding sites, house roofing and wall materials and bednet use [7]. In Rwanda, however, there is paucity of systematic HH studies on malaria burden or associated risk factors with most reported data being aggregated routine health facility data. Despite its tendency to underestimate malaria burden, routine data can be helpful in reflecting malaria trends [14], particularly in low malaria incidence settings where the majority of the population access health services from the reporting health facilities. The passively identified health facility cases may reflect area malaria transmission levels in places where malaria cases tend to cluster in time and place. Index cases may also act as entry points to community HHs where identification of hotspots that could be targeted for optimal malaria control. Malaria hotspots may serve to perpetuate residual malaria transmission in low transmission seasons and hinder efforts to eliminate malaria [15].

In this study, we used HC attendees with presumed malaria as entry points for reactive case identification of malaria infections at the HH level. In a two-phase health facility and HH cross-sectional survey, we employed circle surveillance technique to measure malaria burden and evaluate for associated malaria risk factors. We also investigated for spatial malaria clustering using geographical information system (GIS) and spatial statistical techniques [16,17,18].

Materials and Methods

Ethical Statement

Ethical approval was granted by Rwanda National Ethics Committee. Prior to study initiation, sector and community leaders were informed about the study and their support and verbal consent requested. Written consents were obtained from adult participants and parents/guardians of participating children and from heads of HHs or the oldest person present for the HH surveys.

Study area

The complete survey was conducted in Ruhuha Sector, Bugesera district [19], Eastern Rwanda (Figure1). The sector covers 54 km^2, has a population of about 19,606 persons living in 4279 HHs. It is predominantly rural and traditionally a high malaria endemic area. Ruhuha sector, surrounded by lowland marshes and water-streams draining into the Akagera River System, is separated from Burundi by Lake Cyohoha in the south.

Study Design and Participants

A two-phase cross-sectional survey was conducted between April and October 2011. First, a fever survey was conducted among patients presenting at Ruhuha Health Centre (RHC) with a fever or history of fever in the last 24 hours. Patients of all ages were recruited and after signing the informed consent form, malaria diagnosis by microscopy and individual level risk factor data were collected. Thereafter, study participants were invited to participate in a follow-up HH survey where HH level malaria risk factor data was collected and malaria screening for all HH members performed.

Study Procedures

Health centre (HC) fever survey. At the HC, an interviewer-administered questionnaire, adapted from the Measures group Demographic Health Surveys tools and previous studies [20,21], was administered to adult patients or, in the case of minors, to parents/guardians of the children. The pre-tested questionnaire was administered by study-trained personnel. Data collected included personal demographics, fever characteristics, malaria perception, knowledge and practices including malaria preventive measures, and house structural features (walls and roofs).

Preparation of blood films, microscopic examination and quality assurance. To identify malaria among HC attendees, Giemsa stained thick and thin blood films were prepared and read by two independent experienced microscopists at the RHC laboratory. A third microscopist based at National Reference Laboratory (NRL) settled discrepancies between two readings. Parasite negative results were based on screening of 100 microscopic fields at 1000x magnification. Malaria parasites were counted against 200 white blood cells on thick blood films for enumeration of parasite density and thin smears used for species identification. In addition, 10% of all microscopy slides were sent to the NRL for external quality control.

Household survey. HC-recruited study participants (regardless of their malaria diagnosis status) who consented to a home visit and provided HH locator information were visited 1 to 4 months later for a follow-up HH survey. At this visit, all HHs were enumerated and assigned a unique identification number. An interviewer-administered questionnaire was used to collect data on HH level malaria risk factor characteristics including, bednet availability, type, integrity and use, HH water sources and environmental factors.

Rapid diagnostic test (RDT) screening. In addition to the questionnaire data, all HH members were screened for presence of malaria parasites to measure asymptomatic or minimally symptomatic parasitaemia prevalence using RDTs (*First Response® Combo Malaria Ag (pLDH/HRP2) card test, Premier Medical Corporation Ltd, India*). If HH members were not at home at the time of the survey, they were actively sought out and subsequently screened by the field team. RDTs were performed according to the manufacturer's instruction by trained field team members. All RDTs used were from one batch that was directly obtained through the manufacturer and stored according to the manufacturer's recommendations. However, no external quality control was done on these RDTs. Follow-up confirmatory microscopy was provided at the Ruhuha HC for all RDT-positive individuals to confirm accuracy and inform a malaria treatment decision.

Mapping households and geographical features. GIS was used to capture, manage and geographically integrate data from different sources. Location data for each HH and key geographical feature was collected using a handheld GPS receiver, GPSMAP 60CSx *(Garmin etrex legend®, Garmin International Inc. USA)*. Digitized data from pre-existing shapefiles provided base layers (topography, land use, rivers and surface water) on which study data was overlaid into one geo-database compatible with ArcGIS10. Boundaries shapefiles of administrative units ("cells"), wetlands, water bodies and the elevation contour lines for Ruhuha sector were obtained from the GIS Remote Sensing Training and Research Centre of the National University of Rwanda.

Statistical analysis

Statistical analysis was performed using STATA software (*version 12, College Station, TX, USA*). Univariate analysis to assess for malaria risk for all variables was done using logistic regression and variables with possible malaria risk ($p<0.2$) were included in the initial multivariate logistic regression model. HH data was analyzed using generalized estimating equation (GEE) models with adjustment for HH level malaria case clustering. The level of significance for study statistics was $p>0.05$ and Wald tests were used to quantify variable effects in the model. Possible interaction effects were also assessed for.

Spatial clustering. The Kulldorff spatial scan statistic, using SaTScanTM version 9.1.1 software (http://satscan.org), was used to test for spatial clustering of malaria cases and/or to determine whether the cases were distributed randomly over space [Kulldorff & Nagarwalla. [1995]]. HHs, used as the unit of analysis, were located using the Cartesian coordinate system to specify coordinates with the maximum spatial cluster size set at 50% of the population at risk. As in other studies, SatScan generated circular windows of different sizes for detecting clustering [22,23]. The number of cases in each window was compared to the expected number of cases based on the total number of cases and population size. We used purely spatial analyses based on the Bernoulli probability model that is appropriate for 0/1 event data such as cases/controls. The controls represented the background distribution population. The P-value was obtained from a likelihood ratio test based on Monte Carlo simulation replications of the data set. Spatial scans were performed for both HC attendee and HH member cases. A HC case was defined as being microscopy positive with HC controls defined if they were microscopy negative; a HH member was defined as being a case if they were identified as RDT positive with HH controls defined if they were RDT negative.

Results

In total, 769 HC attendees who presented with fever or with a history of fever in last 24 hours at the outpatient clinic were screened. Of the 769; 175 (22.8%) were diagnosed with malaria, 458 (59.6%) were female, 277 (36.0%) were aged <5 years, 147 (19.1%) aged 5–15 years and 345 (44.9) aged >15 years. A flow chart of study participant enrolment, malaria screening and participation is shown in Figure 2. HH visits were planned for all 769 HC attendees. However, because of the long period between HC case enrolment and HH survey (1–4 months versus the planned 2–4 weeks) and the inaccurate location data reported by study participants, the HH survey was not conducted in HHs of 200 index participants. Among HC attendees, malaria prevalence was comparable between those whose HH were not visited (30.5% (CI. 23.4–38.4) and those visited.

Of the 557 (72.4%) surveyed HHs, 520 HHs had complete data. Only data from these 520 HHs were analysed. In total, 2634 HH members were screened for malaria. Of the 2634, 599 (22.2%) were aged <6 years, 763 (28.3%) aged 6 to 15 years and 1331 (49.4%) aged >15 years. Only 90 (17.3%) HHs had at least one member diagnosed with malaria and the overall malaria prevalence (RDT confirmed) was 5.1% (95% CI 4.34–6.03). All visited HHs had ≥1 bednet and in total, 873 bednets were observed. HH bednet and indoor residual spraying coverage by self-report were 97.1% and 98.2%, respectively. Basic knowledge about malaria was high, with 696 (91%) reporting bednets as the principle malaria preventive measure while 748 (97.3%) reported that fever was the principal malaria symptom. Interestingly, 447 (82.5%) of HHs visited had bednets in their possession but these were not physically hung (Table 1).

Univariate Analysis

Results of univariate analysis for individual and HH (after adjusting for possible house-level clustering of cases) risk factors are displayed in Tables 2 and 3, respectively. Malaria risk among HC attendees was associated with both age and reported bednet ownership. Compared to children ≤5 years, malaria prevalence was three times higher in the 6–15 year olds while a reported ownership of ≥4 bednets was associated with a significant protective effect. HC attendees were evaluated for symptoms predictive of having clinical malaria. Having a measured fever (≥37.5°C) at presentation was associated with higher odds of malaria risk than no fever. Similar to HC cases, malaria risk among their HH members was significantly associated with age and observed bednet coverage. Additionally, HH members living

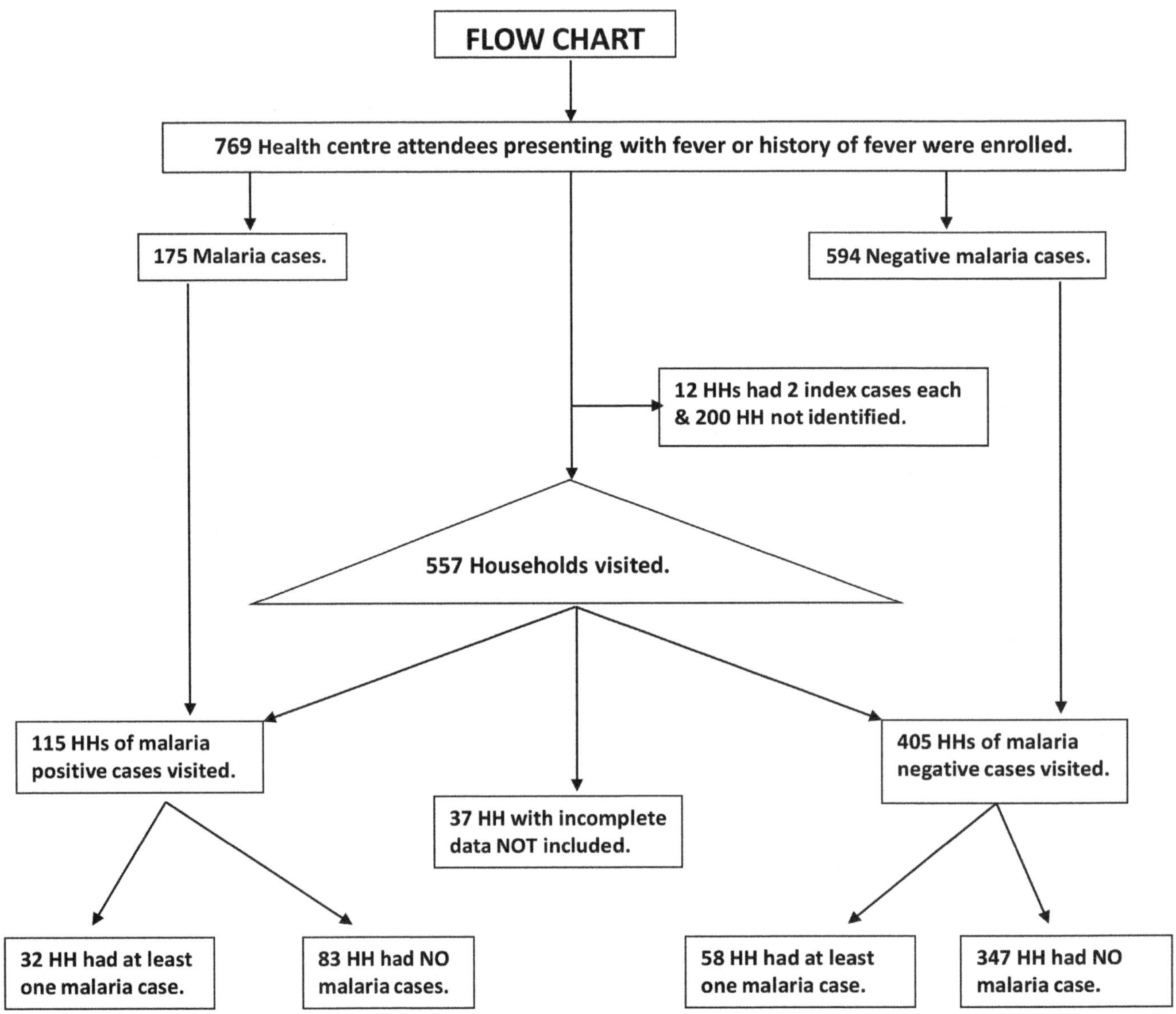

Figure 2. Flow chart of study participant enrolment, malaria screening and participation in a two phase survey.

Table 1. Reported and observed bednet characteristics.

Characteristics of bednets as reported by HC attendees.						Characteristics of bednets observed during HH visits (n = 557 HHs)			
Reported malaria preventive measures used in HH N (%)		*Are bednets in your HH treated? N (%)*		*How many holes are in your bednets? N (%)*		*Observed N (%) of hanged bednets*		*Observed Bednet in HH N (%)*	
Bednets	696 (82.6)	No	21 (2.92)	No holes	649 (91.28)	0	447 (82.47)	0	88 (11.4)
Clear bushes	66 (7.8)	Yes	615 (85.65)	1-10 holes	31 (4.36)	1	93 (17.16)	1	279 (36.3)
No protection	64 (7.6)	Do not know	55 (7.66)	> 10 holes	4 (0.56)	> 1	2 (0.37)	2	240 (31.2)
Others	11 (1.3)	Missing	27 (3.76)	Missing	27 (3.8)			≥ 3	134 (17.4)
Missing	6 (0.7)							Missing	28 (3.7)

in houses made of wood/mud/tent, when compared to those HH members living in dwellings whose walls were made of stone or bricks, and HH ownership of an in-house open water vessel were associated with higher odds of malaria.

Multivariate Analysis

At the individual level, an adjusted multivariate logistic regression model showed significantly higher odds of clinical malaria risk among children aged 5–15 years (OR = 3.02, P value <0.0001) but a protective effective was noted in those with a reported ownership of 4 of more bednets (OR = 0.352, P value 0.003). Having a fever (≥37.5°C) was predictive of having clinical malaria (OR = 1.64, P value 0.011). House level malaria risk remained significantly associated with age, type of material HH dwelling was made of, observed bednet coverage and malaria status of index case after adjusting for malaria case clustering in HHs (Table 3). Compared to the ≤5 year age group, malaria risk was significantly higher among the 6–15 year age group (OR = 2.44, P-value <0.0001) but interestingly lower, albeit with a borderline statistical significance, among the ≥16 year age group (OR = 0.58, P-value 0.047). Living in dwellings made of wood or mud or tent material was associated with a higher malaria risk while an observed ownership of ≥4 more bednets was associated with a protective effect.

Malaria Clustering

Malaria positivity among HC attendees was significantly correlated with a HH having at least one confirmed member (OR = 2.31, P=0.001) but no spatial clustering for HC malaria cases was observed. However, three clusters of HHs with significantly higher risk than expected RDT tested members were identified (Table 4). These HH clusters were located; 1. North East (radius of 2.04 Kilometers (Kms), relative risk of 3.40 and P value 0.0001), 2. South (radius of 0.51 Kms, relative risk of 5.6, (P value 0.0001), and 3. A smaller cluster (not indicated in Figure 3) of only one HH (where 4 of its members tested RDT positive) with a relative risks of 20.8, P value 0.002 (Figure 3). Two of these clusters (1 and 2) were located next to water-based agro-ecosystems.

Discussion

In this study, members of HHs where the index case had clinical malaria showed 1.3 times greater odds of being malaria infected compared to members of HH where the index patient was malaria negative. Comparable findings of a greater risk for malaria infection among HH members of a HC identified clinical malaria case have been shown by Stresman et al. (2010) in Zambia [24]. These findings support the value of circle surveillance as a useful tool for studying HH level malaria burden, risk factors and clustering. In this study, slide/RDT positivity rates of 22.8% and 5.1% among HC malaria presumed cases and HH based asymptomatic cases respectively were found. This demonstrates that circle surveillance can show differences in HH malaria risk and clustering, even in areas of high malaria prevalence as in Ruhuha. A part from living in a HH where the index case had malaria, risk factor analysis identified participant's age and a reported ownership of a ≥4 bednet as variables that, either alone or in unison, significantly influenced malaria risk.

Compared to children aged <5 year, older children and adults had a higher risk of parasite carriage, for both HC attendees and HH members groups. This is in contrast to previous findings of a higher malaria risk in children <5 years [25]. However, a shift to higher malaria risk among older age groups has been reported after the increased coverage with insecticide-treated bednets and

Table 2. Health facility attendee characteristics and malaria risk factors.

Baseline characteristics	N (%)	HC attendees with malaria (n = 175) (%)	HC attendees with No malaria (n = 584) (%)	Univariate Analysis	Multivariate Analysis
Age group					
≤5 years	277 (36.02)	57 (32.6)	220 (37.0)	1.0	1.0
6–15 year	147 (19.12)	57 (32.6)	80 (13.5)	2.444 (1.572–3.801), <0.0001	3.02 (1.890–4.824), <0.0001
≥16 years	345 (44.86)	61 (34.8)	284(49.5)	0.829 (0.555–1.239), 0.36	1.027 (0.663–1.591), 0.906
Gender					
Male	311 (40.44)	73 (41.7)	238(40.1)	0.934 (0.663–1.316), 0.696	
Female	458 (59.56)	102 (58.3)	356(59.9)	1.0	----
Measured temperature at HC					
<37.5	381(49.5)	69(39.4)	312(52.5)	1.0	1.0
≥37.5	388 (50.5)	106(60.6)	282(47.5)	1.700 (1.206–2.396), 0.002	1.636 (1.119–2.392), 0.011
When did fever episode start					
Today	47 (6.11)	10(5.7)	37(6.2)	1.0	----
Yesterday	494 (64.24)	109(62.3)	385(64.8)	1.048 (0.505–2.174), 0.901	
Day before yesterday	162 (21.07)	46(26.3)	116(19.5)	1.467 (0.674–3.193), 0.334	
Long ago	66 (8.58)	10(5.7)	56(9.5)	0.661 (0.250–1.743), 0.402	
Malaria episode in past 12 months					
None	520 (67.62)	105	415	1.0	1.0
1–3 episodes	232 (30.17)	64	168	1.506 (1.052–2.156), 0.025	1.408 (0.959–2.068), 0.081
>3 episodes	17 (2.21)	6	11	2.156 (0.780–5.964), 0.139	2.163 (0.737–6.346), 0.160
Does HH own at least 1 bednet					
Yes	742 (96.61)	167	575	0.653 (0.279–1.530), 0.327	
No	26 (3.39)	8	18	1.0	----
Reported number of bednets in HH					
One	88 (11.88)	24	64	1.0	1.0
Two	279 (37.65)	65	214	0.810 (0.470–1.397), 0.449	0.654 (0.372–1.150), 0.141
Three	240 (32.39)	59	181	0.869 (0.500–1.512), 0.620	0.687 (0.386–1.220), 0.200
≥ Four	134 (18.08)	19	115	0.441 (0.224–0.865), 0.017	0.352 (0.175–0.707), 0.003

the observed follow-up reduced malaria transmission in some communities [26,27]. The reductions in malaria transmission may decrease the risk of malaria inoculation and infections leading to an increase in the age at which malaria infections are first acquired. Additionally, there is a greater likelihood of younger age groups (<5 year olds) using malaria preventive bednets compared to their older siblings, although this data was not collected in this study [26].

In this study, the reported and observed ownership of bednets was associated with significant malaria protective effect. This protective effect of insecticide-treated mosquito net use has also been affirmed in multiple previous studies [28]. Ruhuha sector is a traditionally high transmission setting with high bednet coverage. This high coverage follows the government's massive free bednet distribution after campaigns run between 2009 to 2011 in which government aimed to achieve universal bednet coverage [5,29,30]. Study Participants reported a good level of knowledge of malaria symptoms, transmission and preventive measures with over 82% of respondents reporting use of bednets the night before the survey. However, in only 18% of visited HHs was a bednet found physically hung onto a bed or a sleep space suggesting that bednet use may be sub-optimal. Possible reasons for sub-optimal bednet use may be associated with local house structures and/or sleeping arrangements for the HH members. Most houses in Ruhuha have 1–2 bedrooms with limited structures on which to hang bednets. Additionally, most occupants share sleeping spaces on the floor. These factors may complicate use of available bednets and partially explain the low bednet hanging rates observed and limited bednet protective effects in HHs with bednets. Studies exploring how to optimise bednet usage and effectiveness are recommended.

In this study and others, the quality of housing, apart from being an indicator of HH economic status has been reported, to influence the ease with which mosquitoes can enter and hide in a home and hence contribute to malaria risk [7,31,32]. Occupants of houses with walls made of mud/grass/wood had 1.3 times greater odds (P value 0.016) of having at least one malaria case more than those living in houses with walls made of brick or stone. However, interventions to address type of housing as a malaria risk factor are complex and difficult to achieve and are rarely components of public health programs. A current campaign in Rwanda to phase out grass-thatched houses (locally known as *"nyakatsi"*) and replace them by houses made of brick and iron sheet roofscould impact malaria transmission.

For high transmission countries where essential clinical services are adequately available, the transition from control to elimination

Table 3. Household characteristics and malaria Risk factors.

Variable	Frequency (%)	HH with ≥ one malaria case	HH with No malaria case	Univariate Analysis	Multivariate Analysis
Gender					
Female	1167 (44.3)	64	1,091	1.247 (0.894– 1.740), 0.194	1.191 (0.852–1.667),0.306
Male	1467 (55.7)	66	1,408	1	1.0
Age group					
≤5 years	589 (22.40)	26	563	1	1.0
6–15 years	742 (28.22)	71	671	2.398 (1.528–3.766), <0.0001	2.437 (1.543–3.847), <0.0001
≥16 years	1,298 (49.37)	33	1,265	0.586 (0.350–0.982), 0.042	0.584 (0.344–0.992), 0.047
HH member association with					
Negative index case	2,047 (77.86)	77	1,970	1	1.0
Positive Index case	582 (22.14)	53	529	2.557(1.608–4.066),<0.0001	1.267 (1.068–1.503), 0.007
Observed No of bednets					
One bednet	256 (9.77)	22	234	1	1.0
Two Bednets	2,212 (84.46)	99	2,113	0.490 (0.204–1.179), 0.111	0.456 (0.202–1.029), 0.059
Three Bednets	55 (2.10)	2	53	0.526 (0.217–1.274), 0.155	0.461 (0.207–1.024), 0.057
≥ Four bednets	96 (3.67)	7	89	0.367 (0.143–0.940), 0.037	0.384 (0.165–0.892), 0.026
House wall material					
Bricks and stones	364 (70)	58	306	1	1.0
Wood/mud/tent	156 (30)	32	124	1.324 (1.134–1.546), <0.0001	1.288 (1.082–1.534), 0.004
Type of HH roof material					
Corrugated Iron sheets	457 (87.9)	79	378	1	1.0
Grass thatched/tent/others	63 (12.1)	11	52	0.849 (0.662–1.088), 0.196	0.837 (0.636–1.102), 0.204
Was IRS of HH walls done					
Yes	490 (94.4)	88	402	1	----
No	30 (5.6)	3	27	1.033 (0.742–1.435), 0.852	
Presence of outside water source					
Yes	205 (36.80)	45	150	1	----
No	352 (63.20)	45	280	0.762(0.489–1.190), 0.232	
Have an open water vessel in HH					
Yes	171 (32.88)	35	136	1	1.0
No	349 (67.12)	55	294	0.666(0.425–1.045), 0.077	0.712 (0.351–1.444), 0.347
Green environment around HH					
Very green (grass & trees)	300 (53.86)	47	244	1	1.0
Moderate green (only grass)	174 (31.24)	35	124	1.720 (1.076–2.750), 0.023	1.412 (0.691–2.886), 0.344
No grass at all	83 (14.90)	8	62	0.825(0.383–1.773), 0.622	0.578 (0.218–1.528), 0.269
Does your HH have Electricity?					
Yes	34 (6.10)	4	25	0.512(0.154–1.710), 0.277	0.634 (0.248–1.617), 0.340
No	523 (93.90)	86	405	1	1.0
Have Domestic Animals in HH?					
Yes	376(67.50)	62	291	0.931 (0.564–1.537), 0.779	
No	181 (32.50)	28	139	1	----

is recommended at SPR of <5% [12]. Achieving pre-elimination levels in Ruhuha, given current SPR of >22%, will probably require introduction of novel area-relevant interventions to supplement existing control tools (mainly ITNs and IRS). As malaria transmission declines, a community-based evaluation of transmission intensity and size of infectious reservoir will be required. In this study, malaria prevalence among HH members by RDT was 5.1%. However, since RDTs have a lower sensitivity, as compared for example to molecular tools, the level of true malaria infection prevalence among the predominantly asymptomatic carrier HH members, may have been underestimated [33,34]. In addition, the HH survey was conducted 1 to 4 months, rather than the planned 2–4 weeks, after the initial HC-based fever survey. This delay may have complicated a fair comparison of malaria risk between HC index cases and their HH members. Ruhuha sector is served by only one HC managed primarily by

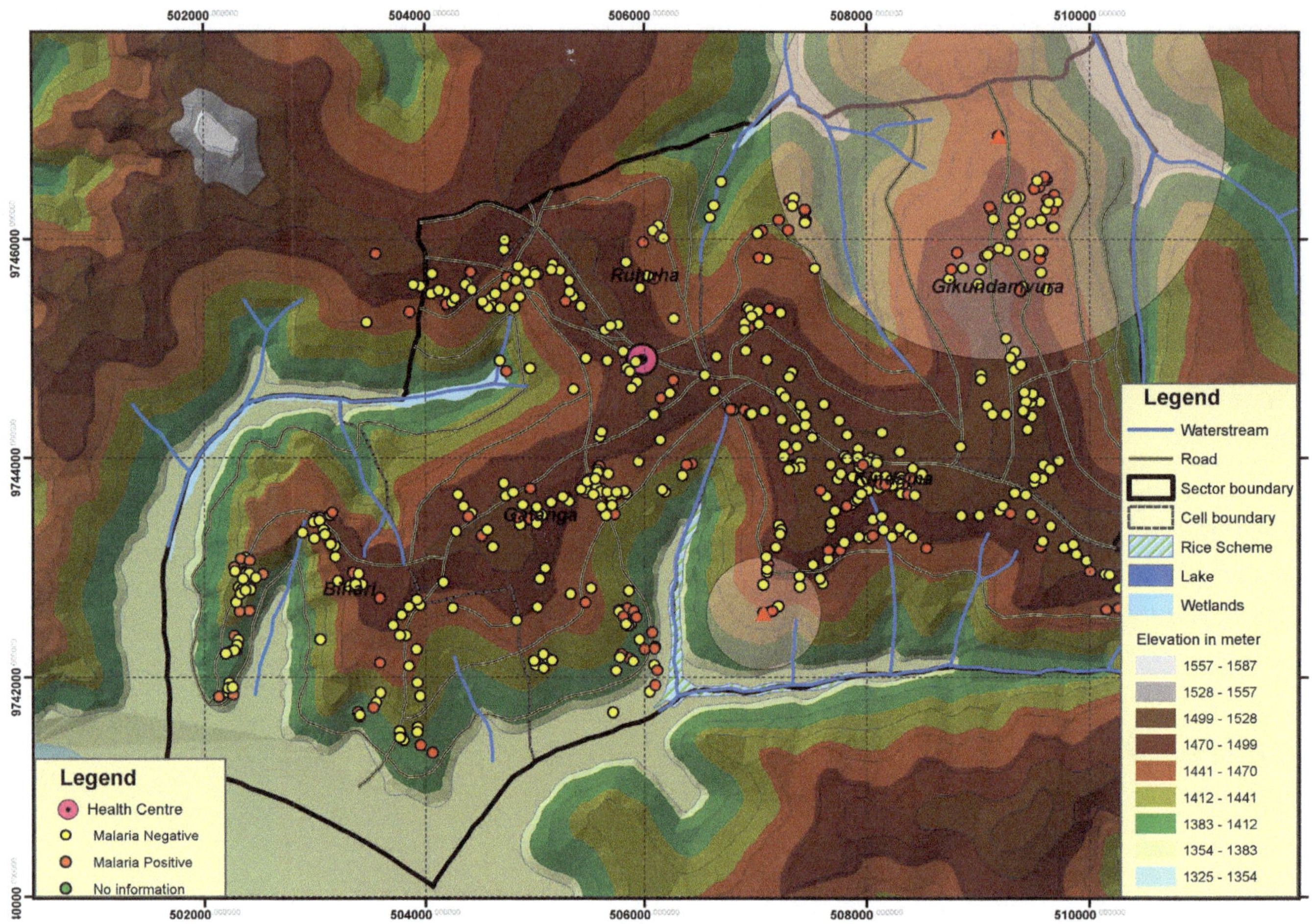

Figure 3. [¥] **Spatial malaria clusters and location of Hhs.** (Yellow dots - control HH with no malaria infected case and small Red dots - case HH with at least one malaria infected case) in Ruhuha sector. [¥]*The used administrative boundaries and geographic features shape files were obtained from the Centre for Research and Training in GIS and Remote Sensing of the National University of Rwanda.*

community health workers with most children ≤5 years. The area population is therefore challenged by inadequate access to health care. Consequently, malaria data reported from this health centre may underestimate the population malaria burden. This further complicates a fair comparison of health centre versus HH level malaria risk [13].

Two hundred HHs could not be identified due to; wrong directions, non-existing HHs and, possibly, out of area study participants who gave wrong data. Given the delay in the follow-up HH surveys and the significant loss to follow-up of index cases, a repeat robust reactive case identification study to assess for clustering, particularly in areas of lower malaria transmission intensity, is recommended [12,15]. In this study, HH cases were RDT confirmed while HC cases were microscopically confirmed in keeping with national malaria guidelines. However, no quality control for used RDTs was conducted. Also, being a cross-sectional survey, malaria burden reported could not reflect seasonal malaria trends and prospective malaria incidence risk.

Apart from the study limitations reported above, this study showed that having malaria among HC attendees was significantly predictive of finding at least one malaria-infected case among his/her HH members (OR = 2.4, P value - 0.001) suggesting that HC-based passive case identification can be a feasible entry point for identifying community hotspots of malaria infection. Guidelines on how to manage asymptomatic and minimally symptomatic RDT positive cases identified through active case detection are

Table 4. Spatial clustering of the more than expected household cases.

Cluster	Year	No of HHs in cluster	P - value	Observed No of HH cases	Expected No of Cases	Relative risk
1	2011	60	0.00002	38	14.064	3.413
2	2011	11	0.00015	16	3.169	5.622
3	2011	1	0.00247	4	0.198	20.808

lacking and would be required in the event that circle surveillance is implemented in the future. The currently recommended first line treatment for uncomplicated malaria in Rwanda is Artemether – Lumefantrine (AL). AL has anti-gametocidal effects and an ability to reduce asexual parasitaemia levels and infectivity among malaria-infected individuals [35,36,37]. It is plausible that AL can be used among asymptomatic and minimally symptomatic cases to clear local reservoir pools and reduce their malaria transmission potential.

Significant spatial clustering for HH cases (but not HC cases) with the clusters located near water-based agro-ecosystems is an interesting finding. The bigger cluster (radius of 5 km) is neighboring marsh lands where traditional rice cultivation is done (North East), while the smaller cluster (0.5 km radius) is located between multiple water streams and Lake Cyohoha in the south where vegetable and other agriculture crops are grown. We speculate that these water agro-ecosystems may provide significant reservoirs for mosquito breeding and hence increased vector intensity for malaria transmission. This finding suggests that future malaria control efforts should consider targeting potential breeding sites and engaging farming communities. To this end, an entomological evaluation of mosquito breeding capacity and endemicity may guide introduction of integrated vector management practices while community-based environmental management approaches for malaria control, as shown to be effective in settings comparable to Ruhuha, may be two potential effective area relevant strategies to employ [32]. To achieve malaria pre-elimination status in Ruhuha, the bednet and IRS strategies, which are principally used, may need to be complimented by interventions that target area breeding sites and malaria risk factors identified through spatial clustering technique as was done in this study may be required [12].

Conclusion

In this study, HC malaria confirmed cases were significantly associated with finding at least one malaria-infected case among their HH members. Reactive case finding, by linking HC-identified passive cases to actively identified HH malaria infection, is a potentially powerful surveillance system for identifying HHs with significant malaria risk and detecting asymptomatic carriers. Especially in low transmission settings, identifying and treating asymptomatic carriers is key in interrupting transmission. Therefore, circle surveillance, when combined with knowledge on the individual, the HH and the environmental malaria risk factors in a given community, can aid detection of hotspots and inform use of targeted malaria control strategies.

Author Contributions

Conceived and designed the experiments: PFM KRB PJdV. Performed the experiments: SR SA JD JdDH LB. Analyzed the data: FK JB PFM PJdV. Wrote the paper: FK JB PFM KRB PJdV.

References

1. Roll Back Malaria Partnership. Global Strategic Plan, 2005–2015. Available: http://www.rollbackmalaria.org/forumV/docs/gsp_en.pdf. Accessed 2012 May 15.
2. WHO: World malaria report 2011. Available: http://www.who.int/malaria/world_malaria_report_2011/9789241564403_eng.pdf. Accessed 2012 May 13.
3. President's Malaria Initiative (PMI). Malaria Operational Plan: Rwanda FY 2012. Washington, DC: PMI, 2011. Available: http://www.pmi.gov/countries/mops/fy12/rwanda_mop_fy12.pdf. Accessed 2012 May 15.
4. Ministry of Health (2009) Health sector strategic plan, July 2009-June 2012 (pp. 82). Kigali, Rwanda.
5. Karema C, Aregawi M, Rukundo A, Kabayiza A, Mulindahabi M, et al. (2012) Trends in malaria cases, hospital admissions and deaths following scale-up of anti-malarial interventions, 2000–2010, Rwanda. Malaria Journal 11: 236.
6. Martens W (1995) Climate change and malaria: exploring the risks. Med War 11: 202–213.
7. Bousema T, Drakeley C, Gesase S, Hashim R, Magesa S, et al. (2010) Identification of hot spots of malaria transmission for targeted malaria control. J Infect Dis 201: 1764–1774.
8. Greenwood BM (1989) The microepidemiology of malaria and its importance to malaria control. Trans R Soc Trop Med Hyg 83 Suppl: 25–29.
9. Greenwood BM, Bradley AK, Greenwood AM, Byass P, Jammeh K, et al. (1987) Mortality and morbidity from malaria among children in a rural area of The Gambia, West Africa. Trans R Soc Trop Med Hyg 81: 478–486.
10. Clark TD, Greenhouse B, Njama-Meya D, Nzarubara B, Maiteki-Sebuguzi C, et al. (2008) Factors Determining the Heterogeneity of Malaria Incidence in Children in Kampala, Uganda. J Infect Dis 198: 393–400.
11. Bousema T, Griffin JT, Sauerwein RW, Smith DL, Churcher TS, et al. (2012) Hitting Hotspots: Spatial Targeting of Malaria for Control and Elimination. PLoS Med 9(1): e1001165. doi: 10.1371/journal.pmed.1001165.
12. Baliraine FN, Afrane YA, Amenya DA, Bonizzoni M, Menge DM, et al. (2009) High Prevalence of Asymptomatic *Plasmodium falciparum* Infections in a Highland Area of Western Kenya: a Cohort Study. J Infect Dis. 200(1): 66–74.
13. World Health Organizatioin (2007) Malaria elimination: a field manual for low and moderate endemic countries. Geneva: WHO.
14. Rowe AK, Kachur SP, Yoon SS, Lynch M, Slutsker L, et al. (2009) Caution is required when using health facility-based data to evaluate the health impact of malaria control efforts in Africa. Malaria Journal 8: 209.
15. Moonen B, Cohen JM, Snow RW, Slutsker L, Drakeley C, et al. (2010) Operational strategies to achieve and maintain malaria elimination. Lancet 376: 1592–1603.
16. Omumbo J, Hay S, Snow R, Tatem A, Rogers D (2005) Modelling malaria risk in East Africa at high-spatial resolution. Trop Med Int Health 10: 557–566.
17. Goovaerts P (2006) Geostatistical analysis of disease data: accounting for spatial support and population density in the isopleth mapping of cancer mortality risk using area-to-point Poisson kriging. Int J Health Geogr 5: 52.
18. Mmbando BP, Kamugisha ML, Lusingu JP, Francis F, Ishengoma DS, et al. (2011) Spatial variation and socioeconomic determinants of Plasmodium falciparum infection in northeastern Tanzania. Malaria Journal 10: 145.
19. Government of Rwanda (GOR) (2010) Bugeseradistrict Available : http://www.bugesera.gov.rw/index.php?option=com_content&view=article&id=73:about-the-district&catid=34:about-the-ditrict&Itemid=27&lang=en.Accessed 2012 May 15.
20. Tsuang A, Lines J, Hanson K (2010) Which family members use the best nets? An analysis of the condition of mosquito nets and their distribution within households in Tanzania. Malaria Journal, 9(1) : 211.
21. Pettifor A, Taylor E, Nku D, Duvall S, Tabala M, et al. (2008) Bed net ownership, use and perceptions among women seeking antenatal care in Kinshasa, Democratic Republic of the Congo (DRC): Opportunities for improved maternal and child health. BMC Public Health, 8(1), 331.
22. Haque U, Hashizume M, Sunahara T, Hossain S, Ahmed SM, et al. (2010) Progress and challenges to control malaria in a remote area of Chittagong hill tracts, Bangladesh. Malar J. 9: 156.
23. Loha E, Lunde TM, Lindtjorn B (2012) Effect of bednets and indoor residual spraying on spatio-temporal clustering of malaria in a village in South Ethiopia: a longitudinal study. PLoS One 7:e47354.
24. Stresman GH, Kamanga A, Moono P, Hamapumbu H, Mharakurwa S, et al. (2010) A method of active case detection to target reservoirs of asymptomatic malaria and gametocyte carriers in a rural area in Southern Province, Zambia. Malaria J. 9: 265.
25. Smith T, Beck HP, Kitua A, Mwankusye S, Felger I, et al. (1999) Age dependence of the multiplicity of Plasmodium falciparum infections and of other malariological indices in an area of high endemicity. Trans R Soc Trop Med Hyg. 93: 15–20.
26. Winskill P, Rowland M, Mtove G, Malima RC, Kirby MJ (2011) Malaria risk factors in north-east Tanzania.Malaria Journal 10: 98.
27. Smith T, Hii JL, Genton B, Müller I, Booth M, et al. (2001) Associations of peak shifts in age-prevalence for human malarias with bednet coverage. Trans R Soc Trop Med Hyg 95: 1–6.
28. Lengeler C (2004) Insecticide-treated bed nets and curtains for preventing malaria. Cochrane Database Syst Rev. CD000363.
29. Ministry of Health -Rwanda. Interim Demographic and Health Survey 2007–08 Kigali, Rwanda, 2009; DHS 2010 (Preliminary Report).
30. President's Malaria Initiative (PMI). Malaria Operational Plan: Rwanda FY 2011. Washington, DC: PMI, 2010. Available: http://pmi.gov/countries/mops/fy11/rwanda_mop-fy11.pdf. Accessed 2012 Sep 15.
31. Coleman M, Coleman M, Mabaso ML, Mabuza AM, Kok G, et al. (2010) Household and microeconomic factors associated with malaria in Mpumalanga, South Africa. Trans R Soc Trop Med Hyg 104: 143–147.
32. Konradsen F, Amerasinghe P, van der Hoek W, Amerasinghe F, Perera D, et al (2003) Strong association between house characteristics and malaria vectors in Sri Lanka. Am J Trop Med Hyg 68: 177–181.

33. Fontecha GA, Mendoza M, Banegas E, Poorak M, De Oliveira AM, et al. (2012) Comparison of molecular tests for the diagnosis of malaria in Honduras. Malar J Apr 18;11: 119.
34. Dal-Bianco MP, Köster KB, Kombila UD, Kun JF, Grobusch MP, et al. (2007) High prevalence of asymptomatic Plasmodium falciparum infection in Gabonese adults. Am J Trop Med Hyg 77: 939–942.
35. Von Seidlein L, Bojang K, Jones P, Jaffar S, Pinder M, et al. (1998) A randomized controlled trial of artemether/benflumetol, a new antimalarial and pyrimethamine/sulfadoxine in the treatment of uncomplicated falciparum malaria in African children. Am J Trop Med Hyg 58: 638–644.
36. White NJ (2004) Antimalarial drug resistance. J Clin Invest 113: 1084–1092.
37. Sutherland CJ, Ord R, Dunyo S, Jawara M, Drakeley CJ, et al. (2005) Reduction of malaria transmission to Anopheles mosquitoes with a six-dose regimen of coartemether. PLoS Med 2:e92.

22

Risk Map of Cholera Infection for Vaccine Deployment: The Eastern Kolkata Case

Young Ae You[1], Mohammad Ali[1]*, Suman Kanungo[2], Binod Sah[1], Byomkesh Manna[2], Mahesh Puri[1], G. Balakrish Nair[2], Sujit Kumar Bhattacharya[2,3], Matteo Convertino[4,5,6], Jacqueline L. Deen[7], Anna Lena Lopez[1,8], Thomas F. Wierzba[1], John Clemens[1,9], Dipika Sur[2]

1 International Vaccine Institute, Seoul, Republic of Korea, **2** National Institute of Cholera and Enteric Diseases, Kolkata, India, **3** Indian Council of Medical Research, New Delhi, India, **4** HumNat Lab, Department of Biological Systems Engineering, Virginia Tech, Blacksburg, Virginia, United States of America, **5** Bioinformatics Institute, Virginia Tech, Blacksburg, Virginia, United States of America, **6** Institute for Critical Technology and Applied Science, Virginia Tech, Blacksburg, Virginia, United States of America, **7** Menzies School of Health Research, Casuarina, Northern Territory, Australia, **8** University of the Philippines Manila, National Institutes of Health, Manila, Philippines, **9** University of California Los Angeles, School of Public Health, Los Angeles, United States of America

Abstract

Background: Despite advancement of our knowledge, cholera remains a public health concern. During March-April 2010, a large cholera outbreak afflicted the eastern part of Kolkata, India. The quantification of importance of socio-environmental factors in the risk of cholera, and the calculation of the risk is fundamental for deploying vaccination strategies. Here we investigate socio-environmental characteristics between high and low risk areas as well as the potential impact of vaccination on the spatial occurrence of the disease.

Methods and Findings: The study area comprised three wards of Kolkata Municipal Corporation. A mass cholera vaccination campaign was conducted in mid-2006 as the part of a clinical trial. Cholera cases and data of the trial to identify high risk areas for cholera were analyzed. We used a generalized additive model (GAM) to detect risk areas, and to evaluate the importance of socio-environmental characteristics between high and low risk areas. During the one-year pre-vaccination and two-year post-vaccination periods, 95 and 183 cholera cases were detected in 111,882 and 121,827 study participants, respectively. The GAM model predicts that high risk areas in the west part of the study area where the outbreak largely occurred. High risk areas in both periods were characterized by poor people, use of unsafe water, and proximity to canals used as the main drainage for rain and waste water. Cholera vaccine uptake was significantly lower in the high risk areas compared to low risk areas.

Conclusion: The study shows that even a parsimonious model like GAM predicts high risk areas where cholera outbreaks largely occurred. This is useful for indicating where interventions would be effective in controlling the disease risk. Data showed that vaccination decreased the risk of infection. Overall, the GAM-based risk map is useful for policymakers, especially those from countries where cholera remains to be endemic with periodic outbreaks.

Editor: Matteo Convertino, University of Florida, United States of America

Funding: This study is supported by the Bill & Melinda Gates Foundation through the Diseases of the Most Impoverished Program and the Cholera Vaccine Initiative. Additional funding is provided by the Swedish International Development Cooperation Agency and the Governments of South Korea, Sweden, and Kuwait. No funding bodies had any role in study design, data collection and analysis, decision to publish, or preparation of the manuscript.

Competing Interests: The authors have declared that no competing interests exist.

* E-mail: mali@ivi.int

Introduction

John Snow's cholera map of 1855 [1] is a disease map on which he plotted the households with cholera deaths in London. A disease map can identify geographic variations of disease incidence, which in turn can be useful to identify areas of unusually high risk requiring preventive action, to formulate etiological hypotheses, and to demonstrate patterns of risk allowing better allocation of resources [2]. As disease variations have a spatial expression, understanding geographical distribution of disease is of considerable importance to public health workers and epidemiologists [3]. Notably, the John Snow's map revealed that families receiving water from the Broad Street hand pump were more likely to have cholera deaths, which helped the health authorities making appropriate area-based interventions for controlling the disease.

Given advancement of our understanding about transmissions of cholera and its control mechanisms, cholera remains a global health problem [4]. More and more countries are being suffered from the scourge of the disease. One reason could be spatial patterns of risk for cholera in an endemic area are not clearly known, thus an effective control mechanism could have never established in a cholera endemic country. During March and April 2010, there was a cholera outbreak in the slums of eastern Kolkata, India, a cholera endemic area. We investigated the data of a geographically referenced population-based surveillance from a cholera vaccine trial conducted in the area, and plotted the cases during the outbreak in the trial area. We used the surveillance data on cholera and applied a generalized additive model (GAM), which combines smoothing with the ability to analyze binary outcomes and adjust for covariates [5], to identify areas of higher and lower risk for cholera over time. We also evaluated the socio-environmental characteristics between the high and low risk areas of the disease and the impact of the vaccination on spatial risk for cholera.

Methods

The Study Area and Data

The study was conducted in urban slum communities in Kolkata, the capital of the state of West Bengal, India (Figure 1). Kolkata is the third largest city in India with 13 million residents into 1,450 km^2 making it is one of the world's most densely populated cities. The Municipal Corporation area is demarcated into 141 administrative wards [6]. The study site comprises three contiguous wards: 29, 30 and 33. In the study area, the streets are narrow with little space between houses, piped municipal water supply is intermittent, and several households share one or two latrines and water taps [7]. There is a canal that runs east of ward 29 and south of ward 33. It is one of the main drainages of rain and waste water in the eastern part of Kolkata.

To register the study population, beginning in January 2003, two censuses were conducted in the original study area that includes all of ward 30 and part of ward 29. Subsequently, the study area was expanded in 2005 in preparation for a cholera vaccine trial. Each census captured the *de jure* population defined as persons who stated their residence in the study area was their regular residence and the residence is legally recognized and a registered slum [8] that excluded people who live in high rise building. We also collected socio-demographic information of each household during census survey. The population database was updated monthly though vital demographic events in the study area. Each household in the study area was identified by its geographic coordinates in the geographic information system (GIS) database. The details on how the GIS database was created have been described elsewhere [9].

Nine project health clinics were established in the community to undertake diarrhea surveillance in addition to the two referral hospitals: Infectious Diseases and B.C. Roy Children. Private medical practitioners were encouraged to refer patients with diarrhea to the project health clinics. Patients from the study area were identified by use of household identification cards and a computerized database. Study physicians recorded pertinent clinical details on a structured clinical data form. Rectal swabs were obtained from all patients presenting with history of loose stools and transported in Cary-Blair media to a laboratory at the National Institute of Cholera and Enteric Diseases (NICED) within 8 hours of specimen collection. At the laboratory, rectal swabs were examined for *Vibrio cholerae* by use of conventional method [10].

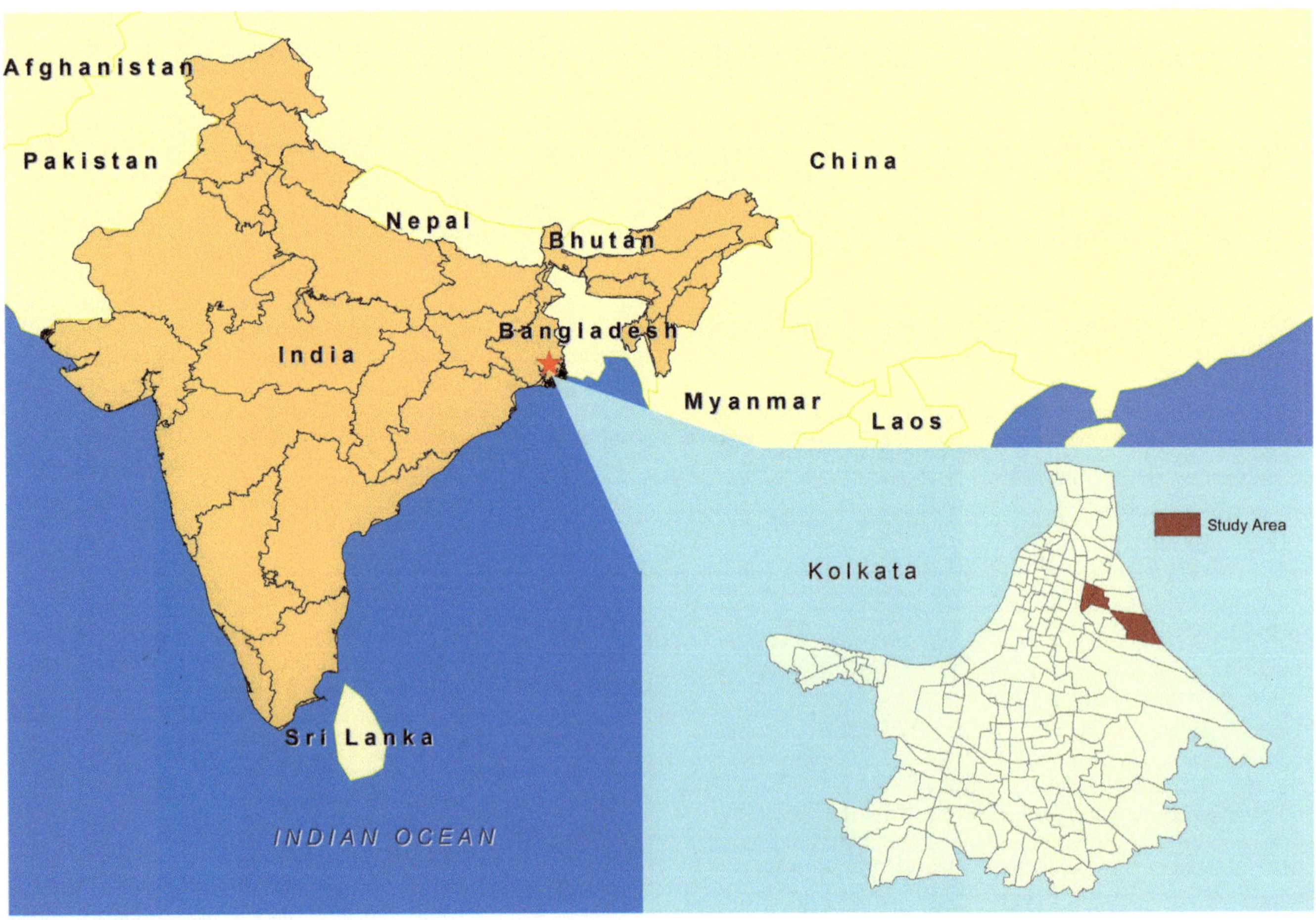

Figure 1. The study area in Kolkata, West Bengal, India.

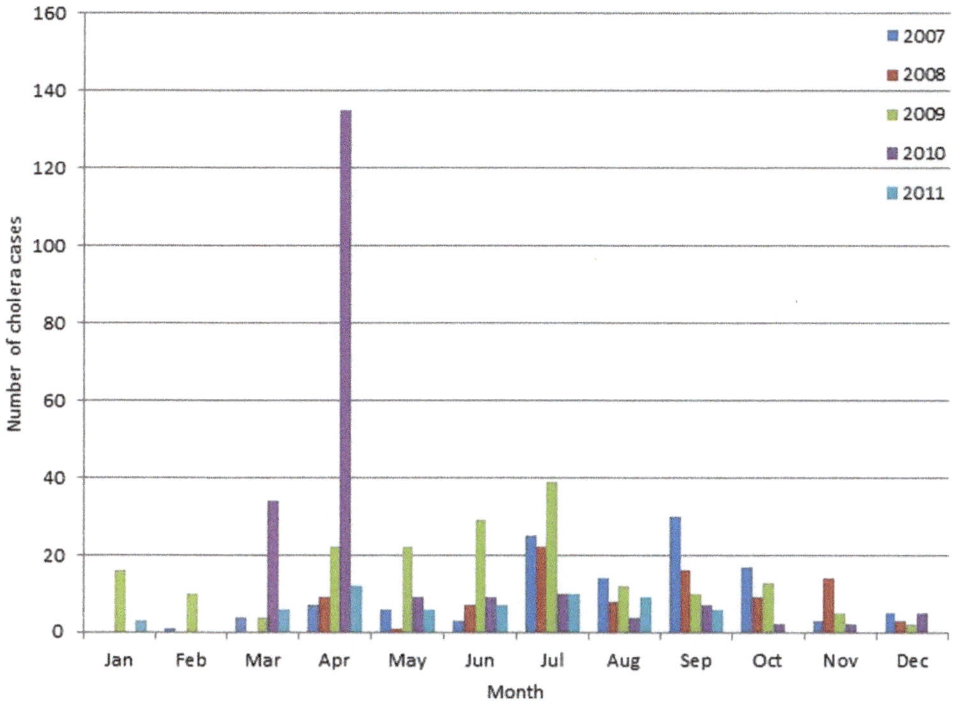

Figure 2. Number of cholera cases by month during 2007–2011 in the study area in Kolkata, India.

Vaccination against cholera followed a two-dose schedule administered in two rounds. Each dose of the vaccine contains inactivated *Vibrio cholerae* O1 cells representing the El Tor and classical biotypes and the Inaba and Ogawa serotypes, as well as serogroup O139 cells. The first round was from July 27 to August 13, 2006, and the second from August 27 to September 10, 2006, to ensure a minimum inter-dose interval of 14 days. Residents were eligible to receive a study intervention if they were aged one year or older and were not pregnant. Each agent was given as a two-dose regimen with an inter-dose interval of at least 14 days, and the coverage for the study intervention was about 68% [11]. We classified the study period as pre-vaccination and post-vaccination. The vaccination period (July 27 to September 10, 2006) was excluded. We aggregated one-year data of the pre- and two-year post-vaccination periods by the geographic points of residence in the GIS database for the spatial analysis. Two-year of post-vaccination data were used, because there were only a few cases in the 1st year of post-vaccination period.

Spatial Models

We used a generalized additive model (GAM) [5] to estimate smoothed log odds as a function of space and converted to odds ratios using whole population as a reference. We modeled location using a bivariate smooth (*S*) of latitude (x) and longitude (y)

$$\mathrm{logit}[\mathrm{p}(\mathrm{x},\mathrm{y})] = S(\mathrm{x},\mathrm{y}) + \gamma \mathrm{z},$$

where logit[p(x,y)] is the log of the disease odds at location (x,y), z is a vector covariates, and γ is a vector of parameters. We required the type of smoother and the span size for estimating smoothing term, *S*(x,y). The span size determines over which averaging takes place. Since the population distribution varied in space, we used a locally-weighted regression smoother to adapt to the changes in population density. The amount of smoothing depends on the percentage of the data points in the neighborhood, referred to as the span size. We chose an optimal span by minimizing the Akaike Information Criterion (AIC) from 1% to 99% of the size of study area increased by 1%. Note that small span sizes reduce bias but increase variance, and large span sizes produce smoother surfaces resulting in increased bias and reduced variability.

We created a rectangular grid covering the study area (100 rows and 100 columns) using the minimum and maximum latitude and longitude coordinates of the study area. Grid points lying outside the study area were clipped. We estimated adjusted odds ratios (ORs) at each grid point using whole study area as the reference, dividing the odds at each grid point by the odds calculated by the reduced model omitting the location smoothing term. We permuted the locations of subjects and reran the GAM model 999 times to estimate the distribution of log odds under the null hypothesis at each point. We define areas of significantly decreased odds ("low risk areas") to include all points that rank in the lower 2.5% of the pointwise permutation distribution and areas of elevated odds ("high risk areas") to include all points that rank in the upper 2.5% of the pointwise permutation distribution [12–14]. We superimpose the 2.5% and 97.5% contour lines on the point estimate map.

In order to make the maps visually comparable, we mapped all results using the blue to red continuous (unclassified) color scale and range of odds ratios, 0.1–2.0. This range covers most but not all of the ORs observed in our analysis. If the ORs is >2.0 we set it to 2.0 and if the ORs is <0.1 we set it to 0.1 for mapping purpose. We used R programming language to run the GAM. Results from the GAM were exported from R into ArcGIS for mapping.

Statistical Analysis

To evaluate the socio-environmental characteristics of the study population between the high and low risk areas, we used Generalized Estimating Equations (GEE) with the logit link

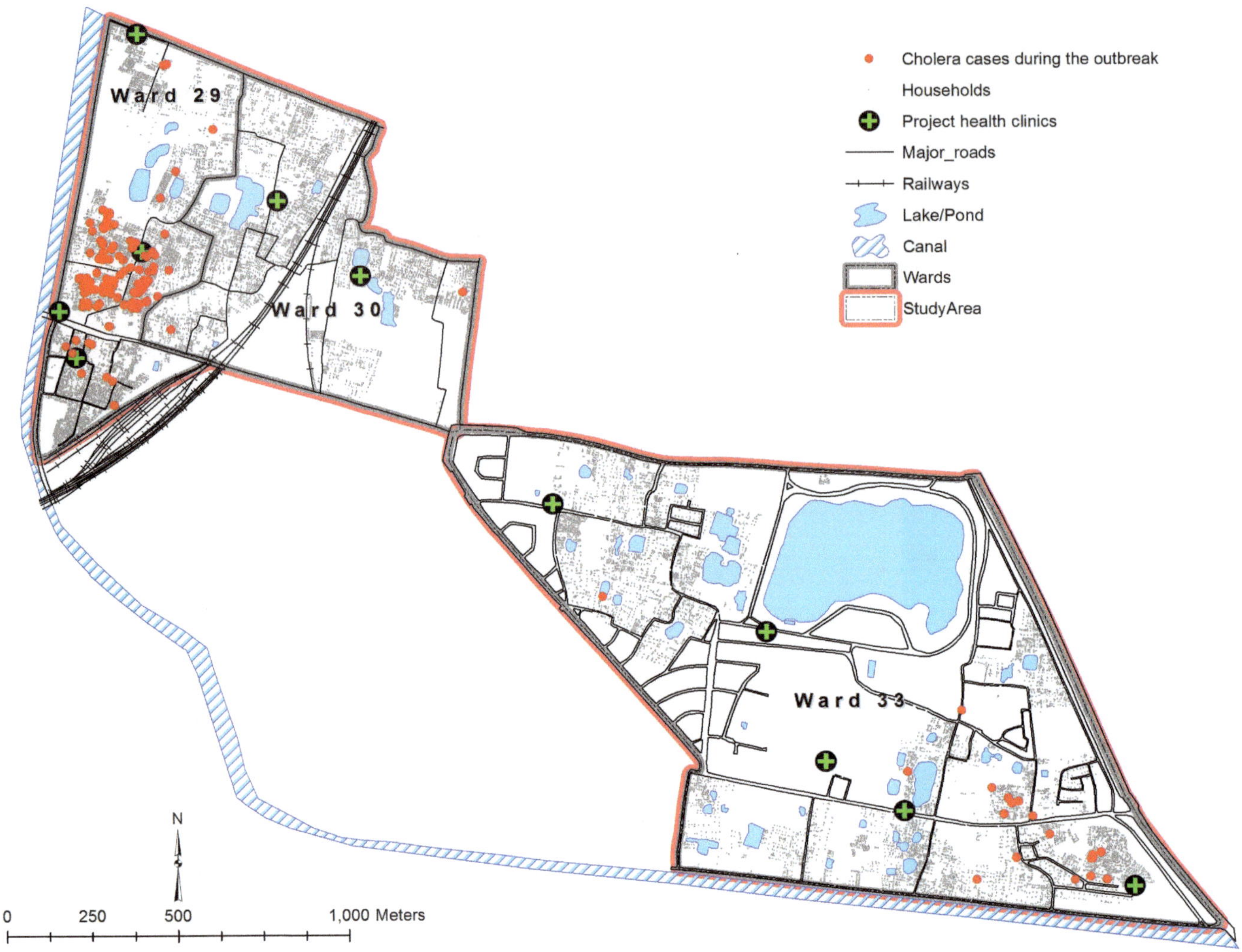

Figure 3. Distribution of the cholera cases in the study area during the large outbreak (March and April 2010) in the study area.

function [15], and adjusted for the household level correlation in the data. The models took living in high or low risk area for each analyzed individual as the dependent variable and fitted several socio-environmental risk factors of cholera detected in earlier studies [11] as the independent variables in models using independent and exchangeable within-household correlation matrices. Coefficients of independent variables in the models were exponentiated to estimate the odds ratio of cholera associated with different levels of the variable. Standard errors for the coefficients were used to estimate p-values and associated 95% confidence intervals (95% CI) for the ORs.

Ethics and Monitoring

The study protocol was approved by the Drugs Controller General of India, the ethics committee of the National Institute of Cholera and Enteric Diseases, the Health Ministry Screening Committee of India, and the International Vaccine Institute Institutional Review Board. Written informed consent was obtained from residents older than 18 years and from the guardians of residents aged 1 to 17 years of age. Additional written assent was obtained from residents aged 12 to 17 years. The trial was registered at ClinicalTrials.gov number, NCT00289224.

Results

During the one-year pre-vaccination and two-year post-vaccination periods, 95 and 183 cholera cases were detected in 111,882 and 121,827 study participants under surveillance, respectively. There were 169 cases during the large outbreak (March and April 2010) in our study area (Figure 2). The residences of those cases are shown in Figure 3.

The results of the GAM model for the pre- and post-vaccination periods are presented in figures 4 and 5. Predicted odds ratios (ORs) during the pre-vaccination period ranged from 0.05 to 4.77, and the global permutation test indicated a statistically significant odds of the disease ($p<.001$) in the west part of the study where most of the cases during the outbreak were occurred (Figure 4). The predicted ORs during the post-vaccination period ranged between 0.23 and 4.70, and the significantly high risk area for cholera was also observed in the same west part (Figure 5). However, there was a change in the alignment of the high risk area for cholera during the post-vaccination period.

The socio-environmental characteristics of study population between high and low risk areas for cholera are shown in Table 1. In both the pre and post-vaccination periods, the high risk areas for cholera were characterized by people using unsafe water

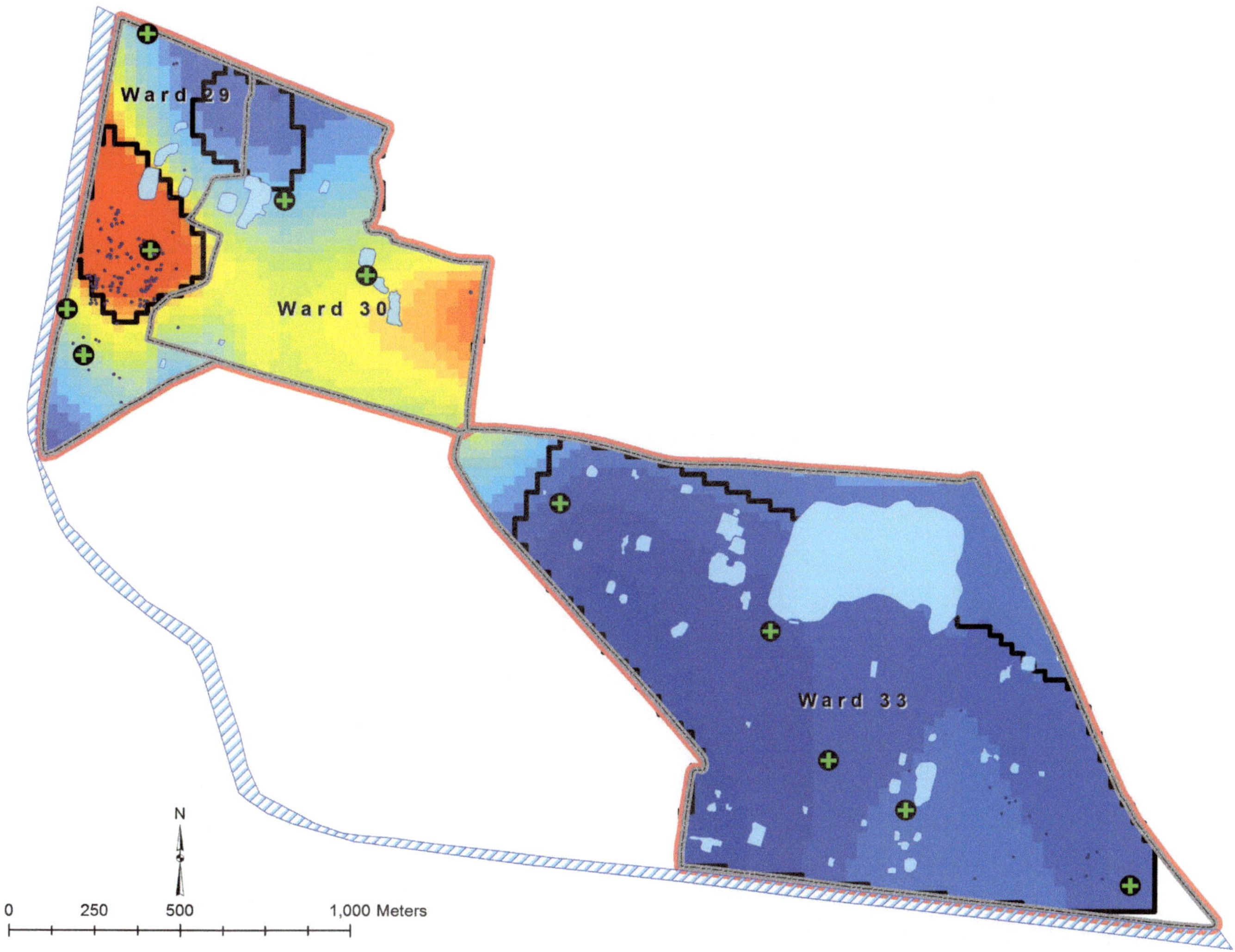

Figure 4. Spatial patterns of risk for cholera during the pre-vaccination period in the study area, Kolkata, India.

sources for drinking, living in poor socioeconomic status (defined by not living in own house), and living close proximity to the canal. During the post-vaccination period, a significantly lower vaccine uptake was observed in the high risk areas for cholera compared to that in the low risk areas for cholera. The high risk areas for cholera were also characterized by people living in close proximity to the health clinics in both the pre- and post-vaccination periods.

Discussion

Our results suggest that the outbreak was largely occurred in the areas of high risk for cholera and among the poorest residents of an area located near the canal, which is used as the main drainage for rain and waste water. Usually, an outbreak occurs in an area with immunologically naïve population. The results of our study indicate that the outbreak can even occur in the areas where immune level of the population is believed to be high. Poverty and living adjacent to the canal were consistent characteristics of the high-risk area population. Following the immunization campaign, the geographic alignment of the high-risk area for cholera was slightly changed. We speculate that the geographic realignment may have been brought about by the vaccination campaign that altered population levels of immunity, which in turn may have altered the disease transmission patterns, ultimately changing the spatial patterns of risk for cholera. Similar changes have been reported earlier [16].

We observed that there was significantly lower vaccine uptake in the high risk versus low risk areas for cholera, and, as reported earlier, only one-third of the population received the cholera vaccine [10]. The poor vaccine uptake in the high risk areas suggests high vaccine coverage is required to reduce disease risk. Close proximity to the canal was a consistent attribute of the high risk area. *Vibrio cholerae* were previously found in the canals of Kolkata [17]. It may be the canal water leaks into damaged pipes that carry drinking water [18], which is common in the slum areas [19], and that could have created increased risk for cholera. Close proximity of the high risk area to the health clinic reflects health care behavior of the people, as observed elsewhere [20–21]. Therefore, it is important to adjust the bias in a risk analysis.

We used the GAM model to determine high and low risk areas for cholera in the slums of Kolkata, because it is an effective approach for modeling spatial distributions of data, combining a number of desirable features, the ability to estimate odds ratios (ORs) while adjusting for confounders and selection of optimum degree of smoothing [13]. An advantage of the GAM method is that theoretical considerations of bias and variance are used to

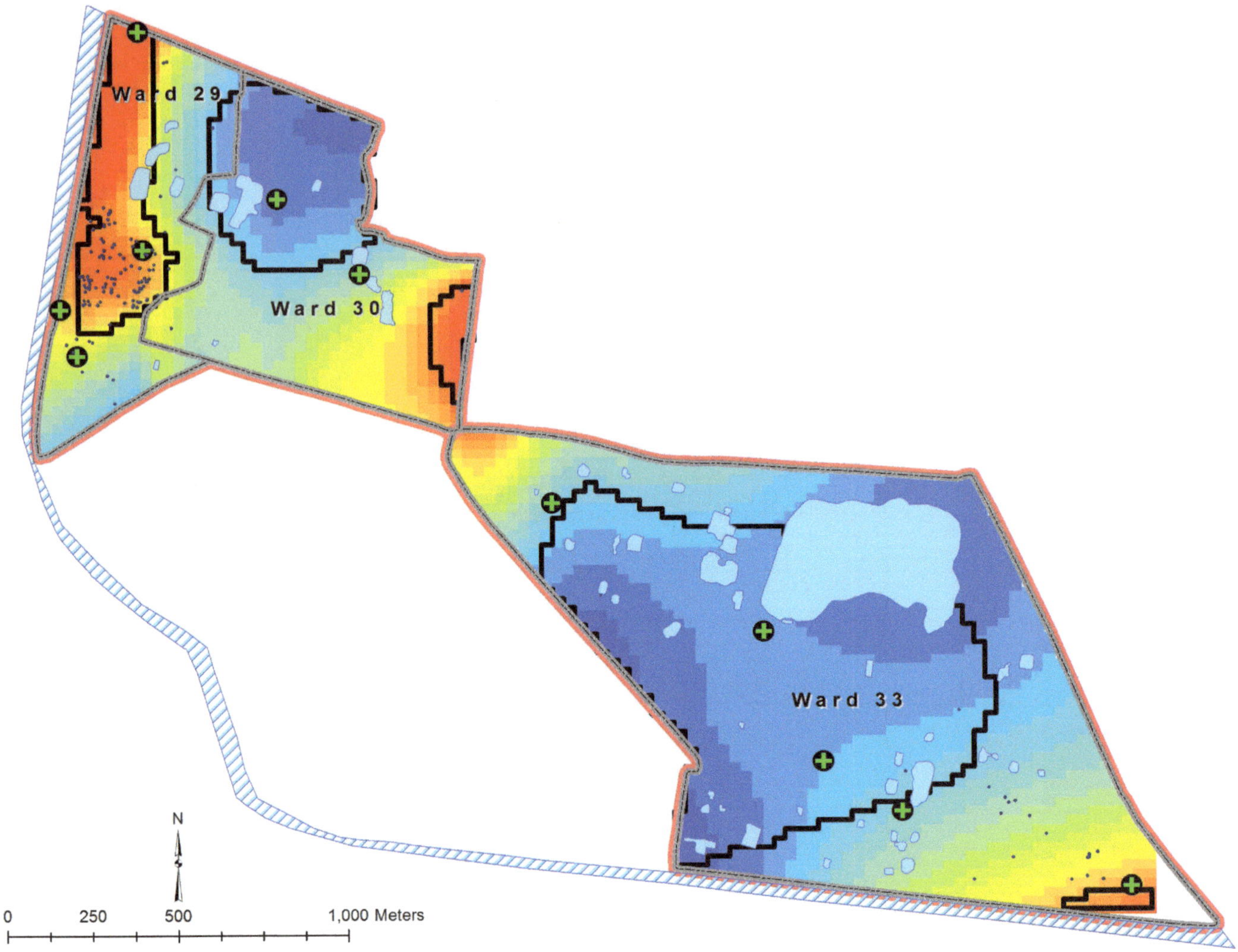

Figure 5. Spatial patterns of risk for cholera during the post-vaccination period in the study area, Kolkata, India.

Table 1. Socio-demographic characteristics between significantly high and low risk areas in the slum of Kolkata, India.

	Pre-vaccination period			Post-vaccination period		
Variables	**High risk (n = 26,354)**	**Low risk (n = 32,532)**	**OR* (95% CI; p-value)**	**High risk (n = 27,191)**	**Low risk (n = 21,777)**	**OR* (95% CI; p-value)**
Cholera vaccine recipients	–	–	–	11,198 (41.2)	10,798 (49.6)	0.525 (0.475–0.581; <.001)
Using safe toilet	948 (3.6)	6,350 (19.5)	0.904 (0.681–1.201; 0.488)	900 (3.3)	3,540 (16.3)	1.017 (0.799–1.294; 0.893)
Using safe water source (own tap)	1,753 (6.7)	10,311 (31.7)	0.754 (0.596–0.955; 0.019)	1,389 (5.1)	6,391 (29.3)	0.374 (0.311–0.452; <.001)
Using boiled or filtered water	1,084 (4.1)	7,575 (23.3)	0.775 (0.598–1.004; 0.054)	875 (3.2)	5,487 (25.2)	0.271 (0.229–0.321; <.001)
Living in own house	4,496 (17.1)	18,334 (56.4)	0.143 (0.123–0.166; <.001)	4,618 (17)	11,654 (53.5)	0.332 (0.293–0.376; <.001)
Distance (meter) from the household to the canal	189.4 (85.6)	482.4 (361.3)	0.9948 (0.9946–0.9951; <.001)	181.8 (205.5)	678.9 (234)	0.994 (0.993–0.994; <.001)
Distance (meter) from the household to the nearest project health clinic	112.8 (45.4)	247.4 (107.7)	0.979 (0.978–0.980; <.001)	112.8 (59.2)	217.1 (99.3)	0.992 (0.991–0.993; <.001)

Note: Number and per cent of population (in parenthesis) are shown for the dichotomous variables and mean and standard deviation (in parenthesis) are shown for the continuous variables (distances).

*The odds ratio for the cited variable, adjusted for all other variables in the table, in a model using Generalized Estimating Equations (GEE) with the logit link function.

choose an optimal smoothing span [22]. The GAM model provided us the magnitude of risk at each point in the study area after adjusting for the population density. However, the GAM model is highly computing intensive. It required more than twenty hours running the model in a state-of-the-art computer.

This study shows that the outbreak was largely occurred in the high risk areas for cholera, and the high risk areas are limited to only a small part of the endemic area. These suggest that a control mechanism does not require setting up in all over in that slum area, but can focus only on the high risk areas where it created increased risk of the disease as well as the outbreak. We call this source drying, and if the source drying is effectively done, it is less likely that the disease will be spread in other part of the area. Implementing such source drying control mechanism is feasible to implement in a resource limited country where cholera is endemic. It is worth mentioning the 64th World Health Assembly urges all member states to strengthen cholera control efforts including the use of oral cholera vaccines together with modalities such as water, sanitation, hygiene and early treatment and detection [23]. Considering the limitation of resources in the cholera affected countries, WHO suggest high-risk area based solution in the control mechanism [24]. The findings of our paper and the kind of this cholera map are helpful for the policymakers, especially those from countries where cholera remains to be endemic with periodic outbreaks, for setting up an appropriate intervention strategies within the limited resource of their countries.

Acknowledgments

We are grateful to the people of Kolkata where our study is being undertaken and to our field staff. We thank Dr. S. K. Niyogi and Dr. B. L. Sarkar for the lab diagnosis of cholera of the project. This manuscript was edited by the IVI Communications & Advocacy Unit.

Author Contributions

Conceived and designed the experiments: YAY MA TFW JC. Performed the experiments: YAY MA. Analyzed the data: YAY MA. Wrote the paper: YAY MA SK BS BM MP GBN SKB MC JLD ALL TFW JC DS.

References

1. Snow J (1855) On the Mode of Communication of Cholera. London: John Churchill, New Burlington Street, England.
2. Kelsall J, Wakefield J (2002) Modeling Spatial Variation in Disease Risk. Journal of the American Statistical Association 97(459): 692–701.
3. Lawson AB (2001) Tutorial in Biostatistics: Disease map reconstruction. Stat Med 20: 2183–2204.
4. Ali M, Lopez AL, You YA, Kim YE, Sah B, Maskery B, Clemens J (2012) Global burden of cholera. Bull World Health Organ 90: 209–218.
5. Wood S (2006) Generalized Additive Models. London: Chapman and Hall.
6. Sur D, Sarkar BL, Manna B, Deen J, Datta S, et al. (2006) Epidemiological, microbiological & electron microscopic study of a cholera outbreak in a Kolkata slum community. Indian J Med Res 123(1): 31–36.
7. Sur D, Deen JL, Manna B, Niyogi SK, Deb AK, et al. (2005) The burden of cholera in the slums of Kolkata, India: data from a prospective, community based study. Arch Dis Child 90(11): 1175–1181.
8. Thomas FC (1997) Calcutta: the human face of poverty. London: Penguin Books Ltd. 3–166.
9. Ali M, Rasool S, Park JK, Saeed S, Ochiai RL, Nizami Q, Acosta CJ, Bhutta Z (2004) Use of satellite imagery in constructing a household GIS database for health studies in Karachi, Pakistan. Int J Health Geogr 3(1): 20.
10. Bopp C, Ries A, Wells J (1999) Laboratory methods for the diagnosis of epidemic dysentery and cholera. Atlanta, GA: Centers for Disease Control and Prevention.
11. Sur D, Lopez AL, Kanungo S, Paisley A, Manna B, et al. (2009) Efficacy and safety of a modified killed-whole-cell oral cholera vaccine in India: an interim analysis of a cluster-randomised, double-blind, placebo-controlled trial. Lancet 374(9702): 1694–1702.
12. Vieira VM, Webster TF, Weinberg JM, Aschengrau A (2008) Spatial-temporal analysis of breast cancer in upper Cape Cod, Massachusetts. Int J Health Geogr 7: 46.
13. Webster T, Vieira V, Weinberg J, Aschengrau A (2006) Method for mapping population-based case-control studies: an application using generalized additive models. Int J Health Geogr 5: 26.
14. Vieira V, Webster T, Weinberg J, Aschengrau A (2009) Spatial analysis of bladder, kidney, and pancreatic cancer on upper Cape Cod: an application of generalized additive models to case-control data. Environ Health 8: 3.
15. Zeger SL, Liang KY, Albert PS (1998) Models for longitudinal data: a generalized estimating equation approach. Biometrics 44(4): 1049–1060.
16. Ali M, Sur D, Kim DR, Kanungo S, Dutta S, Bhattacharya SK, Manna B, Ochiai RL, Clemens J (2011) Impact of Vi vaccination on spatial patterns of typhoid fever in the slums of Kolkata, India. Vaccine 29: 9051–9056.
17. Abou-Gareeb AH (1960) The detection of cholera vibrios in Calcutta waters: the River Hooghly and canals. J. Hyg. 58: 21–33.
18. Hughes JM, Boyce JM, Levine RJ, Khan M, Aziz KMA, et al. (1982) Epidemiology of eltor cholera in rural Bangladesh: importance of surface water in transmission. Bulletin of the World Health Organization 60(3): 395–404.
19. Bhunia R, Ramakrishnan R, Hutin Y, Gupte MD (2009) Cholera outbreak secondary to contaminated pipe water in an urban area, West Bengal, India. Indian J Gastroenterol.28(2): 62–64.
20. Ali M, Emch M, Yunus M, Sack RB (2001) Are the environmental niches of *Vibrio cholerae O139* different from that of *Vibrio cholerae* El Tor? International Journal of Infectious Diseases 5(4): 214–219.
21. Ali M, Canh DG, Clemens JD, Park JK, von Seidlein L, et al. (2005) The use of a computerised database to monitor vaccine safety in Vietnam. Bulletin of the World Health Organization 83(8): 604–610.
22. Hastie T, Tibshirani R (1990) Generalized Additive Models. London: Chapman and Hall.
23. Cholera: mechanism for control and prevention. WHA 64.15 (2011). Sixty-fourth World Health Assembly. Geneva: World Health Organization.
24. WHO (2010) Cholera vaccines: WHO position paper. Wkly Epidemiol Rec 85: 117–28.

23

Nation-Wide, Web-Based, Geographic Information System for the Integrated Surveillance and Control of Dengue Fever in Mexico

Juan Eugenio Hernández-Ávila[1], Mario-Henry Rodríguez[1], René Santos-Luna[1]*, Veronica Sánchez-Castañeda[1], Susana Román-Pérez[1], Víctor Hugo Ríos-Salgado[1], Jesús Alberto Salas-Sarmiento[2]

1 National Institute of Public Health of Mexico, Cuernavaca, Morelos, Mexico, **2** Public Health School of Mexico, Cuernavaca, Morelos, Mexico

Abstract

Dengue fever incidence and its geographical distribution are increasing throughout the world. Quality and timely information is essential for its prevention and control. A web based, geographically enabled, dengue integral surveillance system (Dengue-GIS) was developed for the nation-wide collection, integration, analysis and reporting of geo-referenced epidemiologic, entomologic, and control interventions data. Consensus in the design and practical operation of the system was a key factor for its acceptance. Working with information systems already implemented as a starting point facilitated its acceptance by officials and operative personnel. Dengue-GIS provides the geographical detail needed to plan, asses and evaluate the impact of control activities. The system is beginning to be adopted as a knowledge base by vector control programs. It is used to generate evidence on impact and cost-effectiveness of control activities, promoting the use of information for decision making at all levels of the vector control program. Dengue-GIS has also been used as a hypothesis generator for the academic community. This GIS-based model system for dengue surveillance and the experience gathered during its development and implementation could be useful in other dengue endemic countries and extended to other infectious or chronic diseases.

1

Editor: Shamala Devi Sekaran, University of Malaya, Malaysia

Funding: This study was funded by the National Council for Science and Technology "CONACyT" (http://www.conacyt.gob.mx). The funders had no role in study design, data collection and analysis, decision to publish, or preparation of the manuscript.

Competing Interests: The authors have declared that no competing interests exist.

* E-mail: rsantos@insp.mx

Introduction

Dengue Fever (DF) is caused by the infection with four subtypes of Dengue virus and sometimes progresses to a severe and potentially deadly condition (Dengue Hemorrhagic Fever, DHF). Dengue viruses are transmitted to humans by the bite of infected mosquitoes (*Aedes aegypti* and *Ae. albopictus).* The incidence and geographical dispersion of this disease expanded from nine countries in 1970 to more than 100 at present. More than 2.5 billion people are at risk with 50–100 million infections each year (1.6 million reported in the Americas in 2010) [1], with an estimated average cost of $2.5 billion US dollars per year [2]. Dengue fever is endemic in Mexico [3] where transmission is reported in 28 of its 32 states (Figure 1).

At present, the only public health strategy available for dengue control is directed against the mosquito vector populations. Effective control strategies require the intervention of multiple agencies within the health and other government sectors [4]. Access to timely and reliable epidemiologic and entomologic information is essential to support decision making and facilitate the interaction between the participating stakeholders. The lack of reliable and timely data useful to the different decision and operative components and levels of the control programs is the most important shortcoming in dengue control. For this purpose, national electronic reporting systems were recommended for surveillance, early detection and prediction of outbreaks [5].

Global dengue surveillance systems currently available include DengueNet, the World Health Organization's central data management system, for the global epidemiologic and virologic surveillance [6]. Additionally, DengueMap [7], developed by CDC in collaboration with Health Map, uses Geographical Information System technology (GIS) to monitor DF transmission around the world. The gross detail and delayed data release of these systems limit their use by national control programs. GIS technology, based on the prevalence of suitable mosquito breeding sites and epidemiologic data has been used to assess the distribution and dynamics of mosquito populations or the risk of dengue and DHF in human populations [8–11], but to our knowledge, no national comprehensive electronic surveillance systems with the level of geographical disaggregation as the one presented in this work are currently available.

The Mexican public health system that provides dengue surveillance, preventive and control services is managed by the Federal Ministry of Health (MoH) in collaboration with the 32 decentralized State Health Services (SHS); these in term are organized in local health systems called sanitary jurisdictions. Additionally, Social Security Institutions (organized in delegations) provide services for a limited portion of the population working in the formal economy or state workers [12]. DF surveillance is

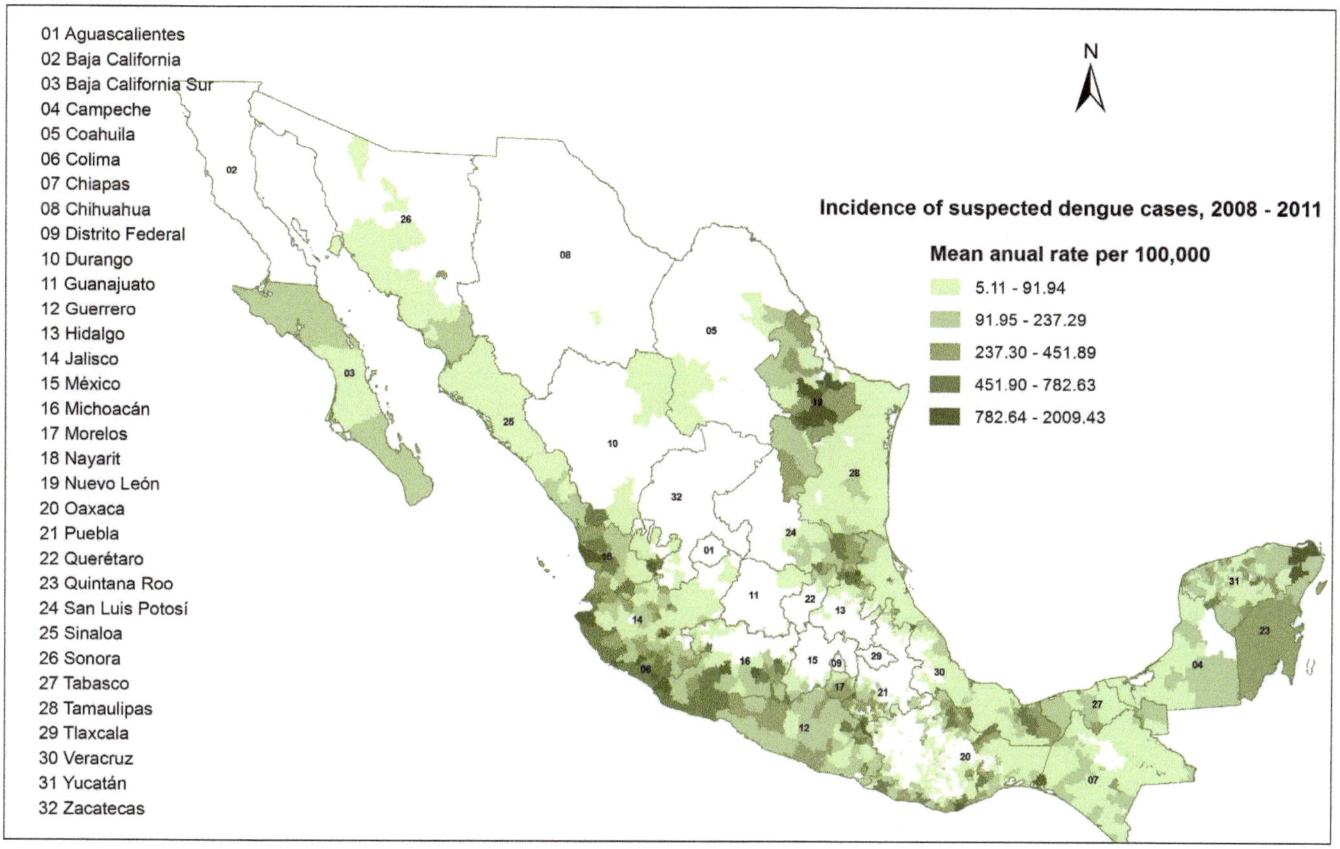

Figure 1. Incidence of dengue cases in Mexico during the period 2008 to 2011. Dengue cases have been reported in 30 of the 32 Mexican states. Geo-referenced case reporting started in 2008 when the EPS began interoperating with Dengue-GIS.

coordinated, at the federal level, by the General Directorate for Epidemiologic Surveillance (GDES) and operated by the SHS through the State Epidemiology Departments[13–15]. A public health laboratory network diagnoses samples from patients with suspicion of dengue (probable DF cases) and a portion of these samples is analyzed for virus isolation (confirmed DF cases). Entomologic surveillance and vector control activities are coordinated by the National Center for Preventive Programs and Disease Control (NCPPDC) [13–15] and carried out by vector control departments in the SHS.

A web-based information system (Epidemiologic Surveillance Platform, ESP) was implemented in 2008 to collect data on suspected cases but it had no systematic means for data sharing and/or utilization; epidemiologic and entomologic surveillance and control information had no spatial disaggregation and flowed vertically to GDES and NCPPDC with little or no feedback to local health levels. This situation caused a proliferation of parallel local systems to support local decision making processes. An analysis of the available epidemiologic surveillance and vector control information systems, as well as data and results from a study to assess the performance of the Mexican Health Information System [16] identified that its main shortcoming were the fragmentation and diversity of non-compatible, electronic and paper-based information systems.

The capacity of web-based GIS applications to facilitate timely registration of data, its integration and analysis could increase the capacity of national programs for dengue surveillance and greatly assist in its prevention and control [17]. We present herein a web-based GIS, with nationwide coverage for dengue surveillance and control based on the conceptual framework presented in Figure 2. It registers data collected at the local level and provides capacities to integrate and analyze information, such as spatial and time aggregation of DF cases, relevant to each level of the health system. The strategy to recruit the participation of health officers at different levels of the system for its design and construction could guide the implementation of similar systems in mid and low income countries.

Methods

Consensus

The format and contents of the new surveillance system were designed in consultation with experts in public health surveillance, disease control and prevention from the federal level and their respective counterparts in the SHS, as well as Information Technology (IT) developers. A detailed analysis of the available epidemiologic surveillance and vector control information systems and data and results from a study to assess the performance of the Mexican Health Information System [16] were used to identify shortcomings.

GIS Basic Platform

Dengue-GIS was constructed as a web based information system with three tier architecture [18]. The Graphic User Interface (GUI) was developed in Microsoft ASP.Net 3.5 © based on MS-ISS 6.0 ©. The Model tier is based on ESRI ArcGis Server [19] and the data tier is based on MS-SQL Server ©. It is hosted at the National Institute of Public Health (NIPH) site on a server

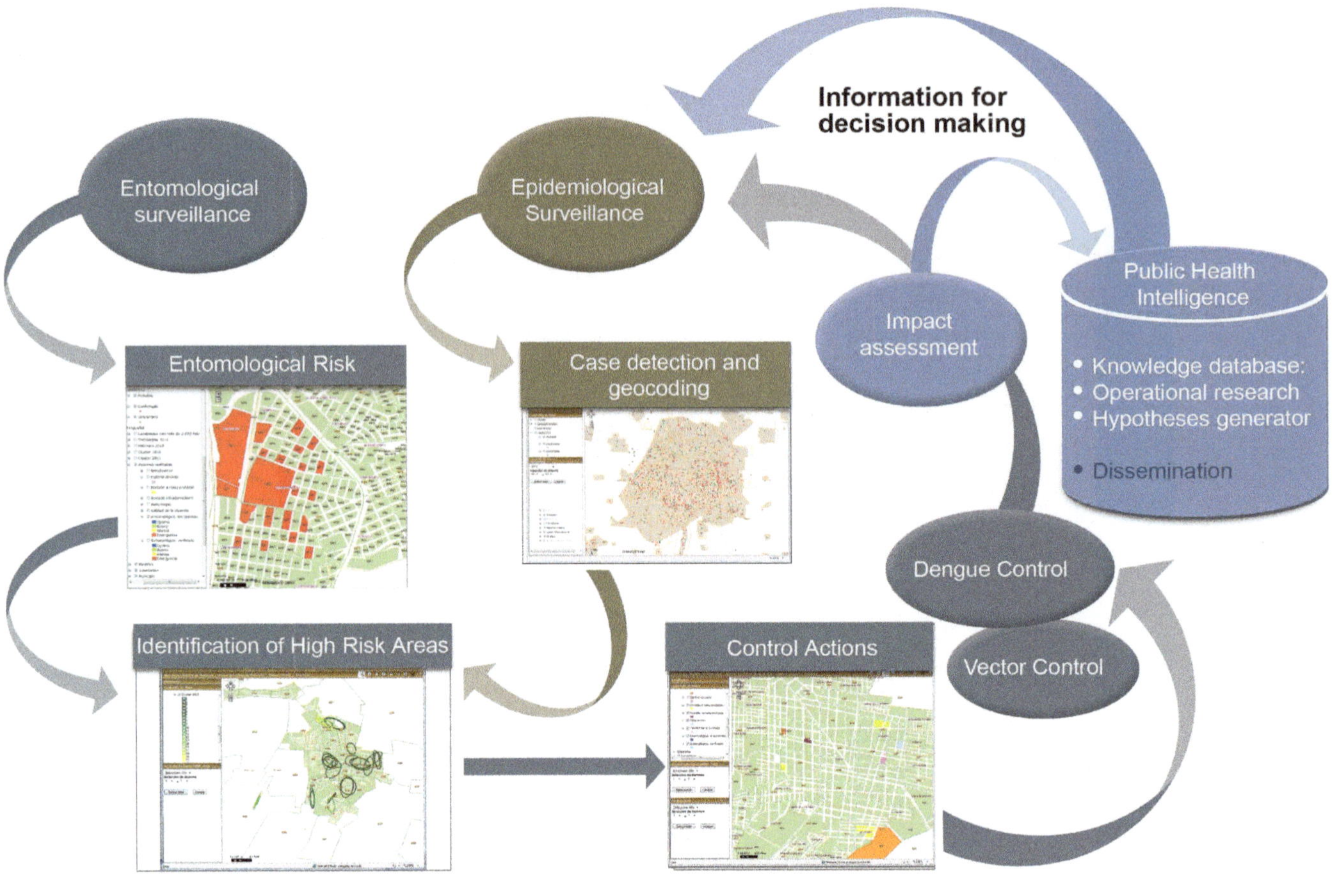

Figure 2. Conceptual Framework. Epidemiological and entomological data are entered and analyzed to produce risk maps which are then used to direct vector control activities. New epidemiological and entomological data are used to assess the impact of control activities thus generating a knowledge database which can be used to evaluate cost-effectiveness of control measures, accountability and operational research. Control interventions are directed according to the risk maps and entered into the system for impact evaluation in the reduction of cases and/or vector population.

farm, totaling 32 cores for map production, eight cores for data base management and four for the internet server; each core with 4 GB ram. The entire site has an internet connection with a bandwidth of 30 Mb/s which is shared with NIPH's institutional web applications.

To provide means for compatibility with the existing electronic information systems and future developments, a location ID-coding system was adopted. This is based on the National Geo-statistical Framework (NGF) from the National Institute for Statistics and Geography (NISG). The NGF is a unique geo-referenced identification system used to reference census data and other statistical data to geographic locations and geo-political features in the Mexican territory [20]. Additional geo-referencing capacities were obtained with the integration of a cartographic database developed by the Federal Electoral Institute (FEI), also based on the NGF, but with greater urban geographic detail. The combined cartography covers small rural towns with attributes, such as public and private schools (based on the Ministry of Education's database) street centers and names and postal code among others. ID codes were assigned to town blocks, town block groups, census tracks, towns, cities, municipalities and states. A unique ID code was adopted to identify and geo-reference health units (Uniformed Code of Health Establishments, UCHE). This ID code was developed by the General Directorate of Health Information of the MoH. The final GIS contains the location of all public and private schools, the location of all health units managed by the public health institutions in Mexico (serving more than 90% of the population) as well as the total population and their socio-demographic characteristics, grouped by block, census tract, town, municipality state, region and nationwide. Other geographical data such as elevation, average temperature and precipitation is also available in the system at different scales.

System Development

The GIS for surveillance and control of DF (Dengue-GIS) was constructed in three phases. In the first phase, an interoperable geocoding web based tool with a mapping application, designed to work with the EPS, was developed to geocode the place of residence (town blocks level) of probable dengue cases. Then, a modification was made on the EPS to enable geocoding through online calls in order to geo-reference the residence of probable cases. The EPS is now operated nationwide by the jurisdictional, state and social security system epidemiologists in coordination with the GDES.

The second phase included the development of a set of modules for entomological surveillance, planning and monitoring control activities, as well as a module for downloading information to the hard drive of local entomologists' and epidemiologists' computers. Vector surveillance is based on ovitrap data; for this purpose, a modification of the geocoding tool was developed to geocode ovitraps. After an ovitrap is geo-referenced, data on egg counts are entered into Dengue-GIS via web based formularies. Another set of web based formularies was developed to allow entering data on vector control activities carried out at the local level. Also, Dengue-GIS allows entering data gathered in entomological surveys, such as the number of water containers per house, the number of positive containers per house and the number of pupae-positive containers per house. These formularies are linked to town blocks using the NGF and the federal electoral identification codes. The modules for control and planning include tools to visualize temporal and geographical accumulation of cases (clusters of DF transmission) in order to direct vector control activities. It also includes a tool to print maps to be used in field activities; these maps have the id code of each block to allow recording of surveillance/control activities.

The third phase was completed with a reporting module and an improved geographic visualization tool that combines epidemiologic, entomological and control data into a single mapping window.

Each of these phases was implemented gradually in the states with most DF incidence, and a group of researchers and IT specialists, along with GDE personnel trained state and jurisdictional epidemiologist in its use.

Statistical Analyses

Identification of areas of probable cases accumulation in time (clusters) is carried out using a nearest neighbor hierarchical clustering algorithm [21]. This statistical methodology is used in epidemiology to detect areas of high accumulation of cases based on the distance between them. For its application in Dengue-GIS, a minimum number of cases to conform a cluster was set to five and the maximum distance between any two cases to 500 meters. The algorithm is run every week using the EPS geo-referenced epidemiological data of the three previous weeks. In this way the entire national territory is scanned every week in search for areas with an accumulation of at least five cases separated by 500 meters or less within a three weeks' time window. The resulting graphical output of the algorithm is an ellipse, centered at the geographical centroid of the set of locations (points) where the probable cases were geocoded. The length of axes of the ellipses is determined by the standard deviation in the north-south and east-west dispersion of the location of the cases. The orientation of the ellipse is based on the directional mean of the set of locations to where the cases were geocoded. The directional mean is determined by an angle that minimizes de distance of every point in the cluster to both axes of the corresponding ellipse. Every week cumulative incidence rates are calculated at a micro-regional scale to help in the detection of high risk areas. Historical information on the yearly incidence per micro-region is also available (shown in File S1 Annex 1). Significance of each week clustering scheme is tested using Monte-Carlo simulation with 100 repetitions.

A risk of transmission index was developed in collaboration with the national vector control personnel, combining the ovitrap contents data and the epidemiological surveillance data. According to the entomological surveillance norm, four ovitraps are to be placed in each city block under surveillance. City blocks are classified into four entomological risk categories, according to the mean number of eggs per ovitrap using the city specific quartiles of the mean counts distribution as cutoff points: low, moderate low, moderate high and high. A 300 meter radius buffer (circle), centered at the geographical centroid of the location of the four ovitraps of each block, is drawn and geographically intersected with the epidemiological surveillance data (points representing the location of cases) to produce the transmission risk index:

- High transmission risk = high entomological risk and presence of dengue probable cases
- Moderate high transmission = Moderate high entomological risk and presence of dengue probable cases
- Moderate low transmission = Moderate low entomological risk and presence of dengue probable cases
- Low transmission risk = no dengue probable cases detected

Results

Epidemiological Surveillance

According to Mexican official norms, epidemiologic data on dengue patients is entered into the surveillance system within one week after the patient's first contact with the health system and it is reported by epidemiological week. New cases detected by primary care providers are notified (using a probable-case-registration form) to the lowest administration level (health jurisdiction) where data is introduced in the EPS/Dengue-GIS. The geocoding system is very intuitive and, when invoked opens a new window displaying a map of the area of residence (based on the address and the NGF) of the probable case. Clicking on a button invokes an instance of the geocoding tool that displays in the computer screen a map of the area based on state, municipality, city, and postal code. To simultaneously register in the EPS and geocode probable dengue cases, the local epidemiologist clicks with the cursor in a town block, using the address information on the probable-case-registry (Figure 3). The probable case information (without any personal data) is then sent to a geo-database for processing and mapping. The screen coordinates are transformed into geographic coordinates and stored at a geocoded data base (at NIPH) and the EPS data base (at GDES) servers. The epidemiologist or health provider has to click only on the screen to record the coordinates and the map window closes to return to the EPS screen. With this simple process the data can be analyzed in a geographical domain with very high resolution. Geo-coding is done at the block level to avoid disclosing the actual place of residence of probable dengue cases.

As part of the regular epidemiological surveillance protocol and according to the Mexican official norms [13,14], blood samples for diagnosis are drawn to a variable portion of dengue probable cases. This blood samples are obtained at the point of care by the health care service providers and sent to the public health laboratory. Data on laboratory case confirmation is entered by the laboratory personnel into the EPS and case status is automatically updated in Dengue-GIS through the internet connection after each registration (Figure 3). Figure 4 shows the location of cases, as color coded dots, in the city of Linares, Nuevo León during the epidemiologic week 22 to 24 of the year 2012. Red dots represent confirmed cases, green dots represent discarded cases (not dengue cases) and yellow dots represent probable cases for which no lab info is available (either because the public health lab has not yet entered their results or because it was not selected for a blood sample according to the surveillance protocol). Every Wednesday, an automated process to detect areas of probable dengue cases concentrations is conducted using the epidemiologic surveillance data of the three previous weeks. Figure 4 shows the graphical results of this process for the weeks 22 to 24 of 2012 as yellow ellipses. In the center of the figure, two areas in which three ellipses intersect indicates that the accumulation of cases occurred, consecutively, during the three weeks showed. For the purpose of display, ellipses are also color coded by epidemiological week in which they appeared: green at the beginning of the year degrading to yellow by mid-year and to purple by the end of the year. This

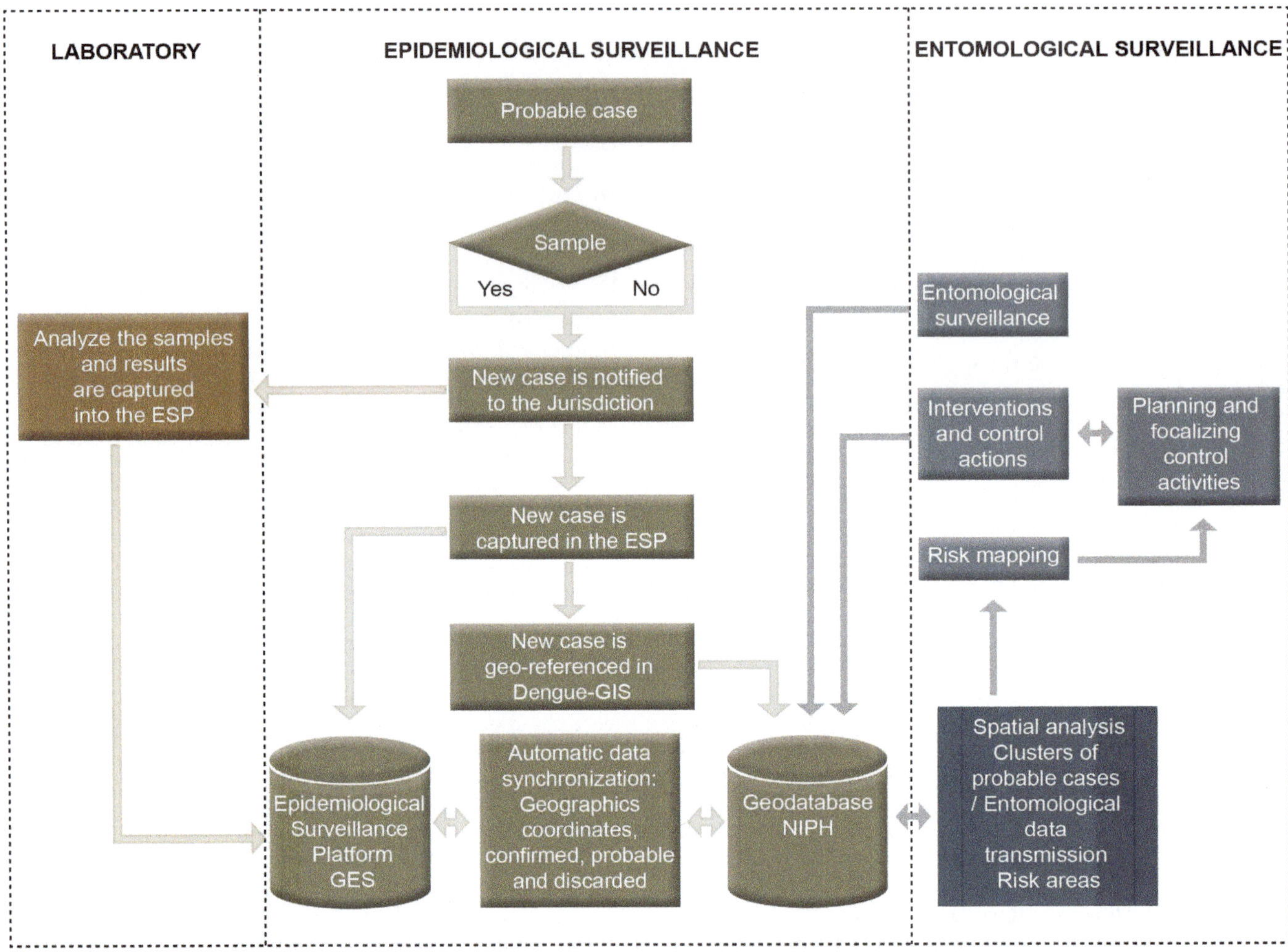

Figure 3. Web-based geographic Information System for integrated dengue epidemiologic surveillance and control data flow diagram. Epidemiological data on probable case is entered into the system at the health center or sent in paper form to the sanitary jurisdiction where is entered into the EPS system. The case is simultaneously geo-coded and data stored in the geodatabase. Laboratory results from blood samples, taken to a portion of the probable cases, enters the system and serves to update the EPS and Dengue GIS data base. Entomological surveillance data and control activities are entered into the Geodatabase by vector control personnel. The integration of these data is used to produce risk maps.

allows the visual analysis of longer periods and the identification of areas where transmission is persistent. File S1 is an animated graphic representation of a longitudinal follow up depicting Merida city blocks with DF cases, the distribution in time and space of clusters of case accumulation in relation to the overall case incidence in 2012, as well as vector control interventions.

Entomological Surveillance and Vector Control

Entomological surveillance is based mainly on mosquito ovitraps, but the system can be used for the inclusion of other parameters, such as numbers and locations of mosquito breeding sites based on the household and/or environmental conditions [13–15,22]. The local dengue program operative personnel perform weekly monitoring of mosquito ovitraps (positivity and number of eggs per trap) and enters into Dengue-GIS the individual results at the jurisdiction administrative offices (Figure 3). Entomological data are geocoded to the respective town block and stored in the Geodatabase (NIPH). Anti-vector interventions are registered in the same manner. Data on entomological surveillance and vector control activities are systematized and displayed in working maps readily available to the local and federal control levels. Figure 4 shows the results of entomological surveys from the east of downtown Linares (center of the figure); the color coded blocks, red and yellow, indicate that those areas are in high risk (alarm and emergency). These areas are precisely where dengue cases clustered for the three consecutive weeks showed in the figure.

Every Tuesday entomological data maps are constructed and used for the estimation of transmission risk. Figure 5 shows the graphic output of this process for the city of Merida, Yucatan for the epidemiological week 12 of 2013. The location of ovitraps with *Ae. aegypti* eggs are represented by red dots. The various degrees of risk are represented by yellow, orange and red circles. For simplicity, low risk areas are not shown in the map.

Dengue-GIS Use

The Information system for epidemiological surveillance (EPS) interoperates on real time with Dengue-GIS (Figure 3). The integration of geo-referenced data on dengue cases, entomological

Figure 4. Screenshot of Dengue-GIS application. Dengue-GIS screenshot showing the distribution of probable cases and case clusters (high transmission areas) during the CDC weeks 22–24 2012 in Linares City, Nuevo Leon. Locations of entomological survey activities during the same period are shown in red. DF cases accumulation in space and time is showed by a set of ellipses which represent the graphical output of the cluster detection algorithm. Two areas, one very close to downtown Linares, and the other to the west of the city show three intersecting ellipses, this indicates that transmission occurred uninterrupted during the three week period showed. Cases are represented by dots; red dots are laboratory confirmed cases, green dots are cases discarded by laboratory tests (other disease) and yellow dots are those for which there is not available laboratory data (either because it was not selected for blood test, according to surveillance protocol or because the public health laboratory has not yet got/entered the respective results). Red squares represent the city block where the entomological surveys were carried out, showing emergency results. Entomological surveys are a separate surveillance activity in which information about the conditions of households is collected. Pictures presented here were modified to translate the text presented to English; the actual Dengue-GIS is an all-Spanish language system.

parameters and control activities makes it possible to evaluate identify and prioritize risk areas (Figures 4 and 5).

Dengue-GIS is available for the federal dengue control program leaders, state program leaders and the operative personnel in all (236 nationwide) sanitary jurisdictions. It is also available to all personnel in charge of epidemiological surveillance at all levels of the health system. Almost two years after starting its implementation Dengue-GIS has more than 302 active users in 31 states, with an average of 204 users during a working day. A total of 147,495 ovitraps have been deployed, geo-referenced and weekly monitored in 526 large cities (where more than 36 million people live) in the states with historically high dengue incidence; covering important cities such as Guadalajara in Jalisco, Monterrey in Nuevo León, Cuernavaca in Morelos., Merida in Yucatán, Cancun in Quintana Roo and Acapulco in Guerrero.

The system has accumulated a total of 6′ 667,385 records. Up to the 48th epidemiological week in 2012 there were 297,797 geo-referenced probable cases in the system (42% of all cases registered in the EPS). The proportion of geo-referenced cases are discretionally decided by control programs on the bases of numbers of suspected cases within geographic locations and the surveillance intensity necessary for the identification of outbreaks, assessment of transmission levels, and control monitoring. With this epidemiological information, 40,687 clusters of probable transmission have been identified scattered in the national territory, since 2010.

According to the laboratory results on patient blood samples (33% of all cases registered in the EPS), the positivity index is 33.1% (26.1% –38.1%) in the period 2008–2012, thus an estimate of 237,748 confirmed cases registered.

Figure 5. Screenshot of Dengue-GIS application showing the transmission risk index. Every Tuesday an evaluation of the abundance of the vector population, estimated by the mean number of eggs per city block and the presence of probable dengue cases is carried-out. The color coded circles represent areas of differential dengue transmission risk: high (red), moderate-high (orange) and moderate-low (yellow). Low risk areas are not shown.

There are 449,295 records with control activities by town block in 8,120 urban and rural areas where more than 54 million people live 48.5% of the country's population, (39.08% in urban areas and 9.45% in rural areas). Currently, this information system registers approximately 84,798 records every week including probable cases and lab confirmation data, ovitrap readings, control interventions (insecticide nebulization, indoor spraying, and larvicide application) as well as the transmission clusters (Figure 4).

Discussion

The main features of Dengue-GIS is its seamless capability to integrate, in an interoperable web based platform, the basic information needed for problem appraisal and control planning, which is readily available to users at all levels of the control program scaffold. Dengue-GIS includes tools for the integrated analyses of epidemiological and entomological surveillance for the identification of increased transmission areas and to monitor vector control activities. The availability of an operable (timely and spatially disaggregated) epidemiological stratification, along with the capability to assess the effect of focalized control interventions, provide local and federal control programs with information for evidence-based decisions within a public health intelligence platform (Figure 2).

Previous to the EPS and Dengue-GIS implementation, epidemiological data, produced locally by health providers, travelled each week in paper forms to their respective sanitary jurisdictions and then to the State Epidemiology Department. The data was then entered into a local electronic system, which in term produced files to be sent to GDE either by disk in the early 90's or more recently by e-mail. GDE compiled all the data into a single database and produced weekly epidemiological reports. Epidemiological and entomological surveillances were unconnected and the information flow was slow. As in most endemic countries, dengue surveillance was of little use in spite of the amount of work required for data collection and registration; the data flow and the aggregation process limited the capacity for local analysis impeding its use for anti-dengue operations. Nonetheless, epidemiological data was collected and at best used for planning next season control operations; while entomological surveillance was mostly used to assess *a posteriori* anti-mosquito interventions.

With Dengue-GIS is now possible to identify at nation-wide scale where and when accumulation of probable dengue cases are occurring (Figure 4) but with a spatial resolution suitable to direct and evaluate vector control activities. The use of cluster detection analyses and risk assessment on a weekly basis produce timely information for decision making at all levels of the dengue control program. For example, the visualization of areas in which accumulation of cases occurred in three consecutive weeks (Figure 4), makes it obvious that some control actions are there needed immediately.

Putting apart the construction of the web site, developing Dengue-GIS implied a complex process. It began with the incorporation in its design of the required components already in use and new ones of possible use by the vector control programs. Operatively, the construction of data input screens required to be as simple as possible and of friendly interaction. Also, the adoption of the system for the routinely registration of data and utilization of integrated surveillance reports required changes in the perception about their usefulness and in the operative culture of its potential users.

The case detection efforts, according to the prevailing local epidemiology, remained the same with the introduction of Dengue-GIS. Basic information related to dengue cases (age, sex, date, address) conserved the traditional usage, but data registration and consultation was facilitated by user-friendly screens. Although weekly epidemiological reports are still produced (in accordance to the specific action program 2007–2012) [22], it has become web based. This reports are now available at http://www.epidemiologia.salud.gob.mx/dgae/panodengue/intd_dengue.html. EPS and Dengue-GIS implementation also introduced other improvements for epidemiological surveillance as the simultaneous registration and availability of the data from local to federal databases and thus the capability of close monitoring in almost real time at all administrative levels of the health system. With several geocoded data layers including environmental, demographic, health system infrastructure and accessibility, Dengue-GIS provides a wider panorama for the identification of the determinants of the disease transmission. Additionally, historical epidemiological data can be easily consulted at different levels of spatial and time aggregations to monitor changes in the intensity of transmission. Laboratory results are now readily available and are mapped in real time. This is a very important improvement for epidemiological surveillance and for the planning and evaluation of vector control activities.

The geo-codification of entomological data required the introduction of new processes in the surveillance activities of the control programs. Data input screens facilitated de registration of the main entomological indicators based on breeding site characterization, and the adoption of a surveillance strategy based on a network of geo-coded ovitraps required new data enter formularies. In addition, the capacity of Dengue-GIS to register anti-vectorial interventions, that can be displayed and analyzed along with the mosquito monitoring information, serves as an input for the analyses of impact and cost-effectiveness of their interventions both in the human and vector populations. Similar to case registration, entomological and control data is simultaneously available to all levels of the control program. The system's capacity to perform spatial-temporal analyses integrating entomological and epidemiological data provides evidence (easily consulted in screenshots) for planning and conducting vector control activities based on risk assessment (Figures 4, 5).

Currently, the national vector control program has adopted the risk based strategy to carry out vector surveillance and control activities, including in its operative guidelines for entomological surveillance the use of the information generated by Dengue-GIS (http://kin.insp.mx/aplicaciones/Plataformadengue/Nacional/Capacitacion/Estrategias%20Dengue/Estrategia_Control_focalizado.pdf accessed 17 april 2013).

The several processes involved in Dengue-GIS development and its implementation present common situations in other dengue endemic countries where traditional surveillance requires modernization to upgrade its usefulness in control and prevention (5). The organizational inertia, built over decades, is probably the most important source of the obstacles faced during the introduction of new procedures [23,24]. To build on ownership, a trans-disciplinary team with experts in dengue surveillance and control from the federal and state levels, as well as local public health workforce. Consensus among all stakeholders on the need to use and share geo-referenced data to facilitate prediction and early outbreaks detection, as well as to evaluate the impact of control activities was achieved. The plasticity of the system allowed the inclusion of new components for compliance to requests made by local epidemiologists and entomologists.

Resistance to change is related to the amount of modifications introduced into current practices [24]. Implementation of Dengue-GIS was facilitated by the adoption of already working information systems as its building blocks. Although the changes introduced in the EPS were minimal, the fragmented health system and the diversity of data management protocols required a great deal of promotion to convince health and vector control personnel about the benefits of geo-coded data surveillance.

The use of Dengue-GIS is now at full scale in Mexico and the data produced by epidemiological and entomological surveillance is used to direct vector control actions. In a self-applied questionnaire answered by the national program leader, nine state leaders and 158 users (operative personnel at the jurisdictional level) of 27 states, 78% found Dengue-GIS amiable, 82% indicated that it provides useful information for the conduction of their activities and 84% uses the transmission clusters to focalize control actions and to visualize coverage of prevention and control activities. Also, 86% considered that Dengue-GIS has been beneficial to the overall management of their activities. State leaders reported that having online access to information helps them to keep track of prevention/control activities and that the system is helping them to strengthen dengue control.

The continuous use of Dengue-GIS for data collection and analyses will eventually generate a population based knowledge database for the development of operative research projects and could be used to identify cost-effective interventions in specific operative settings. This geo-referenced population database can also be useful, in phase III trials of the incoming dengue vaccines [25–28].

Some unexpected uses of the system have begun to appear; the states programs are using Dengue-GIS in different applications for new types of analyses; for example, Jalisco aggregates data to produce risk maps by different geo-political units; Morelos generates maps of insecticide nebulization coverage (as recorded by GPS installed in the fogging machine) which in term could be incorporated into Dengue-GIS. The accessibility to an accumulated set of epidemiological and control assessment data has motivated the scientific community to document and get data for research projects and currently four projects are using data and the spatial capabilities of the system. Its capacity to record health interventions opens the possibility to extend the use of its platform to the surveillance and monitoring and evaluation of other health programs activities.

Integrated Dengue Surveillance systems such as the one presented here are now feasible; they must be part of the prevention and control of this disease. Digital cartography and automated address locators have different development levels in countries where dengue is endemic. Poor urban planning and unregulated urban expansions, along with the lack of national standards to record addresses make it very difficult to automate epidemiological and entomological data geocoding. Nonetheless, these limitations can be solved with tools like Google maps and Google earth and their Application Programing Interfaces (API) [29]. Similarly, open source GIS software such as GRASS and MapServer, a desktop application and web map server respectively [30,31], as well as a growing set of spatial analytical tools, like the *spatial view* in the statistical programming language R [32] could be integrated in these surveillance systems.

Supporting Information

File S1 Flash movie showing the longitudinal nature of the epidemiological, entomological and control data collected through DENGUE-GIS. In this interactive movie the probable case clusters, as well as the control actions are shown by epidemiological week on a micro-spatial scale in the city of Merida Yucatan. The epidemic curve is also shown for reference. The user can play the movie for the 52 epidemiologic weeks and also pause it or play it backwards to better understand the relation of the control actions, the transmission clusters and the epidemic curve.

Acknowledgments

Special thanks to Kenya Valdez, CENAPRECE and GDE working groups for their participation in the implementation of Dengue-GIS. We acknowledge the state coordinators that contributed with ideas and suggestions for its improvement. Most of all, we are grateful to the operative epidemiological and entomological surveillance personnel, as well as the vector control personnel which every day fight dengue in our country and record their activities

Author Contributions

Conceived and designed the experiments: JEHA RSL VSC. Performed the experiments: MHR VHRS SRP JASS. Analyzed the data: SRP VSC VHRS JASS. Contributed reagents/materials/analysis tools: VSC SRP VHRS. Wrote the paper: MHR JHEA RSL SRP.

References

1. WHO (2009) Dengue and dengue haemorrhagic fever. Available: http://www.who.int/mediacentre/factsheets/fs117/en/. Accessed 2011 Nov 19.
2. Shepard DS, Coudeville L, Halasa YA, Zambrano B, Dayan GH (2011) Economic impact of dengue illness in the Americas. The American journal of tropical medicine and hygiene 84: 200–207.
3. Secretaría de Salud (2010) Panorama Nacional de Dengue. Available: www.dgepi.salud.gob.mx. Accessed 2011 Nov 3.
4. San-Martín J, Brathwaite-Dick O (2007) La Estrategia de Gestión Integrada para la Prevención y el Control del Dengue en la Región de las Américas. Rev Panam Salud Publica 21.
5. Beatty ME, Stone A, Fitzsimons DW, Hanna JN, Lam SK, et al. (2010) Best practices in dengue surveillance: a report from the Asia-Pacific and Americas Dengue Prevention Boards. PLoS neglected tropical diseases 4: e890.
6. World Health Organization (2012) Global Health Atlas. Available: http://apps.who.int/globalatlas/. Accessed 2012 Feb 1.
7. Center for Disease Control, Health Map (2012) DengueNet. Available: http://www.healthmap.org/dengue/. Accessed 2012 Mar 3.
8. Sithiprasasna R, Patpoparn S, Attatippaholkun W, Suvannadabba S, Srisuphanunt M (2004) The geographic information system as an epidemiological tool in the surveillance of dengue virus-infected Aedes mosquitos. The Southeast Asian journal of tropical medicine and public health 35: 918–926.
9. Kusnanto H FA, Sunartono. Web-based integrated DHF surveillance system in Sleman district Yogyakarta Indonesia. Available: http://www.amic.org.sg/ict/external/awards/0402A5-L48_Interim_Report1.pdf. Accessed 2011 Nov 25.
10. Cuéllar-Luna LC-RM, Ramírez B, Álvarez-Valdez Á (2009) Geographical Information Systems and its use an Integrated Surveillance System for Prevention of Dengue at a Municipality of Habana City. 166–183.
11. Chalermkittichai S (2006) The application of Web-based Geographic Information System to control dengue Haemorrhagic Fever epidemic: A case study in Amphoe Makham. Amphoe Makham: Mahidol University.
12. OCDE (2005) Estudios de la OCDE sobre los Sistemas de Salud. México. París: OCDE.
13. Secretaría de Salud (1994) Norma Oficial Mexicana NOM-017-SSA2-1994, Para la vigilancia epidemiológica. Diario Oficial de la Federación: Gobierno Federal de México.
14. Secretaría de Salud (2008) Lineamientos para la vigilancia Epidemiológica de Fiebre por Dengue y Fiebre Hemorrágica por Dengue. In: Centro Nacional para la Prevención y Control de Enfermedades (CENAVECE), editor. México: Secretaría de Salud.
15. Secretaría de Salud (2011) NORMA Oficial Mexicana NOM-032-SSA2-2010, Para la vigilancia epidemiológica, prevención y control de las enfermedades transmitidas por vector. Diario Oficial de la Federación: Gobierno Federal de México.
16. Health Metrics Network, WHO (2005) Review of health information systems (HIS) in selected countries IV Mexico,. Available: http://www.who.int/entity/healthmetrics/library/mexico_05apr.doc. Accessed 2011 Oct 21.
17. Duncombe J, Clements A, Hu W, Weinstein P, Ritchie S, et al. (2012) Geographical information systems for dengue surveillance. The American journal of tropical medicine and hygiene 86: 753–755.
18. Fong JHR (1999) Application of middleware in the three tier client/server database design methodology. 6.
19. ESRI (2011) ArcGis Desktop Release 10. Redlands, CA: Environmental Systems Research Institute.
20. Instituto Nacional de Estadística y Geografía (2010) Marco Geoestadístico Nacional. 2010 ed. Mexico: Instituto Nacional de Estadística y Geografía.
21. Levine N (2010) Chapter 6 Hot Spot Analisis I. CrimeStat III: A spatial Statistics Program for the Analysis of Crime Incident Locations (Version 33). Houston, Tx: Ned Levine & Associates and the National Institute of Justice.
22. Secretaría de Salud (2007) Programa de Acción Específico de Dengue 2007–2012. In: Centro Nacional para la Prevención y Control de Enfermedades (CENAVECE), editor. México: Secretaría de Salud.
23. Mihalas GI (2008) Analysis of barriers in implementation of Health Information Systems - EFMI conference introductory address. Studies in health technology and informatics 134: 21–26.
24. Roman-Viñas R (2010) [Factors of success in the implementation of the technologies of the information and the communication in the health systems. The human factor]. Medicina clinica 134 Suppl 1: 39–44.
25. Chao DL, Halstead SB, Halloran ME, Longini Jr IM (2012) Controlling Dengue with Vaccines in Thailand. PLoS neglected tropical diseases 6: 1–11.
26. Johansson MA, Hombach J, Cummings DAT (2011) Models of the impact of dengue vaccines: a review of current research and potential approaches. Vaccine 29: 5860–5868.
27. Letson GW (2009) Dengue surveillance in preparation for field vaccine trials. Journal Of Clinical Virology: The Official Publication Of The Pan American Society For Clinical Virology 46 Suppl 2: S10–S12.
28. Halstead SB, Deen J (2002) The future of dengue vaccines. Lancet 360: 1243–1245.
29. Chang AY, Parrales ME, Jimenez J, Sobieszczyk ME, Hammer SM, et al. (2009) Combining Google Earth and GIS mapping technologies in a dengue surveillance system for developing countries. International journal of health geographics 8: 49.
30. Minnesota Uo (2011) MapServer 6.2.0 Documentation. Available: http://mapserver.org/documentation.html. Accessed 2013 Oct 1.
31. Neteler M, Mitasova H (2004) Open source GIS : a GRASS GIS approach. Boston: Kluwer Academic Publishers. xxvi, 401 p.
32. Bivand R, Pebesma EJ, Gómez-Rubio V (2008) Applied spatial data analysis with R. New York: Springer. xiv, 374 p.

24

GIS-Based Multi-Criteria Analysis for Arabica Coffee Expansion in Rwanda

Innocent Nzeyimana[1]*, Alfred E. Hartemink[2], Violette Geissen[1]

1 Soil Physics and Land Management Group, Wageningen University, Wageningen, The Netherlands, 2 Department of Soil Science, FD Hole Soils Lab, University of Wisconsin, Madison, Madison, Wisconsin, United States of America

Abstract

The Government of Rwanda is implementing policies to increase the area of Arabica coffee production. Information on the suitable areas for sustainably growing Arabica coffee is still scarce. This study aimed to analyze suitable areas for Arabica coffee production. We analyzed the spatial distribution of actual and potential production zones for Arabica coffee, their productivity levels and predicted potential yields. We used a geographic information system (GIS) for a weighted overlay analysis to assess the major production zones of Arabica coffee and their qualitative productivity indices. Actual coffee yields were measured in the field and were used to assess potential productivity zones and yields using ordinary kriging with ArcGIS software. The production of coffee covers about 32 000 ha, or 2.3% of all cultivated land in the country. The major zones of production are the Kivu Lake Borders, Central Plateau, Eastern Plateau, and Mayaga agro-ecological zones, where coffee is mainly cultivated on moderate slopes. In the highlands, coffee is grown on steep slopes that can exceed 55%. About 21% percent of the country has a moderate yield potential, ranging between 1.0 and 1.6 t coffee ha^{-1}, and 70% has a low yield potential (<1.0 t coffee ha^{-1}). Only 9% of the country has a high yield potential of 1.6–2.4 t coffee ha^{-1}. Those areas are found near Lake Kivu where the dominant soil Orders are Inceptisols and Ultisols. Moderate yield potential is found in the Birunga (volcano), Congo-Nile watershed Divide, Impala and Imbo zones. Low-yield regions (<1 t ha^{-1}) occur in the eastern semi-dry lowlands, Central Plateau, Eastern Plateau, Buberuka Highlands, and Mayaga zones. The weighted overlay analysis and ordinary kriging indicated a large spatial variability of potential productivity indices. Increasing the area and productivity of coffee in Rwanda thus has considerable potential.

Editor: Dafeng Hui, Tennessee State University, United States of America

Funding: The authors acknowledge NUFFIC of the Netherlands for funding the PhD training of Innocent NZEYIMANA at Wageningen University. The funders had no role in study design, data collection and analysis, decision to publish, or preparation of the manuscript.

Competing Interests: The authors have declared that no competing interests exist.

* Email: innocent.nzeyimana@wur.nl

Introduction

Coffee is one of the most important tradable commodities in the world and a major foreign-exchange earner in many developing countries [1]. Arabica coffee accounts for two-thirds of the global coffee market [2]. Coffee is a top export commodity and an important source of revenue in Eastern and Central African countries [3]. In some of these countries, such as Burundi, Uganda, Tanzania, Kenya, and the Democratic Republic of Congo, coffee is occasionally grown in association with agroforestry tree species for nitrogen fixation [4].

Rwanda produces mainly Arabica coffee, largely cultivated by smallholder farmers as mono-crop on plots of less than a hectare scattered on hilly slopes. In South and Central America, coffee is mostly grown on large monoculture plantations or under shade [5]. In Rwanda, coffee is predominantly grown along the shores of Lake Kivu in the west, on the plateau in the central part of Rwanda, and in the Mayaga region in the east [6]. Rwanda has ten agro-ecological zones: Imbo, Impara, Kivu Lake Borders, Birunga (volcano), Congo-Nile Watershed Divide, Buberuka Highlands, Central Plateau, Mayaga-Bugesera, Eastern Plateau, and Eastern Savanna. Details of the characteristics of the Rwandan agro-ecological zones can be found in [7].

The total area of arable and permanently cropped land in Rwanda is about 1.45 million ha [8], of which about 30 000 ha was under coffee production in 2005 and it increased to 41 762 ha in 2012 [3]. The total area under coffee production in the tropics is about 10.6 million ha [9]. The expansion of land for the production of coffee depends on three main factors: environmental conditions (e.g. topography, soil type, climate, and elevation), practices of agricultural land management, and genetic resources (i.e. coffee varieties) [10]. The growing conditions for Arabica coffee in Rwanda are characterized by an altitude of 1400–1900 m a.s.l., an annual rainfall of 1500–1600 mm, temperatures of 18–22°C, and an average amount of sunlight of 2200–2400 hours per year. Arabica coffee also requires fine-textured soils of at least one-meter with total porosities of 50–60%, a pH of 4.5–6.0, moderate to high sums of basic cations, and 2–5% organic matter [7].

In Rwanda, as in other developing countries, coffee farming is reserved for steep slopes and soils with low fertility [6]. Most of these lands have been degraded by soil erosion and are under

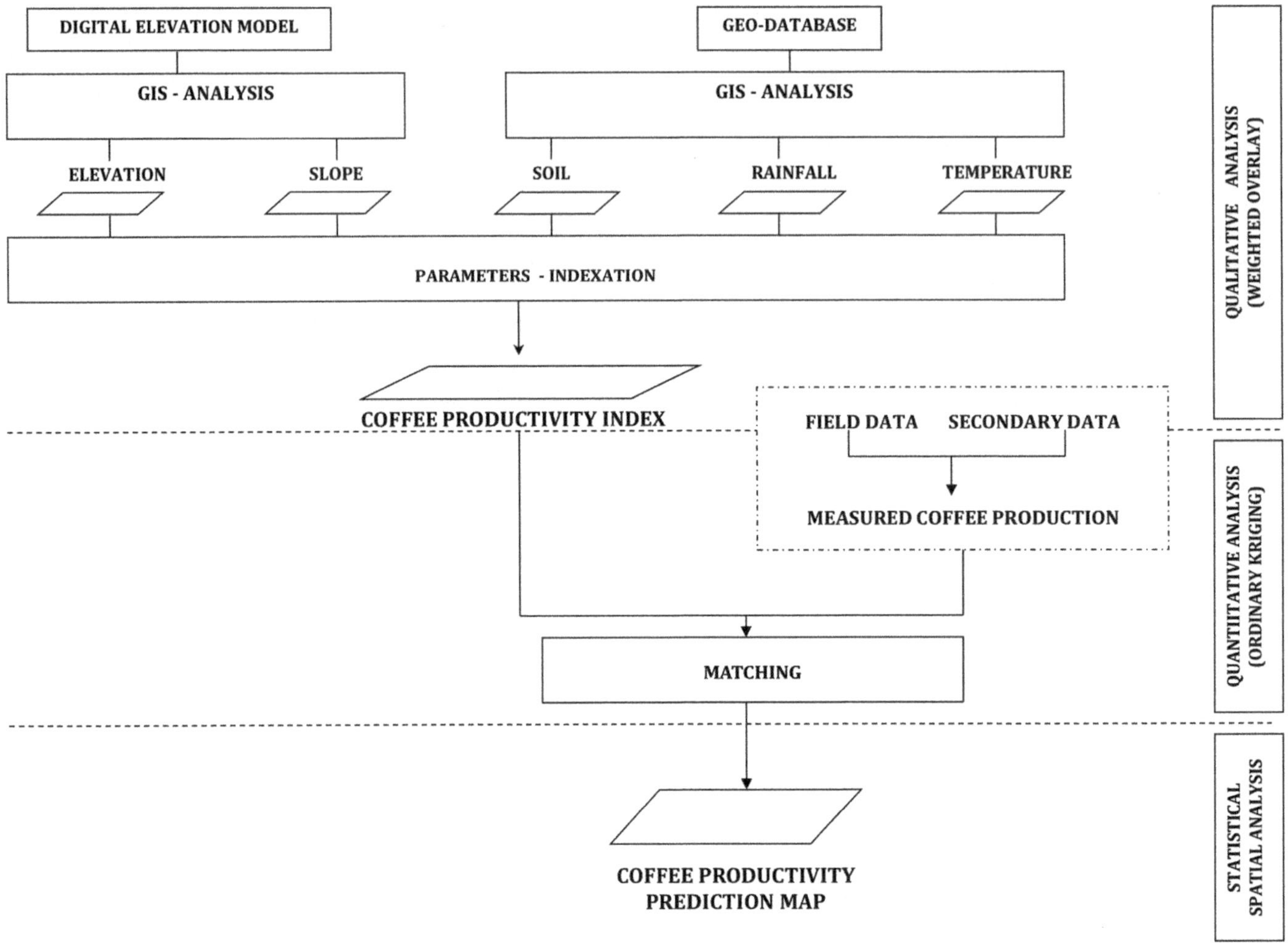

Figure 1. Flow chart of the methodology used to derive coffee productivity indices and predicted Arabica coffee yields.

pressure from intensive cropping by smallholder farmers. Fertile soils are usually reserved for growing staple food crops, which restricts coffee growing to soils of low fertility. In Rwanda, coffee yields above 2.8 t ha^{-1} are rare [6]. The Government of Rwanda has developed a set of policies for improving farmer livelihoods through an increase of sustainable coffee cultivation. A study of agricultural development in Africa has shown that successes are often linked to a cash-crop component and that food crops will profit as a consequence of improved cash income [11]. The identification of potential production zones for expanding coffee production and the prediction of coffee yields are needed to effectively implement these policies.

The evaluation of land is an essential procedure to assess opportunities, potentials, and limitations that a given parcel land can offer for agricultural purposes [12]. Various approaches of land evaluation with specific methodology have been developed to study land-use suitability [12,13]. Geographic information systems (GISs) have been used for mapping and analyzing land-use suitability [14]. Various GIS-based models have been developed by various researchers for land-use planning and suitability analysis [14–21]. The GIS-based models use geo-spatial and geo-statistical tools to assess the land units and to present the results as suitability maps. The models use multi-criteria evaluative approaches and methods by weights, values, or intensities of preference [14,22]. Weighted overlay analysis is one such approach of GIS modeling using spatial multi-criteria evaluation [23,24]. The objectives of this study were: (1) to analyze the spatial distribution of potential production zones for Arabica coffee production and the current productivity levels in the various zones and (2) to predict potential coffee yields and identify potential productivity zones. To achieve the objectives, we developed a model of land evaluation for expanding the production of Arabica coffee in Rwanda based on standard methodologies for land evaluation and geo-spatial analysis.

Material and Methods

Data acquisition

Digitized and tabulated data were assembled for the entire country, including 43 digital soil maps (scale 1:50 000), a digital elevation model (Shuttle Radar Topography Mission (SRTM) at 90×90 m resolution), the coffee production database for 2005,

Table 1. Weighted environmental criteria for evaluating the qualitative Arabica coffee productivity classes and scores (1 to 5 for worst to best)[a].

Environmental coffee productivity criteria							
Elevation (m)	Rainfall (mm)	Temp. (°C)	Soil type[b]	Slope (%)	Qualitative productivity class	Influence value[c] (%)	Score[d]
1600–1800	1400–1600	18–20	MOLL, AND	0–4	Very high	100	5
1400–1600 1800–2000	1200–1400 1600–1800	16–18 20–22	ALF	4–12	High	75	4
1200–1400	1000–1200 1800–2000	15–16 22–24	INCEPT, ULT	12–25	Moderate	50	3
1000–1200>2000	800–1000>2000	14–15 24–26	OX, ENT	25–50	Low	25	2
<1000	<800	<14>26	HIST, VERT	>50	Very low	0	1

[a]The table is a combination matrix that shows the level of productivity if we consider environmental productivity criteria of Arabica coffee production, as guided by [13].
[b]AND, Andisols; ALF, Alfisols; ENT, Entisols; INCEPT, Inceptisols; HIST, Histosols; MOLL, Mollisols; OX, Oxisols; ULT, Ultisols; VERT, Vertisols.
[c]The influence value represents the influence of the raster value compared to the other criteria as a percentage (i.e. 100, 75, 50, 25, or 0%).
[d]Each cell value in each input raster was assigned a new, reclassified score value on an evaluation scale of 1 to 5 (where 5 represented the best score and 1 the worst score 1) The scoring of the environmental productivity criteria of coffee was based on their importance as guided by [13].

general and administrative maps (2006 versions, scale 1:50 000), and a digital agro-climatic database containing temperatures, rainfall, altitudes, and agro-climatic zones. The database contained the amount of coffee produced and the number of trees in each administrative sector. Each sector was divided into cells, which are the lowest administrative units within the Republic of Rwanda.

Data processing: analysis of potential zones for coffee production

The methodology consisted of collecting (on a national scale) data such as soil type and slope gradient and analyzing the spatial distribution of coffee using the spatial-analysis toolset of the ArcGIS software [25]. The methodology aimed at identifying the major production zones based on soil and slope types. The combination of data for coffee distribution and soil was used for identifying the dominant soil types on which coffee is mainly produced and then for estimating the area coverage. The combination of data for coffee distribution and slope was for identifying the dominant slope types on which coffee is mainly produced and then for estimating the area coverage. The spatial distribution of coffee was identified, potential coffee production zones were characterized, and the sizes of the areas of coffee per slope and soil type in the various agro-ecological zones were estimated.

Multi-criteria analysis to estimate a qualitative Arabica coffee productivity index

The assessment of qualitative productivity indices for coffee essentially required the development of a GIS-based database for the optimal use of land resources for coffee. A geo-spatial database of data for elevation, slope, soil parameters, and rainfall and temperature extracted from the digital agro-climatic database, was generated in a GIS multi-criteria model (Figure 1). The landscape characteristics, climatic conditions, and soil parameters of a specific site are the most important determinants of land suitability [17]. The upper part of the flow chart in Figure 1 was thus used to analyze the qualitative productivity indices to indicate the level of productivity in the various agro-ecological zones. The multi-criteria model combined the different layers of data (i.e. elevation, slope, soil parameters, rainfall, and temperature) to identify the major production zones and their current productivity index (CPI). Data for photo synthetically active radiation were not available and so were not included in the model. The multi-criteria analysis used each input raster as a decision variable for sequential GIS interactions between layers. Data were processed using the spatial-analysis tools of ArcGIS [24,25]. The geo-spatial analysis then allowed the combination of the input rasters using weighted overlay analysis in the Model Builder component of ArcGIS to generate output rasters. Each cell value in each input raster was assigned a new, reclassified score value on an evaluation scale of 1 to 5, where 1 represents the lowest suitability and 5 the highest (i.e. scoring of the Arabica coffee requirements over others based on their importance as guided by [13]). Each of the new reclassified score was then weighted by assigning a percentage influence value (i.e. 100, 75, 50, 25, or 0%) (Table 1). This is achieved by multiplying the cell values (i.e. the new reclassified scores) by their percentage influence, and the results are added together to create the output raster. The new output-raster indices were then used as qualitative productivity indices. The weighted Z matrix can have the following form when m input factors and n criteria are considered:

$$Z = \begin{bmatrix} x_{01}..x._{0j}..x_{0n} \\ .... \\ .... \\ x_{i1}..x_{ij}..x_{in} \\ .... \\ .... \\ x_{m1}..x_{mj}..x_{mn} \end{bmatrix}$$

where Z is the combined-factor weighted matrix, m is the number of input factors, n is the number of criteria describing each input factor, x_{ij} is the score representing the level of importance of input factor i based on criterion set j of the Arabica coffee requirements. The score x_{ij} is assigned a percentage influence value according to the importance of the environmental coffee productivity factor within a single input raster as illustrated in Table 1 [24].

The combination of the output rasters is for understanding the influence of the combined environmental factors and/or each factor separately on coffee productivity. The high-productivity class has no significant limitations for sustainable coffee production. The moderate class is characterized by altitudes of 1200–1400 m, annual rainfall below 900 mm or above 2000 mm, and temperatures varying between 18 and 21°C. The low class is characterized by altitudes below 1200 or above 2000 m, annual rainfall below 800 or above 2000 mm, and temperatures below 10 or above 30°C [7,13].

Multi-criteria analysis to estimate the quantitative Arabica coffee productivity index

The qualitative productivity indices (low, moderate, and high) were then quantified with actual yields to generate quantitative Arabica coffee productivity indices using ordinary kriging. The qualitative indices were extrapolated to 121 sampled sites of actual Arabica coffee yields measured at various sites countrywide.

Actual Arabica coffee yield

Actual yields were collected at 121 farms countrywide. Smallholder coffee fields, particularly those near a coffee-washing station, were selected and monitored for yield. The coffee fields are private farms owned by smallholder farmers, technically supported by the National Agricultural Export Development Board (NAEB). In collaboration with the NAEB, the identified farmers participated voluntary in the selection of sample fields. No specific permissions were required for the field activities. In addition, the field studies did not involve any endangered or protected species.

The yields were measured by sampling three branches of coffee trees (low, middle, and high branches). Experimental plots for data collection were approximately 10×10 m and contained 25 coffee trees (i.e. 2500 trees ha^{-1}), each 2 m apart. The coffee trees were predominantly 20–25 years of age and were cropped as monocultures.

All sample sites were independently selected with equal probability. Five randomly selected trees in each plot were sampled by collecting a composite sample of 500 g of good berries from the three branches weekly from April to September 2009. The coffee berries were cleaned, oven-dried at 60°C for 48 h, adjusted to 12% moisture content, and weighed. Grain yield was determined on each randomly selected tree, and a spatial mean plot yield was calculated as:

$$\bar{y} = \frac{1}{n}\sum_{i=1}^{n} y_i \tag{1}$$

where $\bar{y}$ (t ha^{-1}) is the average yield for 2009, y_i (t ha^{-1}) is the yield at sample site i, and n is the number of sample sites.

Predicted Arabica coffee yield

Ordinary kriging analysis was conducted to predict potential yields and to identify potential productivity zones for Arabica coffee [24,25], based on actual yields measured at the study site.

Table 2. Distribution of Arabica coffee areas (ha) and yields (t ha^{-1}) calculated from the coffee database for 2005 for the ten agro-ecological zones of Rwanda.

AEZ No.	Agro-Ecological Zone (AEZ)	Total area (ha)	Area with scattered coffee trees[a] (ha)	Normalized area[b] (ha)	Range of estimated dry yield[c] (t ha^{-1})	Mean dry coffee yield ± SD[d] (t ha^{-1})
1	Imbo (*Lake Kivu region*)	15 832	15 678	804	0.5–2.6	1.2±0.68
2	Impara (*Lake Kivu region*)	64 954	58 532	3376	0.5–2.6	1.2±0.69
3	Kivu Lake Borders (*Lake Kivu region*)	73 593	70 422	2947	0.3–3.5	1.6±0.79
4	Birunga/Volcano	90 887	1952	65	0.5–2.1	1.0±0.74
5	Congo-Nile Watershed Divide	391 930	136 946	4024	0.3–3.5	1.2±0.91
6	Buberuka Highlands	177 154	81 622	1130	0.3–2.8	0.8±0.53
7	Central Plateau & Granitic Ridge	529 772	461 743	10 155	0.3–2.8	0.8±0.47
8	Mayaga-Bugesera (*eastern region*)	223 573	166 085	3328	0.5–1.8	1.0±0.41
9	Eastern Plateau (*eastern region*)	381 367	350 233	5398	0.3–2.2	0.8±0.65
10	Eastern Savana (*eastern region*)	479 761	134 125	692	0.5–2.2	1.1±0.49
	Total	2 564 255	1 162 338	31 921		1.0±0.65

[a]This area is calculated as the area for each sector in the AEZ; only the sector area in the AEZ is extracted by spatial-analysis tools. The 2005 Rwanda coffee database displays only the number of coffee trees and coffee production per sector. Each sector is an administrative entity divided into cells, which are the lowest administrative units within the Republic of Rwanda.
[b]This area is extracted from the area with scattered coffee trees in each AEZ and is calculated using the standard tree density of 2500 trees ha^{-1}, i.e. b = a/2500.
[c]This yield is calculated by averaging the yields for each part of the sector in the AEZ (i.e. sector yield is calculated as the production of each sector divided by the number of trees using the standard spacing of 2× 2 m, or 2500 trees ha^{-1}).
[d]This yield is calculated using SPSS descriptive statistics; the normality of the data was determined using the Kolmogorov-Smirnov test.

Table 3. Distribution of soil types (ha) and areas of Arabica coffee cultivation in the ten agro-ecological zones of Rwanda.

Agro-Ecological Zone (AEZ)	AEZ label No.	Soil/Coffee coverage	Area per soil type/Area covered by coffee per soil type (ha)[a]								Total	
			ALF	AND	ENT	HIST	INCEPT	MOLL	OX	ULT	ha[b]	%[c]
Buberuka Highlands	6	Soil	2000	99	17 838	10 545	45 835	567	10 492	81 901	167277	7
		Coffee	21	-	157	19	385	7	73	456	1118	3
Central Plateau & Granitic Ridge	7	Soil	68 244	373	47 504	2 219	164 855	1634	48 054	195 007	527 890	23
		Coffee	1306	5	876	30	3286	34	778	3722	10 036	31
Mayaga-Bugesera *(eastern region)*	8	Soil	14 564	-	13 650	28 515	34 038	5768	75 141	37 048	208 722	9
		Coffee	288	-	206	392	599	86	1182	569	3322	10
Eastern Plateau *(eastern region)*	9	Soil	22 782	-	66 166	10 652	76 506	20 420	99 118	76 238	371 881	16
		Coffee	311	-	910	161	1111	294	1459	1021	5268	16
Eastern Savana *(eastern region)*	10	Soil	10 555	-	48 331	38 421	89 951	20 235	186 596	26 586	420 676	18
		Coffee	21	-	118	58	303	26	287	57	870	3

Data were extracted from the Rwanda soil dataset and analyzed using the geo-spatial tools of ArcGIS.
[a]AND, Andisols; ALF, Alfisols; ENT, Entisols; INCEPT, Inceptisols; HIST, Histosols; MOLL, Mollisols; OX, Oxisols; ULT, Ultisols; VERT, Vertisols.
[b]Total Rwanda soil and Arabica coffee coverage per agro-ecological zone.
[c]Soil and Arabica coffee coverage in percentage per agro-ecological zone over total Rwanda soil area and Arabica coffee area, respectively.

The ordinary kriging is one of the mostly used geo-statistical methods, quite efficient and accurate for spatial prediction and interpolation [26]. The prediction of yield was based on the qualitative productivity indices validated over the actual yields. A variogram was estimated using Matheron's estimator [27,28]:

$$y^\wedge(h) = \frac{1}{2Mh} \sum_{i=1}^{Mh} \{Z(x_i) - Z(x_i + h)\}^2 \quad (2)$$

where Z (x_i) is the actual yield measured at the study site (x_i), h is the lag, i.e. both distance and direction between the sample sites, M_h is the pair of sample sites separated by lag h, and $\gamma^\wedge(h)$ is the semi-variance at lag h.

To assess the spatial correlation of the yields, prediction accuracy was calculated by comparing expected yields, $Z^\wedge(CYI_j)$, with actual yields measured at the validation sites, *(n)* - $Z^*(CYI_j)$, and to assess a systematic error, calculated as the mean prediction error (MPE) [29]:

$$MPE = \frac{1}{n} \sum_{j=1}^{1} [Z \wedge (CYI_j) - Z(CYI_j)] \quad (3)$$

where CYI is the coffee yield index, $Z^\wedge(CYI_j)$ is the expected yield index generated from the qualitative analysis, and $Z^*(CYI_j)$ is the actual yield measured at the validation sites (n). The validation set accounted for 121 sample sites. The accuracy of prediction was calculated as a root mean square error (RMSE) of prediction [29]:

$$RMSE = \sqrt{\frac{1}{n}} \sum_{j=1}^{1} [Z \wedge (CYI_j) - Z(CYI_j)]^2 \quad (4)$$

The RMSE is a measure of fitness of the prediction curve; the smaller the RMSE, the better the prediction. Ordinary kriging uses and compares different fitting models that perform the analysis, reduce uncertainty, and produce the best prediction map. The RMSE is thus standardized by considering the total variance of the observed values and is then termed the root mean square standardized error (RMSSE) or the mean standardized error (MSE). The RMSSE and the MSE were estimated from the variances between the observed values, i.e. the actual yields measured at the study site [24,29]:

$$RMSSE = \frac{\frac{1}{n} \sum_{j=1}^{1} [Z \wedge (CYI_j) - Z * (CYI_j)]^2}{s^2} \quad (5)$$

where s^2 is the total variance of the CYI at the sample site.

A satisfactory accuracy of prediction has an MSE close to zero and an RMSSE close to unity [24,29]. If the RMSSE exceeds unity, the model underestimates the variability at the validation sites, and thus the prediction is unsatisfactory [24,29,30].

The normality of the measured yield data was determined with Kolmogorov-Smirnov test. All data were normally transformed to meet the assumption of normality by comparing different types of model fitting (exponential and Gaussian) for the analysis, and only the model with the smallest Akaike information criterion (AIC) was adopted. The AIC is a measure of how well a model fits the empirical data; the smaller the AIC, the better the fit [24]. In addition, cross-validation, comparing the predicted values with the measured values, checked the quality of the predicted values [31].

Table 4. Distribution of soil types (ha) and areas of Arabica coffee cultivation in the ten agro-ecological zones of Rwanda (Cont'd).

Agro-Ecological Zone (AEZ)	AEZ label No.	Soil/Coffee coverage	Area per soil type/Area covered by coffee per soil type (ha)[a]								Total	
			ALF	AND	ENT	HIST	INCEPT	MOLL	OX	ULT	ha[b]	%[c]
Imbo (*Lake Kivu region*)	1	Soil	3349	-	268	-	6489	-	-	3910	14 017	0.6
		Coffee	197	-	16	-	382	-	-	230	825	3
Impara (*Lake Kivu region*)	2	Soil	11 831	-	138	1269	10 111	-	363	40 436	64 147	3
		Coffee	657	-	7	75	558	-	1	2109	3407	11
Kivu Lake Borders (*Lake Kivu region*)	3	Soil	8526	91	12 708	35	25 248	258	253	25 469	72 590	3
		Coffee	328	-	539	2	1085	11	11	1045	3020	9
Birunga/Volcano	4	Soil	1782	47 176	15 450	79	2182	4222	-	3174	74 065	3
		Coffee	16	23	-	-	5	7	-	15	65	0.2
Congo-Nile Watershed Divide	5	Soil	8879	11 613	23 396	5388	124 675	2107	11 341	203 960	391 359	17
		Coffee	112	-	330	8	1553	-	17	2004	4024	13
		Subtotal[d]										
		(ha)	152 513	59 352	245 450	97 123	579 890	55 211	431 358	693 728	2 314 625	
		Subtotal[d]										
		(%)	7	3	11	4	25	2	19	30		100
		Subtotal[e]										
		(ha)	3257	27	3159	743	9266	465	3809	11 228	31 954	
		Subtotal[e]										
		(%)	10	0.1	10	2	29	1	12	35		100

Data were extracted from the Rwanda soil dataset and analyzed using the geo-spatial tools of ArcGIS (cont'd); See Table 4 for the notes.

[d]Subtotal of soil area per soil type.

[e]Subtotal of Arabica coffee area per soil type.

Table 5. Distribution of slope classes (%) and areas with Arabica coffee in the ten agro-ecological zones of Rwanda.

Agro-Ecological Zone (AEZ)	AEZ label No.	Slope/Coffee coverage	Area per slope category/Area covered by coffee per slope category (ha)			Total	
			<25%	25–55%	>55%	(ha)	(%)[a]
Imbo (*Lake Kivu region*)	1	Slope	10 863	4652	4705	20 220	1
		Coffee	554	238	11	803	3
Impara (*Lake Kivu region*)	2	Slope	45 321	18 165	1407	64 893	3
		Coffee	2417	871	57	3345	10
Kivu Lake Borders (*Lake Kivu region*)	3	Slope	47 780	25 400	413	73 593	3
		Coffee	1911	1016	17	2944	9
Birunga/Volcano	4	Slope	77 170	12 396	1226	90 792	4
		Coffee	59	6	-	65	0.2
Congo-Nile Watershed Divide	5	Slope	229 902	157 739	4258	391 899	16
		Coffee	2059	1922	46	4027	13
Buberuka Highlands	6	Slope	81 303	89 874	5814	176 991	7
		Coffee	435	633	50	1118	3
Central Plateau & Granitic Ridge	7	Slope	369 316	155 345	5070	529 731	22
		Coffee	7212	2944	105	10 261	32
Mayaga-Bugesera (*eastern region*)	8	Slope	214 946	8122	19	223 087	9
		Coffee	3168	152	-	3320	10
Eastern Plateau (*eastern region*)	9	Slope	316 686	63 065	1586	381 337	16
		Coffee	4462	903	24	5389	17
Eastern Savana (*eastern region*)	10	Slope	453 244	25 987	202	479 433	20
		Coffee	641	57	-	698	2
Rwanda - Slope(ha)		Subtotal	1 846 532	560 746	24 701	2 431 979	
(%)			76	23	1		100
Rwanda - Coffee (ha)		Subtotal	22 917	8743	310	31 970	
(%)			72	27	1		100

Data extracted from the digital elevation model (Shuttle Radar Topography Mission at 90×90 m resolution) and analyzed using the geo-spatial tools of ArcGIS.
[a]Slope and coffee coverage in percentages per agro-ecological zone of the total Rwanda slope and coffee areas, respectively.

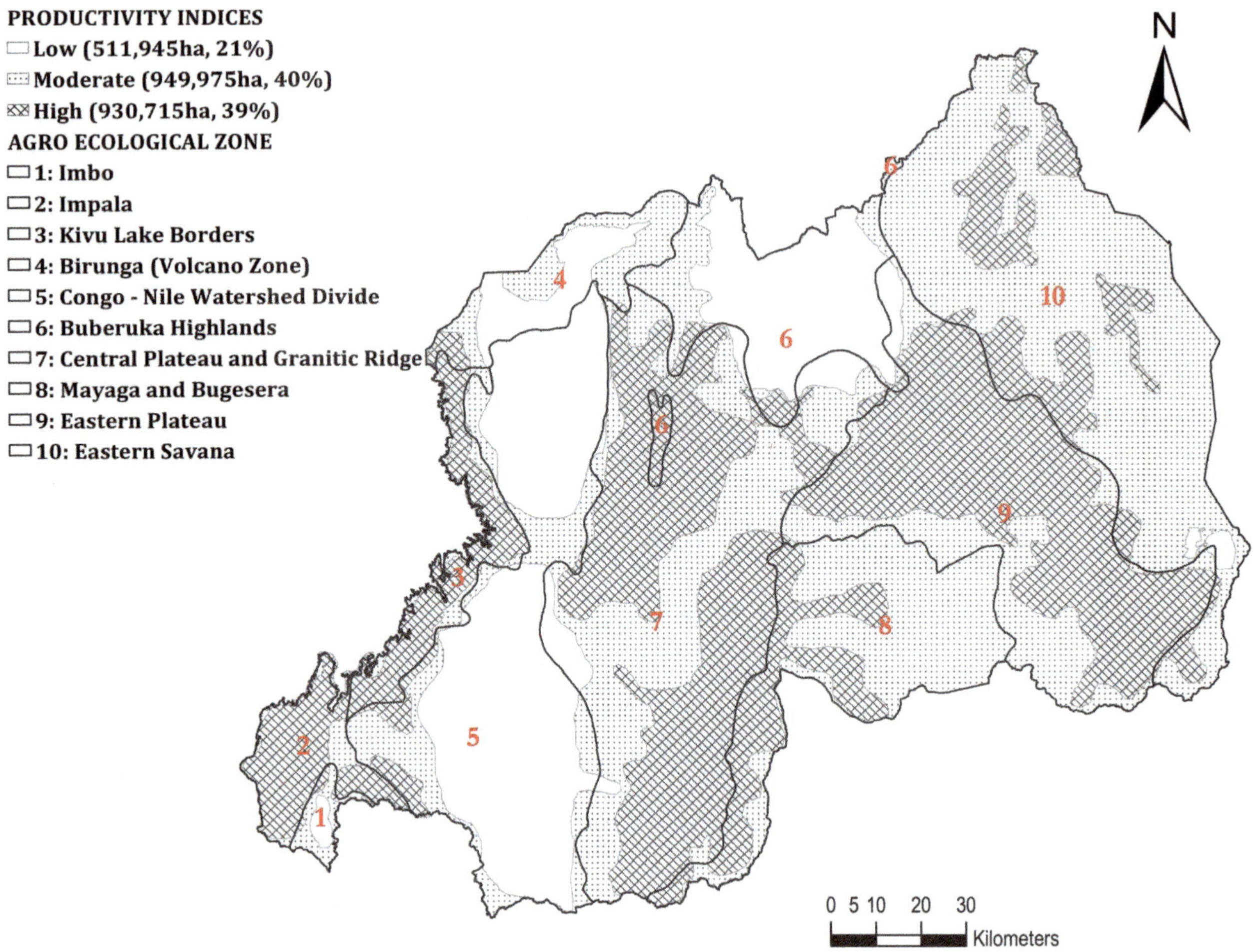

Figure 2. Qualitative Arabica coffee productivity indices (low, moderate, and high) generated by combining factors (elevation, slope, soil type, rainfall, and temperature) using weighted overlay analysis in the ten agro-ecological zones.

Results

Spatial distribution of Arabica coffee and biophysical characterization

The estimated area of Arabica coffee production in Rwanda was about 32 000 ha in 2005, compared to 30 000 ha reported by [3]. This area represents about 2.3% of the total area under agriculture. The Central Plateau had the largest area of coffee production, covering about 32% (10 261 ha) of the total area under coffee cultivation. The Central Plateau was characterized by coffee yields of 0.3–2.8 t ha^{-1} (Table 2). This zone has a wide range of soil types and landscapes. The soils where coffee is cultivated included Alfisols, Inceptisols, and Ultisols, representing 4% (1306 ha), 10% (3286 ha) and 12% (3722 ha), respectively, of the total area under coffee (Tables 3 and 4). Cultivated areas in the Central Plateau and the Granitic Ridge agro-ecological zones are also characterized by moderate (<25%) and steep (25–55%) slopes that cover 23% (7212 ha) and 9% (2944 ha) of the total cultivated area, respectively (Table 5). Coffee productivity is mainly limited by infertile soils derived from schistose and granitic materials. Slopes above 25% affect the productivity of the region due to soil erosion.

The Lake Kivu region (Imbo, Impara, and Kivu Lake Borders zones) in the western province had the highest yields ranging between 0.3 and 3.5 t ha^{-1}, with a mean of 1.6 t ha^{-1}. This region contained 22% (7127 ha) of the total area cropped with coffee (Table 2). The dominant soil types in the region are Inceptisols and Ultisols, representing ~6% (2025 ha) and 11% (3384 ha), respectively, of the total area devoted to coffee production (Table 4). Arabica coffee in the region is dominantly cultivated on moderate (<25%) and steep slopes (25–55%) that cover 15% (4882 ha) and 6% (2125 ha), respectively, of the areas under coffee (Table 5). The region is characterized by environmental conditions favorable to coffee production.

Yields in the Birunga (volcano) agro-ecological zone ranged between 0.5 and 2.1 t ha^{-1}, with a mean of 1.0 t ha^{-1} (Table 2). The extent of coffee in the zone covered only 65 ha of the land, mainly on Alfisols (16 ha), Andisols (23 ha), and Ultisols (15 ha) (Table 4). Andisols are fertile and productive soils, so the farmers will prefer annual crops over perennial crops such as coffee. Coffee is mainly grown on moderate slopes (Table 5). The effective depth of the soil, dominated by Andisols, is the main factor limiting coffee productivity in the zone.

The Eastern Plateau, Eastern Savanna, Mayaga, and Bugesera zones (i.e. the eastern region) together covered ~30% (9418 ha) of the total area of coffee production. Yields in the eastern region ranged between 0.3 and 2.2 t ha^{-1}, with a mean of 1.0 t ha^{-1} (Table 2). The dominant soil types are Inceptisols, Oxisols, and Ultisols, covering ~6% (2013 ha), 9% (2928 ha), and 5% (1647 ha), respectively, of the area under coffee (Table 3). Coffee

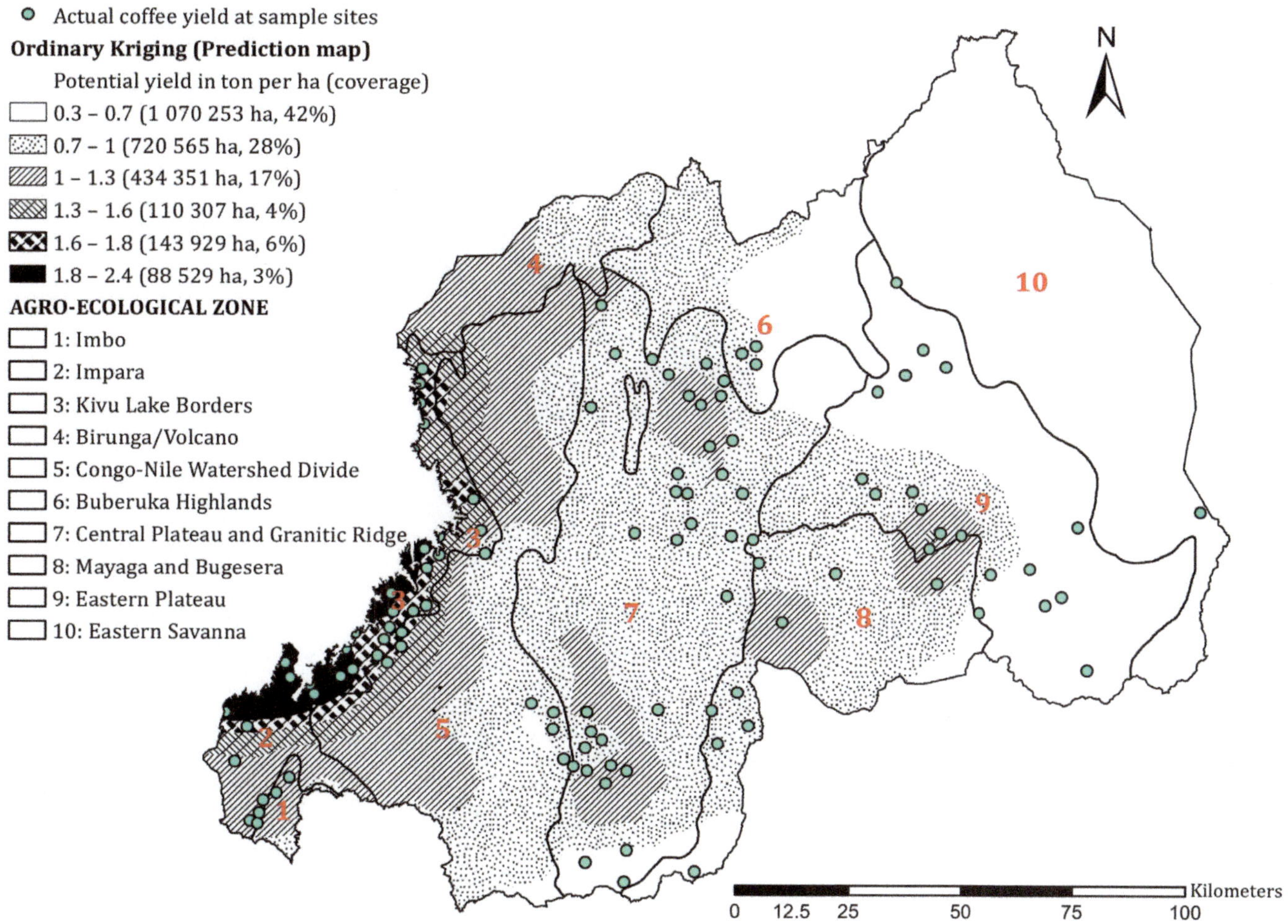

Figure 3. Potential Arabica coffee yield (t ha^{-1}) predicted using ordinary kriging in the ten agro-ecological zones based on actual yields (t ha^{-1}) measured at sample sites.

is cultivated on moderate slopes (<25%) that cover more than 26% (8271 ha) of the area with coffee cultivation (Table 5). The dominant infertile soils of the region, very high temperatures, and low rainfall offer limited conditions for coffee productivity.

The Buberuka Highlands and the Congo-Nile Watershed Divide agro-ecological zones are classified as the highlands of the country (above 2000 m a.s.l.). In both zones, coffee was cultivated on ~16% (5154 ha) of the total coffee area, mainly on moderate (2494 ha, ~8%) and steep (2555 ha, ~8%) (Table 5). Yields in the highlands ranged between 0.3 and 3.5 t ha^{-1}, with a mean of 1.2 t ha^{-1} (Table 2). Inceptisols and Ultisols are the main soil types, representing ~6% (1938 ha) and 8% (2460 ha), respectively, of the total area under coffee cultivation (Table 3). Very low temperatures and heavy rainfall limit the productivity of coffee cultivation in the highlands.

Coffee productivity indices

Qualitative productivity indices were generated based on soil type, elevation, slope, rainfall, and temperature using weighted overlay analysis (Figure 2). The analysis identified three zones with high, moderate, and low productivity indices representing ~930 715 (39%), 949 975 (40%), and 511 945 ha (21%), respectively. Approximately 80% of the total area of the country had moderate to high production potential for Arabica coffee. Zones with high potential productivity indices had fertile soils, moderate slopes and altitudes, and favorable climates. The zones with low productivity indices were mainly at high altitudes with high rainfall and low temperatures. The semi-dry eastern regions, where Oxisols and Ultisols are the dominant soil types (Figure 5), have zones with low indices.

High predicted yields ranged between 1.6 and 2.4 t ha^{-1} along the shores of Lake Kivu and in the Imbo zone (Figure 3). The calculated yields varied between 0.3 and 3.5 t ha^{-1} (Table 2). The prediction map for the country (Figure 3) shows coffee yields varying between 0.3 and 2.4 t ha^{-1}. Eighty percent of the country had low yield potentials of 0.3–1 t ha^{-1}, whereas 21% of the country had moderate yield potentials of 1.0–1.6 t ha^{-1}. The national average yield was predicted to be 1.12 t ha^{-1}, and the measured yield (n = 121 sampled sites) was 1.1 t ha^{-1} y^{-1}. The correlation between the measured and predicted yields indicated that the prediction model was satisfactory (Coefficient of determination $R^2 = 0.73$) (Figure 4).

Discussion

Some general information on the suitability of coffee cultivation in Rwanda was available. The spatial variation of coffee production can be explained by the growing conditions, which include biophysical factors such as soil type and properties, parent material, altitude, slope, and climatic conditions. For example, the

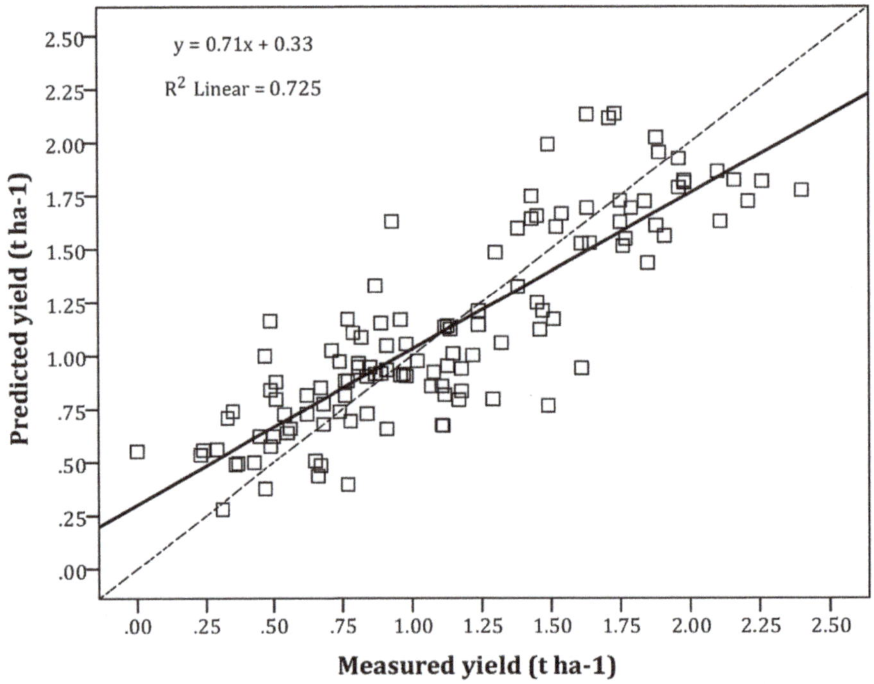

Figure 4. Relationship between measured and predicted Arabica coffee yields – cross validation using ordinary kriging (Predicted Arabica coffee yield index – CYI (t ha^{-1}) = 0.71x + 0.33; Mean Prediction Error – MPE = 0.0187; Root Mean Square Prediction Error – RMSE = 0.278; Root Mean Square Standardized Prediction Error – RMSSE = 0.99; Mean Standardized Prediction Error - MSE = 0.036; Coefficient of determination – R^2 = 0.73; Average Standard Error –Avg. SE = 0.291; Sample points, n = 121).

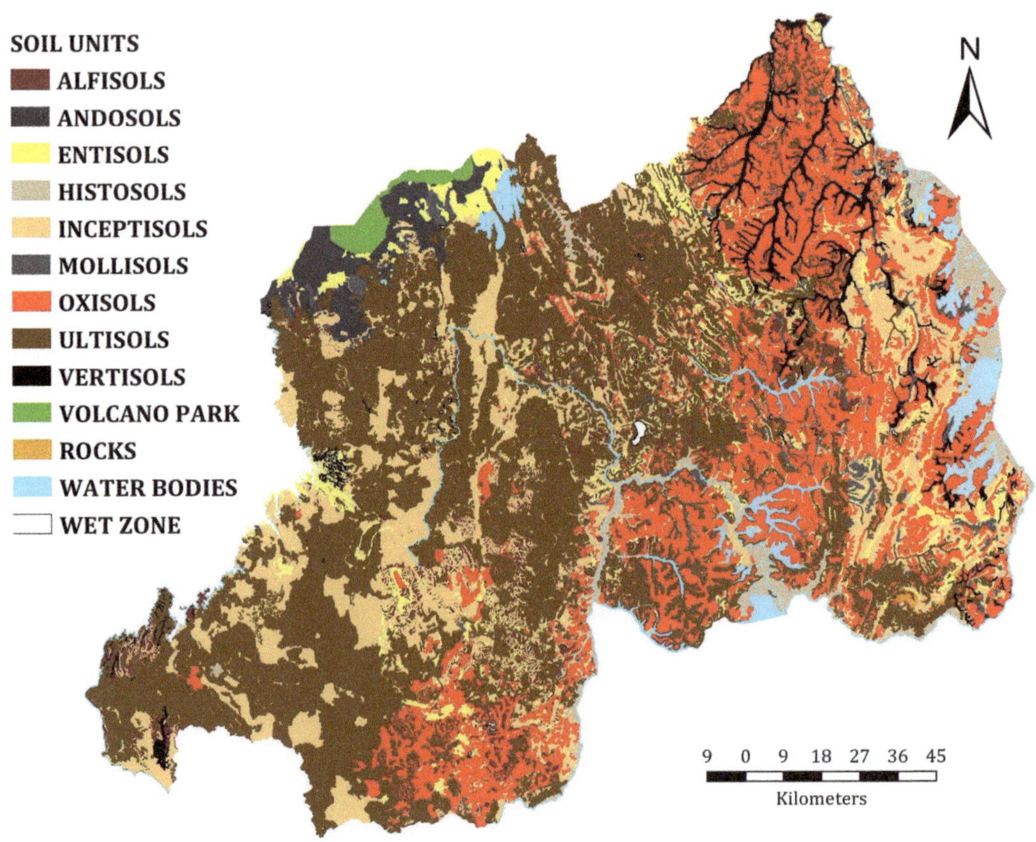

Figure 5. Soil map of Rwanda. Soils are classified using the USDA Soil taxonomy (Source: Data collected from the Ministry of Agriculture and Animal Resources, using the Rwanda soil database) after [32].

yield in the Central Plateau and Granitic Ridge zones was limited by soil acidity and the gravel as these soils developed on granitic and schistose materials. On such soils, erosion on slopes above 25% affects the productivity. Similar observations have been made by [33,34] who indicated that topography can combine with various environmental factors to influence productivity. Topography affects the climate (e.g. variations in temperature and humidity), distribution of soil moisture, soil organic-matter content, soil nutrients, soil textural composition, and soil physical properties, which affect crop growth and yield in a field. The spatial variability of environmental factors can contribute to variability in crop performance, and topography is a vital variable in predicting the spatial variability of crop yields [35].

The multi-criteria overlay analysis used to assess spatial variation in coffee productivity revealed that the Lake Kivu region is an area of high productivity. Similarly, ordinary kriging indicated expected high yields. The Lake Kivu region has favorable soil types and slopes with abundant rainfall and moderate temperatures for the optimal production of coffee. The region has alluvial and very fine clayey soils, developed from basalt, and a high agricultural potential [7].

The Birunga (volcano), Central Plateau, and Buberuka Highlands' zones, the highlands of the Congo-Nile Watershed Divide zone, and the eastern region are areas with both moderate and low productivity indices. The productivity in the Central Plateau zone is mainly limited by infertile soils derived from schistose and granitic materials on the moderately sloped and eroded hillsides. The productivity in the Birunga (volcano) zone is mainly limited by the soil depth (<50 cm) and low temperatures that are sub-optimal for coffee production. High yields were expected in this zone based on the high fertile volcanic soils. Instead, low potential yields were predicted by the kriging model, due to the unsuitable climatic conditions that affect the development and maturity of the berries [36]. In the highlands, mainly in the Buberuka Highlands and Congo-Nile Watershed Divide zones, the production of coffee is limited by very low temperatures, heavy rainfall, and steep slopes that could influence the depletion of soil fertility and reduce yields due to water erosion on the hills. Productivity in the eastern regions is limited by infertile soils and the very high temperatures and low rainfall. Our study demonstrated a decrease in yield in very dry conditions that coincided with lower elevations. Coffee is also constrained by very cold temperatures in the highlands that are often cloudy with low solar irradiation and heavy rainfall. Similar trends of the influence of topography on potato yields have also been reported by [37]. Both the ordinary kriging analysis and the multi-criteria factor analysis thus performed well in assessing and predicting potential yields of coffee. The performance of weighted overlay analysis has also been assessed in cotton by [21]. The relationships of soil, elevation, slope, aspect, and curvature with the stability of crop yields were assessed by [38], who deemed ordinary kriging the best method to estimate crop yield as a function of topography and landscape positions.

A wide range of yields of coffee in Rwanda, varying between 0.8 and 2.8 t ha^{-1} of dry parchment coffee, has also been reported by [6]. The low yields were attributed to coffee variety, agro-ecological conditions, the lack of mineral and organic fertilization, and limited mulching [6]. The national average yield of coffee is estimated at 1.1 t ha^{-1} y^{-1} (this study), but yields above 2.8 t ha^{-1} for dry coffee are rare even with adequate fertilization and sustained crop management [6]. In Uganda, 1.2 t ha^{-1} y^{-1} of dry coffee were recorded for mono-cropped coffee and coffee-banana intercropping systems [39]. The spatial variation in coffee production and productivity are thus mainly influenced by soil properties, soil management, farming practices, and climatic conditions. Similar trends have also been reported by [34,35].

Conclusions

The multi-criteria analysis used to assess spatial variation in potential production zones and the productivity of coffee revealed that agro-ecological factors are largely determined suitable zones of coffee productivity. The spatial variation of coffee productivity in the agro-ecological zones was considerable and was influenced by soil properties, soil management, farming practices, and climatic conditions. High production potentials indicated that smallholder farmers could generate income from coffee and could thus improve their livelihoods. In addition, this may provide an opportunity for farmers to purchase more land and extend the area for the production of coffee.

This study demonstrated that both ordinary kriging analysis and multi-criteria weighted overlay analysis performed well for analyzing the spatial distribution and productivity of coffee and for predicting yield. The depletion of soil fertility due to the lack of erosion control in scattered coffee plots on steep slopes is a major factor limiting coffee productivity in Rwanda. The sustainability of coffee productivity could be ensured by intensifying the use of fertilizers, mainly a well-balanced combination of lime, nitrogen, phosphorus, potassium, zinc, and boron. Limited access to financial resources restricts the purchase of these inputs and the use of different types of mulches that can improve soil properties and reduce the erodibility of the soil is recommended.

Acknowledgments

We thank the Ministry of Agriculture and Animal Resources for providing the digital soil datasets and the National Agricultural Export Development Board (NAEB), former OCIR – CAFÉ, for providing the coffee database and facilitation for field sampling. We also thank the GIS Centre of the National University of Rwanda for providing various spatial datasets and office space. We are very grateful to the NUFFIC for funding the PhD training of Innocent NZEYIMANA at Wageningen University, the Netherlands, and to Mr. Huting of ISRIC and Mrs Mukashema of the GIS Centre for their support in the analysis of the spatial data.

Author Contributions

Conceived and designed the experiments: IN. Performed the experiments: IN. Analyzed the data: IN. Contributed reagents/materials/analysis tools: IN. Wrote the paper: IN AEH VG. Supervision of the research project: AEH VG.

References

1. Ponte S (2002) The "latte revolution"? Regulation, markets and consumption in the global coffee chain. World development 30: 1099–1122.
2. Labouisse JP, Bellachew B, Kotecha S, Bertrand B (2008) Current status of coffee (*coffea Arabica L.*) genetic resources in Ethiopia: implications for conservation. Genetic Resource Crop Evolution 55: 1079–1093.
3. FAO (2014) FAOSTAT - Statistical database. Food and agriculture Organization of the United Nations. Available: http://faostat.fao.org/site/567/DesktopDefault.aspx?PageID=567#ancor. Accessed 2014 Apr 02.
4. Boffa JM, Turyomurugyendo L, Barnekow-Lilleso JP, Kindt R (2009) Enhancing farm tree diversity as a means of conserving landscape-based biodiversity. Mountain Research and Development 25: 212–217.
5. Van Oijen M, Dauzat J, Harmand JM, Gerry L, Vaast P (2010) Coffee agroforestry systems in Central America: II. Development of a simple process-based model and preliminary results. Agroforestry Systems 80.
6. Nzeyimana I, Hartemink AE, de Graaff J (2013) Coffee farming and soil management in Rwanda. Outlook on Agriculture 42 (1): 47–52. Doi 10.5367/oa.2013.0118.

7. Verdoodt A, Van Ranst E (2003) Land evaluation for agricultural production in the tropics: A large – scale land suitability classification for Rwanda. Ghent: Laboratory of soil science, Ghent University. ISBN: 90-76769-89-3.
8. FAO (2014b) FAOSTAT - Statistical database. Food and Agriculture Organization of the United Nations. Available: http://faostat.fao.org/site/666/default.aspx. Accessed 2014 Apr 02.
9. Clay J (2004) Coffee. In: Clay J, editor. World Agriculture and the Environment. Washington: Island press.
10. Bosselmann AS, Dons K, Oberthur T, Olsen CS, Ræbild A, et al. (2009) The influence of shade trees on coffee quality in small holder coffee agroforestry systems in southern Colombia. Agriculture, Ecosystems and Environment 129: 253–260.
11. Gabre-Madhin EZ, Haggblade S (2004) Successes in African Agriculture: results of an expert survey. World development 32: 745–766.
12. Rossiter DG (1996) A theoretical framework for land evaluation. Geoderma 72: 165–190.
13. FAO (1976) A Framework for Land Evaluation. Rome: Soil Resources Development, Land & Water Development Division. FAO Soils Bulletin No 32. 220 p. ISBN: 92-5-100111-1.
14. Malczewski J (2004) GIS-Based land use suitability analysis: a critical overview. Progress in Planning 62: 3–65.
15. De La Rosa D, Moreno JA, Garcia LV, Almorza J (1992) MicroLEIS: a Micro Computer-Based Mediterranean Land Evaluation Information System. Soil Use and Management 8: 89–96.
16. Maji AK, Krishna NDR, Challa O (1998) Geographical Information System in analysis and interpretation of soil resources data for land use planning. Journal of Indian Society of Soil Science 46: 260–263.
17. Coleman AL, Galbraith JM (2000) Using GIS as an Agricultural Land Use Planning Tool. Blacksburg. Department of Crop and Soil Environmental Science, Virginia Tech. Available: http://scholar.lib.vt.edu/ejournals/vaes/00-2.pdf. Accessed 2014 Feb 11.
18. Maji AK, Nayak DC, Krishna NDR, Srinivas CV, Kamble K, et al. (2001) Soil information system of Arunachal Pradesh in a GIS environmental for land use planning. International Journal of Applied Earth Observation and Geo-Information 3 (1): 69–77.
19. Joerin F, Theriault M, Musy A (2001) Using GIS and outranking multi-criteria analysis for land use suitability assessment. International Journal of Geographical Information Science 10 (8): 321–339.
20. Li B, Zhang F, Zhang L-W, Huang J-F, Jin Z-F, et al. (2012) Comprehensive Suitability Evaluation of Tea Crops Using GIS and a Modified Land Ecological Suitability Evaluation Model. Pedosphere 22 (1): 122–130.
21. Walke N, Obi Reddy G, Maji A, Thayalan S (2012) GIS-based multicriteria overlay analysis in soil-suitability evaluation for cotton (Gossypium spp.): A case study in the black soil region of Central India. Computers & Geosciences 41: 108–118.
22. Malczewski J (1996) A GIS-based approach to multiple criteria group decision-making. International Journal of Geographical Information Systems 10 (8): 955–971.
23. Janssen R, Rietveld P (1990) Multi-criteria analysis and geographical information systems. An application to agricultural land use in the Netherlands. In: Scholten HJ, Stillwell JCH, editors. Geographical Information Systems for Urban and Regional Planning. Dordrecht: Kluwer Academic Publishers. pp.129–139. ISBN: 0792307933.
24. ESRI (2011) ArcGIS Desktop: Realease 10. Redlands, CA: Environmental Systems Research Institute.
25. ESRI (2001) ArcGIS Spatial Analyst: Advanced GIS Spatial Analysis Using Raster and Vector Data. Available: http://www.esri.com/software/arcgis/extensions/spatialanalyst/. Accessed 2014 April 02.
26. Diggle PJ, Ribeiro PJJ (2007) Model-Based Geostatistics. New York: Springer.
27. Genton MG (1998) Highly robust variogram estimation. Mathematical Geology 30 (2).
28. Genton MG (2000) The correlation structure of Matheron's classical variogram estimator under elliptically contoured distributions. Mathematical Geology 32 (1).
29. Park SJ, Vlek PLG (2002) Environmental correlation of three-dimensional soil spatial variability: a comparison of three adaptive techniques. Geoderma 109 (1–2): 117–140.
30. Hengl T, Heuvelink GBM, Stein A (2004) A generic framework for spatial prediction of soil variables based on regression-kriging. Geoderma 120 (1–2): 75–93.
31. Mueller TG, Pusuluri NB, Mathias KK, Cornelius PL, Barnhisel RI, et al. (2004) Map Quality for ordinary Kriging and inverse distance weighted interpolation. Soil Science Society of America Journal 68: 2042–2047.
32. Birasa EC, Bizimana I, Bouckaert W, Delflandre A, Chapelle J, et al. (1990) Les Sols du Rwanda: méthodologie, légende et classification. Carte Pédologique du Rwanda. Kigali - Rwanda (Unpublished book): CTB et MINAGRI.
33. Moore ID, Gessler PE, Nielson GA (1993) Soil attribute prediction using terrain analysis. Soil Science Society of American Journal 57: 443–452.
34. Changere A, Lal R (1997) Slope position and erosional effects on soil properties and corn production on a Miamian soil in Central Ohio. Journal of Sustainable Agriculture 11: 5–21.
35. Kravchenko A, Bullock DG (2000) Correlation of corn and soybean grain yield with topography and soil properties. Agronomy Journal 92: 75–83.
36. Descroix F, Snoeck J (2004) Coffee: Growing, Processing, Sustainable Production. In: Wintgens JN, editor. A Guidebook for Growers, Processors, Traders, and Researchers. Switzerland. pp.164–177. ISBN: 352730307311.
37. Soltani A, Stoorvogel JJ, Veldkamp A (2013) Model suitability to assess regional potato yield patterns in northern Ecuador. European Journal of Agronomy 48: 101–108.
38. McKinion JM, Willers JL, Jenkins JN (2010) Spatial analyses to evaluate multi-crop yield stability for a field. Computers and Electronics in Agriculture 70: 187–198.
39. Van Asten PJA, Wairegi LWI, Mukasa D, Uringi NO (2011) Agronomic and economic benefits of coffee–banana intercropping in Uganda's smallholder farming systems. Agricultural Systems 104: 326–334.

25

Moving GIS Research Indoors: Spatiotemporal Analysis of Agricultural Animals

Courtney L. Daigle[1], Debasmit Banerjee[2], Robert A. Montgomery[3], Subir Biswas[2], Janice M. Siegford[1]*

1 Department of Animal Science, Michigan State University, East Lansing, Michigan, United States of America, **2** Department of Electrical and Computer Engineering, Michigan State University, East Lansing, Michigan, United States of America, **3** Department of Fisheries and Wildlife, Michigan State University, East Lansing, Michigan, United States of America

Abstract

A proof of concept applying wildlife ecology techniques to animal welfare science in intensive agricultural environments was conducted using non-cage laying hens. Studies of wildlife ecology regularly use Geographic Information Systems (GIS) to assess wild animal movement and behavior within environments with relatively unlimited space and finite resources. However, rather than depicting landscapes, a GIS could be developed in animal production environments to provide insight into animal behavior as an indicator of animal welfare. We developed a GIS-based approach for studying agricultural animal behavior in an environment with finite space and unlimited resources. Concurrent data from wireless body-worn location tracking sensor and video-recording systems, which depicted spatially-explicit behavior of hens (135 hens/room) in two identical indoor enclosures, were collected. The spatial configuration of specific hen behaviors, variation in home range patterns, and variation in home range overlap show that individual hens respond to the same environment differently. Such information could catalyze management practice adjustments (e.g., modifying feeder design and/or location). Genetically-similar hens exhibited diverse behavioral and spatial patterns via a proof of concept approach enabling detailed examinations of individual non-cage laying hen behavior and welfare.

Editor: Cédric Sueur, Institut Pluridisciplinaire Hubert Curien, France

Funding: Funding came from the United States Department of Agriculture, Agricultural and Food Research Initiative Award #2009-65120-05752 to Dr. Janice Siegford. The funders had no role in study design, data collection and analysis, decision to publish, or preparation of the manuscript.

Competing Interests: The authors have declared that no competing interests exist.

* Email: siegford@msu.edu

Introduction

Many of the fundamental concepts of animal welfare derive from efforts to understand the response of individual animals to prevailing circumstances, and most theories of welfare emphasize a holistic representation of the animal's experience that encompasses such elements as emotional state, physical health, and ability to perform natural behaviors. Thus, the physical and psychological well-being of animals are frequently examined by assessing behavior, health, and emotional states in a single study, which can be complex and time consuming and require multidisciplinary expertise and invasive measurement. Fortunately, correlations between the behavior of the animal and its internal physiology and emotional state have been demonstrated [1]. Therefore, studying behavior can be a simple and non-invasive way of assessing an animal's current welfare state.

Individual animals alone, not aggregated groups of animals, experience the characteristics that make welfare better or worse, such as emotional states and health. Thereby, the concept of animal welfare inherently emphasizes the importance of individual experience and variation [2–4]. Additionally, individual variation with regards to genetics, experiences, and temperament can impact how an individual animal perceives its current situation, the choices it makes, and ultimately its welfare. However, issues of practicality limit the ability of welfare researchers to gather behavioral data for individual farm animals, because most livestock and poultry in modern production systems are housed in large groups. Further, welfare assessments that include behavioral measures require lengthy observation periods, which are often impractical. Because of these considerations, animal welfare tends to be assessed at an aggregate level and behavioral measures are often employed in a limited or indirect fashion. However, the condition of individuals within a group of animals may vary widely, thus group-level averages may not accurately reflect the condition of specific individual animals within that group. Thus, by using the average condition, researchers may be ignoring those individuals in very poor or very good welfare condition. Combined, these issues highlight the need to develop a method for gathering behavioral information at the level of the individual and interpreting the meaning of this behavior with regard to the welfare of the animal.

Currently, social pressure and political mandates are pushing changes in animal agriculture that are resulting in animals being housed in even larger groups and environments, with the goal of improving welfare and allowing animals more behavioral freedom. For example, the egg industry is transitioning from housing laying hens in groups of 4–10 hens in small, simple cages to housing hens in groups of hundreds or thousands in large (mostly) indoor enclosures that contain a variety of environmental features such as nest boxes and perches. The hens housed in these emerging systems likely face different welfare challenges than conventionally-caged hens, which must be recognized and addressed to ensure that alternative housing systems do in fact improve welfare.

Yet current assessments only describe group welfare, possibly masking individual problems.

Information on how individual non-cage laying hens use space and the spatio-temporal variation in their space use is poorly understood. This dearth of information results from the methodological challenges inherent to describing individual animal behavior in group settings. By implementing technological advancements (e.g., wireless sensors) to a representative proportion of the population and adopting analytical techniques used by other disciplines (e.g., Geographic Information Systems (GIS)), we may be able to better capture the responses of individuals housed in large groups. Modeling the spatial configuration of hen behaviors can provide insight into the general welfare of the individual. For example, feeding (here defined as consuming food from a feeder) and foraging (i.e., searching for and/or consuming food found in litter using the feet and beak) are both behaviors that hens are motivated to perform and are required for survival [5,6]. Preening (i.e., a maintenance and comfort behavior where feathers are cleaned with the beak) can be considered a comfort behavior and has been observed to be performed more often in the presence of familiar conspecifics [7]. However, based on previous studies [8,9], it is clear that hens make different choices with regard to where and when they choose to perform these welfare-relevant behaviors based upon their perception of the environment.

Hen behavior is changing in alternative housing systems, compared to what was expected in conventional systems and these changes must be recognized so that management practices and physical housing designs can also be changed to complement the behavioral changes in ways that improve hen welfare or maintain it at a high level. Furthermore, the application of this technology would allow animal managers to identify what should be expected, and this approach would provide them with a new tool to monitor flock movement and identify events when changes in behavior could indicate welfare concerns (e.g., increased activity levels associated with the development of feather pecking [10]). Production animal systems are beginning to mirror some of the large group phenomena that are observed of animals found in natural environments, such as colonial breeding groups of sea birds that face competition for space and resources due to the high concentration of animals in a single area [11]. Thus, management practices and scientific approaches should begin to mirror those applied in wildlife behavioral ecology, as laying hens housed in large groups with limited space and finite space to access resources may face similar challenges to wild species.

Here, we examined spatial patterns in hen behavior and movement in an indoor agricultural setting by combining wireless sensor technology with GIS. Via these techniques, we provide the first characterization of individual hen home ranges in a non-cage environment and the first illustration of spatiotemporal variability in individual hen behavior. This analysis represents the first effort, to our knowledge, to combine wildlife ecology techniques and wireless sensor technology in the study of animal behavior in an intensive indoor, animal production environment.

Methods

Animals and housing

We collected data from laying hens housed in an experimental non-cage system at the Michigan State University Poultry Teaching and Research Center (MSU-PTRC). Prior to the start of the study, all protocols were submitted to and approved by the Michigan State University Institutional Animal Care and Use Committee (AUF: 04/08-060-00). Two identical rooms (6 m×4.5 m) at MSU-PTRC were used. Each room was furnished in the same configuration with nest boxes, perches, tube feeders, and a water line with nipples (Fig. 1). Sixteen nest boxes (each 0.4 m long×0.3 m wide×0.3 m high) in an 8×2 configuration were mounted 0.3 m above the ground on one wall. Perches consisted of a three-level wooden rail structure (with each rail 6 m long and ~5 cm in diameter with a flat top and rounded sides and bottom) and mounted over a 1 m×6 m slatted area at a height of 0.53, 0.76, and 0.99 m from the ground. The perches were mounted to the wall at a slope of 45° with a 40 cm distance between each wooden rail. Room floors were covered with ~8 cm of wood shavings at time of data collection. Food and water were provided daily ad libitum. Daily care, including egg collection, feeding, and hen inspection, occurred at least once a day. Two incandescent light bulbs (60 lux at bulb level) on an automatic timer provided light 15 h per day in each room. Temperature was maintained between 16°C–22°C using a ventilation fan and forced air heating.

Hy-Line Brown laying hen pullets (n = 270, 135 hens/room) were reared in each of the rooms as described above with accommodation made for their size (i.e., smaller perches, which were removed at 6 weeks of age (wk)) and immaturity (i.e., access to nest boxes was not granted until 10 wk of age). Each room provided 0.21 m^2 floor space, 17.8 cm of perch space, 0.01 m^2 of nest box space, and 4.83 cm feeder space per hen (Fig. 1). Thirteen nipples provided drinking access at a ratio of 10.3 hens per nipple. All room parameters met or exceeded United Egg Producers and the Federation for Animal Science Society's housing requirements for non-cage laying hens [12].

At 10 wk, we weighed all hens and fitted them with uniquely numbered leg bands. At 11 wk, 10 hens per room were selected and fitted with a sensor (Fig. 2). We selected hens from different part of the spectrum of body weights because body weight contributes to hen social hierarchy. For instance, heavier hens are more likely to have won a recent fight and perform more double attacks when establishing a social hierarchy [13]. Thus, this step enabled us to evaluate behavior of hens from a variety of social ranks. Collection of data from the sensor to monitor the hen's behavior prior to and immediately after the onset of lay (approximately 18 wk) as part of a separate experiment was the motivation for fitting the hens with sensors prior to comb development. Therefore, even though combs are important to hen social hierarchy [14], it was not possible to include comb size as a parameter indicative of social status to select hens to wear sensors. The remaining 125 hens in each room did not receive sensors. Previous research indicated that wearing the sensor did not have any long-lasting effects on hen resource use or agonistic interactions [8]. The sensor has been shown to not impact hens' resource use or agonistic interaction [15] and can collect locational data (i.e., proximity to stationary sensor) from multiple hens simultaneously through a time division multiple polling access (TDMA) protocol as illustrated in Figure 3. One individual died between the two data collection points, so the final analysis includes data from nine individual hens.

Wireless sensor network

We attached a mobile sensor to each of nine individual hens housed in two separate, identical rooms (five hens in one room, four in the other) using a nylon figure-eight harness. Each mobile node (~10 g) was placed inside a plastic casing and mounted on a hen's back with figure eight nylon harness. The casing was colored to match feather color and painted with a unique number for easy visual identification (Fig. 2). After experimentation, mounting the sensor on the back of the hen was determined to yield maximum sensor stability while maintaining sufficient signal quality for

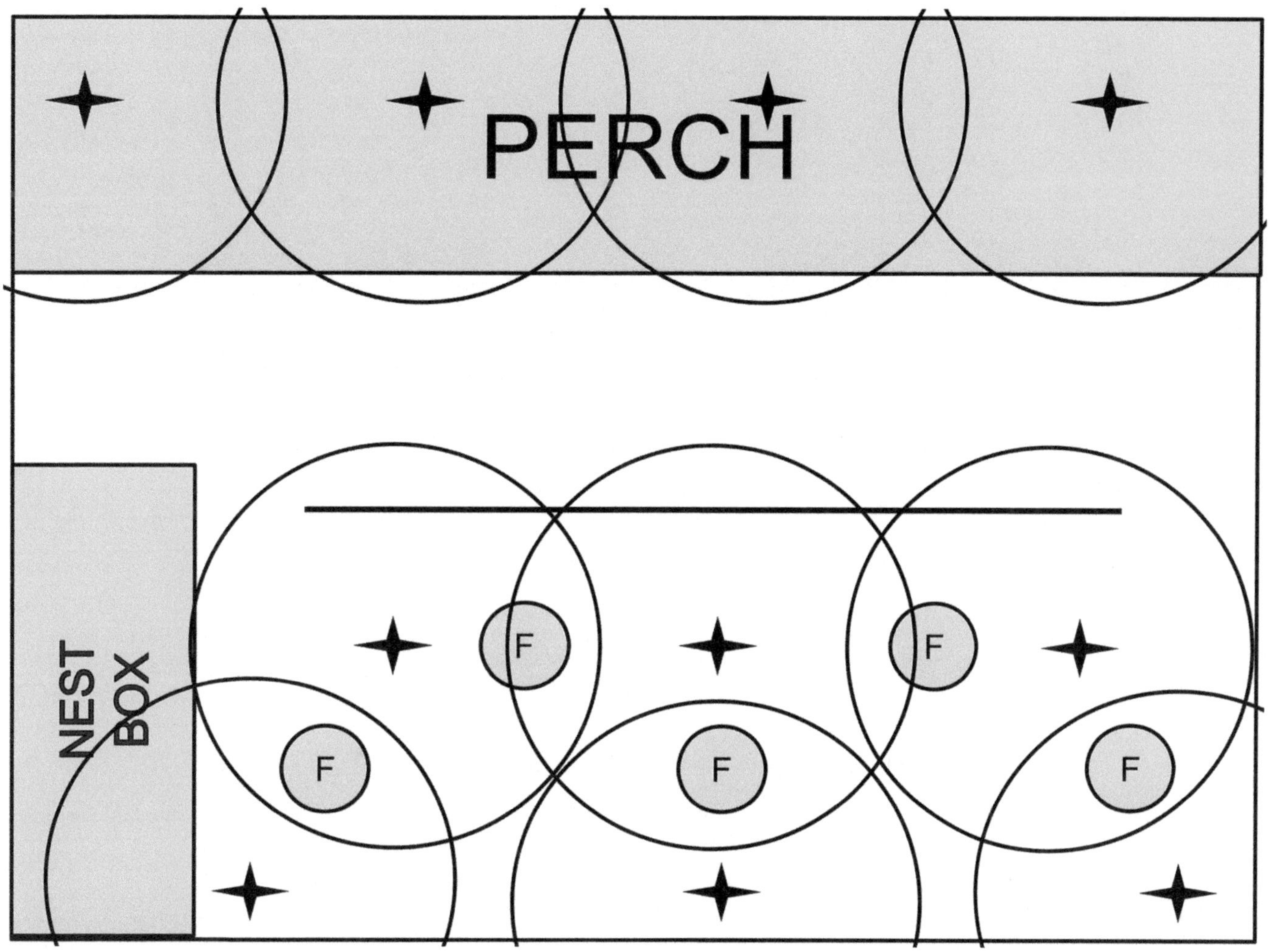

Figure 1. Diagram of room set up and stationary node placement. F represents a feeder. The water line is represented by the thick solid line in the middle of the room.

proper wireless communication with stationary nodes and avoiding tissue damage to the hen. These same hens wore the sensor casing throughout the lay cycle. The base station was strategically placed outside of the room to maintain communication with all stationary nodes. The mobile sensors communicated with a network of stationary sensors strategically placed throughout the hens' environment (Fig. 1).

The sensor system consisted of three components – (i) a Mica2Dot mote radio mobile mounted on hens, (ii) Mica2 mote stationary radio nodes acting as beacons for the proximity detection process, and (iii) a base station that collected the data wirelessly from stationary nodes and stored the data in a laptop PC. Each component consisted of a processor and a radio subsystem, running a Tinyos operating system. All network components operated via a 900 MHz wireless radio link. We strategically placed the ten stationary nodes throughout the room (Fig. 1). To minimize communication interference between stationary and mobile nodes, the stationary nodes were hung one meter above the ground with ceiling-mounted PVC pipes to point their antennas downwards. A polling-based time division multiple access (TDMA) protocol (Fig. 3) administered communication between the nodes in the wireless network at a rate of 0.125 Hz (8 sec sampling interval) to maximize battery life and maintain network synchronization. Furthermore, we implemented an energy-aware sleep schedule to minimize battery consumption and extend the mobile node operating life which allowed the system to run for 48–50 h.

Behavioral data collection

We video-monitored hens in each room continuously across 48-h intervals at 48 and 66 wk using ceiling mounted cameras. The detection of behavior using video was necessary because technology development to detect hen location and behavior simultaneously is in the nascent stages [16]. By associating these observations with the output of the sensor network we made our behavioral data spatially explicit. For instance, observations of feeding, foraging, and preening collected from video were made spatially explicit by time-syncing them with locational data produced from this sensor network. Though 10 hens per room were fitted with sensors, video data were only decoded for the five individual hens per room that were actively transmitting sensor data, as five mobile nodes was the maximum the sensor network could currently poll at that stage of development and retain network synchronization. Therefore, the individual behavioral data collection for this research was restricted to the same five individuals per room.

Figure 2. A laying hen wearing a wireless sensor. The wireless sensor is packaged in a plastic case to prevent entry of dust and moisture. We mounted the sensor the on the back of the hen using a figure-eight nylon harness. Both the sensor case and the harness were colored to blend in with the hen's feathers to avoid attracting the attention of other hens in the experimental room. This picture was taken several days after the hen was fitted with the sensor.

Continuous observation of individual hen posture, behavior, and resource use (Table 1) was made over a 30 min period every hour and a half (06:00–06:30, 07:30–08:00, 09:00–09:30, 10:30–11:00, 12:00–12:30, 13:30–14:00, 15:00–15:30, 16:30–17:00, 18:00–18:30, and 19:30–20:00). Data were collected only during the lights on period following a validated sub-sampling procedure

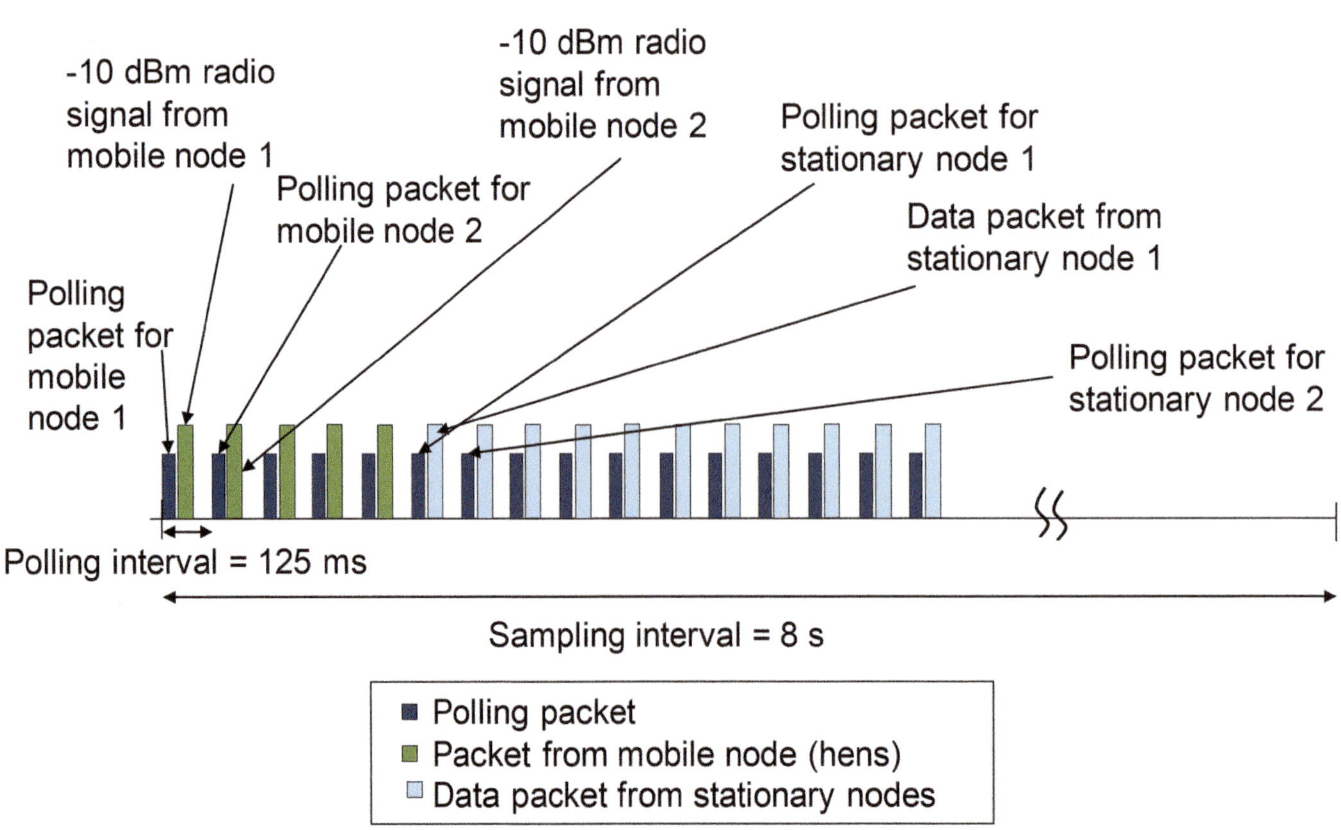

Figure 3. Illustration of the polling-based time division multiple access (TDMA) protocol.

for hens housed in this way and the specific behaviors of interest [17]. Data were collected across a 48 h time period with the dark period (21:00–06:00) omitted because even though infrared cameras were utilized to observe night time behavior, substantial amounts of movement were not observed (most movement during lights off was due to hens transitioning between standing and sitting as they readjusted during the night). Therefore data collection efforts were focused on the lights on period only. Posture, behavior, and resource use were recorded in mutually exclusive categories and are reported in duration of time spent (sec) in that state across the 48 h period and paired t-tests (PROC TTEST, SAS version 9.2) were employed to identify statistical differences between the two ages in the amount of time performing each behavior.

Combining sensor network output and behavioral observations into GIS

We used the spatially-explicit behavioral data to depict hen home range distributions and the spatial distribution of welfare-revealing behaviors. In this respect, we focused our assessment on feeding, foraging, and preening behaviors which can be associated with physical health (e.g., nutrient acquisition or foot condition), natural behavior expression (e.g., grooming and maintenance behaviors) and emotional state (e.g., frustration due to hunger or comfort due to the presence of familiar conspecifics).

Expression of home range and behavior across two periods (48 and 66 wks) allowed us to examine whether these processes varied by individual hen and across time. Specifically, data depicted home range patterns within the non-cage environment, identified the size and amount of overlap in home ranges of hens within these enclosures, and illustrated whether spatial segregation by behavior was apparent among the hens [18–20].

To assess this spatiotemporal variation we needed to develop a GIS depicting physical properties of the indoor enclosures. We georectified and virtually represented the exact dimensions of the room where hens were housed in ArcMap 10.0 (Environmental Systems Research Institute, Redlands, CA). Gaps in the sensor network required that probabilistic movement paths were interpolated between missing locations for each individual hen. Therefore, we fitted a continuous-time correlated random walk model to the sensor data [21]. We developed this model in R (R statistical software version 2.15.1, <www.cran.r-project.org>, accessed 1 May 2013) using the package CRAWL which employed a Kalman-Filter to predict locations from the existing point pattern based on a continuous-time stochastic movement process. In this way, locations that were missed because of gaps in the sensor network were imputed to make the database of hen locations complete.

We used this locational database to delineate home ranges for each individual hen in both rooms by time period (48 and 66 wk). Utilization distributions, widely used in the wildlife sciences, depict the relative probability of animal occurrence in space. In this capacity, we developed utilization distributions (UDs) [22] in R using the Kernel Density Estimator library. We fitted these UDs using a bivariate plug-in matrix that calculated bandwidth along rotated axes for each individual hen [19,20,22]. We subsequently calculated the amount of overlap among the hen UDs for each time period.

To map the spatial configuration of hen behaviors, we developed hotspot maps depicting foraging, preening, and feeding behaviors. Thiessen polygons, which are well-suited to binary data [18] (i.e., the presence or absence of the behavior of interest), were utilized to develop the spatially-explicit maps of these behaviors. The precision of the sensor network (8-sec intervals) meant that hens were recorded performing multiple behaviors in the same

Table 1. Ethogram of behaviors developed to identify posture, behavior and resource use.

Posture	**Description**
Walk	Walking more than 3 steps in succession with head up or when walking hen has not been standing, drinking, feeding, or foraging for the previous 5 s
Stand	Hen is upright and supported off of the ground or perch by legs
Sit	Hen is upright with body touching the ground or perch
Behavior	
feed	Hen pecks at feed in the feeder. Recording starts at first peck
drink	Head is turned upwards towards water source, and hen uses beak to peck at one of the nipples, apparently consuming water
preen	Hen may be sitting or standing. Beak is used to manipulate, rearrange, pull, or smooth body feathers on self. Beak is often run along the length of the feather, starting at the base and moving out towards the tip of the feather
dust bathe	While squatting or laying, hen performs dust bathing activities including vertical wing shaking, bill raking, scratching, ground pecking, movements of the feet and wings to raise dust into the ruffled plumage, rubbing of head and sides in the dust, feather-ruffling and shaking dust out of the feathers. Starts with first wing shake
forage	Hen pecks at substrate while standing or stepping forward with head below rump level. Starts when the hen makes >3 successive pecks at substrate, or when foraging hen has not been standing or walking with head up, or feeding, for the previous 5 s
rest	Starts the hen lies down (sternum resting on substrate) from an upright position or when lying bird has made no dust bathing or preening movements for the previous 8 s
Resource	
feeder	Hen has head in feeder and pecks at feed in the feeder
drinker	Head is turned upwards towards water source, and hen uses beak to peck at one of the nipples, apparently consuming water
perch	Hen is standing, walking, or resting on perch, the rail in front of the nest boxes, or black base of slats underneath raised perches
nest box	Hen is standing or resting inside a nest box

A hen was considered to be performing a new posture, behavior, or resource use when she stopped performing the previous behavior for >5 s or began performing a new behavior for >5 s.

location. Thus, the Thiessen polygons present the mean count of the behavior of interest within each individual's home range. For presentation purposes, these data are divided into equal intervals representing space use for foraging, preening, and feeding in increments of high, medium, low, and none, which correspond to whether the animal was observed performing the behavior more than 75% of the time (high), 74–25% of the time (medium), less than 25% of the time, or not at all (none).

Results

Behavioral observations

Hens varied in the amount of time spent performing specific behaviors (Table 2), and paired t-test determined that hens tended to spend a smaller proportion of their time feeding at 66 wk compared to 48 wk ($t_8 = 1.91$, $P = 0.09$). No significant differences were observed the percentage of time the hens spent performing the rest of the behaviors between the two ages.

Hen home ranges and conspecific overlap

Hens exhibited large differences in home range size and amount of overlap with conspecifics (Fig. 4). On average, home range size (m^2) at 48 wk was larger (5.48 ± 1.13) than at 66 wk (5.17 ± 1.48), while the proportion of overlap (%) with conspecifics decreased from 0.88 ± 0.05 at 48 wk to 0.84 ± 0.07 at 66 wk. Hens that used a smaller proportion of the room were observed to have more of their home range overlap with their conspecifics (i.e., Table 3, Y05), while hens with larger home ranges had less home range overlap (i.e., Table 3, Y08).

Substantial variation was observed among the hens with regard to overlapping home ranges. Five hens (Table 3, B05, B08, Y07, Y08, and Y10) showed a relative decrease in the amount of overlap as they aged; while three (i.e., Table 3, B01, B2, and B10) showed an increase and one remained relatively consistent (i.e., Table 3, Y05). Consequently, this was reflected in changes observed in pairwise associations (m^2) describing degree of overlap between pairs of hens (Table 4).

Feeding, foraging, and preening

We observed a wide variety of feeding, foraging, and preening patterns among the nine hens across time (Fig. 5). Several hens (Y10 and B05) were relatively consistent in the locations they used to perform feeding, foraging and preening. B08 and B10 both performed more foraging than feeding, and B02 appeared to make different choices with regard to foraging and preening at 48 wk compared to 66 wk.

The average percentage of the hens' time devoted to feeding and foraging decreased with age (Table 2). There was also a wide diversity in how much and where preening occurred. Two hens (B08 and B10) within the same room were observed to exhibit similar preening and foraging patterns across time. Y07 was observed to not spend any time foraging or feeding at either age.

Associations among the amount of time spent feeding, foraging, and preening revealed that the amount of time spent foraging was positively correlated with the amount of time spent preening ($R_{18} = 0.653$, $P = 0.0033$), yet was not associated with the amount of time feeding ($R_{18} = -0.021$, $P = 0.935$) even though this relationship was negative. A weak, negative relationship was observed between feeding and preening ($R_{18} = -0.106$, $P = 0.676$).

Discussion

Concerns about animal welfare, including the ability of animals in agricultural production systems to perform natural behaviors, demand that tools are developed to better understand individual animal behavior. Here we provide an example of analysis using data obtained from behavioral observations and a wireless sensor network using GIS that could be used to dynamically assess animal welfare. By combining locational data from the sensor with behavioral observations, spatiotemporal differences among individual laying hens in a non-cage environment were identified across time. Furthermore, these results highlight the large degree of individual variation exhibited among laying hens with regard to their overall and behavior-specific space use. Changes in the location and frequency of behaviors could provide insight into individual hen condition, while variability among hens within the same environment can be used to help identify areas important for performance of specific behaviors and evaluate whether space is a limiting factor in the performance of behavior. This information can improve housing design and animal management practices with regard to animal welfare, not only for laying hens, but for other poultry and livestock species, or, in fact, any confined animals, but particularly those managed in large groups.

In the present study, hen movement within the non-cage system was restricted by the confines of the room. When wild animals are observed to have overlapping utilization distributions, these overlaps can identify areas of optimal habitat for reproduction, foraging, or resting. However, it is likely that hens in a closed system are more likely to show overlap for non-traditional reasons. Hens may exhibit more or less fidelity to specific areas of the room based upon personal preferences, or hens may have a high degree of overlap due to social interactions or resource availability. For example, hens may also have a high degree of overlap simply due to the stocking density of the room and have no choice but to inhabit one another's space. However, further investigation is needed to understand similarities and differences in space use by confined animals compared to wild animals that are free ranging.

Remote monitoring of animals and the development of a GIS can be used to assess the behavior and welfare of agricultural animals in indoor spaces as much as it can for wildlife living in natural environments. In the present study, analysis of home range size, home range overlap, and behavior revealed considerable variability among individual hens. This individual variation provided a window with which we could better view the behavioral differences of certain hens in their enclosures. When we applied this tool to our group of hens, our analysis revealed vast individual variation. Information from such utilization distributions can have implications for hen welfare because they can give insight into where hens are choosing to perform certain behaviors, which may result in altered management practices or facility design. Further, this approach can be used to address multiple types of welfare related concerns across multiple species. For example, future studies applying this approach could identify the impact of sex ratios on the behavior of breeding flocks; the impact of mixing and social dynamics on group-housed sows; or environmental preferences of dairy cattle.

Animal behavior can provide insight into the welfare of the animal. For example, animals behave differently when they are ill (e.g., they may exhibit reduced feeding behavior), injured (e.g., they may exhibit reduced locomotor behaviors), or distressed (e.g., laying hens perform more vocalizations when they are frustrated and unable to perform behaviors such as feeding and drinking [23]). Therefore, changes in their behavioral patterns can be used as a proxy measure for how the animal may be feeling and its overall welfare state. A marked difference in space use was observed for Y07, in which the hen used a smaller proportion of the room at 66 wk compared to 48 wk. The reason for this change in use is unknown, but changes such as this could be indicative of a

Table 2. Percent of time spent performing behaviors for hens at 48 and 66 weeks of age.

Hen ID	Age	Feed	Drink	Preen	Dust bathe	Forage	Rest	Other
B01	48	40.62	0.00	8.59	0.00	16.03	1.78	32.97
	66	22.00	6.48	7.93	2.61	15.62	2.87	42.49
B10	48	12.10	2.69	19.36	0.18	15.47	1.98	48.23
	66	2.91	3.42	1.44	0.00	2.79	0.00	89.44
B02	48	31.46	1.85	12.15	2.73	9.51	11.09	31.21
	66	17.07	3.87	12.91	1.69	17.48	1.95	45.04
B05	48	32.35	4.92	9.18	2.32	15.11	4.22	31.90
	66	20.77	3.79	9.66	1.33	17.03	1.93	45.49
B08	48	1.71	3.54	19.82	2.63	22.45	14.14	35.71
	66	0.45	3.09	15.39	2.18	27.53	0.00	51.36
Y10	48	17.28	4.27	7.72	0.00	20.77	6.55	43.40
	66	10.41	6.22	21.36	4.22	19.14	4.89	33.77
Y05	48	18.45	2.19	16.78	1.80	17.68	0.75	42.34
	66	24.10	2.46	16.02	4.31	8.00	4.06	41.05
Y07	48	0.09	1.44	5.45	0.00	0.00	0.00	93.02
	66	0.03	1.33	2.58	0.00	0.00	0.00	96.06
Y08	48	1.33	2.68	18.75	0.24	30.05	3.39	43.57
	66	7.46	2.27	23.29	0.00	19.01	11.55	36.42
Average	48	17.26	2.62	13.09	1.10	16.34	4.88	44.71
SEM		4.99	0.50	1.88	0.41	2.80	1.62	6.37
Average	66	11.69	3.66	12.29	1.82	14.07	3.03	53.46
SEM		3.19	0.57	2.54	0.56	2.92	1.22	7.64

Results are presented for hens from two separate rooms (Blue, as indicated with a B and Yellow as indicated with a Y) at two different ages (48 and 66 wk).

Table 3. Hen home ranges and home range overlap at two different ages.

Animal ID	Age (wk)	Range size (m^2)	Proportion of room used	Proportion of overlap with conspecifics
B01	48	4.82	0.17	0.97
	66	1.65	0.06	1
B02	48	1.52	0.05	1
	66	7.64	0.27	0.94
B05	48	6	0.22	0.89
	66	9.11	0.33	0.84
B08	48	6.83	0.25	0.94
	66	7.68	0.28	0.85
B10	48	11.28	0.41	0.76
	66	1.66	0.06	0.99
Y05	48	2.97	0.11	1
	66	0.37	0.01	1
Y07	48	2.8	0.1	0.9
	66	1.82	0.07	0.76
Y08	48	10.04	0.36	0.5
	66	13.35	0.48	0.27
Y10	48	3.14	0.11	0.96
	66	3.26	0.12	0.9

Home range area (m^2), proportion of the room that the home range covers (%), and proportion of individual hen's home range that overlaps with her conspecifics (%). Results are presented for hens from two separate rooms (Blue, as indicated with a B and Yellow as indicated with a Y) at two different ages (48 and 66 wk).

physical (i.e., the hen may choose to move less due to an injured foot) or psychological issue (i.e., the hen may choose to move less due to an increase in agonistic social interactions). However, for other species different behaviors may be of more interest and provide more insight based upon the biological relevance of the behavior.

In managed conditions, food is only provided at a feeder(s) and the location of feeding behavior is largely restricted to feeder locations. Thus, for hens in production systems, feeding requires access to a highly valued and spatially concentrated resource. Some hens may be restricted from freely accessing the feeder due to social interactions (i.e., competitive exclusion). When multiple feeders are provided, it is unknown whether hens consistently use the same feeder or whether some feeder locations are more attractive than others. In the present study, one hen appeared to perform feeding behavior in a single area (e.g., Y10), while other hens distributed their feeding behavior evenly throughout the room in which feeders were located.

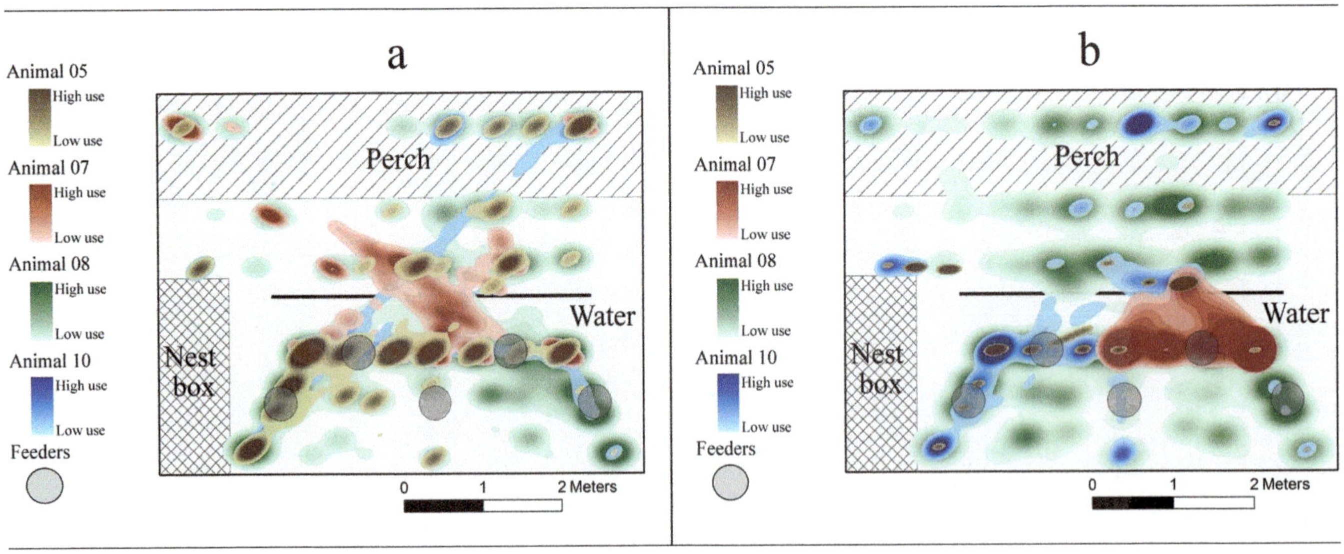

Figure 4. Overall utilization distribution for hens housed within Room Y at 48 (a) and 66 (b) wk. The utilization distribution for Room B is similar and shows a similar amount of variation among hens and across time. Room B is not pictured here due to clarity of image in black and white.

Table 4. Total area (m^2) of overlap between two hens at two different ages (48 and 66 wk).

Pairwise overlap between	48 wk	66 wk
B10–B01	4.39	0.51
B10–B02	1.49	1.53
B10–B05	5.09	1.59
B10–B08	6.21	1.15
B01–B02	1.27	1.64
B01–B05	3.65	1.64
B01–B08	3.10	1.57
B02–B05	1.09	6.57
B02–B08	1.43	5.49
B05–B08	2.98	5.87
Y05–Y07	0.08	1.01
Y05–Y08	0.33	2.87
Y05–Y10	0.03	2.07
Y07–Y08	1.37	2.48
Y07–Y10	0.77	1.22
Y08–Y10	2.88	2.88

Some hens may choose to not use the feeder, however, acquiring their nutrients from feed that has been spilled on the ground by other hens. These hens may be avoiding the feeder because they prefer to acquire their food via foraging in litter, they may be thwarted from accessing the feeder due to social interactions, or they may forage due to inherent behavioral drives. Based on the results of this study, hens may fall into any one of those categories, as they appear to adopt resource acquisition

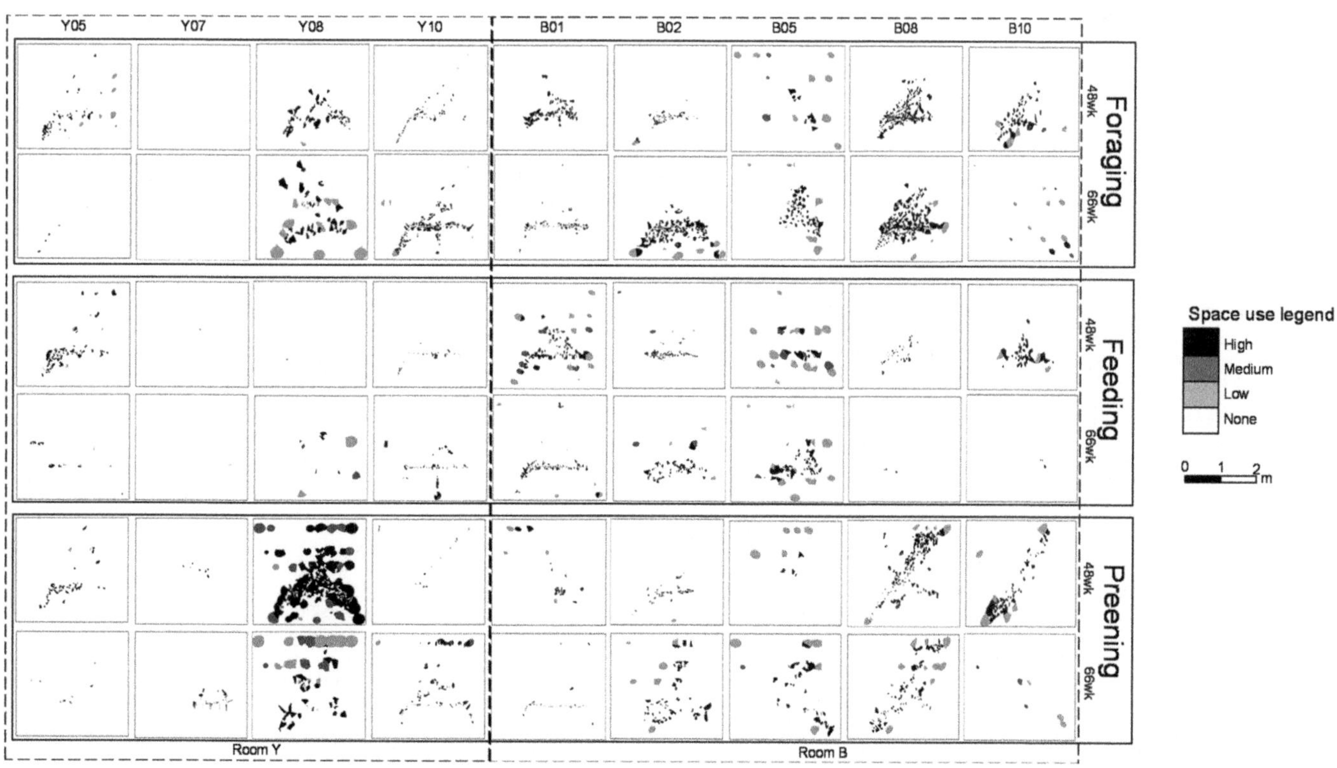

Figure 5. Hotspot maps of feeding, foraging, and preening behavior at 48 and 66 wk. The labels at the top of each panel indicate which hen is represented. The legend to the right illustrated the amount of time (high, medium, low, none) the hen spent performing feeding behavior in that area, which correspond to whether the animal was observed performing the behavior more than 75% of the time (high), 74–25% of the time (medium), less than 25% of the time, or not at all (none)".

strategies based upon current physical and/or social constraints, and do not appear to develop preferences for specific feeders or feeder locations. Beyond using such information to better manage and design systems, changes in feeding behavior could be indicative of illness [24], injury [25,26], or change in social dynamics [27], which could be better studied using the type of data presented here.

However, little is known about the degree to which individual hens may forage versus feed or whether such preferences or patterns change over time. Foraging is not as spatially constrained by the presence or absence or location of specific resources as feeding from a feeder would be, for example. Therefore, the location of hens' foraging behavior may also help producers identify environmental conditions that are favorable for foraging, such as shavings or straw on the floor or provide insight into the amount of foraging space needed. Two hens (B08 and B10) were observed to use the feeder very little, but spent a large proportion of their time budget foraging. Hens must be mobile to forage, able to move around in the enclosure as they search for and consume food by this method. Therefore, changes in the amount of time spent foraging or in the area in which foraging occurs could be indicative of several health problems including illness, painful foot infections [28], neuromas resulting from improper beak trimming [29], broken claws, or broken bones – all of which are relevant to welfare.

Patterns of preening behavior can provide insight into where hens choose to rest, groom, and spend time with familiar conspecifics, and may thus provide insight into the hen's mood or emotional state. Preening can also be a response to stress [30] or be performed as a displacement behavior when motivation or behavioral drives of hens are thwarted [31]. Preening can occur at any time of the day and is not constrained by posture (e.g., sitting or standing) or location of the hen in the environment [32]. Therefore, variation in preening behavior may be indicative of differences in health, maintenance, stress response, preference, or comfort among different hens. Thus, future investigations using dynamic modeling could be used to better elucidate social-spatiotemporal relationships.

Here, a proof of concept approach is presented using GIS to understanding the relationship between non-cage laying hen behavior and space use. This approach provides an opportunity to ask many types of questions important to animal welfare in human-managed environments. With this in mind, concessions must be made with regards to needed improvements to the system as research efforts continue. Duration of data collection (approx. 48 hours) was constrained by battery life. Future research efforts should address how the kinetic movement of the animal can power the technologies used to track their movement and provide welfare-relevant information to the producer and consumer. Data were collected from only five hens per room. As this type of analysis becomes more commonplace, it will be important to identify an optimal sample size for large groups of animals that can provide information important to their management. Also, implementing this technology currently requires expertise in computer programming and data analysis. A generalized user interface is under development for the sensor system used in this study, and implementation of this interface would make the collection and analysis of data less time consuming and more accessible.

Commercially-housed laying hens have been genetically selected to be similar with regards to growth, nutritional needs, productivity, and longevity. Despite this genetic similarity, and even despite identical rearing environments, the mature hens housed within the same environment in this study exhibited very different patterns of space use and had home ranges of varying sizes. Furthermore, some hens exhibited changes in their behavioral patterns over time, while others remained relatively consistent – which may be reflective of different levels of behavioral plasticity or changes in social status or health as the hens aged. This work emphasizes that even though hens may be genetically similar, they still exhibit individual behavioral patterns, meaning that designing housing or management practices for the 'average' animal will not ensure good welfare for all individuals in the system. Therefore, there is a need to provide environmental conditions that can accommodate the diverse spatial needs and preferences for all of the hens housed within. Understanding the diversity of behavioral patterns will enable animal managers to develop best practices and provide resources so animals of all social ranks, personalities, and preferences can acquire the necessary resources and perform the necessary behaviors indicative of a positive welfare state.

Acknowledgments

We extend gratitude towards Angelo Napolitano and the MSU Poultry Teaching and Research Center staff (East Lansing, MI) for their assistance with this project. We also thank Sarah Castle, Kelly Jaynes and Kaitlin Wurtz of the MSU Animal Behavior & Welfare Group (East Lansing, MI) for their help collecting data.

Author Contributions

Conceived and designed the experiments: CLD DB RAM JMS SB. Performed the experiments: CLD DB RAM. Analyzed the data: CLD DB RAM. Contributed reagents/materials/analysis tools: JMS SB. Wrote the paper: CLD.

References

1. Sih A, Bell A, Johnson J, Ziemba R (2004) Behavioral syndromes: an integrative overview. The Quarterly Review of Biology 79: 241–277.
2. Dawkins M (2003) Behaviour as a tool in the assessment of animal welfare. Zoology 106: 383–387.
3. Duncan IJH, Mench JA (1993) Behaviour as an indicator of welfare in various systems. Basic Biology and Welfare. Alternative Housing Systems 7: 69–76.
4. Fraser D (2008) Understanding Animal Welfare. Acta Agriculturae Scandinavica 50: S1.
5. Nicol CJ, Caplen G, Statham P, Browne WJ (2011) Decisions about foraging and risk trade-offs in chickens are associated with individual somatic response profiles. Animal Behaviour 82: 255–262.
6. Duncan IJH, Wood-Gush D (1972) Thwarting of feeding behaviour in the domestic fowl. Animal Behaviour 20: 444–451.
7. Nicol CJ (1989) Social influences on the comfort behaviour of laying hens. Applied Animal Behaviour Science 22: 75–81.
8. Kristensen H, Burgess L, Demmers T, Wathes C (2000) The preferences of laying hens for different concentrations of atmostpheric ammonia. Applied Animal Behaviour Science 68: 307–318.
9. Browne WJ, Caplen G, Statham P, Nicol CJ (2011) Mild environmental aversion is detected by a discrete-choice preference testing method but not by a free-access method. Applied Animal Behaviour Science 134: 152–163.
10. Lee H-j, Roberts SJ, Drake KA, Dawkins MS (2011) Prediction of feather damage in laying hens using optical flows and Markov models. Journal of The Royal Society Interface 8: 489–499.
11. Neil J Buckley (1997) Spatial-Concentration Effects and the Importance of Local Enhancement in the Evolution of Colonial Breeding in Seabirds. The American Naturalist 149: 1091–1112.
12. FASS (2010) Guide for the Care and Use of Agricultural Animals in Research and Teaching. Champaign, IL.
13. Cloutier S, Newberry R (2000) Recent social experience, body weight and initial patterns of attack predict the social status attained by unfamilar hens in a new group. Behaviour 137: 705–726.
14. O'Connor E, Suanders J, Grist H, McLeman M, Wathes CM, et al. (2011) The relationship between the comb and social behaviour in laying hens. Applied Animal Behaviour Science 135: 293–299.

15. Daigle C, Banerjee D, Biswas S, Siegford J (2012) Noncaged laying hens remain unflappable while wearing body-mounted sensors: levels of agonistic behaviors remain unchanged and resource use is not reduced after habituation. Poultry Science 91: 2415–2423.
16. Banerjee D, Daigle C, Biswas S, Siegford J (2012) Remote activity classification of hens using wireless body mounted sensors. Body Sensor Networks; London, United Kingdom. 107–112.
17. Daigle C, Siegford J (2014) When continuous observations just won't do: developing accurate and efficient sampling strategies for the laying hen. Behavioral Processes 103: 58–66.
18. Farris CA, Baisan CH, Falk DA, Yool SR, Swetnam TW (2010) Spatial and temporal corroboration of a fire-scar-based fire history in a frequently burned ponderosa pine forest. Ecological Applications 20: 1598–1614.
19. Gitzen RA, Millspaugh JJ, Kernohan BJ (2006) Bandwidth Selection for Fixed-Kernel Analysis of Animal Utilization Distributions. The Journal of Wildlife Management 70: 1334–1344.
20. Gitzen RA, Millspaugh JJ (2003) Comparison of Least-Squares Cross-Validation Bandwidth Options for Kernel Home-Range Estimation. Wildlife Society Bulletin 31: 823–831.
21. Johnson DS, London JM, Lea M-A, Durban JW (2008) Continuous-Time Correlated Random Walk Model for Animal Telemetry Data. Ecology 89: 1208–1215.
22. Kernohan B, Gitzen RA, Millspaugh JJ (2001) Analysis of animal space use and movements. In: Millspaugh JJ, Marzluff J, editors. Radiotracking and Animal Populations. San Diego, CA, USA: Academic Press.
23. Zimmerman P, Koene P, van Hooff J (2000) The vocal expression of feeding motivation and frustration in the domestic laying hen, *Gallus gallus domesticus*. Applied Animal Behaviour Science 69: 265–273.
24. Weary DM, Huzzey JM, von Keyserlingk MAG (2009) BOARD-INVITED REVIEW: Using behavior to predict and identify ill health in animals. Journal of Animal Science 87: 770–777.
25. Gentle MJ, Waddington D, Hunter LN, Jones RB (1990) Behavioural evidence for persistent pain following partial beak amputation in chickens. Applied Animal Behaviour Science 27: 149–157.
26. Weeks CA, Danbury TD, Davies HC, Hunt P, Kestin SC (2000) The behaviour of broiler chickens and its modification by lameness. Applied Animal Behaviour Science 67: 111–125.
27. Grant RJ, Albright JL (2001) Effect of Animal Grouping on Feeding Behavior and Intake of Dairy Cattle. Journal of Dairy Science 84, Supplement: E156–E163.
28. Hester PY (1994) The role of environment and management on leg abnormalities in meat-type fowl. Poultry Science 73: 904–915.
29. Kuenzel WJ (2007) Neurobiological basis of sensory perception: Welfare implications of beak trimming. Poultry Science 86: 1273–1282.
30. Williams NS, Scampoli DL (1984) Handling, ACTH, ACTH1−24, and naloxone effects on preening behavior in domestic chickens. Pharmacology Biochemistry and Behavior 20: 681–682.
31. Duncan IJH, Wood-Gush D (1972) An analysis of displacement preening in the domestic fowl. Animal Behaviour 20: 68–71.
32. Mishra A, Koene P, Schouten W, Spruijt B, van Beek P, et al. (2005) Temporal and sequential structure of behavior and facility usage of laying hens in an enriched environment. Poultry Science 84: 979–991.

26

What Role Do Local Grocery Stores Play in Urban Food Environments? A Case Study of Hartford-Connecticut

Katie S. Martin[1], Debarchana Ghosh[2]*, Martha Page[3], Michele Wolff[1], Kate McMinimee[4], Mengyao Zhang[2]

1 Department of Nutrition and Public Health, University of Saint Joseph, West Hartford, Connecticut, United States of America, **2** Geography, University of Connecticut, Storrs, Connecticut, United States of America, **3** Hartford Food System, Hartford, Connecticut, United States of America, **4** Department of Allied Health Sciences, University of Connecticut, Storrs, Connecticut, United States of America

Abstract

Introduction: Research on urban food environments emphasizes limited access to healthy food, with fewer large supermarkets and higher food prices. Many residents of Hartford, Connecticut, which is often considered a food desert, buy most of their food from small and medium-sized grocery stores. We examined the food environment in greater Hartford, comparing stores in Hartford to those in the surrounding suburbs, and by store size (small, medium, and large).

Methods: We surveyed all small (over 1,000 ft^2), medium, and large-sized supermarkets within a 2-mile radius of Hartford (36 total stores). We measured the distance to stores, availability, price and quality of a market basket of 25 items, and rated each store on internal and external appearance. Geographic Information System (GIS) was used for mapping distance to the stores and variation of food availability, quality, and appearance.

Results: Contrary to common literature, no significant differences were found in food availability and price between Hartford and suburban stores. However, produce quality, internal, and external store appearance were significantly lower in Hartford compared to suburban stores (all $p<0.05$). Medium-sized stores had significantly lower prices than small or large supermarkets ($p<0.05$). Large stores had better scores for internal ($p<0.05$), external, and produce quality ($p<0.01$). Most Hartford residents live within 0.5 to 1 mile distance to a grocery store.

Discussion: Classifying urban areas with few large supermarkets as 'food deserts' may overlook the availability of healthy foods and low prices that exist within small and medium-sized groceries common in inner cities. Improving produce quality and store appearance can potentially impact the food purchasing decisions of low-income residents in Hartford.

Editor: Rachel A. Nugent, University of Washington, United States of America

Funding: The authors have no support or funding to report.

Competing Interests: The authors have declared that no competing interests exist.

* E-mail: debarchana.ghosh@uconn.edu

Introduction

The availability of nutritious and affordable food can greatly impact chronic disease rates and other critical individual health outcomes in a community [1–4]. For example, insufficient access to healthy and affordable food may adversely affect dietary intake and eventually lead to nutrition related negative health outcomes such as obesity, diabetes, and cardiovascular diseases [5,6]. In urban inner cities, the situation is often thought to be even more critical due to nutritional imbalance caused by easy access to abundant fast-food restaurants on one hand and lack of large supermarket grocery stores on the other hand [7,8].

Dietary intake is determined by a number of factors, including what types of food items are available (*availability*), at what prices (*affordability*), and proximity to grocery stores (*accessibility*).

Researchers typically describe areas with lack of access to healthy and nutritious food as *food deserts* [4,9–14]. Among the several definitions, the United States Department of Agriculture's (USDA) description of food desert is the most commonly used, where, census tracts are identified as food deserts if they satisfy the following two conditions of: 1) "*low-income communities*", based on having a poverty rate of 20% or greater or a median family income at or below 80% of the area median family income; and 2) "*low-access communities*", based on the determination that at least 500 persons and/or at least 33% of the census tract's population live more than *one mile* from a supermarket or a large grocery store [15]. There have also been prior attempts to expand the definition and methodology for delineating food deserts for rural areas [16–18], urban areas [16,19–23], and by incorporating different measures of accessibility determined by distance, time, and modes of transportation to grocery stores [1,24–26] [20,27,28] [29–32]. Residents living in food deserts often experience food insecurity, which is defined as limited food access or uncertain ability to acquire acceptable foods in socially acceptable ways [33]. Similar to food deserts, several researchers have also associated food insecurity with a greater prevalence of chronic diseases and health disparities in urban, low-income areas [1,34]. Recent estimates indicate that approximately 14.5% of all US households experienced food insecurity in 2012 [35].

Few recent studies on food environment, however, have indicated the need to further expand the definition of food deserts by looking more critically at small and medium-sized grocery stores where inner-city residents typically shop for their food items [36,37]. A primary concern for urban areas, in regards to food accessibility, involves the lack of large-scale retail supermarkets with stocks of nutritious food at affordable prices, compared to predominately white or more affluent suburban neighborhoods [25,26,38–42]. Due to 'supermarket redlining' [43,44], large chain supermarkets are often disinclined to locate their stores in inner cities and usually pull their existing stores out and relocate them to suburbs. As the stores close, urban residents have to travel further to purchase nutritious and affordable groceries or shop at local medium and small sized grocery stores that may lack healthy food or offer it at higher prices than the larger stores [45,46]. The situation is often thought to be even more critical for residents with limited access to automobiles or well-connected public transportation. People living in urban communities with restricted access to larger stores pay significantly more (3%–37%) for groceries compared to those living in the suburbs buying the same products at supermarkets [3,25,46–49]. A study by Raja et al. [37], on the contrary, indicated that small-sized grocery stores sometimes do play an important role in providing healthy, affordable, and culturally appropriate foods in minority neighborhoods.

In addition to cost and availability, ethnically diverse shoppers in low-income areas also value variety and express concerns about the quality of fresh produce in smaller stores [50]. Product variety is often defined as the range of product options available to the consumer, enabling consumer choice [51–53]. Even when affordable food sources are present, they have been found to be of poorer quality than those found in wealthy or predominantly white neighborhoods [47,54–58]. The quality of fruits and vegetables is assessed by whether they appear damaged or look unappealing, and often they are less likely to be purchased [52]. The quality of fresh produce and perceived quality of one's food environment directly affect dietary intake [55,59]. Therefore a closer inspection of medium to small-sized stores as an alternate source of healthy, good quality, and affordable food items is required in an area where large supermarkets are hard to find.

This study attempts to contribute to the research of nutrition, food access, and urban food systems by including medium and small sized grocery stores in the definition of food deserts and expanding the concept of accessibility to a multidimensional construct [60]. Rather than defining food access as simply physical distance to a large supermarket, here we examine accessibility (distance to stores), availability (variety of food items), affordability (price), quality of fresh produce, and appearance of small, medium, and larger grocery stores in greater Hartford, Connecticut (CT), to provide a more robust and accurate measure of the urban food environment. Defining access to only large supermarkets can overlook the contribution that small to medium-sized groceries can play in an urban food systems [61]. Therefore we also seek to determine how small and medium-sized grocery stores compare with larger supermarkets in regard to the different dimensions of accessibility mentioned above.

Study Setting: City of Hartford, Connecticut

The City of Hartford in Connecticut (CT), with its diverse demographic, socioeconomic, and health disparity indicators provides a unique opportunity to explore the multidimensional construct of accessibility to not only healthy food but also 'affordable' and 'quality' food. With a population of 124,893, Hartford has an estimated poverty rate of 32.9%, more than double the US poverty rate of 15% [62,63]. Almost one-half of children in Hartford live below the poverty line (47.9% for a total of 14,814 children), compared to the US child poverty rate of 21.8%. The unemployment rate in Hartford in April 2013 was 14.8% [64], compared to approximately 7% nationally [65]. The 2011 median household income was estimated at $29,169, which is less than half of the estimated median household income for the county of Hartford, as well below the median for the United States, which was $50,502 [66,67]. Almost one-third (31.8%) of Hartford households have no vehicle [67].

The youngest members of the Hartford community are at increased risk of diet-related diseases due to nutritionally imbalanced access to foods in their neighborhoods. A 2012 study found that 37% of preschool children in Hartford are overweight or obese, making the prevalence of childhood obesity among preschoolers more than twice as high as Centers for Disease Control and Prevention age and gender body mass index guidelines [68].

The City of Hartford Advisory Commission on Food Policy has historically focused on ensuring that a wide variety of safe and nutritious foods are available for City residents. The Advisory Commission, which was formed in 1991, was one of the first commissions in the US to make nutritious and affordable food a priority. One of the four goals of the Commission is "To ensure that the price of food in the city remains reasonably close to the average price existing in the balance of the state" (Bylaws section 2–327 part b, goal 3 Hartford, CT, 1991) [69]. The Commission has conducted food price surveys in grocery stores on an ad-hoc basis since 1996 and has endorsed this research as supportive of Commission goals. Two of this study's researchers currently serve as Commissioners. The ultimate goal of this empirical research is to inform policies related to healthy food access and food insecurity in urban communities such as Hartford.

Study Goals, Objectives, and Research Questions

The objectives of this empirical study were to examine the *accessibility* (proximity to grocery stores), *affordability* (food prices) and *quality* (whether fruits and vegetables appear damaged or look unappealing and overall appearance of the store) of healthy food sold at large, medium, and small-sized grocery stores in Hartford and within a two-mile radius of the city. The specific research questions are:

1. Do Hartford residents have access to large, medium, and small sized grocery stores?
2. Do availability, price and quality differ by urban/suburban location?
3. Do availability, price and quality differ by store size?

The overall goal was to assess whether low-income households in Hartford who do not own or have limited access to a car would be able to access stores with quality and affordable food. The hypothesis was that small and medium-sized stores that are more common in urban areas could be a potential alternative to scarce presence of large supermarkets. Results can then help inform policy recommendations by the Advisory Commission on Food Policy to City leaders regarding where city resources should be allocated to increase access to healthy foods.

Methodology

Between October 2012 and March 2013, surveys were conducted in 36 stores in Hartford and within a 2-mile radius of

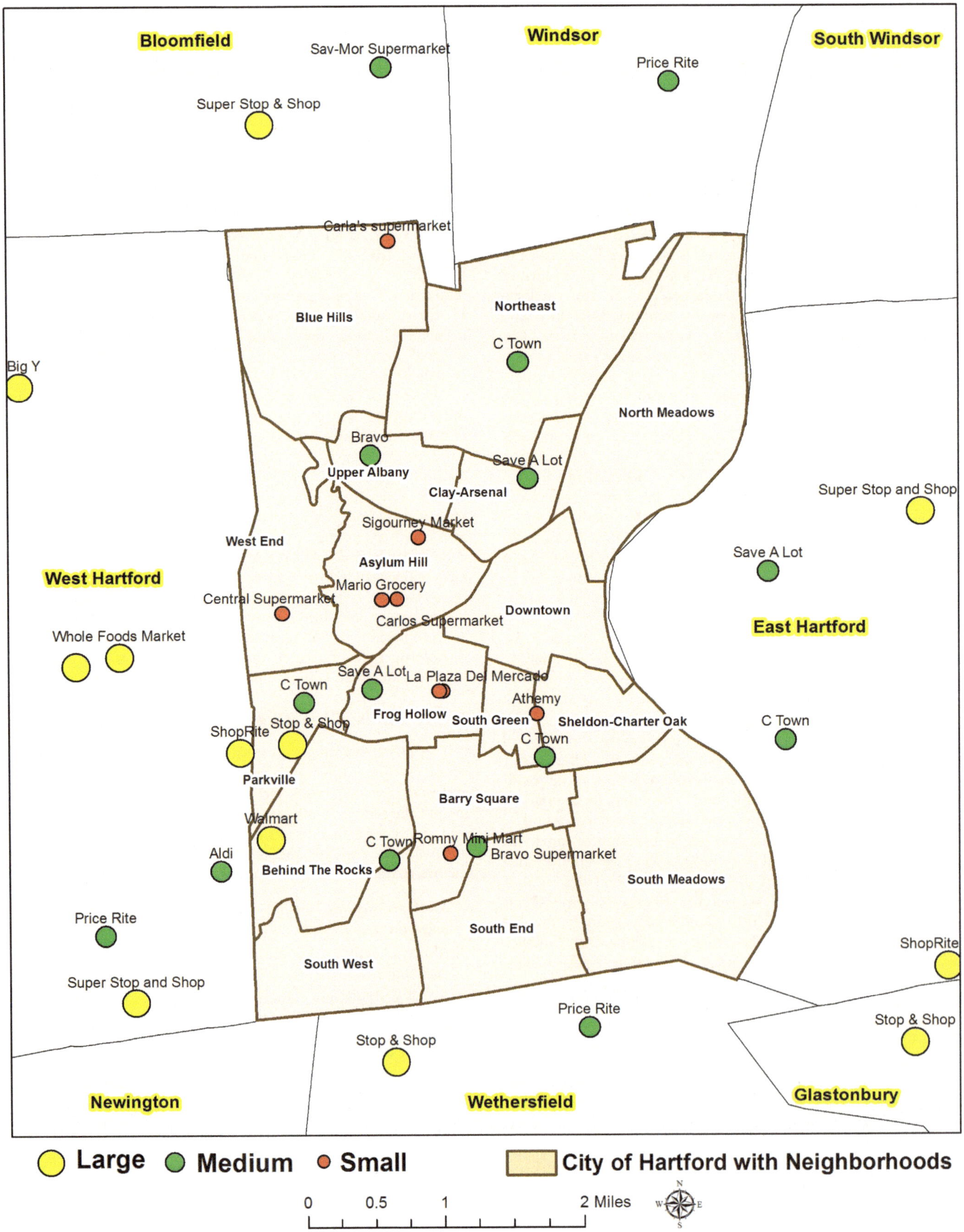

Figure 1. Location of surveyed stores in the City of Hartford, Connecticut.

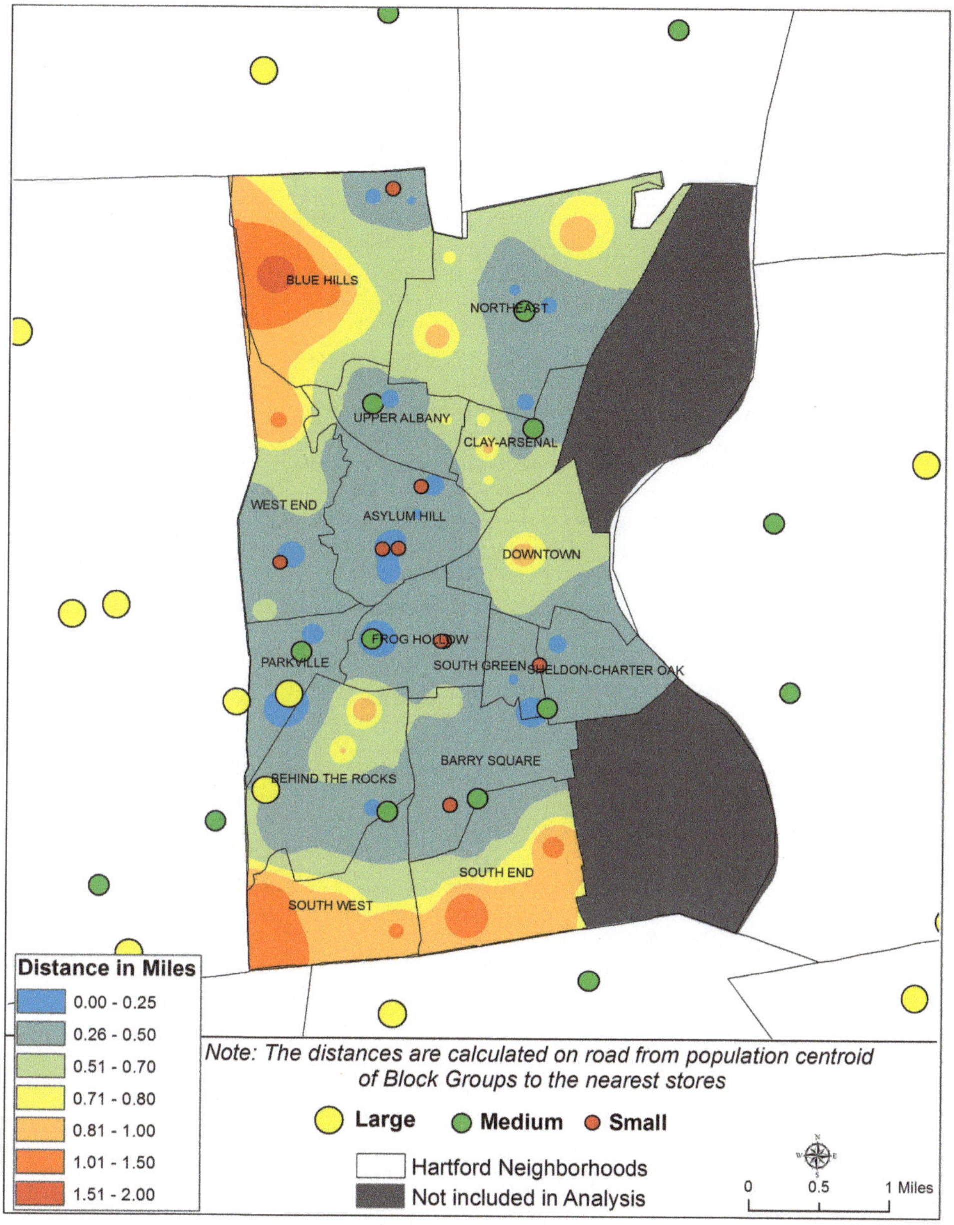

Figure 2. Distance (in miles) to stores from the population centroids of US Census Block Groups.

Table 1. Distance (in miles) from the center of the US census block groups to the nearest grocery store.

Distance Groups (in miles)	Number of Block Groups (N = 93) (%)
0.00–0.25	17 (18.3)
0.26–0.50	41 (44.1)
0.51–0.70	13 (14.0)
0.71–0.80	6 (6.5)
0.81–1.00	8 (8.6)
1.01–1.50	7 (7.5)
1.51–2.00	1 (1.1)

Hartford including parts of several adjacent suburban towns (See Figure 1). The 2-mile radius was drawn from the outline of the City of Hartford. We used a two-mile radius for two reasons: first, the residents of Hartford often shop outside the city limits, and second, to minimize errors from edge effects in the subsequent mapping and spatial analysis. This represents an inventory of all small, medium, and large groceries. The details of selection criteria, size, characteristics, and example of each type of stores are described later. We did not seek formal (or written) permission to conduct these surveys for the following reasons: a) the selected grocery stores are open to the public and do not belong to any private land or property; b) the researchers followed a participant observation data collection technique where they only observed and took notes and did not interact with any human subjects (other shoppers or store workers) during the survey; c) as detailed later, the survey instrument required taking notes of the sale prices and the quality of various food items, which are public information (Refer to Price Survey Instrument); and d) all the researchers have training in conducting ethical research. This survey did not involve endangered or protected species.

Stores were identified based on previous research on the Hartford food environment [61,70] and using a variety of methods to ensure sample completeness, including online yellow pages, business listings, and more importantly "ground-truthing" of driving through neighborhoods to identify stores. This represents an inventory of all medium, and large size grocery stores in the study area, as well as several smaller stores that, based on inventory and selection, operate as grocery stores rather than convenience stores. The City of Hartford, like many other urban settings, has a number of small and medium-sized grocery stores that stock a full array of groceries and are larger than typical corner stores, but smaller than supermarkets.

These "small" stores are defined as independent food stores between 1,000 and 2,500 ft^2 based on field measurements from previous research in Hartford [70]. Medium-sized grocery stores are approximately 15,000–39,999 ft^2 and generally stock a limited number of custom-brand high-volume food items at discount [71]. Examples of stores in this category include Save-A-Lot and Bravo. Large supermarkets range from 40,000 to 80,000 ft^2 and typically include delicatessens, bakeries, pharmacies, and general merchandise in addition to groceries. Examples of stores in this category include ShopRite, Big Y, and Stop and Shop. Stores were also categorized based on whether they were certified to accept vouchers for the Special Supplemental Nutrition Program for Women, Infants and Children (known as WIC) based on data from the Connecticut Department of Public Health. WIC certified stores are required to stock a limited number of fruits and vegetables so they may be more inclined to have produce available. Two of the grocery stores that fit the study criteria were no longer in business and were removed from the sample. Convenience stores, such as bodegas and corner stores, were not included in the study group. Trained researchers participated in a practice survey at several grocery stores prior to the start of data collection. Researchers subsequently completed the data collection in all stores in the study group.

Physical Accessibility and Mapping

To address research question #1 (access to large, medium, and small size grocery stores in Hartford), we used Geographic Information System (GIS) for mapping and spatial analysis. To begin, the 36 surveyed stores were geocoded. Geocoding is the process by which a postal address or any other locational information is associated with geographic units such as points (latitude and longitude coordinates), postal zip codes, or US Census block groups or census tracts [72]. For this study the postal addresses of the surveyed stores were geocoded to points (latitude and longitude).

To measure accessibility (or distance to stores), we used ancillary GIS data sets downloaded from various sources. These data sets are: a) Boundaries of US Census block groups [73], b) Boundaries of neighborhoods in Hartford [74], c) Detailed road network data [73], and d) Location of bus Stops [74]. The transit data was limited and had only locations of bus stops with no additional information of time schedule, routes, or transfers.

Next, we calculated distances on road network from the population centroid of a block group in Hartford to the nearest store using the 'closest facility' function of the Network Analyst Extension in ArcGIS 10.1 [75]. This function measures the distance (sometimes shortest route) of traveling between incidents (population centroid of a block group in our study) and facilities (surveyed store) and determines which are nearest to one another [75]. This measurement is different than calculating Euclidean distances or "as the crow flies" distance between two points [76] and is a better reflection of how people travel and perceive distance. Two Hartford neighborhoods (North and South Meadows) that are non-residential were excluded from the analysis.

Once the distances were calculated from each of the centroids, we then created a continuous distance surface by Inverse Distance Weighting interpolation technique to calculate distances to the nearest stores [72]. Using a spider-diagram function we calculated how many stores have at least one bus stop within a distance of 0.25 miles [77]. This distance cut-off was determined based on the inconvenience of walking over a quarter mile with groceries. Spider Diagram function creates a line representing the shortest distance between centers (surveyed stores in our study area) and the closest destinations (locations of bus stops in our study area). Finally, the variability of prices and quality of available food by store size (large, medium, and small) and location (Hartford versus suburban) were also mapped for better visualization and understanding.

Survey Instrument

To address the above-mentioned research questions #2 (do availability, price and quality differ by store size?) and #3 (do availability, price and quality differ by urban/suburban location) data were collected using a grocery store survey of a market basket of food. The survey instrument is based on previous price surveys conducted by the Hartford Advisory Commission on Food Policy [78]. (Refer to File S1 - Price Survey Instrument). The instrument included a list of 25 grocery items, categorized by milk, cheese, fresh produce, protein, bread, canned and staple items, and beverages. Items were chosen to represent a standard basket of healthy food items that could be used to prepare breakfast, lunch, and dinner and included produce items that would be most likely available during all seasons. The standard size of each item was listed on the survey. The least expensive brand or option available in the store was recorded as an item's price; sale price was not used unless it was the only price posted, and if so, the sale was noted in the comments field. Produce price was calculated by an item's price per weight and standard measures for each item were noted on the survey instrument. If an item was not present in the store at the time of the survey, researchers noted it as unavailable.

Quality of fresh produce was measured on a 1 to 4 Likert scale. A score of 1 indicated most of the item was of poor quality, such as brown, bruised, overripe or wilted. A score of 2 indicated more poor than good; 3 indicated more good than poor. A score of 4 meant most of the item was of good quality, very fresh, no soft

Table 2. Characteristics of Stores: Comparison between Hartford and Suburban location.

Characteristic	Subcategory	Hartford Stores Number (%)	Suburban Stores Number (%)
Total Number of Stores		19 (52.8)	17 (47.2)
Store Size			
	Small	9 (47.4)	1 (5.9)
	Medium	8 (42.1)	7 (41.2)
	Large	2 (10.5)	9 (52.9)**
WIC Certified		13 (68.4)	10 (58.8)
Quality and Appearance			
	High internal quality score of 4	6 (31.6)	12 (70.6)*
	High external quality score of 4	2 (10.5)	11 (64.7)**
	High Produce quality scores from 36–40	8 (42.1)	14 (82.4)*
		Average (SD)	**Average (SD)**
	Internal quality (scale of 1–4)	0.1 (0.8)	3.7 (0.5) *
	External quality (scale of 1–4)	2.5 (0.7)	3.6 (0.6) **
	Produce quality (total score of 40)	34.9 (3.8)	37.9 (3.5) *
Availability of 25 food items		22.8 (3.6)	24.7 (0.6)
Total Price of 25 food items		$47.77 (5.33)	$48.46 (10.5)

SD = Standard Deviation.
* = $0.01 < p < 0.05$, ** = $P < 0.01$.

spots, and of excellent color. Each store was also rated on internal and external appearance, and was scored from 1–4 as poor, fair, acceptable, or good respectively. Internal quality included the overall appearance, lighting, cleanliness, and organization inside the store. External quality was based on the appearance, lighting, perceived safety, and parking of the store's exterior. Researchers assessed several stores as a team and discussed the scoring as a group to ensure standardization for measuring and evaluating store appearance and food quality.

Data Analysis

Data were analyzed using PASW v.18. Some variables were not normally distributed including internal and external appearance, produce quality, and availability of food items. For these variables, non-parametric tests were used. Price was normally distributed and continuous, so t-tests and ANOVA were run to compare means. Significance was determined at $p<0.05$. Comparisons between Hartford and suburban stores were made using Fisher's exact chi-square test for small cell sizes and Mann-Whitney U test for non-parametric data. Quality measures were included as continuous variables and also dichotomized for ease of interpretation. Scores for internal and external appearance were dichotomized between low (1–3) and high (4) scores, and produce quality between low (0–35) and high scores (36–40) based on the distribution of the data. When comparing stores by the three store sizes, comparisons were made using Kruskal-Wallis test for independent samples for non-parametric data.

Results

Distribution of Grocery Stores in Greater Hartford

The sample includes 36 large, medium, and small-sized grocery stores, including 20 stores (55%) within the city limit of Hartford and 16 stores (45%) in the adjacent suburban towns, located in the 2-mile radius from the city boundary. The majority of stores (64%) are certified to accept vouchers for WIC. No significant differences exist in the availability of WIC certified stores, neither between the City of Hartford and the suburban stores, nor by store size.

Accessibility to Grocery Stores in Hartford

Distance in miles from the population centroid of the census block groups to the nearest large, medium, and small sized grocery stores in the study area is shown in Figure 2. Based on the location of the stores, residents living in the central and south central part of the City of Hartford are on average 0.25 to 0.50 miles distance away from stores. Within this area there are a few pockets of lesser accessibility, where residents have to travel more than 0.5 mile to a grocery store. Residents living in Downtown Hartford are at a distance of 0.75 to 1 mile from a store indicating a situation of urban 'food desert'. Moving farther away from the south and central areas of the City, spatial accessibility to any grocery stores decreases. In some areas, especially in the neighborhoods of Blue Hills in the North, and South West and South End in the South there is not a single grocery store within a distance of 2 miles. Out of 93 block groups in the City of Hartford, 44% are at an average distance of 0.25 to 0.50 miles from a store (Table 1). The maximum distance from the center of a block group to the nearest surveyed store is 1.98 miles. The minimum is 0.05 miles and the average distance is 0.57 miles.

We further explored the accessibility to grocery stores in terms of public transportation in the City of Hartford, using the location of bus stops. All 36 stores have at least one bus stop (very often more than one) within a distance of 0.25 miles from the location of the stores. This is true for all the store sizes (large, medium, and small) and locations (Hartford versus Suburban). Four stores have no bus stop within a distance cut-off of 0.1 miles, all of which are

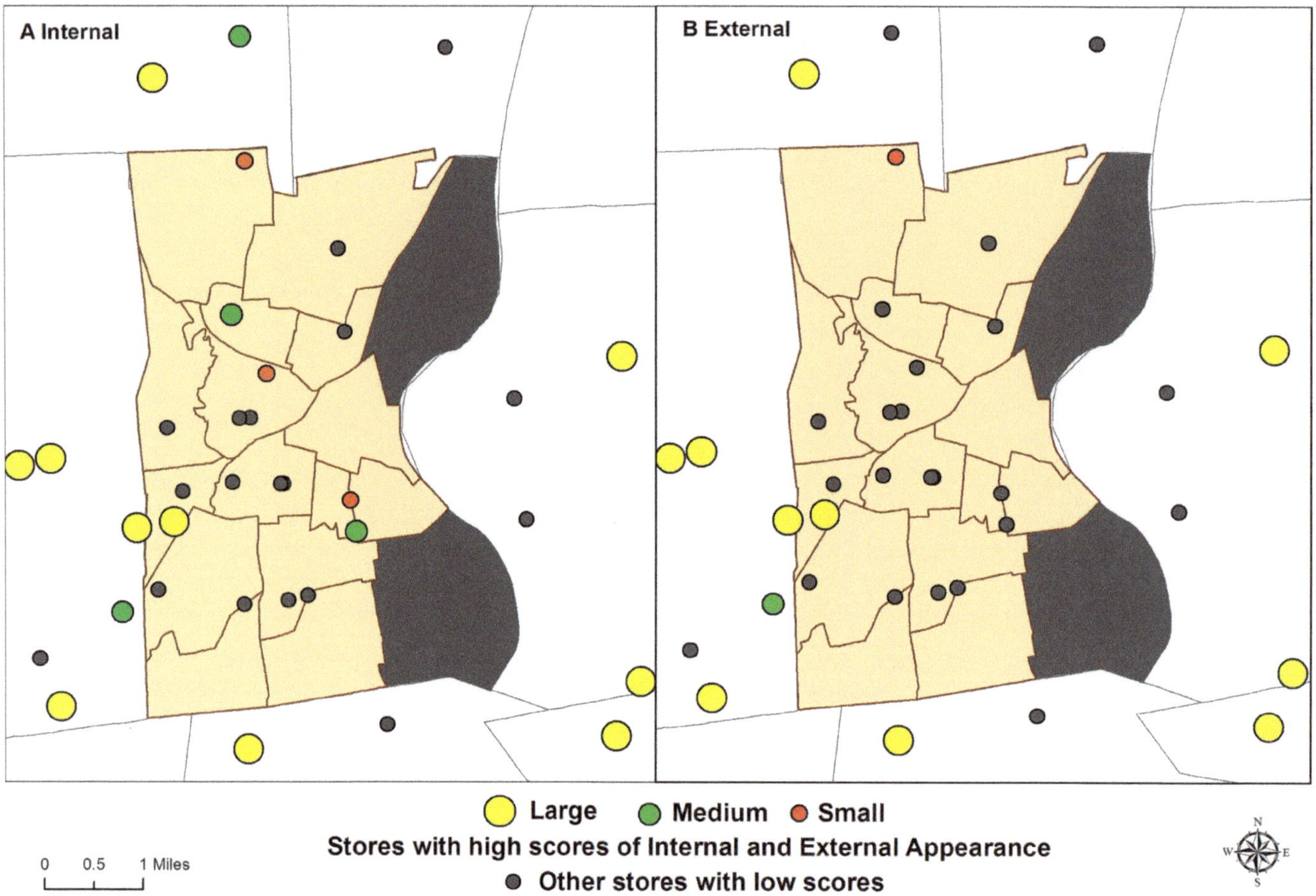

Figure 3. Stores with high scores (Score = 4) of Internal and External appearance.

large sized grocery stores. This could be due to larger parking lots for large-sized grocery stores, which restricted the locations of bus stops beyond 0.1 miles.

Internal and External Appearance/Quality

Significant differences exist between stores in Hartford compared to the suburbs regarding quality of fruits and vegetables, and the internal and external appearance of the stores. Hartford stores were significantly more likely to have low internal store quality, a measure of appearance, lighting, cleanliness and organization. Only six stores (32%) in Hartford had high internal quality scores compared to 12 suburban stores (71%, $p = 0.02$). Of the six stores in Hartford with high internal quality, one was large, two were medium, and three were small sized stores (Table 2).

Similarly, Hartford stores scored much lower on external store quality, a measure of lighting, perceived safety and parking outside the store. Hartford stores were significantly less likely to have high external appearance scores, with only two Hartford stores (11%) receiving high scores, compared to 11 suburban stores (65%, $p < 0.01$). Only one small sized grocery store located in the northern neighborhood of Hartford had high scores for both internal and external quality or appearance. This is notable because this area has a high proportion of low-income minority population with limited access to large supermarkets. Figure 3 displays the variation of internal and external appearance scores by location (Hartford versus suburban) and size of stores (large, medium, small). A clear spatial pattern exists where stores, irrespective of their size, had higher scores for both internal and external appearance in the suburban towns.

Produce Quality

Quality of fresh fruits and vegetables also differed significantly between Hartford and suburban stores. Eight Hartford stores (42%) received high produce quality scores, compared with 14 suburban stores (82%, $p = 0.02$), (Figure 4). In the City, two small grocery stores had high quality produce. Three medium suburban grocery stores had low scores for produce quality, which is contrary to current literature and assumptions that suburban neighborhoods will carry high quality healthy food (Table 2).

Price and Availability of Food Items

No significant differences were found for the price and availability of food items between stores in Hartford and surrounding suburbs. The average price of the 25 items for the overall sample was $48.10; with an average price of $47.77 in Hartford stores, and $48.46 in the suburbs. The majority of stores (64%) carried all items, 22% were missing 1–2 items, and 14% had fewer items. A noticeable exception was two small stores in Hartford that only carried 14 of the 25 items.

Comparisons by Store Size

Significant differences existed for quality, availability, and price, between small to medium, medium to large and between small

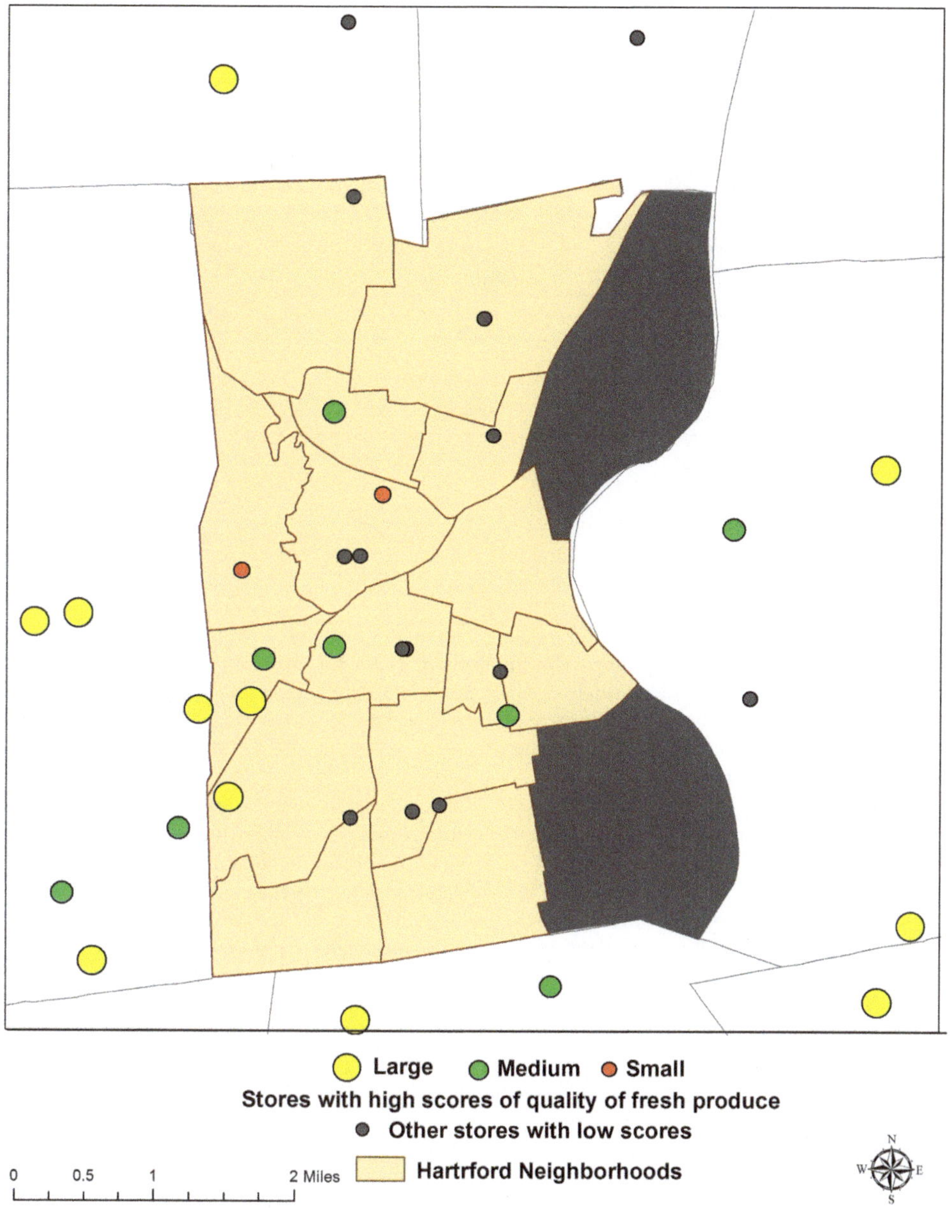

Figure 4. Stores with high scores (Score = 4) of produce quality.

and large stores (Table 3). Large stores had better scores for internal, external and produce quality variables. Medium-sized stores had greater availability compared to small stores, and had lower prices than either small or large stores.

Limitations

The analysis of transit accessibility to stores using just the location of bus stops has limitations and does not capture the complexity of transit network. We only considered the number of bus stops within a distance of 0.25 miles from a store and did not have other supporting information such as number of buses, routes, frequency, schedules, or transfers. This additional information would provide a better measure of accessibility of the grocery stores by public transport.

The scale used to assess internal, external, and produce quality may not have captured the full range of variation among items or store quality. The scale was a subjective measure of quality, although clear guidance on how to rate store quality and fresh produce was given to the survey team. Data was collected from stores in one urban area as a case study; therefore results may not be generalizable to other communities. Due to the small sample, it is possible that some insignificant findings may be significant with a larger number of stores. For example, a larger percentage of medium-sized stores were WIC certified compared to small stores, yet this was not statistically significant. Surveys were administered at one point in time. We assume that quality, availability, and price are fairly stable over time, although produce supply may vary seasonally.

Discussion and Conclusion

Conventional literature and groundwork on urban food systems emphasizes limited access to healthy food in inner-city low-income

Table 3. Characteristics of Stores: Comparison Between Small, Medium and Large sized stores.

Characteristic	Small Stores	Medium Stores	Large Stores
	Number (%)	Number (%)	Number (%)
Total number of stores	10 (27.8)	15(41.7)	11 (30.6)
WIC Certified	4 (40)	11 (73.3)	8 (72.7)
	Average (SD)	**Average (SD)**	**Average (SD)**
Internal quality (scale of 1–4)	3 (1.1)	3.3 (0.5)* b	3.9 (0.3)* c
External quality (scale of 1–4)	2.6 (0.8)	2.7 (0.6)** b	3.8 (0.6)** c
Produce quality (total score of 40)	33.4 (3.7)	35.7 (3.7)* b	39.7 (0.5)** c
Availability of 25 food items	21.7 (4.3)	24.5 (0.9)* a	24.5 (1.8)
Total Price of 25 food items	52.47 (5.0)	41.94 (5.5)** ab	52.52 (8.1)

SD = Standard Deviation.
* = 0.01 < p < 0.05, ** = P < 0.01.
a = significant difference between small and medium stores.
b = significant difference between medium and large stores.
c = significant differences between small and large stores.

neighborhoods compared to surrounding suburbs, with fewer large supermarkets and higher food prices [42,79]. In contrast, results from this study, based in Hartford, Connecticut, found that healthy foods are equally available and sometimes less expensive in local stores in the City compared to suburban stores. Although physical accessibility to grocery stores is not uniform in all Hartford neighborhoods, most Hartford residents live within a 0.5 to 1 mile distance to a grocery store. However, there are a few pockets that would be considered *food deserts* – with no grocery store within a 1-mile or even 2 miles radius.

One of the major findings is the significant variability of quality of produce and store appearance, with Hartford stores faring much worse than the suburban stores. The lower quality may impact customers' willingness to shop in these stores or to purchase fruits and vegetables that contribute to a healthy diet, which can help prevent or mitigate chronic diseases [50]. Improving food quality and appearance of smaller stores in urban inner cities can potentially impact the food purchasing decisions of low-income residents in Hartford and a potential improvement to food insecurity [59].

Focusing solely on large supermarkets may underestimate food access in urban, low-income cities where medium and small-sized grocery stores are more prevalent [26]. Classifying areas with few large supermarkets as food deserts may overlook the availability of healthy foods that exist within small and medium-sized groceries found in inner cities [61]. Results from this study document the availability and lower prices found in the City of Hartford and shed light on the important role that medium-sized grocery stores play in the accessibility, availability and affordability of healthy food in urban areas.

Many of the store owners from small and medium-sized markets in Hartford live locally. Therefore, efforts to improve the business infrastructure and sales of these markets will help support the local economy, which is in line with the principles of healthy, sustainable food systems. In comparison, large supermarkets tend to be owned by national or often international companies where revenues are not reinvested into the city. Studies have shown that store owners' personality and established friendships between owners and patrons fosters store loyalty, especially in neighborhoods without a large supermarket [80,81]. City leadership may want to consider investing in façade improvements, such as internal and external store appearance, and ways to improve produce quality to boost the local economy and potentially improve food purchasing and consumption patterns of low-income residents. These improvements would make the role of small to medium-sized local grocery stores in the urban food system critical in addressing the issues of food deserts in the underdeveloped parts of the city.

Polices and logistics surrounding the opening of large chain supermarkets in impoverished neighborhoods are not easy to implement partly due to a lack of a stable markets and partly due to lack of infrastructure related to easy access to highways, large loading docks for large trucks to unload, and distribution networks. Recent research also indicates that simply placing a large supermarket in a low-income urban community did not improve dietary intake [84]. Large stores are often disinclined to locate their stores in inner cities and usually pull their existing stores out and relocate them to suburbs [43,44]. As the stores close, impacted urban residents have to travel further to purchase nutritious and affordable groceries or turn to small and medium-sized local grocers. In the City of Hartford, eleven of thirteen chain supermarket stores (almost 85% of the stores) have left the city between 1968 and 1984 [82]. Recently, Russell and Heidkamp [83] found that a food desert was created when Shaw's (http://www.shaws.com) was closed in the city of New Haven, CT, which has similar socioeconomic, demographic, and health disparity characteristics to that of Hartford.

This research contributes to nutrition, access to healthy food, and sustainable food systems literature by emphasizing differences based on location and store size. Results from this study can help influence recommendations for the Hartford Advisory Commission on Food Policy and promote action from other city agencies to improve food quality and store appearance in Hartford stores. In addition, this research may contribute to decisions by policymakers in other urban communities to allocate resources to improve existing small and medium sized markets before creating new large stores. Improving the quality of food, and store appearance in urban stores are relatively more plausible than locating a large chain supermarket. These improvements may influence purchasing decisions of low-income households who lack adequate transportation, and improve poor dietary intake linked to health disparities and food insecurity.

Author Contributions

Conceived and designed the experiments: K. Martin DG MP. Performed the experiments: K. Martin DG MP MW K. McMinimee. Analyzed the data: K. Martin DG MZ. Contributed reagents/materials/analysis tools: K. Martin DG MZ. Wrote the paper: K. Martin DG MW.

References

1. Block D, Kouba J (2006) A comparison of the availability and affordability of a market basket in two communities in the Chicago area. Public Health Nutrition 9: 837–845.
2. Baker E, Schootman M, Barnidge E, Kelly C (2006) The Role of Race and Poverty in Access to Foods that Enable Individuals to Adhere to Dietary Guidelines. Preventing Chronic Disease 3: 1–7.
3. Jetter KM, Cassady DL (2006) The availability and cost of healthier food alternatives. American Journal of Preventive Medicine 30: 38–44.
4. Larson NI, Story MT, Nelson MC (2009) Neighborhood Environments Disparities in Access to Healthy Foods in the US. American Journal of Preventive Medicine 36: 74–81.
5. Treuhaft S, Karpyn A (2010) The grocery gap: who has access to healthy food and why it matters. Oakland, CA and Philadelphia, PA: PolicyLink and the Food Trust.
6. Ploeg M (2010) Access to affordable, nutritious food is limited in "Food Deserts". Washington, DC: Amber Waves.: Economic Research Service, United States Department of Agriculture (USDA).
7. Gallagher M (2010) Examining the impact of Food Deserts and Food Imbalance on Public Health in Birmingham, Alabama. Chicago, IL: The Mari Gallagher Research & Consulting Group.
8. Gallagher M (2012) Food and Health in Washington, DC. Chicago, IL: The Mari Gallagher Research & Consulting Group.
9. Avilés-Vázquez KR, Bussmann RW (2009) A Review of "The Future Control of Food: A Guide to International Negotiations and Rules on Intellectual Property, Biodiversity and Food Security". The Professional Geographer 61: 423–424.
10. Beaulac J, Kristjansson E, Cummins S (2009) A Systematic Review of Food Deserts, 1966–2007. Preventing Chronic Disease 6(3): A105. Available: http://www.cdc.gov/pcd/issues/2009/jul/08_0163.htm. Accessed 2013 May 22.
11. McKinnon RA, Reedy J, Morrissette MA, Lytle LA, Yaroch AL (2009) Measures of the food environment: a compilation of the literature, 1990–2007. Am J Prev Med 36: S124–133.
12. Walker RE, Keane CR, Burke JG (2010) Disparities and access to healthy food in the United States: A review of food deserts literature. Health & Place 16: 876–884.
13. Pearson T, Russell J, Campbell MJ, Barker M (2005) Do 'food deserts' influence fruit and vegetable consumption? - A cross-sectional study. Appetite 45: 195–197.
14. Morland K, Diez Roux AV, Wing S (2006) Supermarkets, Other Food Stores, and Obesity: The Atherosclerosis Risk in Communities Study. American Journal of Preventive Medicine 30: 333–339.
15. United States Department of Agriculture (USDA) Economic Research Service (2012) Food Access Research Atlas. Available: http://www.ers.usda.gov/topics/food-nutrition-assistance/food-security-in-the-us.aspx -.UkQNLhYTHzI. Accessed 2013 Mar 22.
16. Hendrickson D, Smith C, Eikenberry N (2006) Fruit and vegetable access in four low-income food deserts communities in Minnesota. Agriculture and Human Values 23: 371–383.
17. Hubley TA (2011) Assessing the proximity of healthy food options and food deserts in a rural area in Maine. Applied Geography 31: 1224–1231.
18. Smith C, Morton LW (2009) Rural Food Deserts: Low/income Perspectives on Food Access in Minnesota and Iowa. Journal of Nutrition Education and Behavior 41: 176–187.
19. Gallagher M (2006) Examining the impact of food deserts on public health in Chicago. Chicago, IL: Mari Gallagher Research & Consulting Group.
20. Hallett LF, McDermott D (2011) Quantifying the extent and cost of food deserts in Lawrence, Kansas, USA. Applied Geography 31: 1210–1215.
21. Larsen K, Gilliland J (2008) Mapping the evolution of 'food deserts' in a Canadian city: Supermarket accessibility in London, Ontario, 1961–2005. International Journal of Health Geographics 7.
22. Whelan A, Wrigley N, Warm D, Cannings E (2002) Life in a 'food desert'. Urban Studies 39: 2083–2100.
23. Wrigley N (2002) 'Food deserts' in British cities: Policy context and research priorities. Urban Studies 39: 2029–2040.
24. Cummins S, Macintyre S (2002) "Food deserts" - evidence and assumption in health policy making. British Medical Journal 325: 436–438.
25. Morland K, Wing S, Roux AD, Poole C (2002) Neighborhood characteristics associated with the location of food stores and food service places. American Journal of Preventive Medicine 22: 23–29.
26. Moore LV, Roux AVD (2006) Associations of neighborhood characteristics with the location and type of food stores. American Journal of Public Health 96: 325–331.
27. Sparks AL, Bania N, Leete L (2011) Comparative Approaches to Measuring Food Access in Urban Areas: The Case of Portland, Oregon. Urban Studies 48: 1715–1737.
28. Sadler RC, Gilliland JA, Arku G (2011) An application of the edge effect in measuring accessibility to multiple food retailer types in Southwestern Ontario, Canada. International Journal of Health Geographics 10.
29. Widener MJ, Metcalf SS, Bar-Yam Y (2011) Dynamic urban food environments: a temporal analysis of access to healthy foods. American journal of preventive medicine 4: 439–441.
30. Burgoine T, Monsivais P (2013) Characterising food environment exposure at home, at work, and along commuting journeys using data on adults in the UK. International Journal of Behavioral Nutrition and Physical Activity 10(1): 85.
31. Widener MJ, Farber S, Neutens T, Horner MW (2013) Using urban commuting data to calculate a spatiotemporal accessibility measure for food environment studies. Journal of Health and Place 21: 1–9.
32. Fuller D, Cummings S, Mathews SA (2013) Does transportation mode modify associations between distance to food store, fruit and vegetable consumption, and BMI in low-income neighborhoods? The American journal of clinical nutrition 97: 167–172.
33. Anderson S (1990) Core indicators of nutritional status for difficult to sample populations. Journal of Nutrition 20: 1557–1600.
34. Seligman HK LB, Kushnel MB (2010) Food insecurity is associated with chronic disease among low-income NHANES participants. J Nutr 140: 304–310.
35. United States of Department of Agriculture (USDA) (2012) Economic Research Service: Food Security in the United States Briefing Room. Available: http://www.ers.usda.gov/briefing/foodsecurity/stats_graphs.htm. Accessed 2013 Mar 22.
36. Cummins S, Macintyre S (2002) A systematic study of an urban foodscape: The price and availability of food in Greater Glasgow. Urban Studies 39: 2115–2130.
37. Raja S, Ma CX, Yadav P (2008) Beyond food deserts - Measuring and mapping racial disparities in neighborhood food environments. Journal of Planning Education and Research 27: 469–482.
38. LeDoux T, Vojnovic I (2013) Going outside the neighborhood: the shopping patterns and adaptations of disadvantaged consumers living in the Lower eastside neighborhoods of Detroit, Michigan. Health & Place 19: 1–14.
39. Cotterill R, Franklin A (1995) The urban grocery store gap: An issue paper. Storrs CT: Department of Agriculture and Resource Economics, University of Connecticut.
40. Powell LM, Auld MC, Chaloupka FJ, O'Malley PM, Johnston LD (2007) Associations between access to food stores and adolescent body mass index. American Journal of Preventive Medicine 33: S301–S307.
41. Lisabeth L, Sa'nchez BN, Escobar J, Hughes R, Meurer WJ, et al. (2010) The food environment in an urban Mexican American community. Health and Place 16: 598–605.
42. Gordon C, Purciel-Hill M, Ghai NR, Kaufman L, Graham R, et al. (2011) Measuring food deserts in New York City's low-income neighborhoods. Health Place 17: 696–700.
43. Eisenhauer E (2001) In poor health: Supermarket redlining and urban nutrition. GeoJournal: 125–133.
44. D'Rozario D (2005) Retail Redlining: Definition, Theory, Typology, and Measurement. Journal of Macromarketing 25: 175–186.
45. Alwitt L, Donley T (1997) Retail stores in poor urban neighborhoods. The Journal of Consumer Affairs 31: 139–164.
46. Chung C, Myers S (1999) Do the poor pay more for food? An analysis of grocery store availability and food price disparities. The Journal of Consumer Affairs 33: 276–296.
47. Hendrickson D, Smith C, Eikenberry N (2006) Fruit and vegetable access in four low-income food deserts communities in Minnesota. Agriculture and Human Values 23: 371–383.
48. Sheldon M, Gans KM, Tai R, George T, Lawson E, et al. (2010) Availability, Affordability, and Accessibility of a Healthful Diet in a Low Income Community. Preventing Chronic Disease 7.
49. Smith DM, Cummins S, Taylor M, Dawson J, Marshall D, et al. (2010) Neighbourhood food environment and area deprivation: spatial accessibility to grocery stores selling fresh fruit and vegetables in urban and rural settings. International Journal of Epidemiology 39: 277–284.
50. Webber CB, Sobal J, Dollahite JS (2010) Shopping for fruits and vegetables: food and retail qualities of importance to low-income households at the grocery store. Appetite 54: 297–303.
51. Ball K, Timperio A, Crawford D (2009) Neighbourhood socioeconomic inequalities in food access and affordability. Health & Place 15: 578–585.
52. Cummings S, Smith DM, Taylor M, Dawson J, Marshall D, et al. (2009) Variations in fresh fruit and vegetable quality by store type, urban-rural setting and neighbourhood deprivation in Scotland. Public Health Nutrition 1: 2044–2050.

53. Thorton LE, Crawford DA, Ball K (2010) Neighbourhood-socioeconomic variation in women's diet: the role of the nutrition environments. European Journal of Clinical Nutrition 64: 1423–1432.
54. Hall B (1983) Neighborhood differences in retail food stores: income versus race and age of population. Economic Geography 59: 282–295.
55. Zenk SN, Schulz AJ, Israel BA, James SA, Bao SM, et al. (2005) Neighborhood racial composition, neighborhood poverty, and the spatial accessibility of supermarkets in metropolitan Detroit. American Journal of Public Health 95: 660–667.
56. Zenk SN, Schulz AJ, Hollis-Neely T, Campbell RT, Holmes N, et al. (2005) Fruit and vegetable intake in African Americans: income and store characteristics. American Journal of Preventive Medicine 29: 1–9.
57. Zenk SN, Schulz AJ, Israel BA, James SA, Bao S, et al. (2006) Fruit and vegetable access differs by community racial composition and socioeconomic position in Detroit, Michigan. Ethnicity and Disease 16: 275–280.
58. Glanz K, Sallis J, Saelens B, Frank L (2007) Nutrition environment measures survey in stores (NEMS-S): development and evaluation. American Journal of Preventive Medicine 32: 282–289.
59. Blitstein J, Snider J, Evans D (2012) Perceptions of the food-shopping environment are associated with greater consumption of fruits and vegetables. Public Health Nutrition 15: 1124–1129.
60. Penchansky R, Thomas JW (1981) The concept of access: definition and relationship to consumer satisfaction. Med Care 19(2): 127–140.
61. Martin K, Havens E, Boyle K, Matthews G, Schilling E, et al. (2012) If you stock it, will they buy it? Healthy food availability and customer purchasing behavior within corner stores in Hartford, CT. Public Health Nutrition 15: 1973–1978.
62. United States Census Bureau (2013) Quick Facts. Available: http://quickfacts.census.gov/qfd/states/09/0937000.html Accessed 2014 Jan 22.
63. United States Census Bureau (2013) Current Population Survey, 2013. Available: http://www.census.gov/hhes/www/poverty/about/overview/index.html. Annual Social and Economic Supplement. Accessed 2014 Jan 22.
64. Connecticut Department of Labor (2013) Current labor force data for Connecticut towns. Available: http://www1.ctdol.state.ct.us/lmi/laus/lmi123.asp. Accessed 2014 Feb 22.
65. US Bureau of Labor Statistics (2014) Labor force statistics from the Current Population Survey. Available: http://data.bls.gov/timeseries/LNS14000000. Accessed 2014 Feb 22.
66. United States Census Bureau (2013) American Fact Finder. Available: http://factfinder2.census.gov/faces/nav/jsf/pages/community_facts.xhtml - none. Accessed 2014 Feb 22.
67. United States Census Bureau (2012) Household Income for States: 2010 and 2011, Available: http://www.census.gov/prod/2012pubs/acsbr11-02.pdf.American Community Survey Briefs 2012. Accessed 2013 Feb 22.
68. University of Connecticut's Center for Public Health and Health Policy (2012) Child weight surveillance in preschool in Hartford, Connecticut. University of Connecticut's Center for Public Health and Health Policy.
69. Advisory Commission on Food Policy (1991) Hartford, Connecticut, Code of Ordinances, Part II, Municipal Code, Chapter 2, Administration, Article V. Boards Commissions and Committees, Division 13.
70. Havens E, Martin K, Yan J, Dauser-Forrest D, Ferris A (2012) Federal Nutrition Program Changes and Healthy Food Availability. American Journal of Preventive Medicine 43: 419–422.
71. SUPERVALU I (2011) Annual Report Fiscal 2011. Available: http://www.supervalu.com/sv-webapp/downloads/SUPERVALU_Fiscal_2011_Annual_Report.pdf Accessed 2014 Feb 22.
72. Chang K-T (2011) Introduction to Geographic Information Systems: McGraw-Hill Science/Engineering/Math.
73. ESRI (2012) ESRI Free Data. Available: http://www.esri.com/data/free-data. ESRI 2012. Accessed 2013 Feb 22.
74. MAGIC (2013) Map and Geographic Information Center (MAGIC). Homer Babbidge Library, University of Connecticut, Storrs, Connecticut: University of Connecticut.
75. ESRI Network Analyst Extension. Available: http://www.esri.com/software/arcgis/extensions/networkanalyst. ESRI. Accessed 2013 Feb 22.
76. Sanders H, Ghosh D, Van Riper D, Manson SM (2010) How do you measure distance in spatial models? An example using open-space valuation. Environment and Planning B: Planning and Design 37: 874–894.
77. Techniques E-S (2012) ET-Spatial Techniques 2012. Available: http://www.ian-ko.com/. ET-Spatial Techniques 2012. Accessed 2013 Feb 22.
78. Hartford Food Policy (2004) Hartford Area Food Stores Price Survey, 2004. Available: http://www.hartfordinfo.org/issues/wsd/familiesandchildren/food_survey_final.pdf.
79. Larson N, Story M, Nelson M (2009) Neighborhood environments: disparities in access to healthy foods in the US. American Journal of Preventive Medicine 36: 74–81.
80. Bloemer J, de Ruyter K (1998) On the relationship between store image, store satisfaction and store loyalty. European Journal of Marketing 32: 499–513.
81. Walker R, Block J, Kawachi I (2012) Do residents of food deserts express different food buying preferences compared to residents of food oases? A mixed-methods analysis. International Journal of Behavioral Nutrition and Physical Activity 9: 1–13.
82. Kane J (1984) The supermarket shuffle. Mother Jones 9: 7.
83. Russell SE, Heidkamp CP (2011) 'Food desertification': The loss of a major supermarket in New Haven, Connecticut. Applied Geography 31: 1197–1209.
84. Cummins S, Flint E, Mathews SA (2014) New neighborhood grocery store increased awareness of food access but did not alter dietary habits or obesity. Health Affairs 33(2): 283–91.

27

Use of Wild Bird Surveillance, Human Case Data and GIS Spatial Analysis for Predicting Spatial Distributions of West Nile Virus in Greece

George Valiakos[1,2], Konstantinos Papaspyropoulos[1,3], Alexios Giannakopoulos[1], Periklis Birtsas[4], Sotirios Tsiodras[5,6], Michael R. Hutchings[7], Vassiliki Spyrou[4], Danai Pervanidou[5], Labrini V. Athanasiou[1], Nikolaos Papadopoulos[8], Constantina Tsokana[1], Agoritsa Baka[5], Katerina Manolakou[1], Dimitrios Chatzopoulos[1], Marc Artois[9], Lisa Yon[10], Duncan Hannant[10], Liljana Petrovska[11], Christos Hadjichristodoulou[12], Charalambos Billinis[1,2]*

1 Faculty of Veterinary Medicine, University of Thessaly, Karditsa, Greece, **2** Department of Biomedicine, Institute for Research and Technology of Thessaly, Centre For Research and Technology Hellas, Larissa, Greece, **3** Research Division, Hunting Federation of Macedonia and Thrace, Thessaloniki, Greece, **4** Technological Education Institute of Thessaly, Larissa, Greece, **5** Hellenic Center for Disease Control and Prevention (HCDCP) Ministry of Health, Athens, Greece, **6** National and Kapodistrian University of Athens, Faculty of Medicine, Athens, Greece, **7** Disease Systems, SRUC, Edinburgh, United Kingdom, **8** Department of Agriculture Crop Production and Rural Environment, University of Thessaly, Volos, Greece, **9** University of Lyon, VetAgro Sup, France, **10** School of Veterinary Medicine & Science, University of Nottingham, Nottingham, United Kingdom, **11** Animal Health and Veterinary Laboratories Agency - Weybridge, United Kingdom, **12** Faculty of Medicine, University of Thessaly, Larissa, Greece

Abstract

West Nile Virus (WNV) is the causative agent of a vector-borne, zoonotic disease with a worldwide distribution. Recent expansion and introduction of WNV into new areas, including southern Europe, has been associated with severe disease in humans and equids, and has increased concerns regarding the need to prevent and control future WNV outbreaks. Since 2010, 524 confirmed human cases of the disease have been reported in Greece with greater than 10% mortality. Infected mosquitoes, wild birds, equids, and chickens have been detected and associated with human disease. The aim of our study was to establish a monitoring system with wild birds and reported human cases data using Geographical Information System (GIS). Potential distribution of WNV was modelled by combining wild bird serological surveillance data with environmental factors (e.g. elevation, slope, land use, vegetation density, temperature, precipitation indices, and population density). Local factors including areas of low altitude and proximity to water were important predictors of appearance of both human and wild bird cases (Odds Ratio = 1,001 95%CI = 0,723–1,386). Using GIS analysis, the identified risk factors were applied across Greece identifying the northern part of Greece (Macedonia, Thrace) western Greece and a number of Greek islands as being at highest risk of future outbreaks. The results of the analysis were evaluated and confirmed using the 161 reported human cases of the 2012 outbreak predicting correctly (Odds = 130/31 = 4,194 95%CI = 2,841–6,189) and more areas were identified for potential dispersion in the following years. Our approach verified that WNV risk can be modelled in a fast cost-effective way indicating high risk areas where prevention measures should be implemented in order to reduce the disease incidence.

Editor: Yury E. Khudyakov, Centers for Disease Control and Prevention, United States of America

Funding: The research leading to these results received partial funding from the European Union Seventh Framework Programme (2007–2013) under grant agreement no. 222633 (WildTech). This project was also co-funded by the «Integrated Surveillance and control programme for West Nile Virus and malaria in Greece» funded by NSRF (National Strategic Reference Framework) 2007–2013. The funders had no role in study design, data collection and analysis, decision to publish, or preparation of the manuscript.

Competing Interests: The authors have declared that no competing interests exist.

* E-mail: billinis@vet.uth.gr

Introduction

West Nile virus (WNV) is a mosquito-borne flavivirus with increasing numbers of reported human disease cases worldwide. In Europe, cases of WNV associated diseases have been reported in several countries in the European Union and in bordering Non-E.U. countries. The largest ongoing European outbreak has been observed in Greece, with more than 524 confirmed cases of human infection and 60 deaths reported since 2010 [1] (Figure 1).

Many studies have associated the presence of specific environmental factors with areas at high-risk for WNV transmission in the USA [2–5] and Europe [6,7]. Tachiiri et al. (2006) developed a model using basic geographic and temperature data to assess WNV risk in British Columbia [8]. Ruiz et al. (2004) used several factors related to the physical environment such as elevation range, physiographic region, and percentage of vegetation cover to determine WNV risk during an outbreak in the Chicago area in 2002 [9]. Methods that have been used in WNV risk modeling include non-linear discriminant analysis [10], logistic [11] or

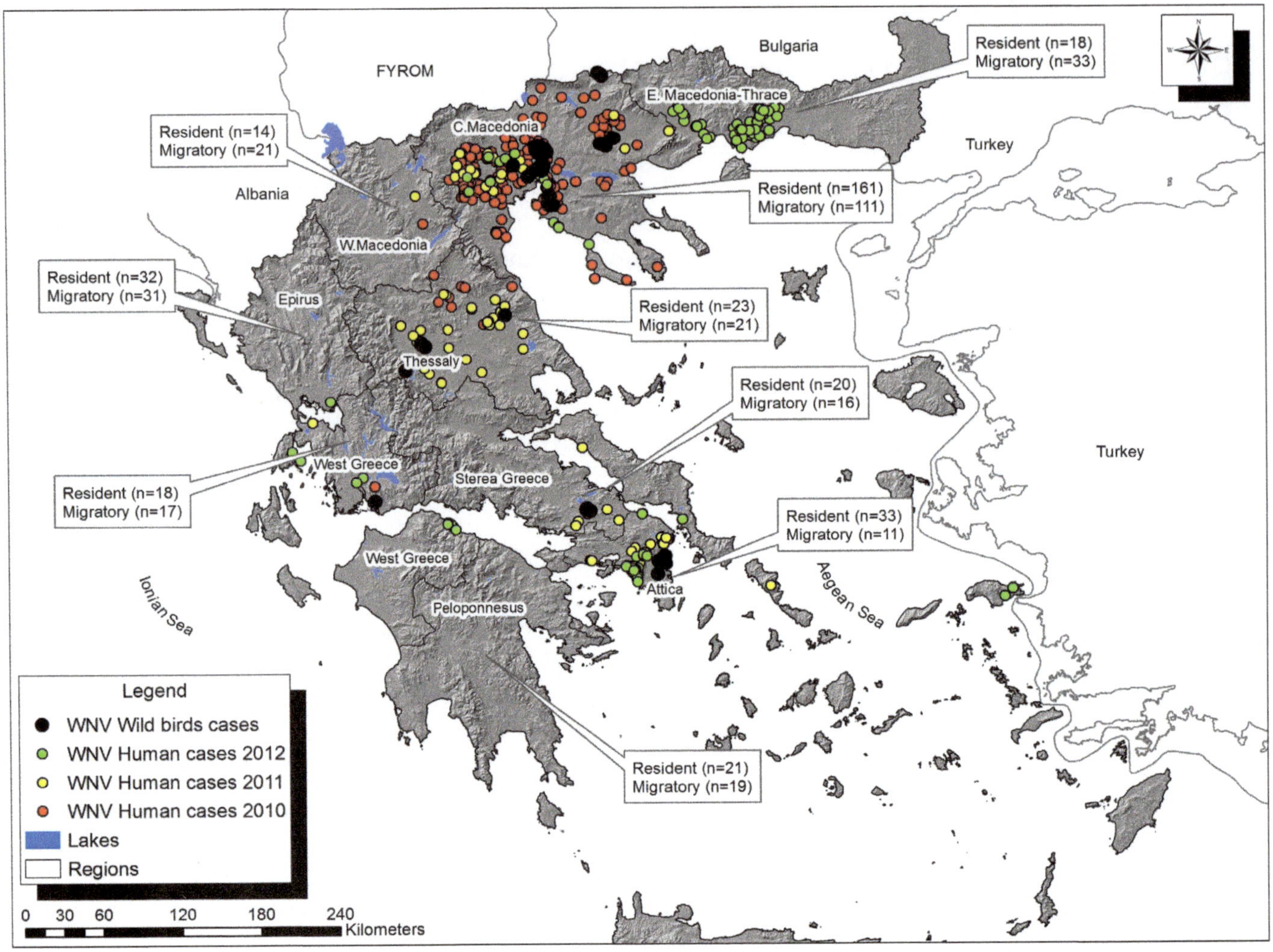

Figure 1. Map of Greece showing WNV laboratory-confirmed human cases and seropositive resident wild birds, 2010–2012. Map of Greece showing the distribution of WNV laboratory-confirmed human cases and seropositive resident wild bird samples for the 2010–2012 period. Red, yellow and green dots indicate human cases reported in 2010, 2011 and 2012 respectively. Black dots indicate seropositive resident wild birds detected during the same period. Text boxes refer to available avian samples (resident and migratory) per each region.

multiple regression models [12] (differential and difference equation modeling [13,14] and cluster analysis [15].

Predictive modeling with Geographic Information System (GIS) can be used to analyze environmental determinants of WNV transmission and determine high risk areas. Most previous WNV risk analyses utilized spatial statistical techniques (mapping clusters, geographic distribution, spatial relationships-regression models) to correlate environmental, climatic and socioeconomic factors with WNV prevalence [2,4,5,9,11,16].

The geographical position of Greece in the Mediterranean peninsula makes it an important transit zone for migratory birds [17]. Greece hosts a wealth of biological diversity, one of the richest in Europe and the Mediterranean.

The main objective of this study was to correlate serological data of exposure of wild birds to WNV and reported human cases data during the Greek outbreak with potential environmental risk factors within a GIS, in order to construct predictive maps identifying areas at risk from further spread. We test the predictive power of the models against recent outbreak data and identify high risk areas for the application of targeted, timely and cost-effective prevention measures such as surveillance, mosquitoes control and campaigns to increase public awareness of the disease.

Materials and Methods

Ethics Statement

This study was co-funded by the «Integrated Surveillance and control programme for West Nile Virus and malaria in Greece» (National Strategic Reference Framework, 2007–2013) and was also part of European Union Seventh Framework Programme (2007–2013) large collaboration project under grant agreement no. 222633 (Novel Technologies for Surveillance of Emerging and Re-emerging Infections of Wildlife - WildTech). University of Nottingham is the project coordinator, and has received signed cooperation agreements by all the project partners, for providing animal samples to the project. All samples used in this project represent material collected by partners and other organisations for other purposes than this project as specified in deliverable D4.5/5.5 entitled "Guidelines for ethical sample collection" submitted to European Commission (26/02/2010, Dissemination Level: PP, Restricted to other programme participants, including

Commission Services). The avian samples were collected opportunistically (no active capture, killing and sampling of wild animals specifically for this study was performed) from animals hunter-harvested by members of Greek Hunting Federation of Macedonia and Thrace, from species considered quarry and during the hunting seasons, according to the prerequisites of the Greek Legislation (FEK 1611/B′/2009, FEK 1183/6-8-2010, FEK 1763/4-8-2011), thus special approval was not necessary and steps to ameliorate suffering were not applicable in this study. Research on animals as defined in the EU Ethics for Researchers document (European Commission, 2007, Ethics for Researchers - Facilitating Research Excellence in FP7, Luxembourg: Office for Official Publications of the European Communities, ISBN 978-92-79-05474-7) is not a part of the study.

The human data were part of the ongoing surveillance of human cases performed by the Hellenic Center for Disease Control and Prevention (HCDCP) and were reported from the treating physicians. Data on human cases is maintained on a database kept by the HCDCP that was completely anonymized to the authors, without being publicly available. Since residential addresses could be potentially identifying, the security of this database was maintained according to the national regulation for the confidentiality of human data. The use of the residential addresses has been approved by the Institutional Review Board (IRB) of the Public Health and Environmental Hygiene post graduate course of the Laboratory of Hygiene and Epidemiology, Faculty of Medicine, University of Thessaly, Greece. The IRB waived the need for written informed consent to use the residential addresses, with the restriction that the analysis would be carried out at the municipality level.

Study Area

The study area comprised the entire country of Greece. Greece occupies the southeastern part of Europe with a total area of 131,990 km^2. Eighty percent of Greece consists of mountains; the country is characterized by a large climatic diversity (29 climatic zones according to the Thorn Waite classification), by its extensive coastline of about 15,000 km and many island complexes in the Archipelagos of Aegean Sea and the Ionian Sea. Climatic conditions of the country are typical Mediterranean: Summer is hot and dry while winter is usually mild. Rain mostly falls in autumn and winter.

WNV Human Cases Data

Reported human WNV cases in Greece (2010–2012) were provided by HCDCP. Most cases were serologically confirmed by the presence of IgM antibodies in the serum and/or the cerebrospinal fluid. Residential address of each human case was used for geocoding and mapping the cases.

Wild Birds Surveillance

A total of 620 avian serum samples were obtained from wild birds hunter-harvested by members of the Greek Hunting Federation of Macedonia and Thrace, from species considered quarry during the 2009/2010, 2010/2011 and 2011/2012 hunting seasons (from 20 August until 28 February the following year), according to the prerequisites of the Greek Legislation. All available samples were obtained from mainland Greece, opportunistically collected during regular hunting activities; samples were available from all 9 mainland regions of Greece (Table 1). Sampling effort was distributed in mainland Greece, avoiding cluster sampling biases, with the exception of the Central Macedonia region, the epicenter of the outbreak, during which a large number of samples were provided. Data on bird specimens that tested positive for WNV during the study were located in the field using handheld Global Positioning System (GPS) units or located by means of longitude and latitude information provided by samplers. Serological screening was performed as already reported [18–20]; a total of 64 resident wild birds were found positive for WNV antibodies, and were used in the current study (migratory wild birds were also found seropositive, but relevant data was excluded from the analysis, see Discussion).

Environmental Variables

Environmental variables for this study were derived from three main database categories: climate, elevation and land cover data.

WorldClim version 1.4 climate data [21] was obtained from the WorldClim website (http://www.worldclim.org). WorldClim is a set of global climate layers (climate grids) with a spatial resolution of 1 square kilometer. Topographic variables including altitude, aspect and slope were extracted from a digital elevation model (DEM) with a spatial resolution of 1 square kilometer (http://srtm.csi.cgiar.org/Index.asp). Land uses were derived from the Corine Land Cover 2000 database (European Environment Agency – EEA, http://www.eea.europa.eu/data-and-maps).

Village and vegetation corrections were digitized from 2007 and 2009 color orthophotos that were available through Web Mapping Service (WMS) (http://gis.ktimanet.gr). To create environmental layers (n = 37) for the analysis (Table 2), ArcGIS 10.1 GIS software (ESRI, Redlands, CA, USA) was used. GIS layers were created to represent factors like the locations of towns and villages, distance to the nearest village, distance from water presence etc. For many of the above parameters, we calculated neighborhood statistics for radii of 100, 200, 500 and 1000 m to determine which spatial scale affects the presence of cases most strongly. These data sets were converted to a common projection, map extent and resolution prior to use in the modeling program.

Statistical Analysis

We used data on 2010 and 2011 human cases for the statistical analysis and model building and kept the 2012 cases for verification. A total of 363 human WNV cases have been reported in Greece for the years 2010 and 2011 (262 cases in 2010 and 101 cases in 2011). The available dataset consisted of presence only data (presence: people infected by the virus). For this dataset, as well as the wild birds seroprevalence dataset a number of explanatory variables (n = 37) were collected and constructed, as mentioned previously (Table 2).

Instead of constructing a number of pseudo-absence controls, a methodology which according to the literature has some significant disadvantages for the prediction modeling [22,23], we decided to search for within the presence data variation of the explanatory variables. We clustered the cases using the agglomerative method of Two Step Cluster Analysis, a method which allows for the utilization of both continuous and categorical variables and clusters the cases by measuring the log-likelihood distance among them [24]. The Two Step Cluster Analysis allowed us to check for a pattern of the virus among the infected people in 2010, in 2011, and in total. The optimal number of clusters was chosen using the Silhouette coefficient, a measure proposed by Kaufman and Rousseeuw (1990) [25]. The coefficient ranges from -1 to 1 and when its value is closer to 1, the clustering is considered efficient.

Before applying the above cluster method, we checked a number of descriptive statistical measures which describe our data. Although Two Step Cluster Analysis is robust to non-normality [24] we used Factor Analysis in order to reduce the number of available variables and to achieve normality and zero-correlation

Table 1. Available avian samples: species, migratory status and number of samples per region.

Species	Migratory Status	E. Macedonia & Thrace	Central Macedonia	Thessaly	Epirus	West Greece	Peloponnesus	Sterea Greece	Attica	West Macedonia
Pica pica	Resident	11	125	17	20	12	14	13	29	9
Streptopelia turtur	Migratory	8	35	4	11	7	8	7	5	8
Corvus cornix	Resident	4	18	5	8	4	4	6	4	3
Anas platyrhynchos	Migratory	5	26	3	0	0	0	2	0	0
Scolopax rusticola	Migratory	4	12	5	9	3	6	2	0	5
Turdus philomelos	Migratory	7	27	6	8	7	5	4	6	8
Corvus monendula	Resident	3	18	1	4	2	3	1	0	2
Anas crecca	Migratory	9	11	3	3	0	0	1	0	0
Total		51	272	44	63	35	40	36	44	35

among explanatory continuous variables. We used the Principal Component Analysis (PCA) as a method of components extraction with rotation method the Varimax method with Kaiser Normalization [26]. Two Step Cluster Analysis was iterated several times using as clustering variables either the components which were extracted by the Principal Component Analysis, or the original variables which were highly correlated with the components. The extracted clusters for humans and the extracted clusters for birds were compared in terms of the variables that are important for clustering.

GIS Analysis

Two significant environmental variables were recognized from the statistical analysis (see Results) and were used to measure environmental conditions for the WNV locations of the seropositive wild birds and the human cases dataset. Mahalanobis distance (MD) [27] was used to develop a distance measure model for wild birds and predict WNV potential distribution prior to the expansion/outbreak of the 2012 period. We calculated MD with ArcGIS software, based on the values of the two significant variables, allowing us to identify suitable areas for WNV potential distribution and occurrence. Model performance evaluation was conducted with the 2012 reported WNV human cases, as provided by HCDCP.

Results

Data analysis demonstrates differences between 2010 and 2011 in terms of positive cases in Greece (Table 3). Fewer cases ($n = 101$) occurred in 2011 and the average case age was 5 years younger compared to 2010 (p-value = 0.024). There was a statistically different distribution in terms of the prefecture of residency of the positive cases, which were found in more southern areas compared to 2010, indicating the pathogen's continued spread in mainland Greece.

Finally, the distribution of the infected individuals in terms of date of infection was different in 2011, where more positive cases were found in July and September, compared to 2010 where the majority of the cases were found in August (64%).

Factor Analysis and Two Step Cluster Analysis revealed that altitude and distance from water were the two variables, among the 37 under study, which clustered significantly the cases. Both variables played a significant role in the clustering procedure. The two variables clustered in a similar way for both humans and birds. For the clustering of human cases, the average Silhouette coefficient was 0.5 which is considered a good clustering value [25]. The same value was achieved for the clustering of seropositive wild birds (Odds Ratio = 1,001 95%CI = 0,723–1,386). Three clusters were created for humans and birds (Figure 2), sharing the same attributes. In particular, humans' Cluster A and birds' Cluster B share the majority of the positive cases in humans and birds respectively (60%). There seems to be a pattern of WNV in Greece in places with low altitude and small distance from water. There are also two other clusters with lower percentages of cases which show that positive cases are also found in places with low altitude and big distance from water (23–24%) and in places with high altitude and small distance from water (almost 17%).

Relevant box-plots (Figure 3) show how well the two variables discriminate in each cluster. A clear separation of the three clusters is seen in both groups of cases.

Regarding the 2010 human cases, clustering showed that low altitude and small distance from water were associated with the majority of the positive human cases as well. A total of 86.6% of

Table 2. Environmental variables used in the analysis.

Variable	Value	Source
Slope (degrees) at 100-, 200-, 500- and 1000-m radii	X, SD, min, max	DEM
Topographic position index (4 classes)	binary	DEM
Altitude	continuous	DEM
Aspect	N,W,S,E	DEM
Distance from nearest village (m)	continuous	ArcGIS-WMS
Distance from water (m)	continuous	ArcGIS-DEM
Habitat types (4 classes: Forests, cultivations, etc.)	Categorical transform to continuous	ArcGIS-Corine LC (EEA)
19 Climatic variables (Temperature 11 indices, Precipitation 8 indices)	continuous	World Clim Database
NDVI (Normalized Difference Vegetation Index) 12 indices	continuous	World Clim Database
Population density	continuous	GEoDatabase

the human cases were grouped in cluster A (Figure 2), which shares similar attributes with cluster A of human 2011 cases and cluster B of birds.

The potential geographic distribution of WNV, predicted by GIS and MD based on the attributes of the major clusters of reported human cases of 2011 and seropositive wild birds is displayed in Figure 4. Fragmented high-risk areas were recognized: Most were concentrated in the Macedonian prefecture, in western Greece as well as in Thessaly. Other suitable high risk areas were located along the coast line of the Peloponnese peninsula and Crete. Moreover, many Greek islands have suitable environmental characteristics such as Rhodes, Mytilene, Chios, Samos etc.

In the early transmission period (June 2012) we reported the high–risk areas recognised throughout this study to the Ministry of Public Health and to HCDCP. As already reported, in 2012, a total of 161 laboratory-confirmed human cases were reported. Out of these 161 cases, only 31 occurred far from WNV high-risk areas recognised by our model (Odds = 130/31 = 4,194 95%CI = 2,841–6,189); four (4) human cases out of 5 were reported in recognised high-risk areas while only 1 out of 5 was not. New areas of potential dispersion of the virus are also suggested for the following years in the areas of Thrace, the Peloponnese peninsula and several Greek Islands (Figure 4).

Discussion

Humans and other mammals, particularly horses, are alternative hosts for WNV; the main route of infection is through the bite of an infected mosquito. Most human infections remain asymptomatic with WNV fever developing in approximately 20% of infected people and West Nile neuroinvasive disease in $<1\%$ [28]. Horses and humans develop low viremic loads ($<10^5$ PFU/ml) of short duration and thus are considered dead-end hosts for WNV [29]. In contrast, various migratory and resident avian species develop high viremic loads, sufficient to infect feeding ornithophilic mosquitoes [30]. Hence, the WNV life cycle is maintained with birds being the main amplifying hosts and mosquitoes the main vectors. Moreover, local movements of resident birds and long-range travel of migratory birds may contribute to pathogen dispersion [31,32]. In southern France, WNV was detected in late summer of 2000 and 2004. Migratory passerines were found with higher prevalence of WNV neutralizing antibodies (7.0%) than resident and short-distance migratory passerines (0.8%), suggesting exposure to WNV or a related flavivirus during overwintering in Africa [33]. Additionally in Spain it was found that Trans-Saharan migrant species had both higher prevalence and antibody titres than resident and short-distance migrants [34].

Table 3. Descriptive statistics of positive human cases 2010–2011.

Variable	2010	2011	Total
Positive cases	262	101	363
Age*	67.5 (sd = 16.9)	62.4 (sd = 19.9)	66 (sd = 17.9)
Sex	47.7% women	36.6% women	44.6% women
Prefecture of residency⁺	31% Thessaloniki	30% Attiki	25% Thessaloniki
	22% Pella	15% Larissa	18% Pella
	19% Imathia	10% Thessaloniki	15% Imathia
Date of infection	11.8% July	20.8% July	
	63.8% August	45.5% August	
	24.0% September	30.7% September	
	0.4% October	3% October	

**Statistically significant reduction in 2011 (5 years) (p-value = 0.024<0.05).*
⁺Independence between 2010 and 2011 (p-value = 0.000<0.001).

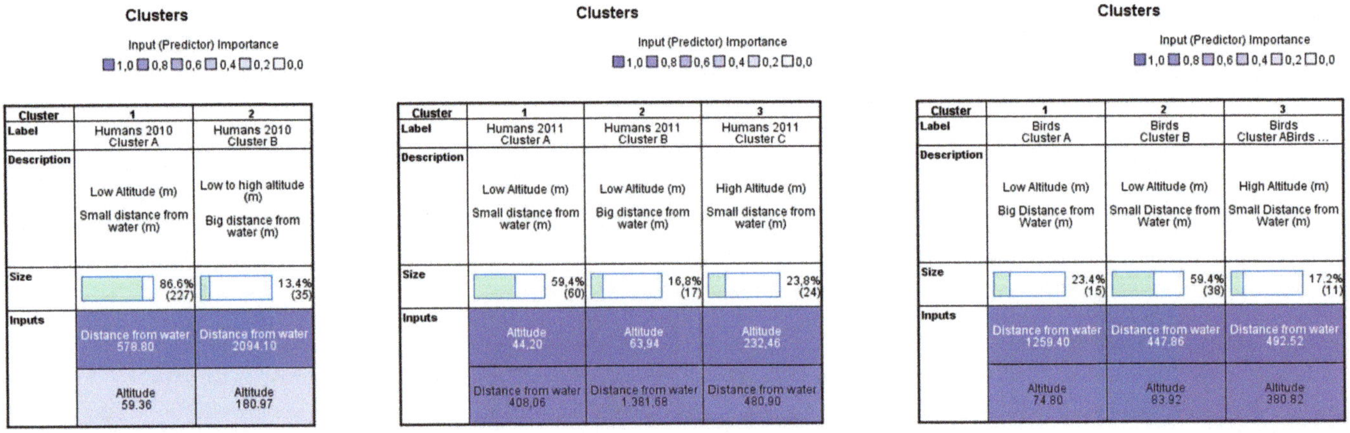

Figure 2. Clusters of human cases of seropositive wild birds and reported human cases of 2010–2011. Clusters of WNV reported human cases of 2010–2011 and seropositive wild birds, according to attributes of altitude and distance from water. Mean values of the two variables are presented under each cluster.

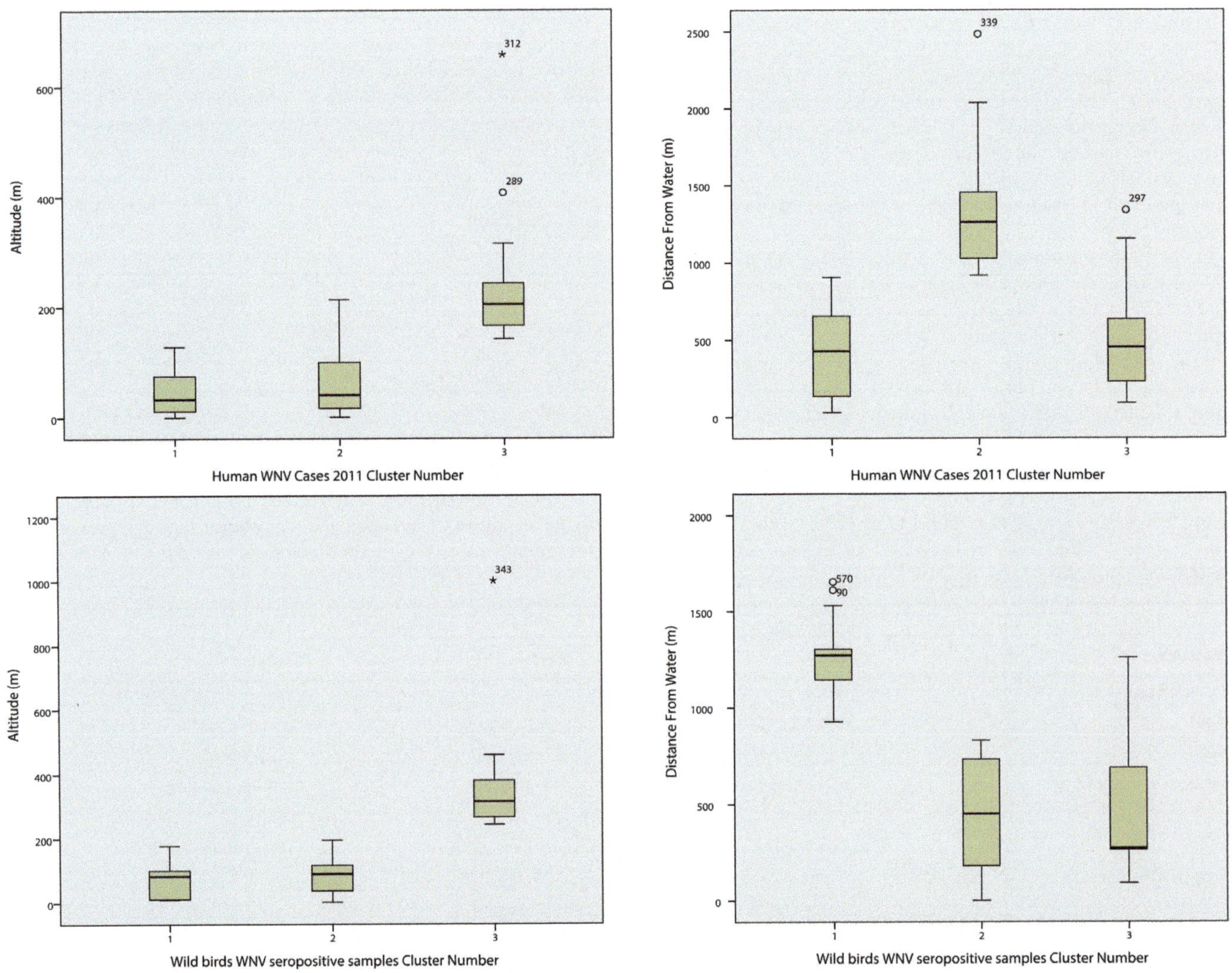

Figure 3. Box-plots of range and altitude. Box-plots displaying range of altitude (left) and distance from water (right) in the three clusters of humans 2011 WNV positive cases and seropositive wild birds.

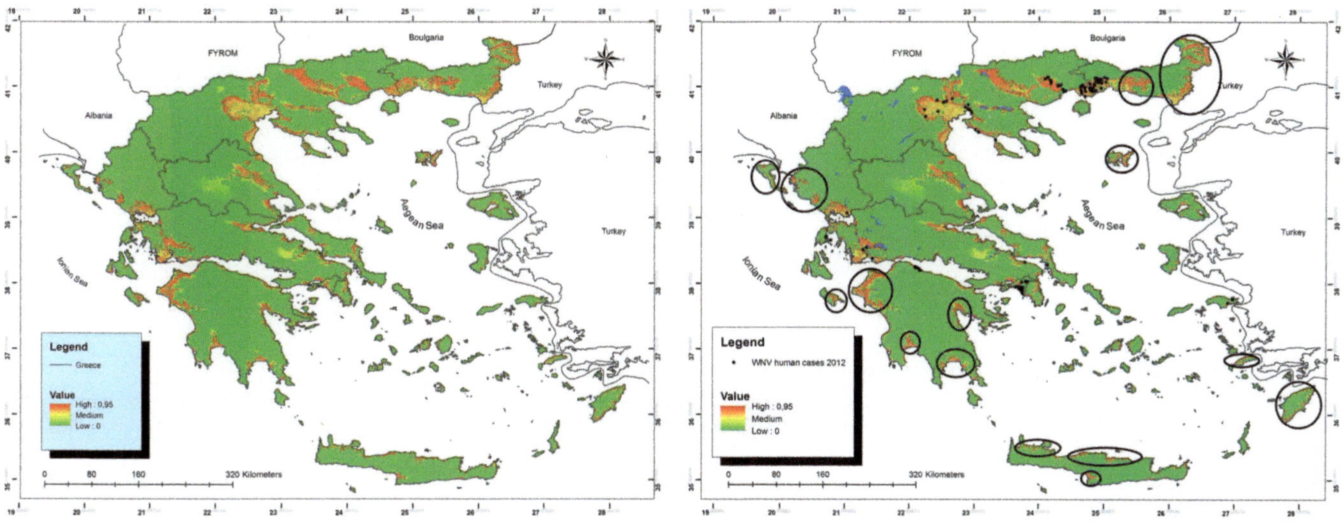

Figure 4. Map of Greece showing potential geographic distribution of WNV. Map of Greece showing potential geographic distribution of WNV, predicted by GIS and MD based on the attributes of the major clusters of reported human cases of 2011 and seropositive wild birds (low altitude, small distance from water). Black dots indicate reported human cases in 2012. Black circles indicate suggested high-risk areas for WNV further dispersion in following years.

In Greece, the disease first appeared in Macedonia prefecture in 2010, with 262 confirmed cases and 35 deaths, and it subsequently spread through mainland Greece in the following years. More specifically, in 2011, the outbreak expanded southwards to central Greece with 101 confirmed cases and 9 deaths, while in 2012, a total of 161 confirmed cases and 18 deaths were reported mainly in Attica and northeastern Greece. A strain of lineage 2 was detected in 2010 in pools of Culex mosquitoes [35] and in wild birds [20]. In this study we correlate various environmental factors with WNV maintenance, amplification and potential for future spread in Greece.

Various public health studies have used Geographical Information System technologies as a tool for data analysis [2,9,15]. Previous studies [9,10] found that certain social and environmental factors were correlated with WNV dissemination patterns: The presence of vegetation, distance to a WNV positive dead bird, the intensity of mosquito abatement, demographic factors such as population age, race and financial status. Low precipitation and warm temperature were also found to associate with WNV cases. On the other hand, spread of WNV has shown some unique distribution patterns in different regions [15,30].

Before reaching the aforementioned results, we undertook several efforts to find out a pattern, or a distinguishable attribute of the WNV positive cases in Greece. Although we used a significant number of explanatory variables for describing each positive case (a mix of both continuous and categorical variables), there was no indication that these altogether could show the pattern in question. Therefore we tried to reduce this dataset by using Factor Analysis. We run PCA once for the temperature variables and once for the precipitation variables. Two components (93% of the variation was explained) for the temperature and two components (96% of the variation) for the precipitation variables were extracted, which means that the fit was very good. These four variables with the rest demographic and environmental variables were used in the Two Step Cluster method. This method was preferred compared to other clustering techniques because it can handle both categorical and continuous variables. However, we also run hierarchical cluster analysis using only the continuous variables, but no pattern of the positive human cases was revealed. Therefore, we used the two step cluster technique in a backward selection way. Initially we used all the explanatory variables together, and we removed one at a time if the Silhouette indicator was not considered good. We used the log-likelihood distance instead of Euclidean distance, because there were initially categorical variables in the dataset. However, when only the two continuous variables were left ("altitude" and "distance from water"), we checked also if the Euclidean distance could reveal the same pattern with the log-likelihood but it didn't. We believe the fact that some of the variables were not significant for the clustering procedure was due to similar environmental conditions existing in Greece during summer. For example, there is no significant variation in terms of temperature or precipitation. This is why more stable variables like distance from water and altitude were responsible for the form of the clusters. After we formed the three clusters with these two variables, this pattern was revealed for both human cases and resident wild birds seroprevalence data.

Distance to water and altitude have both been previously shown to be negatively correlated with mosquito larval presence [36]; mosquitoes are the main biological vectors of WNV and transmission of this arthropod-borne virus is highly dependent on the density of mosquitoes. Low lying areas in close proximity to water include wetland habitats that are used as resting and breeding areas for various migratory and resident birds, allowing the long-distance introduction of the virus via migration routes as well as the rapid local amplification of the virus in a mosquito-bird cycle. In this study, apart from statistically identifying proximity to water and altitude as risk factors of spread of WNV in Greece, we were able to determine specific mean values for these habitat variables that allowed us to predict areas at high-risk for further disease incursion.

WNV positive birds are considered important environmental predictors of WNV human risk and are used in surveillance and risk assessment [2,9,37]. Whilst viremic birds are likely to represent the highest risk to humans, the viremic phase is extremely short, restricting data richness and thus statistical power. Hence we focused our analysis on longer lived serological measures of exposure. Moreover, the use of only the resident WNV seropositive wild birds from all hunter-harvested samples available, even though samples from migratory birds were also found positive, increased the reliability of our analysis, avoiding biases regarding area of exposure e.g. migratory birds travel long distances so the origin of exposure is hard to be determined. Hence, this is a good example of a case in which surveillance regarding exposure and other similar biological data derived from nature, regarding a zoonosis, can be used as an indicator for predicting high-risk areas. This fact was confirmed by the good fit that our model showed for the 2012 positive WNV human cases in Greece.

Conclusions

Modelling results indicated that positive resident wild bird occurrences are correlated with human WNV risk and can facilitate the assessment of environmental variables that contribute to that risk, recognising new high-risk areas where the disease could further spread. Our approach allowed us to create a risk based mapping system to assist and guide WNV disease surveillance, monitoring and control. This risk based approach offers a way to stratify surveillance efforts and resources to improve the efficiency of surveillance for new outbreaks and monitoring existing outbreaks. Furthermore, it could proactively enhance other preventive efforts and educational campaigns for the general public in the not yet "affected" areas. Most importantly, early warning and identification of outbreaks is critical to limiting the animal and human losses to this disease. An active surveillance program undertaken on resident wild birds could be added to active and passive surveillance focused on humans, horses and mosquitoes greatly helping in evaluating and dealing with future outbreaks linked to flaviviruses.

Author Contributions

Conceived and designed the experiments: GV KP AG CB. Performed the experiments: GV LA CT KM DC. Analyzed the data: GV KP AG DP NP CH CT AB DC MRH MA CB LY. Contributed reagents/materials/analysis tools: PB ST VS DP LA NP AB KM MA DH LP LY. Wrote the paper: GV KP AG PB ST VS CH MRH DH LP CB.

References

1. Hellenic Centre for Disease Control and Prevention – HCDCP (2013) West Nile Virus Epidemic in Greece. Greece.
2. Cooke WH 3rd, Grala K, Wallis RC (2006) Avian GIS models signal human risk for West Nile virus in Mississippi. Int J Health Geogr 5: 36.
3. Gibbs SE, Wimberly MC, Madden M, Masour J, Yabsley MJ, et al. (2006) Factors affecting the geographic distribution of West Nile virus in Georgia, USA: 2002–2004. Vector Borne Zoonotic Dis 6: 73–82.
4. Liu H, Weng Q, Gaines D (2008) Spatio-temporal analysis of the relationship between WNV dissemination and environmental variables in Indianapolis, USA. Int J Health Geogr 7: 66.
5. Rochlin I, Turbow D, Gomez F, Ninivaggi DV, Campbell SR (2011) Predictive mapping of human risk for West Nile virus (WNV) based on environmental and socioeconomic factors. PLoS One 6: e23280.
6. Lelli R, Calistri P, Bruno R, Monaco F, Savini G, et al. (2012) West Nile transmission in resident birds in Italy. Transbound Emerg Dis 59: 421–428.
7. Rodriguez-Prieto V, Martinez-Lopez B, Martinez M, Munoz MJ, Sanchez-Vizcaino JM (2012) Identification of suitable areas for West Nile virus outbreaks in equid populations for application in surveillance plans: the example of the Castile and Leon region of Spain. Epidemiol Infect 140: 1617–1631.
8. Tachiiri K, Klinkenberg B, Mak S, Kazmi J (2006) Predicting outbreaks: a spatial risk assessment of West Nile virus in British Columbia. Int J Health Geogr 5: 21.
9. Ruiz MO, Tedesco C, McTighe TJ, Austin C, Kitron U (2004) Environmental and social determinants of human risk during a West Nile virus outbreak in the greater Chicago area, 2002. Int J Health Geogr 3: 8.
10. Liu H, Weng Q (2008) Seasonal variations in the relationship between landscape pattern and land surface temperature in Indianapolis, USA. Environmental Monitoring and Assessment 144: 199–219.
11. Brownstein JS, Rosen H, Purdy D, Miller JR, Merlino M, et al. (2002) Spatial analysis of West Nile virus: rapid risk assessment of an introduced vector-borne zoonosis. Vector Borne Zoonotic Dis 2: 157–164.
12. Crowder DW, Dykstra EA, Brauner JM, Duffy A, Reed C, et al. (2013) West nile virus prevalence across landscapes is mediated by local effects of agriculture on vector and host communities. PLoS One 8: e55006.
13. Bowman C, Gumel AB, van den Driessche P, Wu J, Zhu H (2005) A mathematical model for assessing control strategies against West Nile virus. Bull Math Biol 67: 1107–1133.
14. Cruz-Pacheco G, Esteva L, Montano-Hirose JA, Vargas C (2005) Modelling the dynamics of West Nile Virus. Bull Math Biol 67: 1157–1172.
15. Ruiz MO, Walker ED, Foster ES, Haramis LD, Kitron UD (2007) Association of West Nile virus illness and urban landscapes in Chicago and Detroit. Int J Health Geogr 6: 10.
16. Konrad SK, Miller SN, Reeves WK, Tietze NS (2009) Spatially explicit West Nile virus risk modeling in Santa Clara County, California. Vector Borne Zoonotic Dis 9: 267–274.
17. Handrinos G and Akriotis T (1997) The Birds of Greece. London: C. Helm.
18. Valiakos G, Touloudi A, Athanasiou L, Giannakopoulos A, Iacovakis C, et al. (2012) Exposure of Eurasian magpies and turtle doves to West Nile virus during a major human outbreak, Greece, 2011. European Journal of Wildlife Research 58: 749–753.
19. Valiakos G, Touloudi A, Athanasiou L, Giannakopoulos A, Iacovakis C, et al. (2012) Serological and molecular investigation into the role of wild birds in the epidemiology of West Nile virus in Greece. Virology Journal 9: 266.
20. Valiakos G, Touloudi A, Iacovakis C, Athanasiou L, Birtsas P, et al. (2011) Molecular detection and phylogenetic analysis of West Nile virus lineage 2 in sedentary wild birds (Eurasian magpie), Greece, 2010. Euro Surveill 16.
21. Hijmans RJ, Cameron SE, Parra JL, Jones PG, Jarvis A (2005) Very high resolution interpolated climate surfaces for global land areas. International Journal of Climatology 25: 1965–1978.
22. VanDerWal J, Shoo LP, Graham C, Williams SE (2009) Selecting pseudo-absence data for presence-only distribution modeling: How far should you stray from what you know? Ecological Modelling 220: 589–594.
23. Warton D, Shepherd L (2010) Poisson point process models solve the "pseudo-absence problem" for presence-only data in ecology. The Annals of Applied Statistics 4: 1383–1402.
24. Norusis M (2011) SPSS 19.0 Base. USA: SPSS Inc.
25. Kaufman L, Rousseeuw PJ (1990) Finding Groups in Data: An Introduction to Cluster Analysis. New York: John Wiley & Sons.
26. Hair JF, Black B, Babin B, Anderson R, Tatham R (2006) Multivariate Data Analysis (6th Edition). Upper Saddle River, NJ: Prentice Hall.
27. Ozdenerol E, Bialkowska-Jelinska E, Taff GN (2008) Locating suitable habitats for West Nile Virus-infected mosquitoes through association of environmental characteristics with infected mosquito locations: a case study in Shelby County, Tennessee. Int J Health Geogr 7: 12.
28. Mostashari F, Bunning ML, Kitsutani PT, Singer DA, Nash D, et al. (2001) Epidemic West Nile encephalitis, New York, 1999: results of a household-based seroepidemiological survey. Lancet 358: 261–264.
29. Bunning ML, Bowen RA, Cropp CB, Sullivan KG, Davis BS, et al. (2002) Experimental infection of horses with West Nile virus. Emerg Infect Dis 8: 380–386.
30. Komar N, Langevin S, Hinten S, Nemeth N, Edwards E, et al. (2003) Experimental infection of North American birds with the New York 1999 strain of West Nile virus. Emerg Infect Dis 9: 311–322.
31. Rappole JH, Derrickson SR, Hubalek Z (2000) Migratory birds and spread of West Nile virus in the Western Hemisphere. Emerg Infect Dis 6: 319–328.
32. Malkinson M, Banet C, Weisman Y, Pokamunski S, King R, et al. (2002) Introduction of West Nile virus in the Middle East by migrating white storks. Emerg Infect Dis 8: 392–397.
33. Jourdain E, Zeller HG, Sabatier P, Murri S, Kayser Y, et al. (2008) Prevalence of West Nile virus neutralizing antibodies in wild birds from the Camargue area, southern France. J Wildl Dis 44: 766–771.
34. Lopez G, Jimenez-Clavero MA, Tejedor CG, Soriguer R, Figuerola J (2008) Prevalence of West Nile virus neutralizing antibodies in Spain is related to the behavior of migratory birds. Vector Borne Zoonotic Dis 8: 615–621.
35. Papa A, Xanthopoulou K, Gewehr S, Mourelatos S (2011) Detection of West Nile virus lineage 2 in mosquitoes during a human outbreak in Greece. Clin Microbiol Infect 17: 1176–1180.
36. Bian L, Li L, Yan G (2006) Combining Global and Local Estimates for Spatial Distribution of Mosquito Larval Habitats. GIScience & Remote Sensing 43: 128–141.
37. Patnaik JL, Juliusson L, Vogt RL (2007) Environmental predictors of human West Nile virus infections, Colorado. Emerg Infect Dis 13: 1788–1790.

Permissions

All chapters in this book were first published in PLOS ONE, by The Public Library of Science; hereby published with permission under the Creative Commons Attribution License or equivalent. Every chapter published in this book has been scrutinized by our experts. Their significance has been extensively debated. The topics covered herein carry significant findings which will fuel the growth of the discipline. They may even be implemented as practical applications or may be referred to as a beginning point for another development.

The contributors of this book come from diverse backgrounds, making this book a truly international effort. This book will bring forth new frontiers with its revolutionizing research information and detailed analysis of the nascent developments around the world.

We would like to thank all the contributing authors for lending their expertise to make the book truly unique. They have played a crucial role in the development of this book. Without their invaluable contributions this book wouldn't have been possible. They have made vital efforts to compile up to date information on the varied aspects of this subject to make this book a valuable addition to the collection of many professionals and students.

This book was conceptualized with the vision of imparting up-to-date information and advanced data in this field. To ensure the same, a matchless editorial board was set up. Every individual on the board went through rigorous rounds of assessment to prove their worth. After which they invested a large part of their time researching and compiling the most relevant data for our readers.

The editorial board has been involved in producing this book since its inception. They have spent rigorous hours researching and exploring the diverse topics which have resulted in the successful publishing of this book. They have passed on their knowledge of decades through this book. To expedite this challenging task, the publisher supported the team at every step. A small team of assistant editors was also appointed to further simplify the editing procedure and attain best results for the readers.

Apart from the editorial board, the designing team has also invested a significant amount of their time in understanding the subject and creating the most relevant covers. They scrutinized every image to scout for the most suitable representation of the subject and create an appropriate cover for the book.

The publishing team has been an ardent support to the editorial, designing and production team. Their endless efforts to recruit the best for this project, has resulted in the accomplishment of this book. They are a veteran in the field of academics and their pool of knowledge is as vast as their experience in printing. Their expertise and guidance has proved useful at every step. Their uncompromising quality standards have made this book an exceptional effort. Their encouragement from time to time has been an inspiration for everyone.

The publisher and the editorial board hope that this book will prove to be a valuable piece of knowledge for researchers, students, practitioners and scholars across the globe.

List of Contributors

Sze Hang Fu
Centre for Global Health Research, Li Ka Shing Knowledge Institute, St. Michael's Hospital, University of Toronto, Toronto, Ontario, Canada

Prabhat Jha
Centre for Global Health Research, Li Ka Shing Knowledge Institute, St. Michael's Hospital, University of Toronto, Toronto, Ontario, Canada
Dalla Lana School of Public Health, University of Toronto, Toronto, Ontario, Canada

Prakash C. Gupta
Healis-Sekhsaria Institute of Public Health, Navi Mumbai, India

Rajesh Kumar
School of Public Health, Post Graduate Institute of Medical Education and Research, Chandigarh, India

Rajesh Dikshit
Department of Epidemiology, Tata Memorial Hospital, Mumbai, India

Dhirendra Sinha
South East Asia Regional Office, World Health Organization, New Delhi, India

Dana R. Thomson
Department of Global Health and Social Medicine, Harvard Medical School, Boston, Massachusetts, United States of America

Shrutika Shitole, Tejal Shitole, Kiran Sawant and Anita Patil-Deshmukh
Partners for Urban Knowledge, Action, and Research, Mumbai, India

Ramnath Subbaraman
Partners for Urban Knowledge, Action, and Research, Mumbai, India
Division of Infectious Diseases, Brigham and Women's Hospital, Boston, Massachusetts, United States of America

David E. Bloom
Department of Global Health and Population, Harvard School of Public Health, Boston Massachusetts, United States of America

Ni Huang, Li Wang, Pengyu Hao and Zheng Niu
The State Key Laboratory of Remote Sensing Science, Institute of Remote Sensing and Digital Earth, Chinese Academy of Sciences, Beijing, China

Yiqiang Guo
Land Consolidation and Rehabilitation Center, Ministry of Land and Resources, Beijing, China

Peige Song and Lin An
Department of Child, Adolescent and Women's Health, School of Public Health, Peking University Health Science Center, Beijing, China

Yajie Zhu and Qi Li
Institute of Remote Sensing and GIS, School of Earth and Space Science, Peking University, Beijing, China

Xi Mao
Institute of Cartography and GIS, Chinese Academy of Surveying and Mapping, Beijing, China

Eric Pante and Benoit Simon-Bouhet
Littoral, Environnement et Sociétés Joint Research Unit 7266 Centre National de la Recherche Scientifique, Université de La Rochelle, La Rochelle, France

Eleonore M. Veldhuizen
Department of Human Geography, Planning & International Development Studies, Faculty of Social and Behavioural Sciences, University of Amsterdam, Amsterdam, The Netherlands

Karien Stronks and Anton E. Kunst
Department of Public Health, Academic Medical Center, University of Amsterdam, Amsterdam, The Netherlands

Hsuan-Cheng Huang
Institute of Biomedical Informatics, Center for Systems and Synthetic Biology, National Yang-Ming University, Taipei, Taiwan

Chen-Ching Lin
Institute of Biomedical Informatics, Center for Systems and Synthetic Biology, National Yang-Ming University, Taipei, Taiwan
Graduate Institute of Biomedical Electronics and Bioinformatics, National Taiwan University, Taipei, Taiwan

Chia-Hsien Lee and Chiou-Shann Fuh
Graduate Institute of Biomedical Electronics and Bioinformatics, National Taiwan University, Taipei, Taiwan

Hsueh-Fen Juan
Graduate Institute of Biomedical Electronics and Bioinformatics, National Taiwan University, Taipei, Taiwan
Department of Life Science, Institute of Molecular and Cellular Biology, Center for Systems
Biology, National Taiwan University, Taipei, Taiwan

Salvatore Eugenio Pappalardo
Department of Agronomy, Food, Natural Resources, Animals and the Environment, University of Padova, Padova, Italy

Massimo De Marchi
Department of Civil, Environmental Architectural Engineering, University of Padova, Padova, Italy
Prometeo Project, Universidad Nacional de Chimborazo, Riobamba, Ecuador

Francesco Ferrarese
GIS Laboratory, Department of Historical, Geographical and Antiquity Sciences, University of Padova, Padova, Italy

Zhongzhi Xu., Li Sun, Junjie Wang and Pu Wang
School of Traffic and Transportation Engineering, Central South University, Changsha, Hunan, P.R. China

David K. Wright
Department of Archaeology and Art History, Seoul National University, Seoul, Republic of Korea

Scott MacEachern
Department of Anthropology, Bowdoin College, Brunswick, Maine, United States of America

Jaeyong Lee
Department of Statistics, Seoul National University, Seoul, Republic of Korea

Yuejia Cheng, Zhihang Peng, Rongbin Yu, Hongbing Shen and Feng Chen
Department of Epidemiology & Biostatistics, School of Public Health, Nanjing Medical University, Nanjing, Jiangsu, China

Changjun Bao, Jianli Hu, Wendong Liu, Qi Liang, Ying Wu and Fenyang Tang
Jiangsu Province Center for Disease Control and Prevention, Nanjing, Jiangsu, China

Jessie Norris
National Center for AIDS/STD Control and Prevention, Chinese Center for Disease Control and Prevention, Beijing, China

Alice M. Dalton
School of Environmental Sciences, University of East Anglia, Norwich, United Kingdom
UKCRC Centre for Diet and Activity Research (CEDAR), Institute of Public Health, Cambridge, United Kingdom

Andrew P. Jones
Norwich Medical School, University of East Anglia, Norwich, United Kingdom
UKCRC Centre for Diet and Activity Research (CEDAR), Institute of Public Health, Cambridge, United Kingdom

Jenna R. Panter and David Ogilvie
UKCRC Centre for Diet and Activity Research (CEDAR), Institute of Public Health, Cambridge, United Kingdom
Medical Research Council Epidemiology Unit, Institute of Public Health, Cambridge, United Kingdom

Gavin M. Rudge, Sally C. Fillingham, Alan Girling and Andrew J. Stevens
School of Health and Population Sciences, University of Birmingham, Birmingham, United Kingdom

Mohammed A. Mohammed
University of Bradford, Bradford, United Kingdom

Khesh Sidhu
Welsh Health Specialised Services Team, Caerphilly, Wales, United Kingdom

Sara Varela
Department of Ecology, Faculty of Science, Charles University, Prague, Czech Republic,
Departamento de Ecologia, Instituto de Ciencias Biologicas, Universidade Federal de Goiás, CxP 131, Goiania, Goiás, Brasil,

Javier Gonzá lez-Hernández
Toledo, Spain

Eduardo Casabella
ACCONTE Group, Actividades de Consultoría y Telecomunicaciones, Tres Cantos, Spain

Rafael Barrientos
Departamento de Ciencias Ambientales, Facultad de Medio Ambiente, Universidad de Castilla-La Mancha, Toledo, Spain

Roger Hillson and Joel D. Alejandre
Information Technology Division, Naval Research Laboratory, Washington, District of Columbia, United States of America

Kathryn H. Jacobsen
Department of Global and Community Health, George Mason University, Fairfax, Virginia, United States of America

Rashid Ansumana and Alfred S. Bockarie
Njala University, Bo, Sierra Leone
Mercy Hospital Research Laboratory, Bo, Sierra Leone

Umaru Bangura and Joseph M. Lamin
Mercy Hospital Research Laboratory, Bo, Sierra Leone

Anthony P. Malanoski and David A. Stenger
Center for Bio/Molecular Science and Engineering, Naval Research Laboratory, Washington, District of Columbia, United States of America

Guttorm Raknes
National Centre for Emergency Primary Health Care, Uni Health, Uni Research, Bergen, Norway

Steinar Hunskaar
National Centre for Emergency Primary Health Care, Uni Health, Uni Research, Bergen, Norway
Department of Global Public Health and Primary Care, University of Bergen, Bergen, Norway

Tiberiu C. Sahlean
Faculty of Biology, University of Bucharest, Bucharest, Romania
Department of Terrestrial Fauna, "Grigore Antipa" National Museum of Natural History, Bucharest, Romania

Iulian Gherghel and Monica Papeş
Department of Zoology, Oklahoma State University, Stillwater, Oklahoma, United States of America

Alexandru Strugariu and S¸ tefan R. Zamfirescu
Faculty of Biology, "Alexandru Ioan Cuza" University, Iaşi, Romania

Erin E. Poor
Department of Fish and Wildlife Conservation, Virginia Tech, Blacksburg, Virginia, United States of America
Conservation Science Program, World Wildlife Fund, Washington D.C., United States of America

Colby Loucks
Conservation Science Program, World Wildlife Fund, Washington D.C., United States of America

Andrew Jakes
Faculty of Environmental Design, University of Calgary, Calgary, Canada, 4 Environment Yukon, Dawson City, Yukon, Canada

Mike Suitor
Faculty of Environmental Design, University of Calgary, Calgary, Canada, 4 Environment Yukon, Dawson City, Yukon, Canada

Guanghua Yin, Zhenying Wang and Zuoxin Liu
Institute of Applied Ecology, Chinese Academy of Sciences, Shenyang, China

Fasheng Zhang
Institute of Applied Ecology, Chinese Academy of Sciences, Shenyang, China
Eastern Cereal and Oilseed Research Centre, Agriculture and Agri-Food Canada, Ottawa
Canada

Neil McLaughlin and Xiaoyuan Geng
Eastern Cereal and Oilseed Research Centre, Agriculture and Agri-Food Canada, Ottawa, Canada

Suchithra Naish and Shilu Tong
School of Public Health and Social Work & Institute of Health and Biomedical Innovation, Queensland University of Technology, Kelvin Grove campus, Brisbane,Queensland, Australia

Pat Dale
Environmental Futures Centre, Australian Rivers Institute, Griffith School of Environment Griffith University, Brisbane, Queensland, Australia

John S. Mackenzie
Australian Biosecurity CRC and Faculty of Health Sciences, Curtin University, Perth, Western Australia, Australia

John McBride
School of Medicine and Dentistry, James Cook University, Cairns, Queensland, Australia

Kerrie Mengersen
Mathematical Sciences, Queensland University of Technology, Gardens Point campus, Brisbane, Queensland, Australia

Stephen Rulisa
University Teaching Hospital of Kigali, National University of Rwanda, Kigali, Rwanda
Amsterdam Institute for Global Health and Development, INTERACT Project, Kigali, Rwanda

Steven Agaba, Javier Dukuzumuremyi, Lisette Baas and Jean de Dieu Harelimana
Amsterdam Institute for Global Health and Development, INTERACT Project, Kigali, Rwanda

Fredrick Kateera
Academic Medical Center, Division of Infectious Diseases, Tropical Medicine and AIDS, Amsterdam, The Netherlands
Medical Research Centre, Rwanda Biomedical Centre, Kigali, Rwanda

Petra F. Mens
Academic Medical Center, Division of Infectious Diseases, Tropical Medicine and AIDS, Amsterdam, The Netherlands
Amsterdam Institute for Global Health and Development, INTERACT Project, Kigali, Rwanda
Royal Tropical Institute/Koninklijk Instituutvoor de Tropen (KIT), KIT Biomedical Research, Amsterdam, The Netherlands

Kimberly R Boer
Amsterdam Institute for Global Health and Development, INTERACT Project, Kigali, Rwanda
Royal Tropical Institute/Koninklijk Instituutvoor de Tropen (KIT), KIT Biomedical Research, Amsterdam, The Netherlands

Peter J de Vries
Academic Medical Center, Division of Infectious Diseases, Tropical Medicine and AIDS, Amsterdam, The Netherlands
Amsterdam Institute for Global Health and Development, INTERACT Project, Kigali, Rwanda
Department of Internal Medicine, Tergooiziekenhuizen, Hilversum, The Netherlands

Jean Pierre Bizimana
Geography Department, Faculty of Science, National University of Rwanda, Butare, Rwanda

Young Ae You, Mohammad Ali, Binod Sah, Thomas F. Wierzba and Mahesh Puri
International Vaccine Institute, Seoul, Republic of Korea

Suman Kanungo, Byomkesh Manna, G. Balakrish Nair and Dipika Sur
National Institute of Cholera and Enteric Diseases, Kolkata, India

Sujit Kumar Bhattacharya
National Institute of Cholera and Enteric Diseases, Kolkata, India
Indian Council of Medical Research, New Delhi, India

Matteo Convertino
HumNat Lab, Department of Biological Systems Engineering, Virginia Tech, Blacksburg, Virginia, United States of America
Bioinformatics Institute, Virginia Tech, Blacksburg, Virginia, United States of America
Institute for Critical Technology and Applied Science, Virginia Tech, Blacksburg, Virginia, United States of America

Jacqueline L. Deen
Menzies School of Health Research, Casuarina, Northern Territory, Australia

Anna Lena Lopez
International Vaccine Institute, Seoul, Republic of Korea
University of the Philippines Manila, National Institutes of Health, Manila, Philippines

John Clemens
International Vaccine Institute, Seoul, Republic of Korea
University of California Los Angeles, School of Public Health, Los Angeles, United States of America

Juan Eugenio Hernández-Ávila, Mario-Henry Rodríguez, RenéSantos-Luna, Veronica Sánchez-Castañeda, Susana Román-Pérez and Víctor Hugo Ríos-Salgado
National Institute of Public Health of Mexico, Cuernavaca, Morelos, Mexico,

Jesús Alberto Salas Sarmiento
Public Health School of Mexico, Cuernavaca, Morelos, Mexico

Innocent Nzeyimana and Violette Geissen
Soil Physics and Land Management Group, Wageningen University, Wageningen, The Netherlands

Alfred E. Hartemink
Department of Soil Science, FD Hole Soils Lab, University of Wisconsin, Madison, Madison, Wisconsin, United States of America

Courtney L. Daigle and Janice M. Siegford
Department of Animal Science, Michigan State University, East Lansing, Michigan, United States of America

Debasmit Banerjee and Subir Biswas
Department of Electrical and Computer Engineering, Michigan State University, East Lansing, Michigan, United States of America

Robert A. Montgomery
Department of Fisheries and Wildlife, Michigan State University, East Lansing, Michigan, United States of America

Katie S. Martin and Michele Wolff
Department of Nutrition and Public Health, University of Saint Joseph, West Hartford, Connecticut, United States of America

Debarchana Ghosh and Mengyao Zhang
Geography, University of Connecticut,Storrs, Connecticut, United States of America

Martha Page
Hartford Food System, Hartford, Connecticut, United States of America

Kate McMinimee,
Department of Allied Health Sciences, Universityof Connecticut, Storrs, Connecticut, United States of America

Alexios Giannakopoulos, Labrini V. Athanasiou, Katerina Manolakou Dimitrios Chatzopoulos and Constantina Tsokana
Faculty of Veterinary Medicine, University of Thessaly, Karditsa, Greece

George Valiakos and Charalambos Billinis
Faculty of Veterinary Medicine, University of Thessaly, Karditsa, Greece
Department of Biomedicine, Institute for Research and Technology of Thessaly, Centre For Research and Technology Hellas, Larissa, Greece

Konstantinos Papaspyropoulos
Faculty of Veterinary Medicine, University of Thessaly, Karditsa, Greece
Research Division, Hunting Federation of Macedonia and Thrace, Thessaloniki, Greece

Periklis Birtsas and Vassiliki Spyrou
Technological Education Institute of Thessaly, Larissa, Greece

Danai Pervanidou and Agoritsa Baka
Hellenic Center for Disease Control and Prevention (HCDCP) Ministry of Health, Athens, Greece

Sotirios Tsiodras
Hellenic Center for Disease Control and Prevention (HCDCP) Ministry of Health, Athens, Greece
National and Kapodistrian University of Athens, Faculty of Medicine, Athens, Greece

Michael R. Hutchings
Disease Systems, SRUC, Edinburgh, United Kingdom

Nikolaos Papadopoulos
Department of Agriculture Crop Production and Rural Environment, University of Thessaly, Volos, Greece

Marc Artois
University of Lyon, VetAgro Sup, France

Lisa Yon and Duncan Hannant
School of Veterinary Medicine & Science, University of Nottingham, Nottingham, United Kingdom

Liljana Petrovska
Animal Health and Veterinary Laboratories Agency - Weybridge, United Kingdom

Christos Hadjichristodoulou
Faculty of Medicine, University of Thessaly, Larissa, Greece

Index